Darwin's Fishes
An Encyclopedia of Ichthyology,
Ecology, and Evolution

In *Darwin's Fishes*, Daniel Pauly presents a unique encyclopedia of ichthyology, ecology, and evolution, based upon everything that Charles Darwin ever wrote about fish. Entries are arranged alphabetically and can be about, for example, a particular fish taxon, an anatomical part, a chemical substance, a scientist, a place, or an evolutionary or ecological concept. Readers can start wherever they like and are then led by a series of cross-references on a fascinating voyage of interconnected entries, each indirectly or directly connected with original writings from Darwin himself. Along the way, the reader is offered interpretation of the historical material put in the context of both Darwin's time and that of contemporary biology and ecology.

This book is intended for anyone interested in fishes, the work of Charles Darwin, evolutionary biology and ecology, and natural history in general.

DANIEL PAULY is the Director of the Fisheries Centre, University of British Columbia, Vancouver, Canada. He has authored over 500 articles, books and papers.

Darwin's Fishes

An Encyclopedia of Ichthyology,
Ecology, and Evolution

DANIEL PAULY

Fisheries Centre, University of British Columbia

PUBLISHED BY THE PRESS SYNDICATE OF THE UNIVERSITY OF CAMBRIDGE
The Pitt Building, Trumpington Street, Cambridge, United Kingdom

CAMBRIDGE UNIVERSITY PRESS
The Edinburgh Building, Cambridge, CB2 2RU, UK
40 West 20th Street, New York, NY 10011–4211, USA
477 Williamstown Road, Port Melbourne, VIC 3207, Australia
Ruiz de Alarcón 13, 28014 Madrid, Spain
Dock House, The Waterfront, Cape Town 8001, South Africa

http://www.cambridge.org

© Cambridge University Press 2004

This book is in copyright. Subject to statutory exception
and to the provisions of relevant collective licensing agreements,
no reproduction of any part may take place without
the written permission of Cambridge University Press.

First published 2004

Printed in the United Kingdom at the University Press, Cambridge

Typefaces Lexicon No. 2 9/13 pt. and Swiss   *System* LATEX 2_ε [TB]

A catalogue record for this book is available from the British Library

Library of Congress Cataloging in Publication data
Pauly, D. (Daniel)
Darwin's fishes: an encyclopedia of ichthyology, ecology, and evolution / by Daniel Pauly.
  p. cm.
Includes bibliographical references (p.).
ISBN 0 521 82777 9
1. Fishes – Encyclopedias. 2. Darwin, Charles, 1809–1882 – Knowledge –
Ichthyology – Encyclopedias. I. Title.
QL614.7.P38 2004
597′.03 – dc22   2003064047

ISBN 0 521 82777 9 hardback

Contents

List of figures *page* vi
Foreword J. Nelson ix
Preface and acknowledgments xiii
Conventions used in the text xvi
Darwin and ichthyology xvii
Darwin's Fishes: a dry run xxiii

Entries (A to ZZZ) 1

Appendix I Fish in Spirits of Wine JACQUELINE McGLADE 213
Appendix II Fish of the *Beagle* in the BMNH 234
Appendix III Checklist of fish specimens, identified as collected by Charles Darwin on the *Beagle* voyage, that ought to be present in the collections of the University Museum of Zoology, Cambridge ADRIAN FRIDAY 237

Bibliography 241
Index to the Fishes 323

Figures

1. The Voyage of the *Beagle* (1831–6), with arrival and departure dates *page* 13
2. Head of female (left) and male (right) Bearded catfish *Pseudancistrus barbatus* 14
3. Blennies, Family Blenniidae, and close relatives 19
4. Characins, Family Characidae 34
5. Clingfishes, Family Gobiesocidae 38
6. Creole perch *Percichthys trucha*, Family Percichthyidae 48
7. Cape Verde gregory *Stegastes imbricatus*, Family Pomacentridae 51
8. Diagram to illustrate the transformation of an annelid into a vertebrate 58
9. Dragonet *Callionymus lyra*, Family Callionymidae, illustrating sexual dimorphism 59
10. Eelpouts, Family Zoarcidae, with inserts showing teeth 63
11. The Fine flounder *Paralichthys adspersus*, Family Paralichthyidae 82
12. Fishes of the Family Galaxiidae, based on lithographs by B. Waterhouse Hawkins 93
13. Gobies, Family Gobiidae, with dorsal views as inserts 96
14. Groupers, Family Serranidae, based on lithographs by B. Waterhouse Hawkins 102
15. Sheephead grunt *Orthopristis cantharinus*, Family Haemulidae 104
16. Galápagos gurnard *Prionotus miles*, Family Triglidae 105
17. Argentine menhaden *Brevoortia pectinata*, Family Clupeidae 109
18. Jacks, Family Carangidae 116
19. One-sided livebearer *Jenynsia lineata*, Subfamily Jenynsiinae 120
20. Livebearers, Family Poeciliidae 127
21. Singapore parrotfish *Scarus prasiognathos*, Family Scaridae 154
22. Pigfish *Congiopodus peruvianus*, Family Congiopodidae 158
23. Pipefishes, Family Syngnathidae 159
24. Poacher *Agonopsis chiloensis*, Family Agonidae 163
25. Galápagos porgy *Calamus taurinus*, Family Sparidae 165
26. Narrow-headed puffer *Sphoeroides angusticeps*, Family Tetraodontidae 167

27 Schematic representation of mechanisms emphasized by three groups of evolutionists   168
28 Roundhead *Acanthoclinus fuscus*, Family Plesiopidae   175
29 Heads of Salmon *Salmo salar*, during the breeding season   178
30 Sandperches, Family Pinguipedidae   179
31 Bandfin scorpionfish *Scorpaena histrio*, Family Scorpaenidae   181
32 Silversides, Family Atherinidae, with inserts showing magnified scales   190
33 Thornfish *Pseudaphritis undulatus*, Family Bovichthyidae   202
34 The eponymous wrasse *Pimelometopon darwini*, Family Labridae   209

Foreword

This book by Professor Daniel Pauly is for people interested in fishes, in Charles Darwin, or just plain interested in natural history. Darwin is known for writing about many things, with superb works on orchids and barnacles and, of course, on natural selection. Many authors have written about him and we often hear reference to 'Darwin's finches'. I suspect few people connect Darwin with fishes: this now will change. Daniel Pauly has done a superb job in this book in showing us the many connections between Darwin and fishes. He does this in a delightful way, mixing subtle, cryptic humour with academic discussions. Pauly gives us a tour in discovering fascinating facts; it's a great way to learn about fishes.

Daniel Pauly is internationally known for his work on fish growth and mortality, tropical fisheries management, and ecosystem modelling. A recognized leader in studies of fish population dynamics, he is also well known for his insights into the historic and socio-economic factors that intervene when fish populations are exploited. These, combined with his wide interests in evolutionary subjects, allowed a masterful treatment of Darwin's contributions to ichthyology, the subject of this book.

The book is arranged like an encyclopedia, with items in alphabetical order. The generous cross-references allow the reader to start with a given term of interest and go on an exciting voyage of discovery, exploring all sorts of worlds. One can start from a given fish taxon (by common or scientific name) and be led through, for example, an anatomical part, to a biologist or other scientist, a scientific phenomenon, an ecological or evolutionary subject, a philosopher, a chemical element, a geographic location, some form of life other than a fish, a museum, and then back to some fish. All the topics, however, lead directly or indirectly to the work of Darwin, perhaps the most influential person in biology. The reader never knows where the journey will lead, perhaps to an old fossil, to Louis Agassiz, or to a species flock undergoing evolution. An exciting mixture of topics enters in as we take off in whatever direction we wish. Readers get into whole organismal biology (and respect for whole organismal biology is under serious threat in many so-called 'Biology' Departments), and it is fun to challenge ourselves in seeing just what we know or do not know on given subjects. Resources of interest to ichthyologists also include an appendix list of Darwin's *Fish in Spirits of Wine* by Jacqueline McGlade and the list of *Beagle* specimens in the Natural History (London) and Zoology (Cambridge University) museums. Fishes came into Darwin's life, just as they do ours, in many fascinating ways, as readers will discover. The book does have limits, though, as Professor Pauly has been conservative in his definition of the term 'fishes'. After all, he could have included us: even we humans are fish derivatives!

Darwin receives special respect from biologists. So should fishes. If it were not for fishes we would not have evolved – which cannot be said for finches and other birds. True, Darwin has his detractors, but then so do fishes. Darwin was not the first to suggest that life evolves: such people as his grandfather, Erasmus Darwin, and Jean-Baptiste Lamarck believed life evolved, but we tend to overlook that. What has captivated us is Darwin's explanation of the driving force of evolution, natural selection. Alfred Russel Wallace had the same explanation for evolution as Darwin, but we tend to overlook that also. One can excuse ichthyologists trying to explain the vast diversity of colour in tropical reef fishes from wondering if something else other than natural selection is at play. But let's remind the creationists that our knowledge that evolution has occurred, and is still occurring, is not based on the theory of what its driving force(s) may or may not be.

Darwin has certainly been a major figure in giving us the impetus to make our classifications reflect evolutionary history. It is through his theory of evolution that we explain similarities between taxa and give a modern rationale to our classifications. Our knowledge of evolution and the historical connections of life with explanations of why life is as it is gives biology a unique place among the sciences. In Darwin's day, about 9000 species of fish were recognized as valid, compared with over 28 000 now. We have a concept of species today, as evolutionary lineages separated by irreversible discontinuities (see Nelson 1999), different from that of Darwin, and yet that is not what explains the different number of species recognized. Active ichthyologists of Darwin's day dealing with higher fish classification included Albert Günther of the British Museum and in America Theodore Gill and Edward Drinker Cope. Ironically I think the latter two employed more evolutionary thinking in their work than did the fellow in the British Museum, who as this book points out, studiously avoided references to evolution.

Not surprisingly, many fish species, starting with *Pimelometopon darwini*, have 'Darwin' as part of their scientific or common name, and this book presents all of these (first-order) eponyms, along with second-order eponyms, various retronyms, and one whimsical 'reverse eponym', the unfortunate Mr Fish. There is also Darwin's bass, but that is a book by Paul Quinnett, entitled *Darwin's Bass, The Evolutionary Psychology of Fishing Man*. One can get easily hooked on both Pauly's and Quinnett's books and come out a winner.

Pauly's book is an adventure in learning. This scholarly work is also a wonderful source of quotes, for research papers, public talks, university classes, student essays, or cocktail parties, and the exact source is given for all quotes, so you can pretend you found them yourself. The book will be of interest to both the old-timer and the beginning student of ichthyology (alas, are we not all ultimately beginners) as well as for those with a fondness for Darwin but knowing little or nothing of fishes. It's a good full meal, from the 'dry-run' introduction, through the main course of entries, to Jacqueline McGlade's wine list, and the dessert of finely annotated references.

We can enjoy fishes in so many wonderful ways and I thank Daniel Pauly for presenting yet another way.

Joseph S. Nelson
Department of Biological Sciences,
University of Alberta

Preface and acknowledgments

The idea for this book emerged in the late 1980s when I attempted to find an interesting Darwin quote to add to a volume I was then editing on the impact of El Niño on Peruvian fishes and fisheries. I remembered from earlier readings that Darwin had been in Peru, where he collected specimens of the species which Leonard Jenyns later described as *Engraulis ringens*, the Peruvian anchoveta.

But I did not find any suitable quote: the indexes of books by, or about, Darwin that I consulted all covered 'finches' but not 'fishes'. Still, the pun was obvious, and I decided to write a short essay on Darwin's work on fishes, to be titled *Darwin's Fishes*, if only to get it out of my system.

However, caught in the iron grip of the Law of Unintended Consequences, I ended up writing a book-size chrestomathy. Fortunately, I had the help of Darwin (who contributed about 45 000 of his words, i.e. almost one third of the entire book) and, as we shall see, the help of friends who provided relevant information and helped verify facts.

The book now completed, I will attempt to cover my tracks, and pretend that this was written to fill the 'major gap in scholarship' that is usually recruited in such cases. Thus, I refer to Gruber (1981, p. xix), who, in an influential review of Darwin's work, pointed out that "[w]e need to fill many gaps in our knowledge of detail, and we need new approaches to synthesis. The details wanting are by no means fussy bits. They are, rather, organized chunks or even macro-chunks – for example, a longitudinal and critical reconstruction of Darwin's half-century of work on earthworms." I shall leave the subterranean task of working through *Darwin's Grubs* to my vermiphile colleagues: it is *Darwin's Fishes* that I herewith offer as a latitudinal 'macro-chunk' and 'critical reconstruction'.

This reconstruction of Darwin's work on fishes (and other aquatic organisms interacting with fishes) is 'critical' in that I checked Darwin's work at three levels: (a) within the body of Darwin's formal publications and other writings, by following the development of his ideas and use of the supporting facts, and probing these for internal consistency; (b) with regard to Darwin's use of source material, by accessing the references he cited and appraising his interpretation of the information therein; and (c) by evaluating Darwin's hypotheses in light of present knowledge.

No continuous narrative would have allowed presentation of this multi-layered material in coherent fashion (I did try, and the section below, on 'Darwin and ichthyology' builds on such an earlier attempt), and hence the decision to present it in the form of two lists: (i) 478 alphabetized entries, forming the bulk of this book, and (ii) 958 (mostly) annotated references, used to present Darwin's marginalia and his other comments on publications he had read, and to put his ecologically oriented work in the taxonomy-dominated context of the ichthyology of his time. This enabled the writing up of this material in the form of small, hopefully digestible items (Gruber may have called them 'micro-chunks'), liberally cross-referenced. Moreover, this allowed the book to become largely self-referential, i.e. it enabled me to create entries for explanatory items normally tucked away, in the form of endnotes or glossaries, into the back matters of a book. The price paid for this convenience of access is that Darwin's own narrative had to be 'chunked' into pieces, especially when he illustrated a topic – e.g. **sexual selection** – through a series of examples, referring to different fish species, as in *Descent of Man*. Here, all I can suggest to the reader interested in these topics is to access the original texts, which is a good thing to do anyway.

Another part of the price paid for attempting to include and reference, in agonizing detail, all I could find which Darwin ever wrote about fishes, is that many sections of this book read like laundry lists. I have attempted to cover this up, mainly through levity, the result being that this book will probably irritate serious scholars, but still bore students to tears. This is aggravated by the fact that I have been unable to resist the frequent use of inordinately long, or otherwise difficult words (e.g. 'chrestomathy'), and smuggling some of my pet ideas into this book, e.g., on oxygen's role in fish growth, even if their link to the specifics of Darwin's writings may be seen as tenuous. The cavillers will have a field day, especially among the historians, to whom the natural scientists' easy judgments ('X was wrong about that!') is anathema, given their difficulty in conceiving of a craft that would legitimately reach that way into times past.

On the other hand, Charles Darwin (CD) passed the tests put to him with flying colours, and even passed the anachronical hurdle in (c) above. (I have emphasized the few errors I found, both because CD is such a worthy target, and because without such emphasis, this book may be perceived as a hagiography.) Thus, my overall conclusion is that CD's writings on fishes reflect a generally judicious selection, correct citation and interpretation of his sources, and based thereon, the formulation of hypotheses that have largely withstood the test of time. Moreover, in the process of digging through the literature of Darwin's time and reading his voluminous *Correspondence*, I discovered a person different from most of his contemporaries – even within his social class – and, as well, from the unimaginative compiler of facts that many, often misled by his own *Autobiography*, still see in Darwin. My impression now is that, like Michelangelo, Darwin so well perceived the best thoughts of his time that they enabled him to see glimpses of the future.

Another impression I now have is that the 'taxon-centred' approach used here would lead to many new insights if applied to other large groups, such as insects or mammals, and to biologists other than CD (e.g. Linnaeus or Huxley), whose work, like his, covered a wide range of taxa.

Now, all that is left to do is to thank the friends and colleagues who helped me with *Darwin's Fishes*. Foremost among these is the team, led by Rainer Froese, that created *FishBase*, the computerized encyclopedia of fishes. Rainer was forced – as I was – by the nature of his project to move into areas he had never dreamt of getting into. In this case, the Law of Unintended Consequences turned him into an expert on fish nomenclature. Whatever nomenclatural consistency this book has is due in part (50%) to Rainer Froese. The other 50% are for William Eschmeyer of the California Academy of Sciences, and an honorary member of the FishBase Team. I could not have done without his extraordinary *Catalog of Fishes*, now incorporated in FishBase, and his answers to my queries.

Thanks also to Jacqueline McGlade, for contributing her transcription of *Fish in Spirits of Wine* to this effort, and to Adrian Friday for a list of the fishes collected by CD still in the University Museum of Zoology, Cambridge. As well, thanks are due to Ms Aque Atanacio, the FishBase artist, who processed all the graphs, while Amy Poon, with the assistance of Ms Elsie Wollaston and her staff at the UBC Library, dug up a huge chunk of the ancient references cited here. Thanks also to the students of my *Darwin's Fishes* course, given since 1995 at the University of British Columbia (lately as 'Bio 445'), and in which many of the epistemological and ethical issues touched upon here are further developed.

Maria-Lourdes 'Deng' Palomares wrote the neat 'macro' that kept my word processing program in check, forcing it to maintain the page layout I wanted. I immensely appreciated this, her draft index to the fishes, and the many other ways she contributed to the material incorporated in this book.

The following persons provided information beyond the call of duty (specific topics, if any, are in parentheses), or read the whole or part of the draft, and must now therefore (if somewhat paradoxically), be absolved of responsibility for any remaining errors (which are obviously all my fault): Patrick Armstrong, Nicolas Bailly, Anthony Chow (typhoid fever); Villy Christensen, Philippe Cury, Jonathan Entwisle, Patti Gilbertson, Rune Hagen (king cod), Kristin Kaschner (fish sounds), Mark Kraulis (sexual selection), Sven Kullander, Jessica Meeuwig, Judith Myers (water beetles) Jørgen Nielsen, Tom Okey, Sandra Wade Pauly (yes, my wife, neither last nor least), Torstein Pedersen (king cod), Tony Pitcher, David Preikshot (more references), Neil Rainer (typhoid fever); Donna Shanley (kid stories); Kostas Stergiou, Ray Symonds, Ann Tautz, Peter Tyedmers, Anne van Dam (things Dutch), Michael Vakily, Maria Helena Vieira (Cape Verde Islands), Joseph Wible (Darwiniana), Cindy Young (image processing), and Dirk Zeller.

Conventions used in the text

Most of the material in *Darwin's Fishes* is presented as an alphabetical list of entries, each relating to a particular term or 'headword' (printed in **bold**). [Note that German characters with umlauts, e.g. ä, ü, are alphabetized as if the 'e' were spelled out, i.e. ae, ue.] The entries are liberally cross-referenced by means of asterisks in front of words that are headwords, and by the use of '*see*' and '*see also*'. Charles Darwin is referred to as 'CD' throughout.

CD's own words are differentiated from mine by being presented in this font. Underlinings within CD quotes are his own, except in the entry on *electric organs (II). My few underlinings correct small errors such as incorrect dates of references.

The editors of CD's *Correspondence*, *Marginalia*, *Notebooks and other material not originally intended for publication did a marvellous job deciphering his handwriting and producing texts suitable for scholarly study. Notably, they provide lists of symbols and fonts allowing the distinction of different 'layers' in CD's writing (e.g. to distinguish original text from later emendations). In this book only a few symbols have been retained from the *Correspondence* and *Notebooks* and also used for the transcription of *Fish in Spirits of Wine*. These are:

<....> indicating a deletion by CD;
<<..>> indicating an insertion by CD; and
[......] indicating editors' (or my) notes within the text.

The superscript numbers indicating foot- or endnotes are largely CD's, with a few provided by the editors of work from which quotes were extracted. The original numbers were changed in only a few cases, to ensure their uniqueness within each alphabetic entry. In the annotations to CD quotes, I use 'n.' to refer to items originally in such notes.

Darwin and ichthyology

This book, consisting mainly of alphabetized entries and of quotes ripped from their context, makes it hard for the reader to gain an overall view of its major topic: Darwin's relationships to ichthyology, both as a user of, and a contributor to, the insights in that discipline.

The brief introduction which follows, adapted from Pauly (2002a), attempts to compensate for this by linking Charles Darwin (CD; 1809–82) to various themes, arranged in four periods, and themselves linked to some of the major entries of this book, where detailed information and the relevant quotes are presented.

The years before the *Beagle*
Contrary to a widespread view, CD's youth and student years prepared him well for the *naturalist role he assumed during the voyage of the *Beagle*.

CD's schooling appears to have been typical of that of boys of his social class and time, even if he recalled, in his *Autobiography* (p. 46) that he was doing no good at school. Far more interesting, at least here, is his extraordinary devotion to *angling, which started at an early age, and apparently lasted until well into the *Beagle* years. An extensive *correspondence attests to this devotion, and related activities and readings, notably of Izaak *Walton's *Compleat Angler*. Indeed, Walton's classic, which identifies (and names!) distinct populations of *Trout and other fish species in the British Isles, may have contributed, a decade or so later, to CD's dawning perception of within-species *variation as a motor of *evolution.

CD's angling years were also a period of avid *beetle collection, and this introduced him to Leonard *Jenyns, who later described the fish CD collected during the voyage of the *Beagle*.

Only two elements are highlighted here of CD's student years in Edinburgh: his dissection of a *Lumpfish under the guidance of his then mentor Robert *Grant, and his relationship with John *Edmonton.

The importance of the written account of the Lumpfish dissection derives from the fact that it is the first bit of scientific writing by CD ever found, and from the profound understanding of the relationship between scientific 'fact' and 'theory' that this account documents. Indeed, the mature way of *'seeing' illustrated by this account, while establishing that CD then was already a keen observer, quick to formulate and test fruitful hypotheses, also establishes that Grant, an early evolutionist, cannot have had on CD as little intellectual influence as claimed in his *Autobiography*.

The relationship with John *Edmonton, a former South American slave of African ancestry established in Edinburgh, from whom CD took private lessons in taxidermy, is also important, as it appears to have opened his mind to respecting people outside his narrow class and ethnic background, thus enabling him to learn from the people he met in his later travels and readings, and, ultimately, to formulate a theory that encompassed all of humanity, and embedded us within the same nature. This contrasts with the divisive schemes propagated by less open-minded contemporaries, e.g. Louis *Agassiz and Richard *Owen, whose religious prejudices, combined with social opportunism, ultimately undermined their science.

CD's years in Cambridge, where he performed rather well as a student, again contrary to a widespread belief fuelled by a misleading account in the *Autobiography*, are documented in various biographies, most of which emphasize the role of his mentor there, the botanist John *Henslow. Nothing peculiar to ichthyology is reported from this period, which ends when Captain Robert *FitzRoy accepted CD as his companion and effective *naturalist on the *Beagle*, after *Jenyns had refused the offer.

The *Beagle* years (1831–6)

CD's plan, when he embarked on the *Beagle*, was to collect enough material and observations to write, on his return, an account similar to von Humboldt's *Voyage aux régions équinoxiales du Nouveau Continent* (1805–39), which he greatly admired. Moreover, as ichthyology, in the early nineteenth century, was completely dominated by French taxonomists, as illustrated by the *Histoire Naturelle des Poissons*, CD also planned to collect fish specimens that would prove to be new species; the more the better. Thus, he concentrated his fish sampling effort in areas not previously, or little, explored by French vessels. Hence the thoroughness of his collecting work in Southern South America, and his more limited samples from the Indo-Pacific.

CD's conservative sampling strategy, dictated in part by the difficulties in preserving and shipping specimens back to England (with Henslow at the receiving end), did pay off, as illustrated by a letter of October 1839 to Jenyns, in which he congratulates himself: I am astonished & glad to hear how many new things you seem to have found – four new genera is something. There would have been more, had not a part of the *collection rotted away.

One important reason why the strategy worked is that Jenyns did a very competent job of describing CD's fishes, successfully navigating the waters between the Scylla of lumping distinct species, and the Charybdis of splitting mere variations into named species. (*FishBase was the main source of updated *names, i.e. the main tool I used to establish the correspondence between his species names and those now considered valid; see also the **Index to the Fishes**).

CD clearly believed, long before he conceived *sexual selection (through which he explained sex-related differences in the *colours of fish and other animals), that the colours of animals matter, and the descriptions of the live colours of most of his specimens, e.g. in *Fish in Spirits of Wine, attest to this. Moreover, he did not let his imagination colour his descriptions, basing them, rather, on the colour-coded charts in a book he took with him for that very purpose (Syme 1821). Thus, we can attribute to CD the first rigorous treatment of colours in biology in general, and in ichthyology in particular.

This attention to details which other naturalists may have overlooked is also evident from other aspects of his field work, for example, by his collection in the *Cape Verde, *Falkland or *Galápagos Islands. Notably, this involved performing simple – we might call them Baconian – *experiments, on the behaviour, *ecology, or physiology of various animals, including fishes. This involved, for example, cutting open a marine *iguana in the Galápagos (try that now!) to settle a longstanding dispute on whether they feed on underwater vegetation or on fish, dropping marine fishes into freshwater to see if they would adapt, and more.

The return of the *Beagle* to the *Foundations of Origin* (1837–44)

Particularly revealing to anyone who ever edited a book is the series of letters CD sent to Jenyns, upon his return from the voyage of the *Beagle*, to convince him to start, then to complete the job of describing the specimens CD called my fishes. These letters are fully documented in Burkhardt *et al.* (1986), with additional context provided in this book.

CD even used nationalism: For the credit of English zoologists, do not despair and give up; for if you do, then will it be said that there was not a person in Great Britain with knowledge sufficient to describe any specimen which may be brought here. (Dec. 4, 1837).

As well, CD pleaded with Jenyns for the incorporation, into the fishes' descriptions, of his field notes on colours and behaviour. Jenyns went along, and this made *Fish* (Jenyns 1840–2) rather lively, at least by the standards of the the taxonomic literature of the time. One example, from p. 87: "In Mr Darwin's notes, it is stated that [*Salarias atlanticus] bites very severely, having driven its teeth through the finger of one of the officers in the ships company. Its two very long sharp canine teeth at the back of the lower jaw are well calculated to inflict such a wound."

The point about CD, though, is not any of this, but that he discovered *natural selection. His post-*Beagle* *notebooks, now available in their entirety (Barrett *et al.* 1987), make clear that this discovery happened in the autumn of 1838, with various scholars even venturing specific dates.

This led to an immediate change in the way CD read: before, he absorbed ideas from a wide range of books, almost haphazardly, with what we might wish to call an 'open mind'. He describes his reading during this period thus: I worked on true Baconian principles and without any theory collected facts on a wholesale scale ...

This statement has misled many – because it describes a period which ended when CD hit on natural selection. Thereafter, his readings became more targeted, with all that he read being evaluated (often through critical *marginalia pencilled right onto the offending pages) in terms of its support – or lack thereof – for the nascent theory.

The role played by fish in this phase of CD's personal evolution is hard to pin down. The distributions of fishes clearly played an important role. Notably, CD expected isolated islands to have a relatively large fraction of endemics among their coastal fishes, and one even gets the impression, with regard to the Galápagos at least, that he expected to be able to document, using fish species distributions, the peculiar role that isolated islands play in generating biodiversity, now commonly illustrated with *Darwin's Finches. This couldn't be done at the time, owing to the state of fish *taxonomy, and CD gradually abandoned this theme. However, he continued to discuss fish *distributions when contrasting freshwater with marine habitats, and insisting that the former served as refuge to ancient *ganoid fishes, which have apparently been saved from fatal competition by having inhabited a protected station. (*Origin*, VI, p. 105).

In *Foundations of Origin*, two manuscripts CD wrote in 1842 and 1844 to ensure that his discovery would not be lost (he had his wife promise to publish them, should he die prematurely), 'fish' also served as CD's shorthand for ancestral *vertebrates, especially in terms of their anatomy, habitat, and perceived tendency toward *hermaphroditism, a feature much emphasized in subsequent writings, and fully documented in this volume.

Also of note is CD's membership of the Strickland Commission, which originated the predecessor to the *International *Code of Zoological Nomenclature* (Strickland *et al.* 1843). None of this, however, added to our knowledge of fishes *per se*.

The mature Darwin (1845–82)

CD's contributions to *ichthyology, for the period from 1845 to his death, were both indirect and direct. The indirect contribution, obviously, is that he developed the evolutionary context within which biology must now be done, if it is to mean anything, notwithstanding *creationism.

That story, culminating in the 1859 publication of the first edition of *Origin of Species*, has been told in uncounted biographies and texts, and does not need rehashing here.

However, *Origin*, which ran in six editions during CD's lifetime, contains a multitude of direct references to fishes, notably on sexual selection, on relict forms (variants of the ganoid story, see above), on the position of fishes along the *complexity *scale, and on various *difficulties of the theory, i.e. the seeming lack of transitory forms to explain through natural selection the evolution of *eyes in fishes and other animals; *flying fishes; *swimbladders, erroneously presented as lung precursors; *electric fishes; the metamorphoses of *flatfishes; the pregnancy of male *seahorses; the sudden appearance of *teleosts in the *fossil record; and more. Also discussed are the impact of sea-surface temperatures and geographic barriers on ichthyofauna formation, including an interesting *volte-face* from the first edition, concerning the impact of the Isthmus of *Panama; and the results of his field experiments on how seeds in fish stomachs are distributed by piscivorous *birds. Overall, *Origin* is a firework of ichthyological ideas.

Many of these ideas are amplified in the books CD later wrote to boost the argument in *Origin*, notably *Variations* (1871, 1877), and *Descent* (1871, 1877), his only works with sections explicitly dedicated to fishes. In *Variations*, a two-page section deals with the origin and forms of *Goldfish. In *Descent*, a section discusses the *sex ratios of *Pike, *salmonids and *cyprinids, and another discusses the secondary sexual characters of a large number of fish *species, from *sharks and *rays to highly derived teleosts.

Thus, CD would have had a strong impact on ichthyology, had he decided to gather his thoughts on this group into a small book, similar, say, to the one he devoted to *Worms. He never assembled that book, however, and his impact on the discipline remained indirect – or absent, as illustrated by *Günther's *Introduction to the Study of Fishes*.

The present volume may be seen as a belated entry on CD's behalf.

Darwin's Fishes: a dry run

Presenting aspects of the work of Charles Darwin (CD) in alphabetic form appears never to have been done so far. Thus, it may be useful to provide examples of how this book can be used to extract structured information from its entries and annotated references. This 'dry run' may then serve as:

1. Another introduction to those who skip prefaces and/or introductory essays (which one should not, as this is where the authors usually best explain themselves); and
2. An instructor's guide for the daring colleague considering using *Darwin's Fishes* as main text or supplementary reading for a course in ichthyology, evolutionary biology, or in the history of science.

First to the structure of *Darwin's Fishes*. This book presents all that CD ever wrote on fishes. The bulk of this material is presented in the form of alphabetized **entries**. Another, if small, part of CD's writing on fishes is presented here as annotation to references. This, plus the fact that most references also include my own annotations (and translations, where appropriate), and that all references are cross-linked to all entries where they are cited, make the bibliography an integral part of *Darwin's Fishes*, and not what it might seem at first sight: a grave for the dead bones of scholarship. This integration with the main entries also applies to CD's field notes on fishes, presented here both as a whole (*see* **Appendix I** and/or '**Fish in Spirits of Wine**') and as short quotes within the relevant entries.

Thus, following up on a topic will usually require starting from its main entry, then linking to both the related entries and the references. This is illustrated here with sequences of quotes and comments dealing with three topics, each documenting one typical entry in *Darwin's Fishes*, and pertaining to:

1. A (group of) fish species CD was interested in;
2. A concept which CD documented with fish examples; and
3. A person with whom CD debated fish-related issues.

Parrotfishes CD sampled parrotfishes in both *Tahiti and the *Cocos Islands. While the species in question were not new to science, CD's thoughts about the ecological role of parrotfishes turned out to anticipate his later work on the slow work of earthworms: he believed that parrotfishes, by consuming corals and defecating calcium carbonate, had created the chalk layers that characterize the *Cretaceous, only to be rebuffed by a naturalist, William Buckland, who was often wrong, but not on this. CD also tested whether parrotfishes contain poison, which they do. He also misspelled parrotfish (genus *Scarus*) to *Sparus*, which earlier editors of his work failed to notice.

Thus, if you turn to the entry on parrotfishes (p. 154) you will see that it is linked through the asterisked entries to:

Ciguatera A form of ichthyotoxicity of reef fishes, increasingly affecting the international fish trade and to which CD exposed himself when he consumed a *barrowcooter (his *spelling again) and, possibly, a parrotfish.

Cocos Islands The atolls, now officially known as 'Cocos (Keeling) Islands,' visited by the *Beagle* in April 1836, where CD tested his just-developed theory of coral reef formation (outlined in *Coral Reefs* and predictably rejected by *Agassiz), sampled eleven species of fish (all asterisked), none new to science and a small fraction of the 533 fish species occurring there.

Porgies Fishes of the genus *Sparus*, a misspelling of *Scarus*. CD sampled a porgy in the *Galápagos.

Shoals Referring to a group of fishes, but differing from 'school.' Shoal also pertains to an area of shallow waters, such as the *Abrolhos.

Tahiti where CD sampled ten species of fish (all asterisked), whose range was later briefly discussed, and where he performed a memorable trip inland, described in further entries (*see* **Food-fish; Otters**).

There are other ways than shown above to follow up on the parrotfishes entry. For example, it would have been possible to first visit some of the references cited in that entry (e.g. Valenciennes (1840) or *Fish*, i.e. Jenyns (1840–2)). This would have led to details on the monumental *Histoire Naturelle des Poissons*, the major source of data on fishes at the time, or to CD's annotation on the book in which the specimens he collected were described.

Sexual dimorphism We take this entry to illustrate how *Darwin's Fishes* deals with concepts, in this case one proposed by CD himself. Here we find a definitional quote by CD, and a cross link to *sexual selection, usually the cause of sexual dimorphism. Also, we find a link to *FishBase, an Internet-accessible global database on fish that can be used to test the hypothesis implicit in the CD quote, i.e. that in fish, the females are always larger than the males.

Contrary to the information CD was given on this by *Günther and others, not all fish species have females that are larger than the males (the *Cichlidae represent one of many exceptions). However, this rule does apply to the majority of fish species, and it is sufficiently true to cast doubt on a related belief CD held, which I call the *reproductive drain hypothesis. The belief here is that growth and reproduction act as 'antagonists', i.e. that fish either grow fast or have a high fecundity, but not both. CD favourably cites *Spencer to that effect, and refers to his 'explanation' of the antagonism. Spencer, however, only asserts its existence, and the few examples he gives (Cod, Stickleback) contradict him: Cod grow faster to larger sizes than Sticklebacks, though they have a much higher fecundity. I refer to the entry on *oxygen to explain this apparent paradox, one of the cases where I smuggled a pet theory into *Darwin's Fishes*.

Here again, tracking the references allows identifying further links. Thus, the annotations to Spencer (1864–7) connects not only back to the *reproductive drain hypothesis, but also to *social Darwinism, to which he contributed the key ingredients, and to *survival of the fittest, the term he invented and passed on to a reluctant CD.

Agassiz Our last example is an entry on a person, Jean-Louis Rodolphe Agassiz. The text of the entry betrays that I do not like the man, despite having grown up in a city (La Chaux-de-Fonds, Switzerland, although I am French) whose main street and one of its schools bear his name. He was just too bigoted to now serve as a role model to anyone. And it did not help that he ended up rejecting just about any scientific advance he encountered, be it evolution by *natural selection, or the elegant theory of reef formation outlined in *Coral Reefs*.

In his own entry, Agassiz is cross-referenced to *cavefishes, *classification, *creationism, *evolution, and *taxonomy. As in the above two examples, these entries lead to other entries, etc., and thus to more material on Agassiz' often weird scientific stands (though he did get his glaciers right). We can also sample the virulence of his racism (even within the context of his time) more directly through the references cited in the main entry, i.e. through the annotations to Morton (1854), in which CD suggests he (Agassiz) should be ashamed of what he wrote.

Finally: while navigating through this book, refer to the section on 'Conventions used in the text', and try the entry on *spelling if you think something looks wrong. And please read the Preface if you have not already done so.

Entries (A to ZZZ)

A The first letter of the Roman alphabet, and hence the place where the systematic reader and the author of an encyclopedia first meet. It is therefore the place where the reader is urged not to judge this book by its first letter(s) – just as it shouldn't be judged by its cover. Rather, continue to read further down, 'alphabetically' as it were, or browse. You can do this randomly, or by following the links connecting the entries in this book.

Thus, you can go from here to *Darwin the person (a.k.a. *CD), or to *a* *darwin, the unit of evolutionary change. (Note the subtle introduction of 'a,' the indefinite article, also much used by CD). Or, if you don't already know, you can find out what a *chrestomathy is, or look at the references, either to see if you are cited (you might be if you are an ichthyologist, or a Darwin scholar), or to read some of the nasty remarks CD penned about authors such as Chambers, or *Lamarck. Or you can check on the epistemological problem posed by CD's often strange *spelling.

In this book, CD's writings are always in this font; *italics* are used for emphasis, for Latin or French expressions, and for scientific names (*see also* **ZZZ**).

Aberrant Forms or groups of animals or plants which deviate in important characters from their nearest allies, so as not to be easily included in the same group with them, are said to be aberrant (*Origin VI*, p. 430; see **Cavefishes, Lungfishes, Seahorses**).

Abnormal contrary to the general rule (*Origin VI*, p. 430; *see, for example*, **Analogous organs; Flatfish controversy (II); Monstrosities**).

Abrolhos Reefs named from the Portuguese 'abre olhos,' i.e. 'open your eyes'. Given the soft nature of ships' hulls relative to reefs, there are several places where opening one's eyes was recommended by Portuguese sailors, and two of these are mentioned by CD.

One of these places is the Abrolhos Archipelago, off Brazil (about 18°S, 38°35′W), visited by the *Beagle* in late March 1831. CD noted, while in the vicinity of the Abrolhos, that since leaving Bahia, the only living things that we have seen were a few *sharks & *Mother Carey's chickens (*Diary*, March 24–6, 1832).

The ecosystems of the Brazilian Abrolhos have been relatively well studied (Telles 1998; Ferreira and Gonçalves 1999), and it is hoped that the establishment of a marine park in the area (Dutra 1999) will help overcome the effects of various stresses, notably overfishing.

Another of these places is Houtman's Abrolhos, located on the northwest coast of Australia (28°S), and briefly described in *Coral Reefs* (pp. 234–5). CD noted that: Dampier also repeatedly talks about the immense quantities of Cuttle fish bones floating on the surface of the ocean, before arriving at the Abrolhos *shoals. (*Notebook R*, p. 23; Dampier 1703). However, cuttlefish are molluscs, and do not belong to *Darwin's fishes as defined in this book. Thus, I don't know how these cuttles smuggled themselves in here. Perhaps because of CD's *spelling.

Acanthoclinus fuscus *See* **Roundheads**.

Acanthopterygians A superorder of fishes, characterized by spiny fins and *ctenoid scales, and which includes the perch-like fishes and closely allied groups, i.e. the majority of the over 28,000 extant species of fish.

It is therefore not surprising that *Jenyns, who described CD's fish *collection, should have had problems with the Acanthopterygians. Thus CD's encouragements to Jenyns: I admire the ingenuity, with which you perceive a fishy smell about my book, my silence, & daresay the very name of me: – Moreover this fishy smell, as far as I remember of it in *Henslow's Museum was not very savoury, so that I fear the very idea of me must disturb your nostril. – Far from thinking you have done little, I am delighted to hear that the Acant. are so nearly ready: with respect to the time could you let me have the fish by the end of November, as the latest, so as to produce a number by the final day of the year, or on the 1st of the ensuing March. [. . .]. Have you any idea of the bulk of your M.S. for the Acant. portion of the Fish? [. . .].

I am really very sorry that you find my fish such a troublesome job – ill luck to them they have caused me trouble & plague also, – but I trust you will eventually be repaid in their having led you to study some of the groups of foreign fish – & I feel sure, that whatever you do in them, as far as it goes, will be good work, & a step in the good science of Natural History (*Correspondence*, July 15, 1839).

***Acanthurus* spp**. *See* **Surgeonfishes**.

Achirus lineatus *See* **Sole, Lined**.

Agassiz, **Jean-Louis Rodolphe** Swiss–American biologist and geologist (1807–73) whose early work on fossil (Agassiz 1833–44) and recent fishes (Spix and Agassiz 1829–31; Agassiz 1846) and on the slow work of glaciers won him enough fame for a ticket to America in 1846, where, after ingratiating himself with the most illiberal part of the local elite (see annotation to Morton 1854), he founded Harvard's Museum of Comparative Zoology (Winsor 1991; Tort 1996, pp. 33–7). From this new base, he undertook various expeditions, notably to collect Amazon fishes (Agassiz and Agassiz 1868), and to describe coral reefs in Florida (Agassiz 1883).

Agassiz corresponded extensively with CD, and the fishy part of this correspondence, e.g. on *cavefishes, is documented in this book. However, Agassiz remained to his dying day a prisoner of religious prejudices. Grove and Lavenberg (1997, p. 8) write with reference to an expedition that Agassiz led to the *Galápagos, in 1878, that "curiously the finding of new, different species did not change Louis Agassiz' vigorous opposition to the Darwinian theory of *evolution". Indeed, Agassiz' research programme was geared toward detecting the working of God's mind in the *taxonomy and *classification of living and extinct organisms (Winsor 1991). Agassiz was the last prominent biologist to hold on to such a dream, now the nightmare of biology teachers in less enlightened parts of the world. (*See also* **Creationism**.)

Agriopus hispidus *See* **Pigfishes**.

Albicore CD's *spelling for 'Albacore,' i.e. *Thunnus alalunga* (Bonnaterre, 1788).

As noted by CD, Albacore feed on *flying fishes, which feed on small *crustaceans (*Journal*, Dec. 6, 1833). What supported the latter, i.e. the basis of pelagic *food webs, eluded him, however, as *phytoplankton had not yet been discovered. (*See also* **Plankton**.)

***Aleuteres* spp**. *See* **Filefishes**.

Algae A class of plants including the ordinary seaweeds and the filamentous freshwater weeds (*Origin VI*, p. 430; *see* **Blennies**; **Damselfishes**; **Kelp**; **Lizards**; **Parrotfishes**; **Plankton**).

Alosa *See* **Herrings**.

Altruism An action by, or feature, of a given individual, appearing to benefit a different and unrelated individual.

Altruism represented a serious problem for CD's theory of *natural selection. Thus, after reading in McClelland (1839, p. 230) that Fishes are bright to be caught, he noted: I must utterly deny this. – If this could be passed – farewell my thesis (*Marginalia* 550). CD then developed this point: It has been asserted that animals are endowed with instincts, not for their own individual good or for that of their own social bodies, but for the good of other species, though leading to their own destruction: it has been said that fishes migrate that birds & other animals may prey on them;[2] this is impossible on our theory of the natural selection of self-profitable modifications of instinct. But I have met with no facts, in support of this belief worthy of consideration. (*Big Species Book* p. 520; n. 2 cites Linnaeus (1762), p. 389, and Alison (1847), pp. 7, 15).

CD rightly saw in altruism a clear test of his theory (*see also* **Difficulties**), which thus meets *Popper's criterion of falsifiability: As in nature selection can act only through the good of the individual, including both sexes, the young, & in social animals the community, no modification can be effected in it for the advantage of other species; & if in any organism structure formed exclusively to profit other species could be shown to exist, it would be fatal to our theory. Yet how often one meets with such statements, as that the fish in the Himalayan rivers are bright-coloured, according [to] an excellent naturalist, that birds may catch them! How the fish came to be bright-coloured I can no more pretend to explain than how the *Gold-fish, which Mr *Blyth <informs me he> believes to be a domestic *variety of a dull-coloured Chinese fish, has gained its golden tints, or than how the *Kingfisher, which preys on these fish, comes to be so brilliantly *coloured, without, as far as we can see, any direct relation to its habits. (*Big Species Book*, p. 382; the excellent naturalist is McClelland, cited above. *See also* **Handicap principle**.)

Strangely enough, a Russian school of self-described 'Darwinian' evolutionists emerged which saw altruism of the kind CD rejected as the motor of *evolution (Todes 1989; Sapp 1994). This school included a noted ichthyologist, Karl Fedorovich Kessler, who interpreted fish reproduction, schooling and migrations as forms of 'mutual aid' (Todes 1989, pp. 109–12).

'Mutual aid' is tempting, though it is not what seems to be happening in nature. Rather, the detailed study by Hamilton (1964) and others, first of social insects, then of other social animals, demonstrated conclusively that 'helping,' for an animal, can lead to increased survival and reproduction of kin, i.e. siblings, cousins, etc. Their increased fitness increases the 'inclusive fitness' of the helper, thus compensating for the cost of helping, which can go, for example in the worker caste among eusocial insects, as far as forgoing reproduction. Or put differently: an animal can opt to spread its genes by helping its relatives reproduce successfully, and thereby spread the shared genes, which can be seen as the ones that 'selfishly' benefit from the whole arrangement (Dawkins 1989).

Thus, CD need not have worried about altruism ultimately undermining his theory. In fact, altruism became one of the exceptions that pro<u>b</u>ed the rule.

Amblyopsis *See* **Cavefishes**.

Amblyrhynchus *See* **Lizards**.

Ammocœtus(-es) Larval form of *lampreys, resembling *lancelets in shape and behaviour, and used once by CD to illustrate forms that lack *complexity.

Amphibians A class of vertebrates which includes the frogs, toads and newts (i.e. the 'batrachians'), sporting a mix of features which renders them in some ways 'intermediate' between primitive fishes and early reptiles (though one must be wary of such linear *scale).

CD suggested once that the *lungfish *Lepidosiren* mediates between fish and amphibians: Unknown form probably intermediate between mammals, Reptiles & Birds as intermediate as Lepidosiren now is between Fish and Batrachians (*Corresp.* to C. *Lyell, Sept. 23, 1860). He did this only once, presumably because the lungfishes, at the time, were perceived to *be* amphibians.

The larval form of amphibians, the tadpole, provides a model for the early chordates. The similarities between larval *seasquirts and tadpoles are particularly striking, and are commented upon by CD. (*See also* **Vertebrate origins; Lancelet; Ontogeny**.)

Amphidromy(-ous) A questionable form of *diadromy, referring to fishes that migrate between freshwater and the sea, but not for the purpose of breeding, as *anadromous and *catadromous species do.

Amphioxus *See* **Lancelet**.

Anadromy(-ous) Refers to fish whose adults leave the sea and ascend rivers to spawn; most *salmon and *shads are anadromous. (*See also* **Diadromy**.)

Analogous organs Organs whose similarity depends upon similarity of function, as in the wings of insects and *birds. Such structures are said to be analogous and to be analogues of each other. (*Origin VI*, p. 430). Analogous organs constrast with 'homologous' organs, derived from the same body parts, such as the wings of *bats and birds, both ultimately derived from the forelegs of ancestral lower vertebrates. *Owen contributed greatly to the differentiation of these two concepts, a necessary step in the understanding of organic *evolution. CD discussed the implications of analogous organs under the heading Similar & peculiar organs in beings far remote in the scale of nature, viz.:– I have already alluded to the remarkable case of *Electric organs occurring in genera of fish, as in the *Torpedo & *Gymnotus almost as remote as possible from each other: but the organs differ not only in position, & in the plates being horizontal in one & vertical in the other, but in the far more important circumstance of their nerves proceeding from widely different sources[1] (*Big Species Book*, pp. 374–5; n. 1 cites Owen (1846) who, on pp. 217–18, describes two types of electric organ. However, Owen does not explicitly mention their nerves having different sources, though it may be implied.)

Also: According to our theory when we see similar organs in allied beings we attribute the similarity to common descent. But it is impossible to extend this doctrine to such cases, as those just given of the [...] Torpedo & Gymnotus, the Echnida & Hedgehog &c, – excepting in so far that community of descent, however remote the common ancestor may have been, would give something in common to the general organisation. Just in the same way [...] we have seen that the occurrence of similar monsters in the most diverse members of the same great class may be attributed to a like organisation from common descent, being acted on by like *abnormal causes of change [...].

It is not, I think, at all surprising that natural selection should have gradually given a fish & a *whale something of the same forms, from fitting them to move through the same element; just as man in a small degree has given by his selection something in common to the form of the grey-hound & race-horse. A similar doctrine, I infer, must be extended to the above given remarkable cases of similar, though very peculiar & complex structures, in beings remote in the *scale of nature. Such cases are not common; & in some of them the parallelism, as we have seen in the electric organs of fishes & in the eye of Cephalopod & mammal is not absolutely strict. (*Big Species Book*, pp. 375–6.)

Androgynous *See* **Hermaphrodite**.

Anelasma squalicola *See* **Barnacles**.

Angling Young CD was an avid angler, and this led his older brother Erasmus Darwin to write to him: "As to the tackle you are quite welcome to have it all except the line whose beauties you don't appear to appreciate properly." (*Correspondence*, June 1825).

Three years later, CD described his passion in letters to his cousin W. D. Fox: The reason I delayed answering is that I have been on an expedition for a few days. For you must know that I am become a 'Brother of the Angle' under the superintendence of Mr Slaney (MP. for our town of Shrewsbury), who pronounces me a very flourishing Pupil. (*Corresp.* Aug. 19, 1828). The term Brother of the Angle, emphasized by CD, is from Izaak *Walton's *Compleat Angler*, first published in 1753. One cannot help but wonder how much this work, and its many references to local variants (or *races) of widely distributed fish *species, such as Brown *trout, may have influenced CD's later thinking on *variation. Also, angling led him to keep a field list of fishes, based on Neill (1808; Browne 1995, p. 78; DAR 5: 28–35).

CD remained devoted to angling until shortly before the voyage of the *Beagle*: I have been intending to write every hour for the last fortnight, but *really* have had no time: I left Shrewsbury this day fortnight ago, & have since that time been working from morning to night in catching fish or *beetles. This is literally the first idle day I have had to myself: for on the rainy days I go fishing, on the good ones Entomologizing. [...] And now I give you some account of our Welch trip. [...] Old E & myself staid a few days longer & had some pretty good trout fishing. (*Correspondence*, Aug. 25, 1830).

Later, CD was to remember his passion: I had a strong taste for angling, and would sit for any number of hours on the bank of a river or pond watching the float; when at Maer[1] I was told that I could kill the worms with salt and water, and from that day I never spitted a living worm, though at the expense, probably, of some loss of success (*Autobiography*, p. 27; n. 1 identifies the house as that of CD's uncle, Josiah Wedgwood). CD's experience with angling can be assumed to have contributed to the skill he displayed during the voyage of the *Beagle*, while acquiring his *collections of fish specimens, and while *fishing to supply the *Beagle* crew with fresh food.

Anguilla *See* **Eel; Eels.**

Annelids (-idae) A class of worms in which the surface of the body exhibits a more or less distinct division into rings or segments, generally provided with appendages for locomotion and with gills. It includes the ordinary marine worms, the earthworms, and the leeches (*Origin* VI, p. 431; by ordinary marine worms, CD means bristle worms, or polychaetes, of which he often uses the genus *Nereis* as a representative. Hence his vision of chalk-making nereidous animals in the *Cretaceous. *See also* **Dohrn.**)

Aphritis spp. *See* **Thornfishes.**

Apistus A genus of scorpionfishes to which *Jenyns (*Fish*, p. 163) assigned a specimen collected by CD at *King George's Sound, Australia. He also felt that this specimen, though sharing a number of features with *Apistus niger*, now *Tetraroge niger* (Cuvier, 1829), Family Tetrarogidae, was 'distinct' from it.

This is confirmed by Gomon *et al.* (1994), who point out that the range of *T. niger* does not extend to southern Australia. This leaves open the true identity of the specimen collected by CD.

Aplochiton spp. *See* **Galaxiidae.**

Aplodactylus punctatus *See* **Marblefishes.**

Arripis georgianus *See* **Ruff.**

Artificial selection The process wherein a human breeder chooses which of the progeny of a plant or animal should survive and reproduce. The long-term results of such selection are preferred *breeds or *races. It was CD who first noted the similarity between such selection and the process he called *natural selection.

In fact, the two processes differ only when seen from our perspective as 'selectors'. From the selectee's (e.g. a *Goldfish's) point of view, we are as much part of its environment as, say, its *parasites. Thus, we can also conceive artificial selection as being, from the selectee's point of view, a way of establishing itself in a new niche: the material culture that humans create (including aquaria in pet shops).

Ascension Island A small island in the Southeastern Atlantic, 1290 km to the northwest of *Saint Helena Island, visited by the *Beagle* on July 19–23, 1836. CD took the opportunity for studying the geology of Ascension Island, but does not appear to have sampled its marine life.

Later, however, CD did comment on the fishes of Ascension Island: Fish of Teneriffe. St. Helena & Ascension most species like & *identical* with S. America. & many very *close*:[5] see full paper.[6] L'Institut 1838. p. 338 (*Notebook E*, p. 406; n. 5 refers to Valenciennes (1838a), p. 338; 6 to Valenciennes (1838b), i.e. a summary of Valenciennes (1837–44)).

According to Edwards (1990, p. 45), Ascension Island has a total of 72 species of bottom-dwelling neritic fish, with the following affinities: widespread warm Atlantic species 24; Western and Central Atlantic 21; Central Atlantic 16; Eastern and Central Atlantic 4; *endemic species: 7 (i.e. 10%, similar to the percentage of endemic fishes in the *Galápagos).

Assuming that 'most' means over 50%, it may perhaps be argued that, indeed, most species [of Ascencion are] like & identical with S. America – but this may stretch CD's description too far.

Ascidians *See* **Seasquirts**.

Aspidophorus chiloensis *See* **Poachers**.

Asterisk The symbol*, used in this book to identify terms with an entry of their own. Often written 'Asterix' by French schoolchildren. Find out why.

Asymmetry The results of differences between the 'sides' of structures with one or more longitudinal axes, such as the body of an animal. Echinoderms and coelenterates – CD's *Zoophites – whose bodies are radially symmetrical, have many opportunies for asymmetries (see, for example, Edwards 1966). Some echinoderms, such as the sand dollars and the sea cucumbers, sport variable mixtures of bilateral and radial symmetry, resulting in various asymmetries, depending on one's standpoint. Hence CD's definition of assymetrical as having two sides unlike (*Origin* VI, p. 431) is incomplete.

The bilaterally symmetrical vertebrates only rarely have genes coding for external asymmetries (although asymmetry of internal organs is the rule). Notable exceptions are the *Cichlidae, the *flatfishes, and the *Jenynsiinae, wherein species and/or populations may be defined by the orientation of their asymmetry. Another exception is the *lancelet, whose slight asymmetry may be vestigial (*see* **Dohrn**).

Except in these groups, externally visible asymmetries can therefore be used, in the context of *sexual selection, to evaluate whether a potential partner has suffered from developmental errors, *parasites, predator attacks, or diseases (see Morris *et al*. 2003; Reimchen 1988, 1992, 1997; Sasal and Pampoulie 2000), all of which invariably generate asymmetric injuries. Indeed, many animals, including fishes, generate colour patterns of intricate symmetry, i.e. in which asymmetries are easily detected. Such a *handicap may help females evaluate the true fitness of males.

Laterality is a form of asymmetry reflected in the preferred use, by animals, of (appendages on) their left or right side. It is known as 'handedness' in humans.

Atherina spp. *See* **Silversides**.

Autobiography Short name of the manuscript initially titled *Recollections of the development of my mind and character* written by CD between 1876 and 1881, initially for the benefit of his family, and of which a bowdlerized version was published after his death by his son Francis, along with a selection of his letters, also expurgated (Darwin 1887). A version, with "original omissions restored" was published by CD's granddaughter (Barlow 1958), but serious damage had already been done in terms of casting CD as a conventional, vaguely religious country squire dabbling in nature studies. This may have been accentuated by CD's description of his seemingly unfocused readings during a brief, specific period, from mid-1837 to the autumn of 1838, on p. 119 of his *Autobiography*, which has misled generations of CD's biographers: I worked on true Baconian principles and without any theory collected facts on a wholesale scale... The point here is that CD not only misrepresents Bacon (1620), but most of his own practice, in which 'facts', at least from the end of 1838 to the very end of his working life, were collected *only* to test clearly formulated hypotheses, notably *natural selection.

Here is one of CD's most famous quotes on this: About thirty year ago there was much talk that geologists ought only to observe and not theorise; and I well remember some one saying that at this rate a man might as well go into a gravel-pit and count the pebbles and describe the colours. How odd it is that anyone should not see that all observations must be for or against some view if it is to be of any service (*Correspondence* to Henry Fawcett, Sept. 18, 1861). Similarly, he told Anton *Dohrn, on September 26, 1870, in response to a question on how he started his various studies: I begin always with a priori solutions, if anything happens to interest me. I have generally hundreds of hypotheses before I know the facts; I apply one after the other, till I find one which covers the whole ground. But I am exceedingly careful and slow in printing (Groeben 1982, p. 94). Indeed, CD was an originator of the 'hypothetico-deductive method' usually attributed to *Popper (Ghiselin 1969). Here we can give only glimpses of this, e.g. in CD's *experiments with fishes, which, indeed, were never printed.

Azores An archipelago consisting of nine small volcanic islands west of the Portuguese mainland, and the last stopover of the *Beagle* before she arrived in Falmouth on October 2, 1836, and completed her voyage. Armstrong (1992c) thus called Terceira, where the *Beagle* anchored, "Charles Darwin's Last Island".

While performing his usual land-based explorations, CD did not collect marine organisms from the Azores. Indeed, during the last phase of the voyage of the *Beagle*, CD sampled very few marine animals in general, and fishes in particular.

Thus, CD did not sample *Scorpaena azorica* Eschmeyer 1969, originally described as an *endemic but since reported from the Mediterranean (Golani 1996), nor any of the many other fishes from that archipelago (Santos *et al.* 1997). Moreover, in his haste to get back home, CD reported, in both the *Diary* and the *Journal*, his arrival in Terceira as having occurred on September 20, and the departure from the Azores on September 25. The more patient *Beagle* log reports these dates as September 19, and 23, respectively (Armstrong 1992c, p. 60).

Bacalao Spanish word for (salted) *cod, appearing in CD's *Notebook Z* (p. 481), i.e. Molina Vol. 1, p. 244. Baccalao. migratory fish. – See *Kings drawings. – for real name.

Spanish-speaking colonists applied the name 'bacalao' to many large fishes unrelated to *cod, but locally abundant in parts of Central and South America, just as English-speaking colonists used 'cod' as common *name for different fishes in North America, Australia, South Africa, etc.

It is thus very difficult to identify which species CD had in mind, especially since all but one of Philip G. *King's drawings have disappeared (*see* **Hippoglossus kingii**). A clue, however, is given in the text referred to by CD: "such was the abundance of bacalao near the Juan Fernandez Islands that the same occurs as what is said about Newfoundland Bank, i.e. throwing a hook on a line always led to a fish being caught" (Molina 1788, p. 244; my translation).

The 'Bacalao' *Polyprion oxygeneios* (Bloch & Schneider, 1801) is one of the most important fish species of the Juan Fernandez Islands (Rojas *et al*. 1985), and may be the species Molina was referring to. Since they also occur on the Chilean mainland, they could have "come close to the beaches of *Valparaiso in great *shoals during the months of October, November and December" (Molina 1788, p. 244).

Catches of *P. oxygeneios* along the Chilean coast have now dwindled to almost nothing (see catch data in www.fao.org, or in *FishBase), and their abundance in the Juan Fernandez Islands had already much declined when Rojas *et al*. (1985) conducted their field work. Hence this species' fate may well continue to parallel that of the Northern cod *Gadus morhua* off Newfoundland, which collapsed in the early 1990s (Hutchings and Myers 1994), owing to overfishing.

Bahia A state in north-eastern Brazil, as well as the previous name of its capital, now renamed 'Salvador,' and which the *Beagle* visited from February 29 to March 18, 1832, and again from August 1 to 6, 1836.

CD took the opportunity of his 1832 visit for a first sampling trip in a tropical forest, and onshore, to collect marine life, notably a *burrfish. In contrast, CD does not appear to have collected anything on his second visit, and indeed, Bahia had lost part of its charms. The novelty & surprise were gone & perhaps our memories had, in the long interval, exaggerated the *colours of the scenery. (*Diary*, August 1, 1836).

Bahia Blanca Bay and city in Northern *Patagonia, Argentina, named after the white salt crystals that form on the exposed flats at low tides (tidal range: 3–4 m).

There, CD collected five fish species described in *Fish as *Batrachus porosissimus* (*see* **Toadfishes**); *Clupea arcuata* (*see* **Herrings**); *Platessa orbignyana* (*see* **Flatfishes**); *Syngnathus crinitus* (*see* **Pipefishes** (**I**)); and *Diodon antennatus* (*see* **Burrfishes** (I)).

Bajada Capital of Entre Rios. In 1825, the town contained 6000 inhabitants, and the province [of Argentina] 30,000; yet, few as they are, none of them have suffered more from bloody and desperate revolutions. They boast of representatives, ministers, a standing army, and governors: so it is no wonders that they have their revolutions (*Journal*, Oct. 5, 1833).

The question is now: what was CD doing in this charming place? The answer is in the same source: I was delayed here for five days, and employed myself in examining the geology of the surrounding country, which was very interesting. We here see beds of sand, clay and limestone, containing sea shells and *sharks' teeth, passing above into an undurated marl, and from that into the red clayey earth of the Pampas, with its calcareous concretions and the bones of terrestrial animals.

And why all this? Because this sheds light on an otherwise obscure line in the *Zoology*

Notes (p. 393, no. 1510), which list Fishes teeth. Liniston. Bajada. Such clarity is what we want, especially as it also establishes that the Liniston of the *Zoology Notes* is in fact *limestone*.

Balistes spp. *See* **Triggerfishes**.

Barbel Common name of *Barbus barbus* (Linnaeus, 1758), a member of the Family Cyprinidae, occurring in the upper, fast-flowing reaches of rivers ('barbel zone') in West and Central Europe, and characterized by poisonous eggs (Maitland and Campbell 1992).

Barbels are mentioned by Edmund Langton in his report of the experiments he conducted for CD, his uncle. *See* **Experiments (IV)**.

Barnacles Highly modified crustaceans of the Order Cirripedia, to which CD devoted eight years of arduous work (Darwin 1851; Crisp 1983; Smith 1968; Southward 1983).

Here, we deal only with one species of barnacle, introduced as follows: A long course of selection might cause a form to become more simple as well as more complicated; thus the adaptation of a crustaceous animal to live attached during its whole life to the body of a fish, might permit with advantage great simplification of structure, and on this view the singular fact of an embryo being more complex than its parents is at once explained (*Foundations*, p. 227; *see also* **Complexity**).

The species CD means here is *Anelasma squalicola*, which is parasitic on northern Shark (*Correspondence* to J. G. Forchhammer, Nov. 12, 1849). Thus, CD mentions The curious Anelasma, which lives buried on the skin of sharks in the northern seas, is said to live in pairs (*Collected Papers II*, 1874, p. 179) in a discussion of *hermaphroditism in barnacles. This species tends to occur in closely clustered patches, or at least in pairs, to ensure cross fertilization through a proboscii-formed penis capable of great elongation.

Anelasma squalicola is formally described in Darwin (1851, Vol. 1, p. 170), with North Sea. Parasitic on *Squalus as its *type locality. (Note that these barnacles may represent cases of commensalism, rather than function as true *parasites.)

Under habits, CD notes that According to Lovén, this *species lives embedded in the skin of Squalus maximus, *and* spinax; and Professor Steenstrup informs me, that from late observations, it appears that this animal always adheres to the shark's body in pairs. (Darwin 1851, Vol. I, pp. 178–9; Lovén 1844; for *S. maximus* and *S. spinax see* **Basking shark** and **Velvet belly**, respectively).

Barracuda Common name of members of the Family Sphyraenidae, all sporting the awesome, pointed teeth befitting piscivorous fishes.

On January 18, 1832, after geologising on Quail Island (*Cape Verde Islands), CD reports, For dinner I had Barrow Cooter for fish & sweet potatoes for vegetables: quite tropical and correct (*Diary*, p. 26; *see also* **Spelling**). CD's Barrow Cooter was most likely the European barracuda *Sphyraena sphyraena* (Linnaeus, 1758).

This species, which *ranges from the Bay of Biscay, France, to Angola, and also occurs in the Mediterranean and Black Seas, reaches 165 cm in length. The European barracuda is reported to attack people (review in De Sylva 1963, pp. 121–7), though the converse, as the above meal illustrates, is far more frequent.

On the other hand, *ciguatera occasionally makes members of this and related species rather toxic (Auerbach 1991; Lewis and Endean 1984; Tosteson *et al.* 1988), thus levelling the field a bit.

Barriers One of the key factors affecting the *distribution and *range of organisms.

CD opines: If we take a general view of distribution, I think we must conclude that barriers, whatever the nature, in regard to powers of passage of organisms is the chief, I sh^d say decidedly the most important element in their distribution. For marine *productions, landing [?] stretching N. & S. is a perfect barrier, if it has long existed, so again a wide space of ocean; now compare the shells on each side of I. of *Panama, only one the same; so with crustacea, so with Fish. – (Isthmus of Suez so low).

Again there is profound ocean, fully as wide as Atlantic ocean, west of S. America, without an island & here there is not a shell in common – but westward in ocean strewn with isl[d] (& with evidence of former isl[d]) the shells & fish extend with very many in common, even to W. coast of Africa, almost exactly an hemisphere. Again land shells of America, correlate with water shells on opposite sides of Alleghanies. (some fish cases of Hooker) (*Big Species Book*, pp. 584–5; *see also* **Squirrelfishes**).

As pointed out by Kay (1994), CD became gradually less convinced of the role of barriers in explaining the distribution of organism. This is why the eponym proposed by *Huxley for one of the barriers in question is '*Wallace's line', in spite of CD's early work on the biogeography of the area in question.

Barrow cooter *See* **Barracuda**.

Basking shark This shark's scientific name, *Cetorhinus maximus* (Gunnerus, 1765), and its junior *synonyms (*Squalus elephas, S. rhinoceros*) emphasize its large size, which can reach 9 m, and provide enough space to support lots of *barnacles.

An interesting feature of the basking shark is its possession of a set of well-developed gill rakers (Steenstrup 1873), whose bony substance resembles the baleen of *whales, and hence its other name of 'Bone-shark.' CD felt that Steenstrup's explanation of the nature of these gill rakers as modified teeth was a most wonderful case (*Calendar, no. 10594).

Wonderful as they might be, these gill rakers are shed in winter, and this must be considered when studying the feeding and growth (Pauly 1978, 2002b), and other aspects of the biology of basking sharks.

Batrachus porosissimus *See* **Toadfishes**.

Bats Flying mammals belonging to two suborders, the Microchiroptera (small bats) and the Macrochiroptera (large bats, i.e. 'flying foxes'), which appear to have evolved independently from non-flying mammals. Flying foxes of the genus *Pteropus,* which tend to consume fruits, are important pollinators and seed dispersers for the forest communities of tropical oceanic islands (Banack 1998; Neuweiler 2000).

CD wondered [w]ho would have ever supposed that at the present day there should be a Bat feeding chiefly on frogs & occasionally on fish[3]; or that the fructivorous Pteropus, when put on a floating raft, should take to the water & 'swim pertinaceously after a boat[4]'. (*Big Species Book*, p. 342; n. 3 states: Mr *Blyth gives an account of these habits in the Magaderma lyra in India; Blyth 1845, p. 463; *M. lyra* is a macrochiropterid; n. 4 refers to Lay 1829).

A fishing bat story is reported by Balon *et al.* (1988, n. 3), who note that *Pteropus*, in the Comoro Islands, "catch fish in the sea in a similar way to fisheagles". They also mention that they could not find any mention of fishing bats in the literature, wondering if they were "witness to *evolution in making?" In a sense, they were: organisms are constantly evolving, notwithstanding creationists (*see also* **Punctuated equilibrium**). Unfortunately, that evolutionary bit has been part of the literature since 1845. The score is Evolution 1 : Balon *et al.* 0.

Bay of Islands Port of call of the *Beagle* in New Zealand, December 21–30, 1835, used by CD, among other things, for sampling of marine organisms (Armstrong 1992b, 1993b); in the nineteenth century "a popular, almost cosmopolitan, place for scientific expeditions" (Roberts and Paulin 1997).

The fish CD collected at this location consisted of specimens assigned in Jenyns' *Fish to *Trigla kumu* (*see* **Gurnards**); *Acanthoclinus fuscus* (*see* **Roundheads**); *Tripterygion capito* (*see* **Blennies**); *Eleotris gobioides* (*see* **Sleepers**); and *Mesites attenuatus* (*see* **Galaxiidae**).

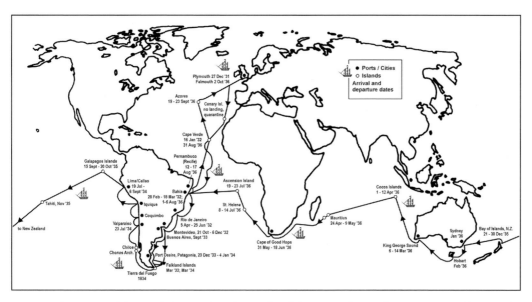

Fig. 1. The voyage of the *Beagle* (1831–6), with arrival and departure dates. The segments north of Recife (Brazil) and south of Montevideo (Uruguay) and Valparaiso (Chile) are simplified; the dates for the Azores differ from standard accounts.

Beagle H.M.S. *Beagle* was launched in 1820 as a sloop rigged as a brig (doesn't that sound wonderful?), measuring 90 ft (about 30 m), and displacing 235 tons (Freeman 1978, p. 30). On her first surveying voyage to South America, from 1826 to 1830, her Captain was Philip Parker King, father of Philip Gidley *King (King *et al.* 1836).

It is the second of her voyages (Fig. 1), captained by R. *FitzRoy, which launched the *Beagle* into immortality: young CD was on board, though not as her *naturalist (Burstyn 1975). Basalla (1963) reviews aspects of that voyage not related to CD, which involved testing the batteries of chronometers used to infer longitudes (Sobel 1995), and testing "the new Beaufort scale to reckon wind forces around the world" (MacLeod and Rehbock 1994). Many books for specialist and/or lay persons have been published which retrace all or parts of that voyage, and give emphasis to one or other of its adventures. Examples are Campbell (1997); Keynes (1979); Marks (1991); or the superb volume by Moorehead (1971). None, however, beats CD's own *Journal of the Beagle*, extracted from his *Diary*.

Hopkins (1969, p. 201) noted that "the *Beagle* outlasted both Darwin and FitzRoy. For some years after the voyage to South America, it stayed in service, continuing survey work through the 1870's, when steam replaced sail. The *Beagle* was finally sold out of the Royal Navy a few years after Darwin's death and ended its sailing as a training ship for the Japanese Navy, finally being broken up in 1888."

The log of the *Beagle*'s years with the Royal Navy is kept in the archives of the Public Record Office at Kew, London, as item ADM 51/504-3055.

Beards Fish have beards, just like people, but these beards have different origins.

Agassiz (1835) reports that: "from the bearded *Carps to the bearded *Siluri there appears to be a natural transition by means of the bearded *Loaches; [but] it is important to distinguish that in these latter, as well as in the

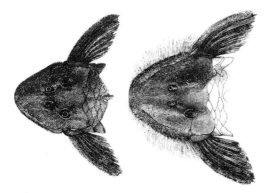

Fig. 2. Head of female (left) and male (right) Bearded catfish *Pseudancistrus barbatus*, Family Loricariidae (*Plecostomus barbatus* in *Descent II*, p. 339). Based on a drawing by G. Ford, under the supervision of A. Günther. Note bearded male, presumably a result of *sexual selection.

"*Carps* and other *Cyprinidæ*, the beards, as they are called, are merely processes of the skin; while in the *Siluri, the *cirri* of the angles of the mouth are actually prolongation of the maxillary bones becoming gradually cartilaginous and tapering into thread-like extremities."

Amazingly, CD knew what to add to this prickly matter: In a siluroid fish, inhabiting the fresh waters of South America, the *Plecostomus barbatus*[18] (Fig. 2), the male has its mouth and inter-operculum fringed with a beard of stiff hairs, of which the female shows hardly a trace. These hairs are of the nature of scales. In another species of the same genus, soft flexible tentacles project from the front part of the head of the male, which are absent in the female. These tentacles are prolongations of the true skin, and therefore are not *homologous with the stiff hairs of the former species; but it can hardly be doubted that both serve the same purpose.

What this purpose may be, it is difficult to conjecture; ornament does not here seem probable, but we can hardly suppose that stiff hairs and flexible filaments can be useful in any ordinary way to the males alone. (*Descent II*, p. 338; n. 18 refers to Günther (1868a), p. 232).

CD's own beard poses similar questions. Why did he get one, given that, as his wife wrote on April 28, 1866, "[h]e was obliged to name himself to almost all of [his old friends], as his beard alters him so much"? (Litchfield 1915, Vol. II, pp. 184–5).

Browne (2002, p. 302) writes "[a] beard like Darwin's was a visual symbol of the real seat of Victorian power, and one of the most obvious outward manifestations of what Darwin would soon be describing as factors involved in *sexual selection among humans."

The Great *Linnaeus had earlier provided a similar answer: "God gave men beards for ornament and to distinguish them from women" (cited in Blunt 1971, p. 157), which is not as silly as it appears: in both cases, catfish and humans, it is sexual selection that did the job Linnaeus assigns to God, i.e. ornament does [...] seem probable. Or put differently: female catfish may share with women a liking for males with beards.

Beauty An attribute of organisms – including fishes – often noted by CD. Here is a typical quote: The males sedulously court the females, and in one case, as we have seen, take pains in displaying their beauty before them. (*Descent II*, p. 342).

Until recently, such a reference – and the many more strewn throughout this book, and CD's books and *Correspondence* – would have been dismissed as hopelessly subjective and/or anthropomorphic. Beauty was perceived as being 'in the eye of the beholder', and not accessible to scientific analysis. Yet, both humans and animals have recently been found to prefer mates with bodies that are highly symmetrical. Symmetry indicates that embryonic development proceeded harmoniously (Reimchen 1997), and remained unaffected by diseases (Haldane 1949b), or *parasites. It further indicates that later interactions with competitors and potential predators did not lead to injuries (Reimchen 1988, 1992; Barlow 2000, p. 131) – all indicative of good genes. Thus, in animals, symmetrical bodies correlate with increased fitness. In humans, facial symmetry is a necessary (though not a sufficient) condition for the perception of beauty (Edcoff 1999).

Except for studies of symmetry, studies devoted to the perception of 'beauty' in fishes do not appear to have been conducted to date. However, its perception is likely, as many species combine a high visual acuity with complexly structured colour patterns on their bodies. (*See also* **Asymmetry; Handicap principle; Reproduction; Sexual selection; Social Darwinism**).

Beetles *See* **Water-beetles**.

Benthos The plants (phytobenthos, e.g. *kelp) and animals (*zoobenthos) living in, on or just above the sea bottom.

CD sampled zoobenthos on many of his seashore trips, e.g. Collected some marine animals at Quail Island & spent most part of the day examining them. (*Diary*, Jan. 25, 1832; *Cape Verde Islands); Walked to the coast West of Quail Island with *King, & collected numerous marine animals, all of extreme interest. I am frequently in the position of the ass between two bundles of hay; so many beautiful animals do I generally bring home with me. (*Diary*, Jan. 30, 1832).

The 'ass' in question is often attributed to the French scholastic philosopher Jean Buridan (1300–58), a disciple of William of *Occam (yes, he of the Razor): placed at an equal distance between two equally tempting bundles of hay, the animal would starve, unable to choose where to start munching (Schischkoff 1961, p. 75). Of course, no real donkey would be such an ass.

Bering pike An unidentified fish, mentioned in passing by CD in *Notebook C* (p. 289): If any one is staggered at feathers & scales. Passing into each other let him look at wing & orbit of *penguin & then he will cease to doubt: Scales into Teeth in Bering Pike (Waterhouse).

I could not find any source identifying this fish; even Jordan and Gilbert (1899), who should have known, failed me. Thus a guess, based on three (weak) arguments: CD's Bering pike is a shark, and specifically the dogfish *Squalus acanthias,* because (i) sharks can be perceived as having their (placoid) scales 'turning into teeth' (Daniel 1922); (ii) *S. acanthias* is called *piked* dogfish in Britain; (iii) it is very common in the Bering Sea.

As for (was it George?) *Waterhouse, he is of no help, because he wrote on mammals and insects, but, as far as I can tell, not on fishes.

Beroe A free-swimming 'comb-jelly' (Ctenophora), lacking the tentacles common in its phylum, and feeding predominantly on other comb-jellies.

Beroe are part of the *zooplankton, and mentioned by CD as food of *flying fishes, in a discussion of pelagic *food webs.

Big Species Book One of CD's names for the book he was planning to write (and for which he was accumulating notes) when he received, in June 1858, a letter from Alfred *Wallace accompanying a little paper (Wallace 1858) neatly summarizing 'his' (Wallace's *and* CD's!) discovery of *natural selection.

CD immediately abandoned his plan of publishing a *Big Species Book*, and, after assembling from pre-existing documents the brief text of Darwin (1858), quickly extracted an abstract from the notes accumulated so far. This 'abstract' is *Origin. The original notes, whose bearing on *ecology was analysed by Stauffer (1960), have now been edited and published (Stauffer 1975), and all their references to fishes incorporated into this volume.

Bioluminescence The emission of light by living organisms, called phosphorescence by CD.

Here is what he wrote about this: Near Fernando Noronha the sea gave out light in flashes. The appearance was very similar to that which might be expected from a large fish moving rapidly through a luminous fluid. To this cause the sailors attributed it; at the time, however, I entertained some doubts, on account of the frequency and rapidity of the flashes.

With respect to any general observations, I have already stated that the display is very much more common in warm than in cold countries. I have sometimes imagined that a disturbed electrical condition of the atmosphere was most favourable to its production. Certainly I think the sea is most luminous after a few days of more calm weather than ordinary, during which time it has swarmed with various animals. Observing that the water charged with gelatinous particles is in an impure state, and that the luminous appearance in all common cases is produced by the agitation of the fluid in contact with the atmosphere, I have always been inclined to consider that the phosphorescence was the result of the decomposition of the organic particles, by which process (one is tempted almost to call it a kind of respiration) the ocean becomes purified. (*Journal*, Dec. 6, 1833).

CD was lucky with most of his guesses, but not here: far too little biochemistry was known in his days for him to have figured out the outline of the luciferin–luciferase reactions underlying most instances of bioluminescence in bacteria and higher organisms. Indeed, their complex evolution (reviews in Herring 1978) would have been too much for comfort when added to CD's compilation of *difficulties for his theory. Moreover, had CD known about planktonic bioluminescence, he would also have needed to find a reason for the organisms in question to advertise their existence and location, and this too would have been exceedingly difficult. Indeed, it is only recently that the hypothesis of Burkenroad (1943) has been developed to the point where it is compatible with *natural selection.

Burkenroad originally suggested that bioluminescent *phytoplankton produce flashes of light when disturbed (usually by herbivorous zooplankton) in order to attract a second-order predator (as do *minnows with their 'Schreckstoff'). As stated, this hypothesis did not make much evolutionary sense, since it postulated a behaviour that would not benefit the specific gobbled-up individuals that displayed it. This would require too much *altruism. Wells (1998, p. 96) noted, however, that copepods (and presumably other small grazers) release, unharmed, any cells that flash: ingesting them would make the copepods' small bodies glow, and render them highly visible to predators. This agrees with the observation of Fitch and Lavenberg (1968, pp. 11–12) that predatory deep-sea fishes have black-lined stomachs, useful, as they suggest "to keep the luminous fish which they swallow whole from 'lighting up' in their death struggle and making the predator an easy meal for a larger fish".

However, alerting second-order predators is not the only, or even the main, reason for bioluminescence in fishes (Hastings 1975). Thus, many species of coastal bioluminescent fishes produce light that is directed downward, and which is thought to match the ambient light, such that their shadows are obliterated, thus foiling first-order predators swimming further below (Hastings 1971).

I once tested this hypothesis, if somewhat crudely, by ranking five species of slipmouth (Leiognathidae), a family of small, coastal bioluminescent fishes, by the size of their light organ, and by the different depths at which they occur, the idea being that large light organs are required to match the stronger light at shallow depths, and conversely. The two ranked lists were similar enough for the resulting paper to be publishable (Pauly 1977), but barely so, and it will not be mentioned again.

Birds A group of vertebrates probably descended from dinosaurs and characterized by feathers, toothless beaks, and egg-laying.

CD reminisced in his *Autobiography* that, From reading White's Shelborne, I took much pleasure in watching the habits of birds, and even made notes on the subject. In my simplicity I remember wondering why every gentleman did not become an ornithologist (p. 45; White 1789, 1843).

CD, ever a gentleman, did become an ornithologist, as well, but we can't follow him there: the birds considered in this book consist only of species mentioned in *Birds, or in CD's Ornithological Notes* (Barlow 1963) and *Big Species Book*, and which feed on, or otherwise relate to fish. (*See also* **Bird's nest soup; Cormorants; Kingfishers; Mother Carey's chickens; Penguins; Petrels; Scissor-beak**). Thus, we shall ignore the fishing owl in Richardson's *Fauna Boreali-Americana* (1829–37) although it was noted in CD's *Marginalia*.

Birds Short title of *The Zoology of H.M.S. Beagle. Part III: Birds* (Gould *et al*. 1841), which presented the bulk of the *birds collected by CD during the voyage of the *Beagle*, notably the finches (Genus *Geospiza*), later called 'Darwin's Finches' (Lack 1953).

Bird's nest soup The question arises: what could this entry possibly have to do with fishes, given that [t]he nest is composed of a brittle white translucent substance very like pure gum-arabic or even glass [...]. This dry mucilaginous matter soon absorbs water & softens: examined under the microscope it exhibits no structure, except traces of lamination & many generally conspicuous in small dry fragments, & some bits looked almost like vesicular larva (*Big Species Book*, p. 499).

The answer is quite simple: Most authors believe that the nest of the esculent swallow is formed of either a *Fucus or of the Roe of fish; others, I believe, have suspected that it is formed of a secretion from the salivary glands of the bird. This latter view I cannot doubt from the preceding observations is the correct one. The inland habits of the Swifts, & the manner in which the substance behaves in flame almost disposes of the supposition of Fucus. Nor can I believe, after having examined the dryed roe of fishes, that we should find no trace of cellular matter in the nests, had they been thus formed. How could our Swifts, the habits of which are so well known, obtain roe, without being detected?

Mr Macgillivray has shown that the salivary crypts of the Swift are largely developed, & he believes that the substance with which the materials of its nest are felted together, is secreted by these glands. I cannot doubt that this is the origin of the similar & more copious & purer substance in the N. American Swifts, & in that of the Collocalia esculenta. We can thus understand its vesicular & laminated structure, & the curious reticulated structure of the Philippine island species. The only change required in the instinct of these several birds is that less & less foreign material should be used. Hence I conclude, that the Chinese make soup of dried saliva! (*Big Species Book*, p. 500; MacGillivray 1840). This is nearly as revolting as eating insect vomit (i.e. honey).

There are modern authors still believing that bird nests are made of Fucus. One is Keys (1963), who writes that "[s]ome kind of seaweed is gathered by a certain species of *petrel of the Family Procellariidae, which builds its nest in the cliffs of the Malayan Archipelago. The seaweed is predigested by the alkaline fluid of the mouths of these birds before being used to construct the nests. The nests are gathered in the fall, for export to China." So let's use this opportunity to clarify the issue. CD was right, again: these nests are made entirely of the saliva of insect-feeding swiftlets (Family Apodidae), notably the White-nest swiftlet *Aerodramus fuciphagus,* the Volcano swiftlet *A. vulconorum,* and other species of the same genus, all abundant around the South China Sea. The nests are harvested, dangerously, from cliffs (Summers and Valli 1990), or more safely and sustainably from 'bird houses' constructed for the purpose, notably in Java (Whitten *et al.* 1996, pp. 234–5), where they have become a profitable, if small, export industry.

However, bird's nest exporters must care about their location. Connoisseurs from Southern China will tell you (at least some have told this author) that bird's nests from coastal areas are far more tasty than those from inland areas. We conclude this by citing Madame Wu (1984, p. 220), who points out that "[b]ird's nest is one of those prized ingredients the Chinese believe add to a woman's femininity."

Blennechis spp. *See* **Blennies**.

Blennies A group of closely related fishes of the Suborder Blennioidei, of which the combtooth blennies, Family Blenniidae, are the most speciose (*see also* **FishBase**). The shallow habitat of the blennies, which protects them from many potential predators (Gibson 1993) may be a cause for their marked *sexual dimorphism, and *territoriality, indicative of *sexual selection.

Thus CD: with some blennies, and in another allied genus,[20] a crest is developed on the head of the male only during the breeding-season, and the body at the same time becomes more brightly-coloured. There can be little doubt that this crest serves as a temporary sexual ornament, for the female does not exhibit a trace of it. In other *species of the same genus both sexes possess a crest, and in at least one species neither sex is thus provided. (*Descent II*, pp. 338–9; n. 20 refers to Günther (1861), i.e. p. 221 [on *Blennius* spp.] and p. 240 [on *Salarias*, the allied genus; [Fig. A]; however, neither page indicates the males' crests to be developed only during the breeding season).

There are many blennies among Darwin's Fishes, and we present them in the order of their appearance in *Fish*. The Rusty blenny *Blennius palmicornis*, now called *Parablennius sanguinolentus* (Pallas, 1814) was described in *Fish* (p. 83), based on a specimen of about 12 cm collected by CD in the *Cape Verde Islands. The Rusty blenny lives in very shallow waters (down to about 1 m) dominated by pebble fields, and rocks exposed to sunlight and covered with filamentous *algae, its main food, and can be observed 'flying' underwater over relatively long distances.

Next are *Blennechis fasciatus* from *Concepcion (pp. 83–85; Fig. B), and *Blennechis ornatus* from *Coquimbo (pp. 85–6; Fig. C), also in Chile, both of which are now identified as *Hypsoblennius sordidus* (Bennett, 1828). Then we have the Redlip blenny *Salarias atlanticus*, now *Ophioblennius atlanticus* (Valenciennes, 1836), described in *Fish* (p. 86) from three specimens collected by CD in the Cape Verde Islands. The adults can be quite a handful. Thus, CD wrote that they bite very severely; having driven teeth through Mr Sullivan's finger (*Fish in Spirits*, no. 19; the victim was a 2nd Lieutenant, later Admiral Sir James Sul̲ivan). Jenyns adds that its "two very long sharp canine teeth at the back of the lower jaw are well calculated to inflict such a wound." (*Fish*, p. 87).

Another species is the Rippled rockskipper, described in *Fish* as *Salarias quadricornis*, based on two specimens exhibiting dull red transverse lines. This fish is now called *Istiblennius edentulus* (Schneider & Forster, 1801), with a specific epithet indicating safety for Mr. Sulivan's fingers. The smaller of these two specimens, also lacking the nuchal crest typical of male blennies, bore a label indicating it had been collected by CD in the *Cocos Islands (*Fish*, p. 87). The other, lacking such a label, but bearing a nuchal crest, was assumed by Jenyns to have been taken at the same location. The Rippled rockskipper lives just under the water line, in rocky or reef areas with slight to moderate wave action. Its common name is due to its habit of jumping or 'skipping' from one pool to another when pursued. Last among the combtooth blennies, we have *Salarias vomerinus*, described in *Fish* (pp. 88–89), based on a specimen collected by CD at Puerto Praya, Cape Verde Islands. This species is now known as *Entomacrodus vomerinus* (Valenciennes, 1836).

Also, representing the Family Labrisomidae, we have *Clinus crinitus*, now *Auchenionchus microcirrhis* (Valenciennes, 1836), based on a specimen from Coquimbo, Chile (*Fish*, pp. 89–91; *Fish in Spirits*, no. 1204; Fig. D). Finally, CD sampled two specimens of the Spotted robust triplefin, originally *Tripterygion capito*, now *Grahamina capito* (Jenyns, 1842), family Tripterygiidae, from tidal rocks in the *Bay of Islands, New Zealand (*Fish*, p. 94; *Fish in Spirits*, no. 1343; Fig. E).

Blennius spp. *See* **Blennies**.

Blyth, Edward English naturalist (1810–73), based in India from 1841 to 1862, and who, during that period, responded to many of CD's queries on plants and animals.

Blyth is much acknowledged as a source of information throughout *Descent and *Variations, notably on the variability of domestic mammals and birds, but also including carp (*Correspondence*, Jan. 23, 1852). In fact, he is one of the authors most cited by CD (Sheets-Pyenson 1981; Tort 1996, p. 346). Thus, it is difficult to accept the notion, put forward by Eiseley (1959), that Mr. Blyth's contribution to CD's work was deliberately obfuscated (see Smith 1960, 1968 for a cogent refutation).

Even more preposterous is the notion that Blyth – who believed in the fixity of species – was the real source of CD's major insights,

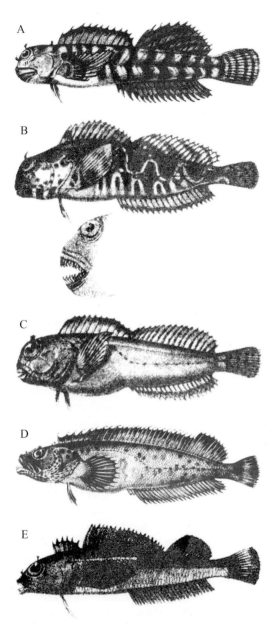

Fig. 3. Blennies and close relatives, Families Blenniidae (A–C), Labrosomidae (D), and Tripterygiidae (E). From lithographs by B. Waterhouse Hawkins. **A** *Entomacrodus vomerinus* (*Salarias vomerinus* in *Fish*, Plate XVII), based on a specimen from Puerto Praya, Cape Verde Islands. **B** *Hypsoblennius sordidus* (*Blennechis fasciatus*, Plate XVII). The insert shows the teeth, similar to those drawn through Mr Sullivan's finger. **C** *Hypsoblennius sordidus* (*Blennechis ornatus*, Plate XVII), specimen from Coquimbo, Chile. **D** *Auchenionchus microcirrhis* (*Clinus crinitus*, Plate XVIII), specimen from Coquimbo, Chile. **E** Spotted robust triplefin *Grahamina capito* (*Tripterygion capito*, Plate XIX), from Bay of Islands, New Zealand.

particularly into *natural selection (Sheets-Pyenson 1981). Indeed, the material assembled by Eiseley to support this notion – justly forgotten articles by Mr Blyth – show that he did not come even close. However, Blyth did provide CD with information relevant to this book, and so we end this entry on a positive note. (*See also* **Altruism; Bats**).

Bonito A common name applied to a number of tuna-like fishes, including Skipjack *Katsuwonus pelamis* (Linnaeus, 1758). As CD mentions 'bonitos' in the context of a *food web involving *flying fishes, we may cite *Günther (1880, p. 458), who confirms that bonito "eagerly pursues the Flying-fish, and affords welcome sport and food to the sailor".

Bony fishes The most diverse and speciose of fishes, earlier reported to appear 'suddenly' in the fossil record (highest fish in Old Red Sandstone; *Notebook C*, p. 257). If true, this would have posed a problem for CD's gradualist theory: The case most frequently insisted on by palaeontologists of the apparently sudden appearance of a whole group of *species, is that of the *teleostean fishes, low down, according to *Agassiz, in the *Chalk period. This group includes the large majority of existing species. But certain *Jurassic and *Triassic forms are now commonly admitted to be teleostean; and even some palaeozoic forms have thus been classed by one high authority.

If the teleosteans had really appeared suddenly in the northern hemisphere at the commencement of the chalk formation, the fact would have been highly remarkable; but it would not have formed an insuperable difficulty, unless it could likewise have been shown that at the same period the species were suddenly and simultaneously developed in other quarters of the world. It is almost superfluous to remark that hardly any fossil fish are known from south of the equator; and by running through Pictet's *Palæontology* it will be seen that very few species are known from several formations in Europe. Some few families of fish now have a confined *range; the teleostean fishes might formerly have had a similarly confined range; and after having been largely developed in some one sea, have spread widely.

Nor have we any right to suppose that the seas of the world have always been so freely open from south to north as they are at present. Even at this day, if the Malay Archipelago were converted into land, the tropical parts of the Indian Ocean would form a large and perfectly enclosed basin, in which any great group of marine animals might be multiplied; and here they would remain confined, until some of the species became adapted to a cooler climate, and were enabled to double the Southern capes of Africa or Australia, and thus reach other and distant seas (*Origin* VI, pp. 285–6; Pictet 1853–4).

The last paragraph shows CD accepting a notion – that continents change positions relative to each other, break up and reassemble – which many geologists bitterly opposed until the advent of the theory of plate tectonics (Wegener 1966; Le Grand 1988). Also, this paragraph captures the broad outlines of how that part of the Tethys Sea overlapping with what we now call Southeast Asia became the world's hottest spot of marine, and especially teleostean, biodiversity (Pauly and Froese 2001). How could we here follow up on all of this?

Boxfishes Fishes of the Family Ostraciidae, characterized by a bony carapace encasing the body, hence their common name.

The family, which occurs throughout the Tropics, is represented here by *Ostracion punctatus* Bloch and Schneider 1801, sampled by CD in *Tahiti, "where it had previously been observed by Captain Cook" (*Fish*, p. 158).

Note that Jenyns initially identified this species as *Ostracion meleagris* Shaw, 1796 (in

Shaw and Nodder 1789–1813), a different species (*Fish in Spirits*, nos. 1325, 1326).

Brains Fish have small, simple brains, especially *Pike (see plot in *FishBase), but this is no reason to agree with Brullé (1844), who suggested that their brains, and other complex organs, develop *before* other organ systems in the course of their ontogeny.

This gave *Huxley an opportunity to write one of his devastating critiques, here in a letter to CD: "The heart of a Fish is very simple as compared with that of a Mammal & a like relation obtains between the brains of the two – if Brullés doctrine were correct therefore the Heart & Brain of [. .] the Fish would appear at a later period relatively to the other organs than those of the Mammals – I do not know that there is the least evidence of anything of the kind – On the contrary the history of development in the Fish & in the Mammal shews that in both the relative time of appearance of these organs is the same or at any rate the difference if such exists is so insignificant as to have escaped notice [. . .]. The animal body is built like a House – when the judicious builder begins with putting the simple rafters – According to Brullés notion of Nature's operations he would begin with the cornices, cupboards & grand piano." (*Correspondence*, July 7, 1857).

To which CD quickly agreed: Your instances of Heart & Brains of Fish seems to me very good (*Correspondence*, July 9, 1857).

Branchiae Gills or organs for respiration in water (*Origin VI*, p. 431). Branchiae work by exposing a respiratory surface to a water flow, this surface being equipped for gas exchanges (O_2 in, CO_2 out).

However, branchial surfaces cannot, for compelling reasons of geometry, grow as fast as the mass (of body tissue) that requires oxygen and produces carbon dioxide, and in water-breathers respiratory surface area per unit of body mass always declines with increasing body mass (Pauly 1981, 1997, 1998). This leads to strong constraints on the life history and size of fish (*see* **Oxygen**).

Bream Common name of *Abramis brama* (Linnaeus, 1758), a member of the Family *Cyprinidae.

Bream, also known as 'Carp bream', can reach over 80 cm and live in Eurasian rivers, from 40° to 75° N. They are mentioned only once by CD, along with the *sex ratios of fishes.

Breed A lineage of domesticated animals. CD used the word *race to refer to what we now call a breed (*see also* **Carp, Goldfish**).

Bull(dog) cod *See* **Cod**.

Bull trout *See* **Trout**.

Burbot Common name of *Lota lota* (Linnaeus, 1758), a freshwater gadid with a circum-arctic distribution, and an affinity for the bottom of deep lakes and large rivers (Cohen *et al.* 1990, p. 53).

CD discussed the eyes and swimbladder of this fish: In the depths of the ocean, & in deep & dark wells some Crustaceans [. . .] are blind.[4] Now though I am not aware that any Fish inhabiting very deep water is normally blind, yet it seems to bear on the above facts, that the Gadus lota[5] at the depth of 100 fathoms has its air-bladder frequently atrophied, often accompanied by total blindness. (*Big Species Book*, p. 296; n. 4 refers to Forbes (1851), p. 254; n. 5 to Jurine (1825), p. 149).

Note that CD defines 'atrophied' as Arrested in development at a very early stage (*Origin VI*, p. 431), which completely overlaps with his definition of 'aborted'; *see* **Lungfishes**.

Burrfishes (I) Fishes of the Family Diodontidae, also known as porcupine fishes, the latter name referring to the spines covering their bodies, which stand erect when their owners are inflated.

Quoy and Gaimard (1824, pp. 201-2) reported that the widely distributed

Indo-Pacific Birdbeak burrfish *Diodon caeruleus* [now *Cychlichthys orbicularis* (Bloch 1785)] was not consumed, and that in Guam, people considered it "venomous". They also speculated that this may be due to its consumption of coral, confirmed by opening the stomach of what must have been a large specimen, containing "one pound of rocky debris". CD summarized their account by stating that it has sometimes been thought (*vide* Quoy in Freycinet's *Voyage*), that coral-eating fish were poisonous, then used a *parrotfish to test the generality of the proposition. We now know that burrfishes are indeed dangerous, owing to toxins in their gonads and intestine (Halstead 1978; *see also* **Puffers**).

CD collected three species of burrfish himself. One was the Globefish *Diodon nicthemerus* (Cuvier, 1818, p. 135), misspelled *nycthemerus* in *Fish* (p. 150). As the label of the specimen was lost, the sampling locality is unknown, but it can be assumed to have been *King George's Sound, as the species is *endemic to southern Australia (Coleman 1980, p. 296), and CD collected Australian fish only at that location.

Next are the Striped burrfish, *Cyclichthys schoepfi* (Walbaum, 1792), formerly *Diodon rivulatus*, collected by CD on the shore of the Rio Plata, at *Maldonado, Uruguay (*Fish*, p. 150; *Fish in Spirits*, no. 723), and the Bridled burrfish *Cyclichthys antennatus* (Cuvier, 1816), formerly *Diodon antennatus*.

C. antennatus is a shallow-water species, found mainly on seagrass beds (Randall 1996). CD collected a specimen of this species in *Bahia, Brazil (*Fish*, p. 151; *Fish in Spirits*, no. 132), and upon his return to England, wanted it identified: Will you ask Leonard *Jenyns whether he can tell me the genus of little fish, which I believe is a Diodon (132). – It is the only fish I care about the name; but I am far from certain, whether it is one of those preserved or whether it was thrown away (*Correspondence* to J. S. *Henslow, July 12/13, 1837).

The reason CD was pressing Jenyns for the name of his '*Diodon*' was his wish to include a story, presented below, in the *Journal of the Beagle*, the book he was then working on. Also, true to his habit of recycling the same story through his various publications (see, for example, entries on **Megatooth shark**, or **Pike**), CD suggested to Jenyns he could incorporate elements of his *Diodon* story into *Fish*: You will find in my Journal p. 13 & 14 some account of the habits of the Diodon (No^r. 132) – & I send the notes, I made at the time, which very briefly relate to *colour. – You can extract shorten, & alter anything you think worthy of insertion. – I would have corrected a copy, but I really do not know whether any, or how much, is worth repeating. – Please use your own judgement. – & bring with you, when you come to town, the three original pages of notes (*Correspondence* Nov. 1841).

Burrfishes (II) Here are the notes CD wanted Jenyns to use:

March 10th a Diodon was caught swimming in its unexpanded form near to the shore. – Length about an inch: above blackish brown, beneath spotted with yellow. On head four soft projections; the upper ones longer like the feelers of a snail. – Eyes with pupils dark blue; iris yellow mottled with black. – The dorsal *caudal & anal fins are so close together that they act as one. These as well as the Pectorals which are placed just before branchial apertures, are in continuous state of tremulous motion even when the animal remains still. – the animal propels its body by using these posterior fins in same manner as a boat is sculled, that is by moving them rapidly from side to side with an oblique surface exposed to the water. – The pectoral fins have great play, which is necessary to enable the animal to swim with its back downwards. –

When handled a considerable quantity of fine "Carmine red" fibrous secretion was emitted from the abdomen & stained paper, ivory &c of a high *colour. – The fish has several means of defence it can bite & can squirt water to some distance from its Mouth, making at the same time a curious noise with its jaw. – After being taken out of the water for a short time, & placed in again, it absorbed by the mouth (perhaps likewise by the branchial apertures) a considerable quantity of water & air, sufficient to distend its body into a perfect globe. – This process is effected by two methods: [swallowing the air and water] & then forcing it into the cavity of the body, its return being prevented by a muscular contraction which is externally visible [and] by the dilatation of the animal producing suction. – The water however I observed entered in a stream through the mouth, which was wide open & motionless; hence this latter action must have been caused by a kind of suction. When the body is thus distended, the papillæ with which it is covered become stiff, the above mentioned tentacula on the head being excepted. – The animal being so much buoyed up, the branchial openings are out of water, but a stream regularly flowed out of them which was constantly replenished by the mouth.

After having remained in this state for a short time, the air & water would be expelled with considerable force from the branchial aperture & the mouth. – The animal at its pleasure could emit a certain portion of the water & I think it is clear that this is taken in partly for the sake of regulating the specific gravity of its body. – The skin about the abdomen is much looser than that on the back & in consequence is most distended; hence the animal swims with its back downwards. Cuvier doubts their being able to swim when in this position; but they clearly can not only swim forward, but also move round. – this they effect, not like other fish by the action of their tails, but collapsing the *caudal fins, they move only by their pectorals. – When placed in fresh water seemed singularly little inconvenienced. (*Zoology Notes*, pp. 25–6, which also give details on CD's corrections of his notes, not reproduced here).

Jenyns did follow CD's suggestion to include some of his field notes, and included in *Fish* (p. 151) a paragraph on the behavior of '*Diodon antennatus*', and a reference to pp. 13–14 of CD's *Journal*, both rare features in taxonomic books, then even more than presently devoted almost exclusively to the cataloguing of pickled corpses.

Note, however, that burrfishes are unlikely to have developed the ability to blow water as a *defence* tactic. [Which shark would be impressed by having water blown at its face?] Rather, this behaviour is a foraging technique, used to excavate invertebrates from sandy or muddy bottoms (Wainwright and Turigan 1997). Moreover, it is this behaviour, and its anatomical correlates, which provided the basis for the evolution of inflation in burrfishes.

Burrfishes (III) And here is the story CD was so eager to tell:

One day I was amused by watching the habits of a Diodon, which was caught swimming near the shore. This fish is well known to possess the singular power of distending itself into a nearly spherical form. After having been taken out of water for a short time, and then again immersed in it, a considerable quantity of both water and air was absorbed by the mouth, and perhaps likewise by the branchial apertures. This process is effected by two methods; the air is swallowed, and is then forced into the cavity of the body, its return being prevented by a muscular contraction which is externally visible; but the water, I observed, entered in a stream through the mouth, which was wide open and motionless: this latter action must, therefore, depend on suction.

The skin about the abdomen is much looser than that of the back; hence, during the inflation, the lower surface becomes far more distended than the upper; and the fish, in consequence, floats with its back downwards. Cuvier doubts whether the Diodon, in this position, is able to swim; but not only can it thus move forward in a straight line, but likewise it can turn round to either side. This latter movement is effected solely by the aid of the pectoral fins; the tail being collapsed, and not used. From the body being buoyed up with so much air, the branchial openings were out of the water; but a stream drawn in by the mouth, constantly flowed through them.

The fish, having remained in this distended state for a short time, generally expelled the air and water with considerable force from the branchial apertures and mouth. It could emit, at will, a certain portion of the water; and it appears, therefore, probable, that this fluid is taken in partly for the sake of regulating its specific gravity. This diodon possessed several means of defence. It could give a severe bite, and could eject water from its mouth to some distance, at the same time it made a curious noise by the movement of its jaws. By the inflation of its body, the papillæ, with which the skin is covered, became erect and pointed.

But the most curious circumstance was, that it emitted from the skin of its belly, when handled, a most beautiful carmine red and fibrous secretion, which stained ivory and paper in so permanent a manner, that the tint is retained with all its brightness to the present day. I am quite ignorant of the nature and use of this secretion. [I have heard from Dr. Allan of Forres, that he has frequently found a Diodon, floating alive and distended, in the stomach of the *shark; and that on several occasions he has known it eat its way not only through the coats of the stomach, but through the sides of the *monster, which has thus been killed. Who would ever have imagined that a little soft fish could have destroyed the great and savage shark?] (*Journal*, Feb. 29, 1832. Cuvier doubts is an understatement: Cuvier (1829) *asserts* that inflated *Diodon* cannot control their movements, but his authority is another author, as unlikely as Cuvier himself to have seen the fish when alive. So we should trust CD's direct observations. Dr J. Allen of Forres provided much information to CD when he completed *Coral Reefs, which cites his MS thesis, based on coral transplantation experiments along the east coast of Madagascar, in 1830–2.)

Randall (1996) describes for this species, as well, "a curious red secretion [which] may colour large regions of the body, especially ventrally" (p. 348). At my request, he amplified his description as follows: "I remember the red secretion of *C. antennatus*, but I do not know the function. If I were to have a live one in my hands today and it produced the secretion, I would taste it. In this way I and Japanese colleagues first determined the skin toxin of soapfishes and clingfishes" (John E. Randall, Bishop Museum, Hawai'i, pers. comm., May 1997). I have not followed up on this matter, as I did not want to put Jack at risk.

Finally, I must confess that I have been so far unable to corroborate the Jonahesque story about a burrfish chomping its way out of a shark's stomach.

Butterfishes Members of the Family Stromateidae, with about a dozen species. These coastal, smooth-looking fishes are very tasty (pers. obs.), which may be the reason for their common name (or maybe it is because their flesh is somewhat soft?). One of two species sampled by CD, *Stromateus stellatus* Cuvier, 1829 was, at *Chiloé, at the extreme southern end of its *range. Jenyns described it as *Stromateus maculatus*, a *synonym (*Fish*, p. 74; *see also Fish in Spirits*, nos. 788, 1146).

Butterflyfishes Members of the Family Chaetodontidae, known for their lively colours, and for the complete dependence of many of their long-snouted species on specific, live corals for their food, just as some butterflies with a long "proboscis" rely on specific flowers for their food (see Darwin 1877a).

The Threadfin butterflyfish, *Chaetodon auriga* Forsskål, 1775, which complements its diet of corals with other small *zoobenthos organisms (Hobson 1974; Sano *et al.* 1984), was collected by CD in April 1836 in the *Cocos Islands, and described as *Chaetodon setifer* in *Fish* (pp. 61–2).

Calendar Short title of *A Calendar of the Correspondence of Charles Darwin, 1821–1882*, a list of all the letters so far identified that were authored or received by CD.

The annotations in the *Calendar*, compiled by Burkhardt *et al*. (1985a) as a first step toward their monumental **Correspondence of Charles Darwin*, were used here to infer the contents of letters not included in the twelve volumes of the *Correspondence* so far published, and the total number of *words in CD's letters. Letters not seen, but cited from the *Calendar*, are identified here by their number.

***Callionymus* spp.** *See* **Dragonets**.

Cape of Good Hope Not the southernmost tip of Africa (which is Cape Agulhas), but close enough for 'good hope' to be justified. (*See also* **Jenyns' questions**.)

The **Beagle* and her crew spent three weeks at the Cape, from May 31 to June 18, 1836.

CD did not collect any fishes at the Cape, and only a few shells among marine organisms, emphasizing instead his land-based collection (Armstrong 1991a).

There is, however, one connection between fish and the Cape in CD's writings: As glacial action extended over whole of Europe, & in Himalaya on *both* sides of N. America & *both* sides of Southern S. America & I believe in N. Zealand [. . .], I consider it probable that some of the warmer temperate plants would spread into the Tropics, whilst the arctic plants reached the foot of the Alps & Pyrenees [. . .].

Some, I consider it possible might cross the Tropics & survive at C. of Good Hope, *T[ierra]. del Fuego & S. Australia. [. . .]. This theory, I conceive, explains certain aquatic *productions in S. hemisphere &c &c. (& European Fish at C. of Good Hope). (*Big Species Book*, p. 529).

And indeed, there are 'European fish' at the Cape of Good Hope, one example being the anchovy *Engraulis encrasicolus* Linnaeus, 1758, a species reaching North all the way to Central Norway (*see* **FishBase**). Its South African population – even if defined as a species of its own (which would be called *E. capensis* Gilchrist, 1913) – is in any case derived from a European population of *E. encrasicolus*.

Cape Verde Islands A small oceanic archipelago off the coast of northwest Africa, visited by the *Beagle* early in her voyage, in mid-January 1831. Lobban (1998) gives a detailed account of CD's activities while in the Cape Verde Islands, for which CD's own **Diary* or **Journal* should be consulted as well.

CD collected fish specimens described in Jenyns' *Fish*: *Serranus goreensis* and *S. aspersus* (*see* **Groupers (I)**); *Upeneus prayensis* (*see* **Goatfishes**); *Stegastes imbricatus* (*see* **Damselfishes**); *Blennius palmicornis*, *Salarias atlanticus* and *S. vomerinus* (*see* **Blennies**); and *Muraena* spp. (*see* **Morays**).

Capelin CD writes of this fish that [t]he males alone of the capelin (*Mallotus villosus*, one of Salmonidae), are provided with a ridge of closely-set, brush-like scales, by the aid of which two males, one on each side, hold the female, whilst she runs with great swiftness on the sandy beach, and there deposits her spawn.[2] (*Descent II*, p. 331; n. 2 cites Anon. 1871, p. 119; CD had written to Günther that he felt this account to be trustworthy [*Calendar, no. 9383], notwithstanding Günther's opinion to the opposite; see letter no. 9316 in the *Calendar*, whose editors erroneously assigned *Mallotus* to the Family *Cyprinidae).

Templeman (1948, p. 34), citing numerous sources, and based on his own observations, confirms that individual female Capelin frequently spawn jointly with two males, though only one male may be involved as well. In the latter case, the male bends around the female, and if she is much smaller than he, "they tend to move in a semi-circle while spawning". Thus here: CD 1 : Günther 0.

Capelin, originally described by Müller (1776), and now reassigned from the

Salmonidae to the Osmeridae (which also includes the *Smelt), are small, abundant, oil-rich fish, widely distributed in subarctic waters (Stergiou 1989), and important as prey for a large number of predators, notably cod-like fishes, seabirds and marine *mammals. Previously used as bait in the great cod fisheries of Atlantic Canada (Anon 1871), or for their roe (the carcasses being thrown away), they have been recently much depleted by industrial fisheries (usually supplying fish meal plants), of which at least one supplied a power plant in which Capelin were burnt to generate electricity (Clover 1991). This is only slightly worse than the proposed use of fish oils as alternative to diesel fuels (Blythe 1996). These practices have led to a shortage of prey for human food fishes and for birds, notably in the North Sea (Robertson *et al.* 1996), prompting calls to phase out such destructive operations.

Capybara A large semi-aquatic *mammal of South America, preyed upon by *jaguar. This large rodent (actually the largest of extant rodents) belongs, as well, to a large group of mammals and birds that were in the past often declared to be 'fish' so they could be eaten during the Lent season by otherwise good Catholics, e.g. in Venezuela and Colombia (Topoff 1997). In the European Middle Ages, this pseudo-fish group also included geese, spuriously connected with *barnacles ('goose barnacles') because of their feeding antennae, which can be construed as resembling feathers. Obviously, this group also includes all manner of marine mammals. CD, whose *Journal* frequently mentions capybara, had nothing to do with this.

***Caranx* spp.** *See* **Jacks.**

Carcharias megalodon *See* **Megatooth shark.**

Carp The common carp *Cyprinus carpio* (Linnaeus, 1758), one of the few fishes that are truly domesticated for aquaculture (Balon 1995a), a reason for their morphological variability.

CD commented on this: A *species may be highly variable, but distinct *races will not be formed, if from any cause selection be not applied. It would be difficult to select slight *variations in fishes from their place of habitation; and though the carp is extremely variable and is much attended to in Germany, only one well-marked race has been formed, as I hear from Lord A. Russell, namely the *spiegel-carpe*; and this is carefully secluded from the common scaly kind. On the other hand, a closely allied species, the *goldfish, from being reared in small vessels, and from having been carefully attended to by the Chinese, has yielded many races. (*Variations* II, p. 222; 'Spiegel-karpfen' [mirror carp], or, as CD put it looking-glass carp, is so named because it has only a few, large, shiny scales on its chest and back; Russell's letter to CD, correcting a statement in the first edition of *Variations*, was sent on May 19, 1873; *Calendar, no. 8915).

Lord Arthur Russell's studies in Germany (Tort 1996, p. 3754) may not have been that deep, as he misled CD. There were, at the beginning of the nineteenth century in Germany, at least two well-established 'races' of carp besides that mentioned above (Balon 1995b), namely the 'line' carp, with a single row of pectoral scales along the body, and the 'leather' or 'naked' carp, entirely wanting scales.

Carp, Crucian Common name of *Carassius carassius* (Linnaeus, 1758), Family *Cyprinidae. CD mentions the ability of Crucian carp to hybridize with a close relative, *C. gibelio* (*see* **Carp, Prussian**).

Carp, Indian Members of the Family Cyprinidae, important in South Asian aquaculture.

CD noted that: in India several species of freshwater fish are only so far treated artificially, that they are reared in great tanks; but this small change is sufficient to induce much variability[13]. (*Variations* II, p. 246: n. 13: McClelland on Indian Cyprinidae, Asiatic Researches, vol. xix, part ii, *1839*, pp. 266, 268, 313).

The relevant quotes, from McClelland (1839), are as follows: "*Cirrhinus rohita* var. attains a large size in Assam, and is probably the true *Ruee* of the natives. That which is figured by Buchanan is as far as I have seen a small fish, though the larger kind which I have figured would seem to be the one he has described. This as well as the preceding species present so many *varieties, probably the result of artificial means resorted to for their propagation; from their value as an article of food, that it is difficult to define their true characters." (p. 266, footnote; for 'Buchanan,' see Hamilton 1822).

"They correspond with the species named by Buchanan, *Cyp. curchius*, *C. cursa* and *C. cursis*, but I cannot altogether reconcile them with his descriptions; they appear to me to be varieties resulting from domestication" (p. 268, referring to what is now *Labeo rohita* (Hamilton, 1822)).

"*Cyp. rohita*, Buch. *Ruee* of the natives; no less celebrated in India than the *Carp in Europe. It is the fish described by Buchanan, though not the one he has figured as the *Ruee*, the principal difference being in the form of the mouth. The various slight modifications of form under which the *Ruee* appears, prove the extent to which this species must have, at one period, been propagated in India. It is one of the largest and most abundant fishes in all parts of the country." (p. 313, in figure legend).

Returning to the CD quote, we note, with Eknath and Doyle (1990), that in India, as in most other parts of the world, traditional aquaculture practices usually lead to inbreeding, i.e. to less variability than occurs in wild stocks . . .

Carp, Prussian Common name of *Carassius gibelio* (Bloch, 1782), a member of the Family *Cyprinidae, originally from East Asia, and now established in Eastern Europe.

CD comments on specimens of this Prussian carp hybridizing with, and also somehow changing into Goldfish (*Carassius auratus*):

I will now give a single case in Fish taken from Bronn;[3] the Cyprinus gibelio & carassius have generally been considered distinct species, for they differ in almost every part in proportion, as shown by the table given by Bronn; but Eckstrom narrates that the offspring of the C. carassius removed from a large lake into a small pond, assumed an intermediate form; & on the other hand the offspring from C. gibelio from a small pond turned into a large lake 40–50 years before, had become changed into C. carassius. (*Big Species Book*, p. 124; Eckström 1840; n. 3, referring to Bronn 1843, reads: Ges[ch]ichte der Natur. B. 2. s. 106; note CD's use of 'B.' for *Band* or volume, and 's' for *Seite* or page, as he usually did for German references. Remember: CD was not xenoglossophobic!)

Given the complex taxonomy of European members of the cyprinids (*see* FishBase), and the fact that Prussian carp is one of the few existing species of fish practising parthenogenesis, their having changed into Crucian carp (*C. carassius*) is probably due to misidentifications. However, their forming *hybrids, with intermediate characteristics, may be possible. Moreover, Prussian carp can be seen as 'changing into' another species, if one accepts, with several authors (*see* FishBase), that Prussian carp is a *subspecies of Goldfish, with the name *Carassius auratus gibelio* (Bloch, 1782).

Cartilaginous fishes A group of fishes also known as 'Chondrichthyes' (meaning the same thing) and including the *sharks, *rays, and *chimaeras, all appearing in CD's writings.

Their lack of bone tissue, previously thought to be a primitive condition, is now seen as secondary, since their ancestors (Devonian placoderms) had bodies covered by a bony armour. Indeed, the fine 'teeth' covering the skin of sharks are remainders of this armour.

In the 1820s, when young CD was a student of medicine in *Edinburgh, the *Lumpfish *Cyclopterus lumpus*, though a *bony fish, was still *lumped with the Chondrichthyes, owing to its cartilaginous body cover (see Cuvier (1828) for a review of early *classifications of fishes). This explains the conundrum faced by the young CD when he dissected this fish (*see also* **Seeing**).

Caspian Sea A large body of brackish water, surrounded by Russia, Turkmenistan, Iran, and Uzbekistan.

A remnant of the ancient Tethys Sea (*see* **Distribution**), the Caspian Sea was repeatedly connected to the Sea of Aral and to the Black Sea, forming a large Pontocaspian Sea, and is now in deep ecological trouble (Stone 2002). The Caspian Sea was subjected to large changes of salinity in the course of its long history, and as a consequence, its *fauna consists of several components: (1) a relict brackish-water fauna (Pontocaspian *endemics, e.g. the Caspian *lamprey *Caspiomyzon wagneri* Kessler 1870); (2) a Mediterranean element (e.g. *mullets); (3) an Arctic–Baltic element (mainly crustaceans); and (4) numerous freshwater species of the Pontacaspian region (Remane 1971, pp. 143–57).

The origin of the Caspian fauna were in CD's time already a matter of considerable debate, and he noted on this: Annals of. Natural. History. p. 135. Natural History of the Caspian. Fresh Water Fish!!?Adapted to salt water? – peculiar species, crabs & molluscs few. – ? are not some same – what is the alliance with the Black sea. – it would be ocean, what is land to continent – Original Paper, worth studying. (*Notebook D*, p. 379; Eichwald 1839, p. 135); Decemb. 21th. – L'Institut 1838. p. 412.[1] M. Eichwald has published Fauna of Caspian. –[2] fishes fresh water kinds. (yet living in the salt?.) – very few animals of any kind – Fauna, must be very curious. – (*Notebook E*, pp. 418–19; n. 1 refers to Eichwald 1838; n. 2 to Eichwald 1834–8).

Some jottings on modifications of the shell of *Cardium* from tertiary strata in the Crimea may be viewed as related to this: Bulletin Geologique April 1837, p. 216 Deshaye on changes in shells from salt & F. Water – on what is species. very good Has not Macculloch written on same changes in Fish? (*Notebook B*, p. 184; Deshaye 1836; Macculloch 1824).

Catadromy(-ous) Referring to fish that migrate from freshwater to the sea to spawn, as do *eels. (*See also* **Diadromy**.)

Catalogue Short name for CD's master list of the plants and animals he collected during the voyage of the **Beagle* (Barlow 1933; Porter 1985).

This list, which may be called 'Darwin's Database,' consists of three parts (i.e. red pocket notebooks with soft covers) labelled Catalogue for specimens in spirits of wine (nos. 1–660; 661–1346; and 1347–1529), and three parts labelled Printed numbers (nos. 1–1425; 1426–3344; and 3345–3907). The numbers, which are the same as in the *Zoology Notes*, refer to the numbers on the labels attached to the specimens, and are arranged in order of their acquisition.

"From this master catalogue and the zoological diary twelve separate classified annotated catalogues were prepared during the summer of 1836 so that they could be turned over to naturalists who would be engaged in naming and describing the specimens in their special fields (Sulloway 1982). Some of them were returned to CD when the work was completed, and they are now in the Darwin Archive: DAR 29.1 contains '*Animals*,' i.e. mammals; '**Fish in Spirits of Wine*'; '*Insecta*'; and '*Shells in Spirits of Wine*';" (Burkhardt *et al.* 1985b, p. 547; the words "zoological diary" refer to the **Zoology Notes*, edited by Keynes 2000).

CD's catalogue of specimens, presently on exhibit at CD's family house (Down House), and the annotated lists derived from it provide further illustration of the meticulousness with which CD assembled his *collection, thus justifying his subsequent authorship, in 1859, of a field manual for geologists and biologists.

Catfishes Members of a set of closely related families, all belonging to the Order Siluriformes, formerly a broad 'Siluridae' family. A few species practise *mouth-brooding (and even lower lip-brooding by male armored catfishes, or Loricariidae), some function as *cuckoo-fish, and all species are whiskered, with some South American species sporting *beards as well.

At least four species of catfish were sampled by CD, and three are described in *Fish (p. 110–14). One, based on a specimen taken from a running brook near *Rio de Janeiro, was *Pimelodus gracilis* Valenciennes 1836, i.e. it had already been described (see Jenyns' note in *Fish in Spirits*, no. 180). The other two were new: *P. exsudans* Jenyns 1840 (also from Rio?) was found in the *collections without labels (*Fish in Spirits*, after no. 181), and *Callichthys paleatus* (*Fish*, p. 113–14; Another specimen from same site; *Fish in Spirits*, nos. 181/182). Of these new species, the former is now called *Rhamdella exsudans* (Jenyns, 1840), and assigned to the long-whiskered catfishes (Family Pimelodidae), along with what is now *Pimelodella gracilis*. The third, now *Carydoras paleatus* (Jenyns, 1842) is presently assigned to the Callichthyidae, a family of armoured catfishes. Additionally, Jenyns noted "One of the Siluridae – very bad & thrown away" (*Fish in Spirits*, no. 1244). And since we are at it, we may mention item no. 1511 in CD's list of Specimens not in Spirits, consisting of the Pectoral bone from the Armado. Fish (*Zoology Notes*, p. 393), sampled somewhere near Buenos Aires in August 1833.

After this boring bit, here is what CD has to say about catfish-ing: Owing to bad weather we remained two days at our moorings. Our only amusement was catching fish for our dinner: there were several kinds, and all good eating. A fish called the 'armado' (a *Silurus), is remarkable from a harsh grating noise it makes when caught by hook and line, and which can be distinctly heard when the fish is beneath the water. This same fish has the power of firmly catching hold of any object, such as the blade of an oar or the fishing-line, with the strong spine both of its pectoral and dorsal fin. (*Journal*, Oct. 12, 1833; also in *Fish in Spirits*, no. 745; *see also* **Sounds**).

CD has more stories about catfishes. Thus, he notes in *Notebook B* (p. 205) that a Fish which emigrates over lands is a siluris, p. 123 (referring to Kirby 1835); On male fishes hatching the ova in their mouths, see a very interesting paper by Professor Wyman, in *Proc. Boston Soc. of Nat. Hist.*, 15 September 1857; also Professor Turner in *Journal of Anat. and Phys.*, 1 November, 1866, p. 78. Dr *Günther has likewise described similar cases. (*Descent I*, p. 163, n. 30, and p. 345, n. 38; Günther 1868a).

The item on male fish hatching ova in their mouths may be found on pp. 268–9 of the *Proceedings of the Boston Society of Natural History*, Vol. VI, pertaining to September 16, 1857, and was presented by its President (i.e. Jeffrey Wyman). There, we read that "some species of fishes from the Surinam River" were exhibited, belonging to a species of 'Bagré' – which indeed practises buccal incubation. However, the original text states that it is the *females* which incubate the *eggs: "[d]uring the months of June, the females have their mouths filled with eggs, and the young may be seen in all stages of formation, if a large number of individuals is examined. There are at least four species of Siluroids which have this habit."

Turner (1866), based on Wyman (1859), provides more details on Wyman's fishes: they were obtained from a market in Paramaribo (Dutch Guyana, now Surinam) and the "eggs were always in the mouth of the males, and were not bruised, and none were found in the stomach". Turner (1866) also cites Günther (1864, Vol. 5) to the effect that male specimens of '*Arius fissus*' – now *Cathorops spixii* (Agassiz, 1829) – sampled in Cayenne, French Guyana, had about 20 pea-sized eggs in their mouth and gill-chambers. Turner then goes on documenting another case of male *mouth-brooding in '*Arius boakeii*' from 'Ceylan' (Sri Lanka), simultaneously describing it as a new species.

Now we should digress a bit: Turner (1866) cites Wyman (1857) as having been presented on the September 15, just as CD cites it. Yet these proceedings are dated September 16, 1857. Hence, it appears that CD did not actually see Wyman (1857), but cited it from Turner (1866). This would explain why CD did not pick up the original (and erroneous) reference to *female* mouth-brooding *Bagré*. Turner's error appears to be due to Wyman (1859, p. 5) having given September 15, 1857 as the date in question. Hence, we may have here a case of a chain of authors (including CD and Tort (1996), p. 4709) citing a paper they have not seen (as I sometimes do in this book, as well). Let's add, to complete this story, that the true publication date of Turner '1866' is actually 1867.

Transcription errors often lead to results of this sort, though their specific nature differs among disciplines. Thus, geneticists now can track mutations (a form of transcription error) when these induce diseases in families with well-established genealogies (one of the few instances where aristocrats' exaggerated sense of themselves serves a useful purpose). Similarly, philologists have developed methods for tracking manuscripts through transcription errors (even monks are not perfect!).

Caudal (fin) Of or belonging to the tail (*Origin VI*, p. 431). In fish, the thing 'of the tail' that comes most readily to mind is the caudal fin, usually homocercal (equal-lobed), but sometimes *heterocercal, as in larval *Gar-pike, or in *sharks.

This entry could end here, given some witty coda, but won't, as there is more to fish tails than their giving a finish to their owner. Rather, I will use this opportunity to mention that caudal fins can be used to infer the metabolic rate of their owners. Thus, the spoon-shaped, broad caudal fins of, for example, *dragonets (Fig. 9) indicate low metabolic rates, whereas more open, forked tails such as in *jacks (Fig. 18) and especially in tuna, indicate high consumption of *oxygen. This difference in shape can be easily quantified (through aspect ratios, as used to describes the wings of airplanes) and used to estimate food consumption and related processes in fish (see Palomares and Pauly 1998; Pauly *et al*. 2000 – DNA box).

Finis.

Cavefishes Both an ecological group (fishes that live in caves) and a Family, the Amblyopsidae.

Both kinds of cavefishes interested CD, who after having read an article by Silliman (1851) asked for further information: What I want to know is, whether any of the Crustacea, spiders, insects (flies*beetles, crickets &c) & Fish belong to the American type (Has not *Agassiz noticed the Fish?) ie to genera or sections of genera, found only on the American continent. – I sh^d be most grateful for any, the least, information on this head (*Correspondence* to J. D. Dana, July 14, 1856).

CD felt that the distribution of cave organisms supported his cause: I have little doubt that like the fish Amblyopsis & like Proteus *in Europe,* these insects are 'wrecks of ancient life' or 'living fossils,' saved from competition & extermination (*Correspondence* to C. *Lyell, Jan. 10, 1860; *see also* **Extinction**).

This opinion later went into print: Far from feeling surprise that some of the cave-animals should be very anomalous, as *Agassiz has remarked in regard to the blind fish, the Amblyopsis, and as is the case with the blind Proteus with reference to the reptiles of Europe, I am only surprised that more wrecks of ancient life have not been preserved, owing to the less severe competition to which the scanty inhabitants of these dark abodes will have been exposed. (*Origin VI*, p. 112; Agassiz 1851).

Moreover, CD felt that, as in the case of island species, the variety of cave organisms on different continents, in spite of the similarity of their habitat, argued directly against *creationism: With respect to the cave animals, reflect on the cave-Rat, the fish Amblyopis & Astacus in America – the Proteus in caves of Europe, & you will admit that on creation doctrine, there has been surprising diversity for such similar habitations (*Correspondence* to A. Murray, May 5, 1860).

The Amblyopsis in question, first described from Mammoth Cave, Kentucky, USA, is the troglodyte *mouth-brooding Northern cavefish *Amblyopsis spelaea* DeKay, 1842, one of about a dozen species of Amblyopsidae, the others consisting of epigean, troglophile and more troglodyte species, which thus represent the whole gamut of adaptations for cave life (Poulsen 1963; or Nelson 1994, pp. 221–2). The Family occurs only in North America, and its troglodyte forms have no predators, as CD suggests above. Moreover, they are not anomalous, i.e. they are not the *aberrant relatives of the *guppies which Agassiz (1851) suggested they were. However, the amblyopsids are not a particularly 'ancient' group, i.e. not wrecks of ancient life. The score here is CD 3 : Agassiz 1.

On the other hand, Agassiz' patronizing call for "young American naturalists" to devote their lives to studying the origin of blindness in cavefishes has been followed up (if not only by 'young Americans'), with rather astounding results. This resolved the inconsistencies in CD's uses of the concepts of 'uselessness' and 'simple disuse,' as occurs in the *Big Species Book* (p. 295): in the caves of Kentucky there are, also, blind insects crustaceans, fish, & a Rat. The various stages of abortion of the eyes in these Kentucky animals is very curious: some have no trace of an eye, some have a rudiment, & the Crustacean has the footstalk for the eye without the organ, – it has the stand for the telescope without the instrument. Now as the existence of useless eyes could hardly be injurious to these animals, I should attribute their blindness to simple disuse.

As it turns out, maintaining functional eyes in an environment where vision provides no selective advantage is injurious, because this costs metabolic energy, which may better be used elsewhere (*see* **Oxygen**), e.g. for olfactory and tactile receptors and the associated parts of the *brain (Poulsen 1963). Thus, it is not simple disuse that causes the eye of successive generations of cavefish to atrophy, but competitions with variants whose fitness is high because of their atrophied eyes. This is also the reason why the degeneration of the eyes, well studied in *Astyanax mexicanus*, a *characin, follows highly predictable steps (Génermont *et al.* 1996), and stops when it starts compromising other, ontogenetically related, and still useful organ systems.

Caviller A British term for a person who raises irritating and trivial objections (to cavil: to find fault where unnecessary, to complain, criticize, carp).

CD had his fill of those (Ellegård 1958; Kogan 1960; see also the many scurrilous reviews in Hull 1973, itself scurrilously reviewed by Rudwick 1974), and this affected his response even to good news: Mr Walsh has evidently been a close & good observer of Fishes. – What he says about the Sea – *Trout in the Loch is very curious; & it would be very desirable that some professed Ichthyologist should examine these Trout. – It would make a good case for me. – But until thus examined & pronounced on, cavillers would simply deny that the fish was a sea-trout (*Correspondence* to G. R. *Waterhouse, Nov. 12, 1861).

Unfortunately, we do not know who Mr Walsh was, nor what his observations were, which some might cavil about.

CD Abbreviation of 'Charles *Darwin', used by the editors of his *Correspondence*, and other scholars, and also used here.

'CD' is also used for 'CD-ROM' (Compact Disc, Read Only Memory), i.e. a medium for mass storage of digitized text and figures that is eminently suited to making important works, such as Darwin's, accessible to a wide audience.

A 'Darwin CD-ROM,' edited by Ghiselin and Goldie (1997), is available which contains the full text of (i) most of CD's major books (*Journal,* *Zoology, *Coral Reefs, Cirripedia, *Origin, Orchids, *Descent, *Expression*), (ii) a few of his shorter papers (including Strickland *et al.* 1843), and (iii) Ghiselin (1969). The CD-ROM also contains a Darwin Biographical Dictionary, a Darwin Bibliography, pictures, and some fluff (details in Provine (1997) and at www.lbin.com).

I used this CD-ROM extensively when completing this book. On the other hand, I contributed to it the file of *Fish, so there was some give and take.

Chaetodon setifer *See* **Butterflyfishes**.

Chalk *See* **Cretaceous**.

Char Members of the genus *Salvelinus*, large trout-like fishes of northern Eurasia and North America. *Salvelinus umbla* (Linnaeus, 1758) is a form whose specific status has been questioned until recently, and most of the biological knowledge on char is thus derived from close relatives, notably the Brook and Arctic char, *S. fontinalis* and *S. alpinus*, respectively.

Char show up only once in CD's writings: the males of the char (S. umbla) are likewise at this season rather brighter in colour than the females[23] (*Descent II*, p. 340; this season refers to the breeding-season and n. 23 refers to Thompson (1841), p. 440).

However, p. 440 in CD's source only states that "the male fish can at a glance be distinguished from the females either by colour or by the many characters which are comprised under 'form'". It is on other pages of that source that the brighter colours of the males are described. But should anyone care?

Characins Members of a very speciose freshwater family, the Characidae, related to the Salmonidae, and ranging from Southern Texas to Argentina.

The most famous members of this family are the piranha, never mentioned by CD, notwithstanding their fearsome, if somewhat ill-deserved reputation (Pauly 1994b) and their occurrence in the upper Parana River, from which some of his fish *collection originates.

In *Fish*, the characins are classified as harmless cousins of the *trout, i.e. among the salmonids, though their pectoral fin causes painful pricks (*Fish in Spirits*, no. 180). They consist of six species, of which *Jenyns assigned five to the genus *Tetragonopterus* (pp. 123–8). For three of these, the only taxonomic updating required is their assignment to the genus *Astyanax*: *A. abramis* (Fig. 4A), *A. scabripinnis* (Fig. 4B), and *A. taeniatus* (Jenyns, 1842). The fourth species, *T. rutilus*, was found to be a synonym of *A. fasciatus* (Cuvier, 1819), while the fifth, *T. interruptus*, is now *Cheirodon interruptus* (Jenyns, 1842; Fig. 4C). The sixth species of characin in *Fish* is *Oligosarcus hepsetus* (Cuvier, 1829), which Jenyns originally assigned to the genus *Hydrocyon* (*Fish*, p. 128; *Fish in Spirits*, no. 661).

CD collected his characins from a running brook at Socego, province of *Rio de Janeiro (*A. scabripinnis*; *Fish in Spirits*, no. 288), near Rosario on the Parana River (*A. abramis*; *A. fasciatus*; *Fish in Spirits*, nos. 747, 748), and a lake near *Maldonado (*C. interruptus*; *O. hepsetus*). He reported on the live *colours of three species: Back bluish silvery, with a silver band on the side: a bluish black spot behind the gills. Fins pale orange; tail with a black central band (*A. abramis*); Back iridescent greenish brown: a silver band on the side. Fins dirty orange: tail with a central black band: above and below the band bright red and orange (*A. fasciatus*; Fig. 4D); Bluish silvery (*O. hepsetus*).

Characins – especially those of the genus *Astyanax* – have often evolved into eyeless *cavefishes (Kullander 1999), a topic of much interest to CD.

Cheilio ramosus *See* **Wrasses**.

Chimaera

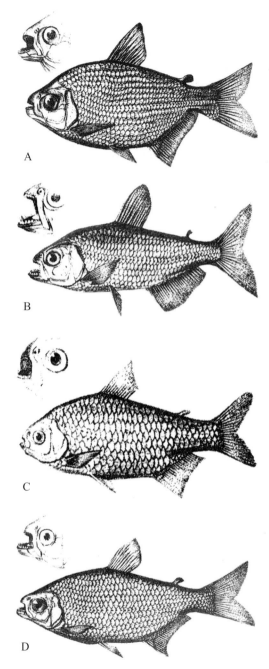

Fig. 4. Characins, Family Characidae, based on lithographs by B. Waterhouse Hawkins, with inserts showing the mouth, much less impressive than that of piranhas, their close relatives. **A** *Astyanax abramis* (*Tetragonopterus abramis* in *Fish*, Plate XXIII), based on a specimen from the Parana River, Argentina. **B** *Astyanax scabripinnis* (*Tetragonopterus scabripinnis*, Plate XXIII), from Rio de Janeiro, Brazil. **C** *Cheirodon interruptus* (*Tetragonopterus interruptus*, Plate XXIII), from Maldonado, Uruguay. **D** *Astyanax fasciatus* (*Tetragonopterus rutilus*; Plate XXIII), from the Parana River, Argentina.

Chiloé Island off southern Chile (about 42°S), north of the Chonos Archipelago. Jenyns refers in *Fish* (pp. v, xiii–xvii) to an "Archipelago of Chiloe", but this is misleading, even though there are a few small islands around Chiloé.

CD visited Chiloé in November 1834, and January–February 1835, and collected the following fish species: *Aspidophorus chiloensis* (see **Poachers**); *Stromateus maculatus* (see **Butterfishes**); *Iluocoetes fimbriatus* (see **Eelpouts**); *Gobius ophicephalus* (see **Gobies**); and *Gobiesox marmoratus* (see **Clingfishes**).

Chimaera Mythical animal, composed of parts from other animals, often a lion's head, a goat's body, and a serpent's tail. Also a group of odd-looking fishes of the Order Chimaeriformes, similar to sharks in many features, and to teleost fishes in others. The first species of this order to be described in the scientific literature is the 'rabbitfish' *Chimaera monstrosa* Linnaeus, 1758, whose common and scientific *names jointly illustrate the ontological confusion they generated – notwithstanding that 'rabbitfish' is also used, and more appropriately, for the seagrass-loving Siganidae (Tsuda and Bryan 1973; Stergiou 1988).

CD jotted p. 71 Chimera – Antarctica [. . .] <<caught>> Chile, Van Diemen's Land & Cape of Good Hope, upon reading Lesson (1830–1), thus referring, if implicitly, to two other chimaeras, *Callorhinchus callorhynchus* (Linnaeus 1758), and *C. capensis* (Duméril, 1865), which jointly have the range given in Lesson, and noted by CD (*Notebook C*, p. 247). However, CD did not follow up on this.

He also wrote that: The males of *Plagiostomous fishes (*sharks, *rays) and of Chimaeroid fishes are provided with claspers which serve to retain the female, like the various structures possessed by many of the lower animals. (*Descent II*, pp. 330–1). And: [i]n that strange *monster, the *Chimaera monstrosa*, the male has a hook-shaped bone on the top of the head, directed forwards, with its end rounded and covered with sharp spines; in the female 'this crown is altogether absent,' but what its use may be to the male is utterly unknown.[19] (*Descent II*, p. 338; n. 19 reads: F. Buckland, in *Land and Water*, July, 1868, p. 377, with a figure. Many other cases could be added of structures peculiar to the male, of which the uses are not known; see Burgess (1967) for details on Buckland's contributions to *Land and Water*).

Finally, Helfman *et al.* (1997, p. 194) inform us that the hook-shaped bone on the top of the head is called a 'tentaculum' and functions as "an additional clasper". The latter bit is slightly misleading, since even male chimaeras (I think) would not dare attempt to inseminate female Chimaeras by using an organ sticking up from their forehead.

Chrestomathy A compilation and rearrangement of selected literary passages, from one or several authors, often meant to assist readers in acquiring a language.

Darwin's Fishes, which quotes copiously from CD's work, is a chrestomathy devoted to the language he used to express his ideas and, equally, to the language of *ichthyology and of science in the nineteenth century.

Chromis(-ids, idae) See **Cichlidae**; **Damselfishes**.
Chromis crusma See **Damselfishes**.
Chrysophrys taurina See **Porgies**.

Cichlidae (I) A very speciose family in fresh and brackish waters of Africa and South America, derived from marine ancestors, and related to the *wrasses and *parrotfishes.

The cichlids share with the *damselfishes the feature of having only one nostril on each side of their snout, and this is the reason why, in earlier classifications, they were considered members of a broadly defined 'Chromidae' family (Berg 1958, pp. 254–5).

Some cichlid species, notably among the lepidophages (*scale-eaters) display a strong *asymmetry in the orientation of their mouth, which may be bent left or right, thus facilitating lateral attacks (Barlow 2000, p. 36).

The only member of the Family Cichlidae sampled by CD is *Cichlasoma facetum* (Jenyns, 1842), formerly *Chromis facetus*, from *Maldonado, Uruguay (*Fish*, pp. 104–5; Above, greenish black; the sides paler; slightly iridescent).

Cichlidae (II) CD discussed cichlids at length, as they provided neat examples for *sexual selection. Thus: In many of the Chromidae, for instance in Geophagus, and especially in Cichla, the males, as I hear from Professor Agassiz,[21] have a conspicuous protuberance on the forehead, which is wholly wanting in the females and in the young males. Professor Agassiz adds, 'I have often observed these fishes at the time of spawning when the protuberance is largest, and at other seasons when it is totally wanting, and the two sexes show no difference whatever in the outline of the profile of the head. I never could ascertain that it subserves any special function, and the Indians on the Amazon know nothing about its use.' These protuberance resemble, in their periodical appearance, the fleshy caruncles on the heads of certain birds; but whether they serve as ornaments must remain at present doubtful. (*Descent II*, p. 340; n. 21 reads: See also A Journey in Brazil, by Professor and Mrs. Agassiz, 1868, p. 220); the protuberance on the forehead is now called a 'nuchal hump' and appears to be caused by a 'local edema', i.e. the hump is filled with fluid, and probably serves to signal maleness; Barlow 2000, p. 142].

With the various species of Chromids, as Professor Agassiz likewise informs me, sexual differences in colour may be observed, 'whether they lay their *eggs in the water among aquatic plants, or deposit them in holes, leaving them to come out without further care, or build shallow nests in the river mud, over which they sit, as our *Pomotis does. It ought also to be observed that these sitters are among the brightest species in their respective families; for instance, Hygrogonus is bright green, with large black ocelli, encircled with the most brilliant red.' Whether with all the species of Chromids it is the male alone which sits on the eggs is not know. It is, however, manifest that the fact of the eggs being protected or unprotected by the parents, had had little or no influence on the differences in colour between the sexes. It is further manifest, in all the cases in which the male take exclusive charge of the nests and young, that the destruction of the brighter-coloured males would be far more influential on the character of the *race, than the destruction of the brighter-coloured females; for the death of the males during the period of incubation or nursing would entail the death of the young, so that they could not inherit his peculiarities; yet, in many of these very cases the males are more conspicuously coloured than the females. (*Descent II*, pp. 345–6; based on a letter from *Agassiz dated July 22, 1868; *Calendar* no. 6286; *see also* **Sexual selection**).

An important group of cichlids are those of the Great Lakes of Africa, felt by many ichthyologists to be the aquatic counterparts of *Darwin's Finches, owing to their rapid *speciation. Hence the title of a recent, if misleadingly titled, book on fish collecting on the shore of Lake Victoria, *Darwin's Dreampond* (Goldschmidt 1996), which does not seriously explore any issue raised by CD's work. These issues, on the other hand, are masterfully handled in Fryer and Iles (1972), Greenwood (1974), Kingdon (1989), and Barlow (2000), and regularly updated taxonomic and biological accounts of the many cichlid and other fish species involved here may be found in *FishBase.

Ciguatera A form of ichthyotoxicity due to the consumption of coral reef fishes containing toxins secreted by a dinoflagellate *Gambierdiscus toxicus* (Lewis and Holmes, 1993). Herbivorous fishes ingest this toxin and remain unaffected by it, but concentrate it in their gut and muscle tissues. Further concentration occurs when predatory fish consume ciguatoxic herbivores.

CD appears to have consumed at least two kinds of fish now known to be often ciguatoxic. In the case of the *barrow cooter (barracuda), he clearly did not know of the potential danger their consumption represents (De Sylva 1963, pp. 128–53). However, in the *Cocos Islands he appears to have deliberately consumed a *parrotfish to test whether it contained poison. This test was dangerous, as parrotfishes are frequently ciguatoxic (Auerbach 1991; Bagnis *et al*. 1974; Chungue *et al*. 1977). Indeed, they contributed 12% of more than 500 cases attributed to fish in the ciguatera database assembled by the Secretariate of the Pacific Communities, Nouméa, New Caledonia (Dalzell 1993), and incorporated in *FishBase.

Cirriped(es, -edia) *See* **Barnacles.**

Classification The process by which entities – including living organisms – are assigned to groups with 'similar' characteristics, the result of that process also being a 'classification'. Traditionally, the similarities used for classifications were in the eyes, or rather in the mind, of the beholder (Lakoff 1987), evolutionary thinking requires another way of *seeing.

Thus, CD warns: how grossly wrong would be the classification, which put close to each other a Marsupial and Placental animal, and two birds with widely different skeletons. Relations, such as in the two latter cases, or as that between the *whale and fishes, are denominated 'analogical,' or are sometimes described as 'relations of adaption.' They are infinitely numerous and often very singular; but are of no use in the classification of the higher groups. (*Foundations*, pp. 199–200; *see also* **Vertebrate origins**).

And: generally, it may be observed in the writings of most *naturalists, that when an organism is described as intermediate between two *great* groups, its relations are not to particular species of either group, but to both groups, as wholes. A little reflection will show how exceptions (as that of the *Lepidosiren, a fish closely related to *particular* reptiles) might occur, namely from a few descendents of those species, which at a very early period branched out from a common parent-stock and so formed the two orders or groups, having survived, in nearly their original state, to the present time. (*Foundations*, p. 212).

I think considerable light can be thrown by the theory of descent on these wonderful embryological facts which are common in a greater or less degree to the whole vegetable kingdom, and in some manner to the animal kingdom: on the fact, for instance, of the arteries in the embryonic mammal, bird, reptile and fish, running and branching in the same courses and nearly in the same manner with the arteries in the full-grown fish; on the fact I may add of the high importance to systematic naturalists of the characters and resemblances in the embryonic state, in ascertaining the true position in the natural system of mature organic beings. (*Foundations*, p. 220).

In spite of the last statement, CD later claimed that his theory has little impact on the *practice* of *taxonomy. Thus, in *Origin*, he suggested that systematists will be able to pursue their labor as at present (p. 484), and this was (and still is) true for most taxonomists (De Queiroz 1988).

Only one ichthyologist – Giovanni Canestrini (1835–1900) – readily comes to mind as having understood, upon reading *Origin*, the need for members of his profession to acknowledge variations within and between populations, i.e. to move away from the typological thinking that had so long dominated European philosophy and the emerging natural sciences (Mayr 1982, pp. 45–7; Sinclair and Solemdal 1988). Hence his remark that "allowing *varieties to be incorrectly called species not only makes classification of our fishes difficult but also curtails *a priori* the investigation of the causes that give rise to many varieties" (Canestrini 1866, cited in Pancaldi 1991, p. 83).

Climbing fish A marvel noted by CD in his *Notebook B* (p. 205; based on Kirby 1835): A climbing fish. p. 122, and again in the *Big Species Book*: Certain fish use their pectoral fins [...] even for climbing trees; if fish had become [...] terrestrial animals, how easily ancient transitional uses of the pectoral fins, might have baffled all conjuncture. (p. 341).

These notes, which refer to the Climbing perch *Anabas testudineus* (Bloch, 1792), named after *testudo*, the turtle (not known to climb much), are interesting in that some creationists complain that key links between early fishes and tetrapods are still "missing", even though that story, as shown by Zimmer (1998), has been nailed down.

Clingfishes Small fishes of the Family Gobiesocidae, so named because they 'cling' to the substrate by means of a sucking disk derived from their pelvic fins.

The family is represented here by *Gobiesox marmoratus* Jenyns, 1842, a Sucking fish collected at *Chiloé, Chile (*Fish in Spirits*, no. 1081; Fig. 5A), and the Red clingfish *Arcos poecilophthalmus* (Jenyns, 1842), from a tidal pool at Chatham (San Christobal) Island, *Galápagos (*Gobiesox poecilophthalmos* in *Fish in Spirits*, no. 1288, and *Fish*, pp. 140–1; Fig. 5B).

Grove and Lavenberg (1997, p. 241, based on Briggs 1955) have suggested that *Arcos poecilophthalmus*, which differs markedly from the two other species of *Arcos* occurring in the Eastern Pacific, colonized the Galápagos from the

Cocos Islands

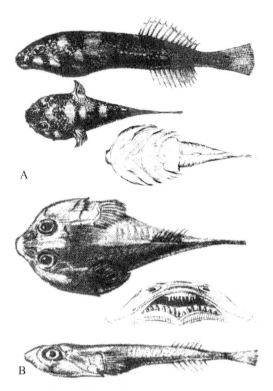

Fig. 5. Clingfishes, Family Gobiesocidae, based on lithographs by B. Waterhouse Hawkins (*Fish*, Plate XXVII). **A** *Gobiesox marmoratus*, sampled off Lemuy Island, southern Chile. **B** *Arcos poecilophthalmus* (*Gobiesox poecilophthalmus*), from a tidal pool in Chatham (San Christobal) Island, Galápagos. The insert shows a magnified view of its teeth.

Atlantic, during a period when the isthmus of *Panama was still open.

Clinus crinitus *See* **Blennies.**
***Clupea* spp.** *See* **Herrings.**
Cobites(-is) *See* **Loaches.**
Cocos Islands These islands, whose official name is 'Cocos (Keeling) Islands', consist of about 27 islands forming two distinct atolls, jointly covering a total area of 14 km². They are now an Australian territory, although located close to the Indonesian island of Java (12°05′S; 96°55′E). Their flora and *fauna have been much modified by human impact (Armstrong 1992d).

The *Beagle visited the Cocos Islands from April 1 to 12, 1836, and this gave CD the opportunity to test his theory of coral reef formation, whose basic outline was thought out well before he had, in the Cocos, his first (and only) opportunity to study a coral reef (Armstrong 1991b; Stoddart 1962; Woodruffe *et al.* 1990). This theory, which, notwithstanding *Agassiz (1883), turned out to be correct, was thus developed well before CD had all his 'facts', in stark contrast to naïve perceptions of how science works.

Here are some of his observations: There is no want of animal food at these Islands, for turtle & fish abound in the *lagoon. (*Diary*, April 1, 1836). Also: I was employed all the day in examining the very interesting yet simple structure & origin of these islands. The water being unusually smooth, I waded in as far as the living mounds of coral on which the swell of the open sea breaks. In some of the gullies & hollows, there were beautiful green & other colored fishes, & the forms & tints of many of the *Zoophites were admirable. It is excusable to grow enthusiastic over the infinite numbers of organic beings with which the sea of the tropics, so prodigal of life, teems; yet I must confess I think those *Naturalists who have described in well known words the submarine grottoes, decked with a thousand beauties, have indulged in rather extravagant language. (*Diary*, April 4, 1836).

Note that it was easy for CD, at this point, to abstain from extravagant language: he already had the key to the simple structure & origin of coral reefs!

A total of 533 fish species are now reported from the Cocos Islands (Allen and Smith-Vaniz 1994), and the specimens sampled by CD in the Cocos Islands were assigned in Jenyns' *Fish* to the following eleven: *Diacope marginata* (*see* **Snappers**); *Upeneus flavolineatus* (*see* **Goatfishes**); *Gerres oyena* (*see* **Mojarras**); *Chaetodon setifer* (*see* **Butterflyfishes**); *Seriola* *bipinnulata* (see **Rainbow runner**); *Acanthurus triostegus* (see **Surgeonfishes**); **Mugil* sp. (see **Mullets**); *Salarias quadricornis* (see **Galaxiidae**); *Scarus chlorodon* and *Scarus* sp. (see **Parrotfishes**), and *Tetrodon implutus* (see **Puffers**).

Referring to agencies working against the growth of corals, CD states that For instance, Mr Liesk informed me, that some years before our visit unusually heavy *rain killed nearly all the fish in the lagoon, and probably the same cause would likewise injure the corals. (*Coral Reefs*, p. 27; Mr. Liesk is, according to the *Diary* (April 1–3, 1836), an English resident [. . . living with Capt. Ross, the master of the islands] in a large barn-like house open at both ends; *see also* **Parrotfishes**).

Cod Common name of *Gadus morhua* (Linnaeus, 1758), commercially the most important species of its genus and family (Gadidae), all frequently referred to as *Gadus* or cod in CD's writing (*see*, for example, *Fish in Spirits*, nos. 751, 877).

Despite its enormous economic importance to nineteenth century England and other countries around the North Atlantic, *Gadus morhua*, "the Fish that Changed the World" (Kurlansky 1997) rates only scattered mentions in CD's writings.

Thus, he notes that their fecundity is very high: Mr F. Buckland found 6,867,840 *eggs in a cod-fish (Land and Water, 1868, p. 62) (*Variations* II, p. 373, n. 48).

In a discussion of the prevalence of 'struggle' in nature, CD mentions that Mr. Couch caught a cod-fish with no eyes, yet in good condition[2] (*Big Species Book*, p. 205; n. 2 refers to Couch (1825), p. 72), and concludes from this and other examples that the struggle for existence is periodical & not incessant (*Big Species Book*, p. 206; *see also* **Punctuated equilibrium**).

CD also mentions change in cod abundance: Codfish were formerly never caught at Cape Hinlopen, but now they are numerous there (*Big Species Book*, p. 573; based on Kalm [1753] 1/294; Cape Hinlopen is in Delaware, USA, 38°48′N, near the southern limit of the range of cod in the Western Atlantic; Scott and Scott 1988, p. 269; Wise 1958).

Also, Jeffries Wyman informed CD of the existence, off Labrador, of a rather common *monstrosity (*Correspondence*, Sept. 15, 1860). CD encouraged him: I am very glad to hear that you are collecting facts on the 'Bull-dog' fish (*Correspondence*, Dec. 3, 1860), but no additional information came, and CD had to use what he had for a short note: Prof. Wyman, of Cambridge, United States, informs me that the common cod-fish presents a [...] *monstrosity, called by the fishermen the 'bulldog cod' (*Variations* I, p. 93, n. 64).

In Norway, a country where cod matter, this monstrosity is considered to bring good luck. There, bulldog cod are called *kongetorsk* (king cod), the upper part of the head being seen as looking like a crown. This obviously implies that kongetorsk should be leading the other cod, and one can find written account of this belief (e.g. in a translation from p. 219 of Lilienskiold 1701): "as the bees have their pathfinders, and the geese have their leaders, so the cod do not lack a leader; they have the Torsche-Kongen, which leads the entire schools toward the country . . ."

Cod liver oil Cod, i.e. *Gadus morhua*, have large livers, whose lipids serve as energy storage.

The oil extracted from cod livers has many uses. One of them is to serve as a basis for paints, as was the case in Newfoundland until the advent of plastic-based paints.

The better-known use of cod liver oil, however, is medicinal, as generations of children will attest who were force-fed a spoonful a day of the "vile-tasting liquid" (Kurlansky 1997, p. 154). In Britain, the use of cod liver oil intensified after the publication of Bennett (1841), though it was already in use as "a remedy for rheumatism, then a catchall diagnosis for aches and pains." (Kurlansky 1997).

In 1860, a daughter of Emma and Charles Darwin, Henrietta, became ill, suffering from severe vomiting attacks and a bout of typhus fever (*Correspondence* to W. E. Darwin, Jan. 24, 1861; note that typhus and typhoid fever, two different diseases, were not well distinguished in CD's time; Herbert 1980a). As Henrietta's illness had led to a hardening of the abdominal tissues that the doctor believes will only gradually subside (*Correspondence* to J. D. Hooker, August 7, 1860), her parents had good reasons to worry: the same year, the 42-year old Albert, Prince Consort of Queen Victoria, died of typhoid fever.

CD's friend J. D. Hooker, a botanist, recommended that Henrietta be given cod liver oil, as this had been newly suggested in France for symptoms such as hers. Henrietta's parents went along with this and so did Henrietta, but her stomach did not: she could not keep the stuff down. Hooker suggested it would also work if rubbed on and CD, ever inquiring, asked him for details: You mentioned in a former letter about rubbing in cod-liver oil. Etty cannot take it internally. I asked our Doctor here & he had never heard such a thing (*Correspondence*, Feb. 8, 1861).

Hooker answered, and CD was grateful: Many & cordial thanks for your admirable & clear letter about rubbing in oil. We have begun (*Correspondence*, Feb. 20, 1861). And behold, a few months shortly thereafter, CD noted: Some gain of strength has *certainly* been coincident with rubbing in the oil, which my wife has steadily continued since February (June 19, 1861); Etty is going on well & continues to absorb Cod-liver oil – we think this best advice of yours of any we have had (*Correspondence*, July 17, 1861); The sea has certainly done Etty great good; but I still pin my faith to your oil, which is never neglected (*Correspondence*, August 13, 1861); and finally Etty goes on splendidly, & still sucks in the Oil (*Correspondence*, Sept. 24, 1861). She must have stopped at some time, since she later got married, and went on to edit her mother's correspondence (Litchfield 1915).

A good guess is that the cod-liver oil application had no effect whatsoever on Henrietta's health. However, just to be sure, two Professors at the Division of Infectious Disease, Department of Medicine, University of British Columbia, were consulted on this. One wrote "I am unaware of any specific medicinal effects of cod liver oil which may either provide microbicidal activity against *Salmonella typhi*, the causative agent of typhoid fever, or by bolstering the cellular immune mechanisms, which are important in host defence against typhoid fever" (Dr A. W. Chow, pers. comm., January 24, 1995). The second opinion was very similar to the first, except that its author added that "[t]he anecdote [. . .] regarding Darwin's treatment of his daughter with cod liver oil is certainly romantic and interesting, but as far as I am aware, this has not been followed in any objective way. I wonder whether the Medical Research Council would be interested in funding such a study?" (Dr N. Rainer, pers. comm., February 2, 1995). This is how Science progresses.

Code Short name for the *International Code of Zoological Nomenclature* (ICZN 1999) which contains the rules and recommendations referring to the availability of scientific *names in the animal kingdom, as used for *taxonomy. The *Code* is largely based on rules initially standardized by the Strickland Commission (Strickland *et al*. 1843), with CD one of the *alia*.

Collected Papers Short title of *The Collected Papers of Charles Darwin*, edited by Barrett (1977), and containing all (152) of his short publications. The first edition of *Collected Papers*, which also includes a few previously unpublished short documents, notably on *Lumpfish, was issued as two separate volumes; the edition used here consists of 'two volumes in one', while retaining the original pagination. Thus, the few references to *Collected Papers* in this book indicate the volume number in addition to the pages cited. Also, the original references to the few short papers of CD's mentioning fish are given in the bibliography of this book.

Collection The ensemble of specimens, notably of fishes, collected by CD during the voyage of the *Beagle*, and that she brought back, or that were earlier sent back to England via other vessels.

These consignments, to J. S. *Henslow, include descriptions of the material sent, e.g. [t]he small cask contains fish; will you open it, to see how the spirit has stood the evaporation of the Tropics [. . .]. I have sent to you by the Duke of York Packet, commanded by Lieu: Snell to Falmouth. – two large casks containing fossil bones. – a small cask with fish (*Corresp.*, Oct./Nov., 1832). By the same packet, which takes this, there will come four barrells: the largest will require opening, as it contains skins, Plants &c &c, & cigar box with pill boxes: the two next in size, only Geological specimens need not be opened, without you like to see them, the smallest & flat barrell, contains fish; with a gimlet, you can easily ascertain how full it is of spirits (*Correspondence*, July 18, 1833); and [t]he Cask is divided into Compartments the upper contains a few skins. – the other a jar of fish, & I am very anxious to hear how the Spirit withstands evaporation (*Correspondence*, Nov. 12, 1833).

What CD feared did in fact happen, as indicated by a letter from Henslow: "I am afraid that I have been rather negligent in not writing sooner to announce the arrival of your last Cargo which came safe to hand excepting a few articles in the Cask of Spirits which are spoiled, owing to the spirit having escaped thro' the bung-hole. [However,] I have popped the various animals that were in the Keg into fresh spirits in jars & placed them in my cellar" (*Correspondence*, Aug. 31, 1833). CD learned from this, and his few piece of advice to collectors, which may be found in his *Journal* (pp. 598–601), and, in reprinted form, in Barlow (1967), emphasizes the need for tight containers: jars should be closed with a bung covered by bladder, twice by common tinfold, and by bladder again; let the bladder soak till half putrid. I found this plan quite worth the trouble it cost.

Still, a sizeable number of CD's fishes reached England in acceptable condition, and could be described by *Jenyns. What then happened to these fishes is well documented: I have sent off per Waggon, (Carr Paid) the fish skins & all the bottles carefully packed up, directed to Mr Crouch Phil. Soc Cambridge (*Correspondence* to Jenyns, May 9, 1842, right after the last number of *Fish* was published). Unsurprisingly, CD's fishes then went to the University Museum of Zoology, Cambridge, where they remained until 1917, when the majority of the specimens (see **Appendix II**) were transferred to the Natural History Museum, London, formerly the British Museum (Natural History). The specimens still in Cambridge are documented in **Appendix III**.

Other zoological specimens collected by CD (mainly crustaceans) are kept at the *Oxford University Museum.

Colours These play a very important role in the writings of the mature CD, notably with regard to *secondary sexual characteristics, and his theory of *sexual selection.

Thus, for example, he asks: What, then, are we to conclude in regard to the many fishes, both sexes of which are splendidly coloured? Mr. *Wallace[30] believes that the species which frequent reefs, where corals and other brightly-coloured organisms abound, are brightly coloured in order to escape detection by their enemies; but according to my recollection they were thus rendered highly conspicuous. In the fresh-waters of the tropics there are no brilliantly-coloured corals or other organisms for the fishes to resemble; yet many species in the Amazons are beautifully coloured, and many of the carnivorous Cyprinidae in India are ornamented (*Descent II*,

Complexity

p. 343; n. 30 refers to Wallace 1867, who emphasized the camouflaging effect of coral reef fish colours, still an active research area (see Marshall 1998); *see also* **Altruism**).

This affinity to colours was noticeable in the young CD as well, who, in a letter to J. S. *Henslow, even mentions these colours in the same context as the labels on his specimens, and behavioural observations, thus stressing their importance: N.B. What I have said about the numbers attached to the fossils, applies to every part of my *collections. – Videlicet. Colors of all the Fish: habits of birds &c & &c. (*Correspondence*, March 1834; '*videlicet*' means 'obviously').

In line with this, CD used standardized colour names to describe specimens, taken from Patrick Syme's second edition of *Werner's Nomenclature of Colours* (Syme 1821), or, as he explained to Jenyns: colours, when given were compared with Pat. Syme's nomenclature book in hand (*Corresp*. Oct. 17, 1839; see annotation to Syme 1821 to see what this implied). Moreover, CD often put the colours' names from Syme's book in quotation marks, many recoverable even from *Covington's copy of his notes (see *Fish in Spirits*). Thus, CD's descriptions of the colours of fish – all reproduced here (the descriptions, not the colours) – are still useful today, contrary to the fanciful wordage of many other naturalists of his time.

Colymbetes *See* **Water-beetles**.

Complexity The state of being intricate, i.e. consisting of a large number of interacting parts or components. Complexity being what it is, this entry is far too long. However, its start, based on CD's comments on p. 27 of the draft of Huxley and Etheridge (1865), is quite appropriate: I think little expansion is wanted in middle paragraph to show how a fish can be morphologically more complex and physiologically less so than the highest mollusc. I do not quite understand. (*Correspondence* to *Huxley, Dec. 16, 1857).

Most educated lay persons, and many trained biologists as well, believe that CD's theory of *evolution by *natural selection necessarily implies that complexity must increase with time. This is probably so, given co-evolution, i.e. the fact that at least some of the competitors, grazers or predators that evolving plants and animals encounter, have 'opted' for complexity, thus requiring complex countermeasures (Wright 1999, and see below). Moreover, what may appear to be morphologically 'simple' or 'retrograde' organisms often evolve incredibly complex life cycles, as pointed out by Gould (1996a). However, Gould (1996b), whose *Full House* does not discuss co-evolution, strongly argues that there is *no* evolutionary trend toward complexity.

CD, like Gould, provides arguments for both sides: The enormous *number* of animals in the world depends, of their varied structure & complexity.- hence as the forms became complicated, they opened fresh means of adding to their complexity.- but yet there is no <<NECESSARY>> tendency in the simple animals to become complicated although all perhaps will have done so from the new relations caused by the advancing complexity of others. – It may be said, why should there not be at ay time as many species tending to dis-developments (some probably always have done so, as the simplest fish &), my answer is because, if we begin with the simplest forms & suppose them to have changed, then very changes <len> tend to give rise to others. – Why then has there been a retrograde development in Cephalopoda & fish & reptiles.? [...] I doubt not if the simplest animals could be destroyed, the more highly organized ones would soon be disorganized to fill their place. [...]

It is quite clear that a large part of the complexity of structure is adaptation. though perhaps differences between *jaguar & tiger may not be so. – Considering the Kingdom of nature as it now is, it would not be possible to simplify the organization of the different beings, (all fishes to the state of the *Ammocœtus, Crustacea to –? &c) without reducing the number of living beings – but there is the strongest possibility of increase them, hence the degree of development is either stationary or more probably increases." (*Notebook E*, pp. 422–3; note that here, CD anticipates most of Gould's case in *Full House*, then goes beyond it as he considers co-evolution).

And: A long course of selection might cause a form to become more simple as well as more complicated; thus the adaptation of a crustaceous animal to live attached during its whole life to the body of a fish, might permit with advantage great simplification of structure, and on this view the singular fact of an embryo being more complex than its parents is at once explained. [...]

I may take this opportunity of remarking that naturalists have observed that in most of the great classes a series exists from very complicated to very simple beings; thus in Fish, what a range there is between the *sand-eel and *shark – in the Articulata, between the common crab and the Daphnia – between the Aphis and butterfly, and between a mite and a spider.[27] Now the observation just made, namely, that selection might tend to simplify, as well as to complicate, explains this; for we can see that during the endless geologico-geographical changes, and consequent isolation of species, a station occupied in other districts by less complicated animals might be left unfilled, and be occupied by a degraded form of a higher or more complicated class; and it would by no means follow that, when the two regions became united, the degraded organism would give way to the aboriginally lower organism.

According to our theory, there is obviously no power tending constantly to exalt species, except the mutual struggle between the different individuals and classes; but from the strong and general hereditary tendency we might expect to find some tendency to progressive complication in the successive production of new organic forms. (*Foundations*, p. 227; **n. 27 reads** Scarcely possible to distinguish between non-development and retrograde development).

But it may be objected that if all organic beings thus tend to rise in the *scale, how is it that throughout the world a multitude of the lowest forms still exist; and how is it that in each great class some forms are far more highly developed than others? Why have not the more highly developed forms everywhere supplanted and exterminated the lower? *Lamarck, who believed in an innate and inevitable tendency towards perfection in all organic beings, seems to have felt this difficulty so strongly, that he was led to suppose that new and simple forms are continually being produced by *spontaneous generation. Science has not as yet proved the truth of this belief, whatever the future may reveal. On our theory the continued existence of lowly organisms offers no difficulty; for *Natural selection, or the *Survival of the Fittest, does not necessarily include progressive development – it only takes advantage of such *variations as arise and are beneficial to each creature under its complex relations of life.

And it may be asked what advantage, as far as we can see, would it be to an infusorian animalcule – to an intestinal worm – or even to an earthworm, to be highly organized. If it were no advantage, these forms would be left, by Natural selection, unimproved or but little improved, and might remain for indefinite ages in their present lowly condition. And geology tells us that some of the lowest forms, as the infusoria and rhizopods, have remained for an enormous period in nearly their present state. But to suppose that most of the many now existing low forms have not in the least advanced since the first dawn of life would be extremely rash; for every naturalist who has dissected some of the beings now ranked as very low in the *scale, must have been struck with their really wondrous and beautiful organization.

Nearly the same remarks are applicable if we look to the different grades of organization within the same great group; for instance, in the vertebrata, to the co-existence of mammals and fish – among mammalia, to the co-existence of man and the ornithorhynchus – amongst fishes, to the co-existence of the shark and the *lancelet (Amphioxus), which latter fish in the extreme simplicity of its structure approaches the invertebrate classes. But mammals and fish hardly come into competition with each other; the advancement of the whole class of mammals, or of certain members in this class, to the highest grade would not lead to their taking the place of fishes.

Physiologists believe that the *brain must be bathed by warm blood to be highly active, and this requires aerial respiration; so that warm-blooded mammals when inhabiting the water lie under a disadvantage in having to come continually to the surface to breathe. With fishes, members of the shark family would not tend to supplant the lancelet; for the lancelet, as I hear from Fritz Müller, has as sole companion and competitor on the barren sandy shore of South Brazil, an anomalous annelid (*Origin VI*, pp. 98–9; an account of Johann Friedrich Theodor Müller's (1822–97) life, travails, and exchanges with CD is given in Tort (1996), pp. 3109–12).

Let's pick the idea that marine mammals lie under a disadvantage in having to come continually to the surface to breathe to conclude this discussion. I think this remark misses out on a key advantage that the marine mammals, which have evolved only recently, have over fishes and other water-breathers: the mammals bring along their own oxygen supply wherever they dive, which often includes waters bereft of this life-supporting element. Thus, the sperm whale (*Physeter macrocephalus*) is able to forage actively within the Deep Scattering Layer, i.e. the oxygen-poor water layer occurring at depths between about 400 and 1000 m in most oceanic areas, where lanternfishes, squids and other water-breathers remain in suspended animation, following daytime feeding bouts in surface waters (Papastavrou *et al.* 1988). Here, clearly, the evolution of mammalian complexity has led to quite a large step in the evolutionary arms race.

There have been many attempts to apprehend and quantify complexity (see, for example, Ulanowicz 1986, or Slobotkin 1992). So far, the most convincing attempt is that of Chaisson (2001), whose measure of complexity is energy density, the rate of energy flow through a system per unit mass (Φ_m, in erg s^{-1} g^{-1}). This enables, among other things, a comparison of our sun ($\Phi_m \sim 2$) with our brains ($\Phi_m \sim 150\,000$) and those of marine mammals, all bathed by warm blood. The Φ_m values of organisms have tended to increase in evolutionary time, and those of human society in history as well, both of these trends thus supporting the notion of *progress.

Concepcion City in Chile where CD collected, in early March 1835, a species described as *Blennechis fasciatus* by Jenyns (*Fish*, p. 84; *Fish in Spirits*, no. 1202; *see* **Blennies**).

Some may argue that there is more than that to Concepcion.

Concordance Alphabetical index of the non-trivial *words of a book (excluding 'a,' 'the,' etc.) with a reference to the passage where the words occur.

The passages are usually single lines in small characters, with the keyword in question (IN CAPITAL LETTERS) put at the centre, e.g.

243c020 f wide *range work this out – L. Jenyns, about my FISH New Zealand & New Holland fish very similar.–

which is the last line of p. 231 in Weinshank *et al.* (1990), and where the alphanumeric code indicates the location of this line in *Notebook C*.

In addition to that for the *Notebooks*, concordances exist for CD's *Origin* (Barrett *et al.* 1981), *Expression* (Barrett *et al.* 1986) and *Descent* (Barrett *et al.* 1987). These concordances were used to verify that all of CD's references to 'fish' and closely related terms were included in this volume.

Conger eels *See* **Eels**.

Conger punctus *See* **Eelpouts**.

Constancy Lack of change, here discussed by CD with reference to the evolution of repetitive patterns in body parts: Whenever any part or organ is repeated many times over in the structure of a species, it is variable in number, the same part or organ becoming numerically constant, either in other parts of the body of the same individual, or in other species, whenever the number is few [...In birds, it] might be thought that the greater importance of the wing and tail feathers would account for their constancy; but I doubt this, for we find the same rule in the vertebrae, which are generally constant in mammals & birds, but in snakes, according to Schlegel, the number varies greatly in the same species. So I believe it is in the teeth of fish & reptiles compared with the teeth of mammals. (*Big Species Book*, p. 567; Schlegel 1843, p. 27).

Coquimbo City in Chile where CD collected the following species, described by Jenyns: *Umbrina ophicephala* (*Fish*, p. 45; *see* **Croakers**); *Blennechis ornatus* (*Fish*, p. 85; *Fish in Spirits*, no. 1211; *see* **Blennies**); and *Clinus crinitus* (p. 90; *C. crinitis* in the index of *Fish*; *see* **Blennies**).

Coral Reefs Short title of "The Geology of the Voyage of H.M.S. Beagle. Part 1. Structure and distribution of Coral Reefs" (Darwin 1842), in which CD presented his revolutionary (and now widely accepted) theory linking the subsidence of land and coral reef formation, formulated before [he] had ever seen a true coral reef (*Autobiography*, p. 97, Stoddart 1962, 1994; *see also* **Cocos Islands**).

The key point of the theory in *Coral Reefs* is that, while individual corals will rapidly settle on the rocks around the emerging, usually conical tip of new volcanic islands, or along emerging coastlines, these structures need to slowly sink for substantial reefs to develop on them. Given subsidence, 'fringing' reefs will be the first to be formed. As volcanic islands sink deeper, and the coral reefs along their coastline grow upward to stay within the lighted zone, atolls (with their internal *lagoons) may then be formed, whose roundish shape reflects their having been formed on top of a cone. Or as put by CD when he began to flesh out the theory later articulated in *Coral Reefs*: In time, the central land would sink benath the level of the sea & disappear, but the coral would have completed its circular wall (*Diary*, April 12, 1836). Also, repeated sea-level changes may gradually shift the locations of fringing reefs away from a coastline, thus forming barrier reefs (such as occur off eastern Australia, and Belize).

Among other things, this theory implies that if an island sinks so fast that coral growth cannot keep up, then its reefs will 'drown.' It is thus appropriate to use the *eponym 'Darwin Point' for the combination of lowest temperature and light (i.e. the highest latitude) at which coral growth can still compensate for subsidence (Grigg 1982).

These concepts will become increasingly better known to the public in the next decades, as increased numbers of bleaching events (due to increasing water temperature and possibly other anthropogenic stresses), and increased coastal turbidity (also anthropogenic) will reduce the ability of coral reefs to match sea-level increases.

Coral Reefs provided material for only two entries in this book, on *parrotfishes, and the effect of strong rain on *Cocos Islands reef fishes.

Cormorants Fish-eating birds of the genus *Phalacrocorax*, of which one species is described in *Birds* (p. 145), based on a specimen from *Patagonia, though without the behavioural observation that CD added to most of John Gould's morphological accounts.

This was the Imperial shag *Phalocrocorax atriceps* (*P. carunculatus* in *Birds*), previously treated as a complex of species, but now considered a single species with variable plumage, and a wide distribution that includes both Patagonia and the *Falkland Islands (Harrison 1987, p. 230). Hence, we can now assign to that species the field observations CD made at Berkeley Sound, East Falkland Islands, where he [s]aw a Cormorant catch a fish and let it go 8 times successively like a cat does a mouse or *otter a fish (*Notebook 1.14*; March 21, 1833, cited in Armstrong 1992a, p. 104; Armstrong 1993a).

Correspondence Short title of the *Correspondence of Charles Darwin*, published since 1985 by Cambridge University Press, in a series of volumes with varying combinations of the same group of (associate) editors.

Twelve volumes were available as source of material for this book: 1 (1821–36; Burkhardt *et al.* 1985b); 2 (1837–43; Burkhardt *et al.* 1986); 3 (1844–46; Burkhardt *et al.* 1987), 4 (1847–50; Burkhardt and Smith 1988); 5 (1851–5; Burkhardt and Smith 1989); 6 (1856–7; Burkhardt and Smith 1990); 7 (1858–9, + supplement for 1821–57; Burkhardt *et al.* 1991); 8 (1860; Burkhardt *et al.* 1993), 9 (1861; Burkhardt *et al.* 1994), 10 (1862; Burkhardt *et al.* 1997), 11 (1863; Burkhardt *et al.* 1999), and 12 (1864; Burkhardt *et al.* 2001; the only volume not to contain anything by CD on fish).

Once fully published, the *Correspondence* will present the text of all letters still extant (about half of all; Browne 2002, p. 11) that were written by CD, or received by him, a total of about 14 000 items (all briefly documented in the *Calendar* of Burkhardt *et al.* (1985a), used here to infer the contents of letters so far unpublished). Also, these magnificent volumes include biographic notes, bibliographies and other material documenting the work and impact of CD.

Previous compilations of CD's correspondence existed, notably by his son Francis (Darwin 1887; Darwin and Seward 1903; *see also* **Autobiography**), meant to cover the entire range of his work through a selection of (heavily expurgated) letters. Other compilations document his interactions with various individuals, e.g. his mentor J. S. *Henslow (Barlow 1967), or the marine biologist A. *Dohrn (Groeben 1982). The *Correspondence* supersedes these previous efforts, which in any case did not exhibit the same, extremely high standard of scholarship.

Corvina adusta See **Croakers**.

Covington, Syms The young sailor who worked as personal assistant to CD during the voyage of the *Beagle (1831–6), and as his secretary to 1839 (*see* **Fish in Spirits of Wine**). Covington (1816?–61) later emigrated to Australia, where he became postmaster.

From there, Covington exchanged ten letters with CD from 1843 to 1859; he also appears to have been the only of CD's correspondents in the Pacific area to have sent him specimens (of *barnacles; Garber 1994). A fictionalised account of his life, based on his diary and correspondence (Weitzel 1995), was published by McDonald (1998). A nice read.

Covington's diary does not include more than a few casual mentions of fishes. One exception is his observations from the *Cocos Islands: "In the small *lagoons or pools on reefs are immense numbers of small fish of different species, and of the most brilliant colours and shapes I ever saw or fancy could paint. Here are great numbers. A green fish, the coral eater", referring to *parrotfishes. The other exception is when he noted that CD "purchased the jaws of an enormous *shark killed by an English whaling crew just as it was about to crunch the men's launch, but what he did with them remains unknown".

FitzRoy (1839, p. 618), on the other hand, tells us where the jaw went: "Mr. Chafter [an Officer on the *Beagle*] obtained the jaw of a huge blue shark, at Hobart Town, which had been killed by the boat's crew of Mr. James Kelly whaling vessel. The extreme length of the *monster was thirty-seven feet. Its jaw is now in the United Service Museum" (and presumably lost, as the Royal United Service Museum, a repository of military objects from 200 years of British campaigns, was closed in the early 1960s).

Creationism The belief that a deity created the world and its organisms as they now are. Once a reasonable view, espoused by the best scientists of the time, creationism has since been convincingly refuted (Eldredge 2000). Even the Pope has officially conceded this (John Paul II 1996), thus beating by 150 years his team's previous world record in foot-dragging (i.e. recognizing Galileo's heliocentric system three centuries after it was proposed). The transition of creationism, from a theory compatible with the best current knowledge (of the early to mid nineteenth century; see Armstrong 2000) into the morass it is now, happened during CD's lifetime, and, of course, he was instrumental in hastening it. However, this transition would have happened anyway, given simultaneous advances in physics, geology, etc., which jointly ended up counting far more than the "Oxford Declaration", asserting that the Bible and its miracles must be taken literally, and which was signed by 11 000 (!) Anglican clergymen (George 1982, p. 99).

This transition, incidentally, also spans the lifetime of Louis *Agassiz, an otherwise competent ichthyologist, whose attempts to deny evidence staring him in the face (e.g. his *Threefold parallelism) would strike one as funny, were it not, as well, such a sad example of social opportunism (*see also* Winsor 1991).

Biologists are often advised to pitch against creationism, which now passes itself as 'Intelligent Design Theory' – *déjà vu* all over again. As my contribution to this noble effort, I reproduce here a table in Wise (1998), which compares our best knowledge of the evolutionary sequence with a text reputed to be divinely inspired:

Actual order	Biblical order (Genesis)
Sun and other stars before Earth	Earth before Sun and other stars
Sun before land plants	Land plants before Sun
Marine organisms as first life forms	Land plants as first life forms
Fish before fruit trees	Fruit trees before fish
Insects before fish	Fish before insects
Reptiles before birds	Birds before land reptiles

The major feature of the natural sciences is that the sequences on the left might change, or involve exceptions. Thus, for example, studies on the Tambaqui, *Colossoma macropopum* (Cuvier 1818), a species of Amazonian frugivorous fish (Saint-Paul 1986) may establish that it evolved *after* the fruit trees it now depends on. [Note subtle choice of authority for the frugivory of *C. macropopum*, to show that there are good people on the left, as well.]

On the other hand, whatever the results of empirical research, the column on the right never will change. It is unfalsifiable, as *Popper would say. Hence, it cannot be taught in science classes.

***Crenilabrus* spp.** *See* **Wrasses**.

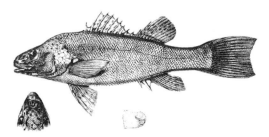

Fig. 6. Creole perch *Percichthys trucha*, Family Percichthyidae, described as *Perca laevis* in *Fish* (Plate I). Hawkin's lithograph is based on a specimen found dead on a bank of the Santa Cruz, River, Patagonia, Argentina.

Creole perch Common name of *Percichthys trucha* (Valenciennes, 1833), a species of temperate bass (Percichthyidae), found in Andean streams in Chile and Argentina.

A specimen pale yellowish brown, with dark mottlings was "found dead by Mr Darwin, high up the river of Santa Cruz, in South Patagonia," and described as *Perca laevis* (*Fish*, pp. 1–4; *Fish in Spirits*, no. 947; Fig. 6). The low productivity of the Santa Cruz River (which flows from the Andes nearly straight eastward and discharges into the Atlantic at 50°S) and its banks was noted by both CD and Captain FitzRoy, following an ascent deep into the Patagonian interior, on board three of the *Beagle's* whale-boats.

Thus, CD wrote that [t]he curse of sterility is on the land. The very water running over the bed of pebbles are stocked with no fish. Hence there are no water-fowl, with the exception of some few geese and ducks (*Diary*, April 22, 1834). This is confirmed by his Captain (FitzRoy 1837, pp. 123, 126), who "could not have believed that the banks of any large freshwater river could have been so devoid of wood, or so unfrequented by man, beast, bird or fish". Indeed, "only one fish was taken, – [. . .] which was similar to a trout. Not more than half a dozen live fish were seen, and none that could be caught, either with hook or nets."

Cretaceous A period of the Mesozoic era, lasting from 146 to 65 million years ago, named after its mighty layers of sedimented 'chalk' (Latin: *creta*) that occur in the United Kingdom, France, and elsewhere in Europe.

In a paper presented before the Geological Society of London on Nov. 1, 1837, CD had suggested that In recent coral formations, the quantity of stone converted into the most impalpable mud, by the excavations of boring shells and of nereidous animals, is very great. Numerous large fishes (of the genus *Sparus*) likewise subsist by browsing on the living branches of coral. Thus, "a large portion of the chalk of Europe was produced from coral, by the digestive action of marine animals, in the same manner as mould has been prepared by the earth-*worm on disintegrated rock" (see Barrett (1977), Vol. I, p. 53, Note 4). This was neatly rephrased by CD's first cousin Elizabeth Wedgwood, as "your hypothesis of chalk being made by fishes- if fish made Chalk Hill I dont see why *worms may not make a meadow" (*Correspondence*, Nov. 10, 1837).

However, William Buckland, in his referee's report to the Geological Society of London, to which CD's paper had been submitted for publication, recommended "that the Author be advised to withdraw the passage relating to the origin of Chalk – as introducing very disputable matter into a paper that is otherwise unexceptionable" (*Correspondence,* March 9, 1838). Though he saw Buckland as a vulgar and almost coarse man (*Autobiography*, p. 102), CD had no option but to drop this hypothesis from the published version of his talk (Darwin 1840). It was a good idea, as chalk consists almost exclusively of foraminiferan shells. However, he left this bit in the first (though not the second) edition of the *Journal* (April 6, 1836; see **Parrotfishes**).

The other error to be noted here is that the fish in question should be *parrotfishes (*Scarus*), not porgies (*Sparus*), an error also popping up elsewhere (*see* **Porgies; Geology**).

Croaker(s) My trusty Random House Dictionary defines a 'croaker' as a person who grumbles, or forebodes evil. Thus CD, when complimenting *Huxley on his editorship of a new series of the *Natural History Review*, noted I am rather a croaker & I do rather fear that the merit of the articles will be above the run of common readers & subscribers. (*Correspondence*, Jan. 3, 1861).

However, this is not the reason why we have croakers here. Rather, our main croakers are fishes of the Family Sciaenidae. CD sampled several species of this widespread group, though all in South America, and we shall take stock of these by moving clockwise around that continent.

First we have *Otolithus guatucupa*, now *Cynoscion striatus* (Cuvier, 1829) and *Corvina adusta*, now *Ophioscion adustus* (Agassiz, 1831), both based on specimens from *Maldonado, Uruguay (*Fish* pp. 41–2; *Fish in Spirits*, nos. 458, 694, 695). This is followed by the Southern king croaker *Umbrina arenata*, from both Maldonado and *Bahia Blanca, Argentina (*Fish in Spirits*, nos. 392, 714). This species ranges from New York to Buenos Aires, and is now identified as *Menticirrhus americanus* (Linnaeus, 1758). CD described its colouring as follows: Body mottled with silver and green: dorsal and *caudal fins lead–colour. (*Fish*, p. 44).

Next is the Snakehead king croaker *Umbrina ophicephala* – now *Menticirrhus ophicephalus* (Jenyns, 1840) – from Coquimbo, Chile (*Fish in Spirits*, nos. 1218, 1220). We conclude with the Peruvian weakfish *Otolithus analis* Jenyns, 1842, based on a specimen from 'Lima' (i.e. Callao), *Peru (*Fish*, p. 164).

Our grand tour ends here, since *Prionodes fasciatus*, sampled in Chatham Island, *Galápagos (*Fish in Spirits*, no. 1284), and originally, if hesitantly, assigned by *Jenyns to the croaker family (*Fish*, pp. 46–7, 164), turned out to be a *grouper.

Crustacea(-ans) A class of articulated animals, having the skin of the body generally more or less hardened by the deposition of calcareous matter, breathing by means of gills. (Examples, crab, lobsters, shrimp, etc.) (*Origin VI*, p. 432).

CD sampled numerous crustaceans during the voyage of the *Beagle*, and devoted several years of his life to the study of a group of crustaceans, the *barnacles.

In this book, crustaceans occur only in the context provided by CD's writing on fishes. (*See* **Entomostraca, Zooplankton**.)

Ctenoid A type of fish *scale, typical of *Acanthopterygians, characterized by spikes, arranged in comb-like fashion (hence the name) along the trailing edge. Previously used to identify a group of fish roughly overlapping with the Acanthopterygians. (*See* **Trout-perch**, and CD's annotations to Agassiz 1850, Pictet 1853–4 and Sedgwick 1850).

Cuckoo-fish Term used by CD to characterize the (then) hypothetical behaviour of a fish that would somehow force another fish to care for its *eggs, as the cuckoo does to other birds.

CD regretted such fish did not seem to exist, especially as J. Wyman (1859) had described buccal gestation in frogs and *catfish: It is quite a pity that there are not fish of the same group with cuckoo-like habits; your fact would so well have explained how the habit might have arisen. (*Correspondence.*, Dec. 3, 1860).

As it turns out, fish of the same group with cuckoo-like habits have evolved. The upside-down catfishes *Synodontis multipunctatus* and *S. petricola* (Family Mochokidae), *endemic to Lake Tanganyika, lay their small eggs at the same time as their target hosts, *mouth-brooding members of the Family *Cichlidae. The hosts pick up the foreign eggs along with theirs, but the eggs of the *parasites hatch faster, owing to their smaller size (Pauly and

Pullin 1988), enabling the newly hatched catfish to consume the host's eggs (Wisenden 1999; Barlow 2000, pp. 205–8).

Moreover, there are many species of fish that parasitize the nests built by other species, though without the eggs of the parasitized species being consumed or ejected. Here, the parasites, often *minnows and other members of the *Cyprinidae, benefit from not having to construct their own nests, nor guard and oxygenate the eggs, activities that lead to exposure to predation (Wisenden 1999).

However, as the review of Wisenden (1999) emphasizes, the evolution of 'alloparental care' in fishes is known only in its broadest outline, and much of the job of explaining how the habit might have arisen is still to be done.

Cusk eels Members of the Family Ophidiidae, whose elongated bodies make them look similar to true *eels.

CD notes, concerning the emission of *sound by fish, the cases of two *species of *Ophidium, in which the males alone are provided with a sound-producing apparatus, consisting of small movable bones, with proper muscles, in connection with the swim-bladder[41]. (*Descent II*, p. 347; n. 41 cites Dufossé (1858a,b, 1862)).

Summarizing earlier writing on this, Günther (1880, p. 145) confirmed that "[a] peculiar mechanism has been observed in the air-bladder of the *Ophidiidae*, the anterior portion of which can be prolonged by the contraction of two muscles attached to its anterior extremity, with or without the addition of a small bone".

Cycloid A type of roundish, thin *scale occurring in various teleost groups, notably the *Cyprinidae and the *catfishes, and earlier *lumped into a single group, the 'Cycloids', defined by their having such scales. The term is used by CD in his annotations to contributions by Agassiz (1850), Pictet (1853–1854), and Sedgwick (1850). (*See also* **Trout-perch**.)

Cyprinidae The family of fish comprising the *Carp and their relatives, the *Goldfish, *Minnow, *Roach, *Tench, etc., and of which many species are used for aquaculture, notably in Central Europe, India and China.

Thus, we have CD's observation that many of the carnivorous Cyprinidae in India are ornamented with 'bright longitudinal lines of various tints'.[31] Mr. M'Clelland, in describing these fishes, goes so far as to suppose that 'the peculiar brilliancy of their colours' serves as 'a better mark for *king-fishers, terns, and other birds which are destined to keep the number of these fishes in check'; but at the present day few naturalists will admit that any animal has been made conspicuous as an aid to its own destruction.

It is possible that certain fishes may have been rendered conspicuous in order to warn birds and beasts of prey that they were unpalatable, as explained when treating of caterpillars; but it is not, I believe, known that any fish, at least any fresh-water fish, is rejected from being distasteful to fish-devouring animals. On the whole, the most probable view in regard to the fishes, of which both sexes are brilliantly coloured, is that their colours were acquired by the males as a sexual ornament, and were transferred equally, or nearly so, to the other sex. (*Descent*, p. 343, with CD's view on this now being widely accepted; n. 31 reads: Indian Cyprinidae, by Mr. J. M'Clelland, Asiatic Researches, vol. xix, part ii, 1839, p. 230). (*See also* **Altruism; Carp, Indian;** *Perilampus perseus*; **Sexual selection**.)

Dace Common name of *Leuciscus leuciscus* (Linnaeus, 1758), a 'white fish' of the family *Cyprinidae.

CD wanted to use Dace for his various *experiments on seed dispersal, and thus he concluded a long list of questions to Thomas Eyton, an ornithologist, with: Lastly (if you are not sick of my enquiries) have you ever examined the stomachs of dace & other white fish? Do they ever eat seeds; I know it is good to bait a place with grains. For like the house which Jack built, a heron might eat a fish with seed of water plant & then fly to another pond.

I have been trying for a year with no success to get some dace &c. Have you any & could you catch some in net. & order your kitchen maid to clean them, & you c^d. send me the whole stomach & I would sow the contents on burnt earth with every proper precaution. If ever your good nature sh^d. lead you to send me any such rubbish; it might be put in bladder or tin foil & sent by Post, & if you will not think me very impertinent I could repay you the shilling or two for postage; as the rubbish w^d. thus come much quicker & cheaper to me (*Corresp.*, August 31, 1856).

The house which Jack built refers to a story where "This is the farmer who sowed the corn/ That fed the Cock that crowed in the morn/ That waked the Priest all shaven and shorn/ That married the Man all tattered and torn/ [. . .] /That killed the Rat/That ate the Malt/That lay in the house that Jack built." (Caldecott 1878), similar to the story of 'The old woman and her pig' (Conner 1813), also mentioned by CD in connection with his fish feeding *experiments.

Dajaus diemensis See **Mullets**.

Damselfishes Fishes of the Family Pomacentridae, formerly a part of the *chromids, particularly abundant on coral reefs, to which they add their own tufts of colours, as flowers do to spring meadows. Some damselfish species actively 'farm' filamentous *algae as a food source, defending their meadows vigorously against competitors and even large predatory fishes (Zeller 1988).

CD collected two species of this family during the voyage of the *Beagle*. The first was the Cape Verde gregory *Stegastes imbricatus* Jenyns, 1840, whose original description is based on a specimen of 3 inches from Quail Island, in the bay of Puerto Praya, *Cape Verde Islands (*Fish*, pp. 63–5, 165; *Fish in Spirits*, after no. 157; Fig. 7). The Cape Verde gregory, which reaches 10 cm, lives on rocky bottoms down to a depth of 25 m, and ranges, along the West African coast, from the Canaries and Senegal in the North to Angola in the South.

The second specimen, which was Above leaden colour, beneath paler, grows considerably larger, was sampled near *Valparaiso, Chile. It belongs to *Chromis crusma* (Valenciennes, 1833), and was described as *Heliases crusma* (*Fish*, pp. 54–6; *Fish in Spirits*, no. 1011).

CD never returned to the damselfishes. Nevertheless, this group includes a Darwin *eponym, *Pomacentrus darwiniensis*, now *Dischistodus darwiniensis* (Whitley, 1928).

DAR Abbreviation for 'Darwin Archive,' used, for example, by the editors of CD's *Correspondence*. The Darwin Archive, kept partly at the Cambridge University Library and partly at the Darwin family's 'Down House', England, has

Fig. 7. Cape Verde gregory *Stegastes imbricatus*, of the Family Pomacentridae, or damselfishes. Based on a specimen taken by CD near Quail Island, Puerto Praya, Cape Verde Islands (*Fish*, Plate IX).

been extensively mined by Darwin scholars, although there are still parts that have not been transcribed, including some that would have been useful for this book. This applies particularly to his *Experimental Book* (DAR 157a), and to various smaller items scattered throughout this volume, which are cited here from secondary sources.

darwin(s) Unit of relative change in the structure of organisms, proposed by Haldane (1949a), with one 'darwin' corresponding to the difference between the natural logarithms of two measurements (x_1, x_2), reflecting change in an attribute, divided by the corresponding time difference in million years (($\ln x_2 - \ln x_1$)/($t_2 - t_1$)). This unit has several deficiencies, but still can be used to illustrate some issues, notably concerning *punctuated equilibrium.

Darwin, Charles Robert The person here referred to as 'CD', born on February 12, 1809, who died on April 19, 1882, and whose published work (listed in Freeman 1977), especially Darwin (1859), changed the way humans see themselves.

There are numerous biographies of CD. Among the recent ones, the best are probably those by Desmond and Moore (1992), and Browne (1995, 2002), and there is no point in this book attempting to substitute for these. For the limited biographic information in this book, see **Autobiography**, **Beagle**, and **Jenyns**.

Darwin's Bass Title of a book by Quinnett (1996), and largely delivering on its ambitious subtitle: *The Evolutionary Psychology of Fishing Man*.

Here Darwin's bass – the Largemouth bass *Micropterus salmoides* (Lacepède, 1802) – is the great survivor, which, having learnt to avoid the lures of the clumsy angler, gives such challenge to the sophisticated one. As this war of wits and technology (carbon fibre rods, plastic lures) escalates over bass generations, continued angling pleasure is guaranteed to successive generations of anglers – given that they do not overfish, and that we humans do not trash our planet.

The depth of thought – and of feelings – and the soundness of the scientific observations playfully presented in *Darwin's Bass* make it the contemporary counterpart of *Walton's *Compleat Angler*, which CD used to consult. Thus, *Darwin's Bass* somehow closes a cycle, at least within this book.

Darwin's Finches A group of small birds of the genus *Geospiza*, whose different morphology, specific to each of the *Galápagos Islands, is often believed to have given CD, *while he was on these islands*, the key to what later became his theory of *natural selection. In reality, CD failed to notice these inter-island differences, and they were brought to his attention much later by John Gould, when the latter was working on the material for *Birds* (Gould *et al.* 1841; Lack 1953).

Still, Darwin's Finches inspired this book.

Darwin's Fishes CD's interest in coral reefs, *barnacles, *worms, and orchids is well documented and, indeed, his work led to monographs now essential to those working on these groups.

Not so for *fishes: although he was interested in this group, as attested by numerous observations scattered throughout his published work, and his *notebooks and *correspondence, CD never authored any book or paper devoted solely, or even mainly to fishes. Thus, ichthyologists and Darwin scholars interested in CD's treatment of this most speciose group among the vertebrates, until now had to content themselves with *Fish*, a book edited, rather than authored, by CD.

The present volume addresses this by covering all the fishes (over 250 species) with which CD can be associated, i.e., 'Darwin's Fishes.' (The pun is intended, and it works also in French: *les poissons de Darwin – les pinsons de Darwin*). Graphically, this is:

$$<°DARWIN >\{$$

or abbreviated, and as a *shoal:

<°CD >{ <°CD>{ <°CD >{ <°CD >{ <°CD >{ <°CD >{

The single version somehow resembles a (European) mackerel (*Scomber scombrus*), while the fish in the shoal resemble Indo-Pacific mackerel (genus *Rastrelliger*), both taxa not mentioned elsewhere in this book.

The first group of Darwin's Fishes ($n = 137$) consist of the species collected by CD during the voyage of the *Beagle*; the overwhelming majority of those were described by *Jenyns in *Fish.

The second group of 'Darwin's Fishes' ($n = 91$) are those mainly European species whose biology CD commented upon, either in his formal publications, or in his notebooks, correspondence or *marginalia. Jointly, these two groups make up 5.7% of the 4152 fish species which, according to *FishBase, had been formally described before 1850 (the year taken here to represent the midpoint of CD's ichthyological work).

The third group are the *eponyms, i.e. the fish species named after CD ($n = 17$), e.g. the *wrasse *Pimelometopon darwini*.

There is very little overlap between these three groups: the biological information on the species he collected during the voyage of the *Beagle* was too limited to illustrate the broader principles CD was interested in. Thus, he relied for the bulk of his post-*Beagle* writings on well-studied European species (e.g. *Pike), though with some interesting exceptions (e.g. *Galaxiidae; *Goldfish). Fortunately, CD never commented on the fishes named after him, or after Port Darwin, thus saving us from having to coin a term for the semantico-epistemological conundrum this would have created.

An additional 72 fish species are introduced to illustrate various issues, bringing the number of fishes mentioned in this book to a total of 320 species, i.e. roughly 1% of the extant fish species, including those still to be described (*see* **FishBase**; **Ichthyology**).

Darwin's roughy Common name of *Gephyroberix darwinii* (Johnson, 1866), also known as Finescale roughy (Gomon *et al*. 1994, pp. 400–1), a member of the Family Trachichthyidae (slimeheads, roughies and sawbellies), previously *Trachichthys darwinii*, and still an *eponym. The famously long-lived (up to 149 years; Fenton *et al*. 1991) Orange roughy *Hoplostethus atlanticus* Collett, 1889 is a close relative.

Deal fish Common name of *Trachipterus arcticus* (Brünnich, 1788), also known as Ribbon fish, and a species that may be seen as providing a model for the evolutionary transition toward the peculiar adaptations of *flatfishes. Thus, CD mentions that certain species, whilst young, habitually fall over and rest on the left side, and other species on the right side. Malm adds, in confirmation of the above view, that the adult *Trachypterus arcticus*, which is not a member of the Pleuronectidae, rests on its left side at the bottom, and swims diagonally through the water; and in this fish, the two sides of the head are said to be somewhat dissimilar. Our great authority on fishes, Dr.*Günther, concludes his abstract of Malm's paper, by remarking that 'the author gives a very simple explanation of the *abnormal condition of the Pleuronectoids.' (*Origin*, VI, p. 187; Malm 1868, abstracted in Günther 1868b).

The deal, here, is that the transition from bilaterally symmetrical fishes to fishes that swim on the side is easy to conceive, notwithstanding the issues that led to what may be called the first *Flatfish controversy.

Deep-sea spiny eels A family of eel-like fishes with a global distribution, the Notacanthidae, whose species feed mainly on *zoobenthos at depths ranging from 125 to 3500 m.

CD, citing Richardson (1846, pp. 189, 191), mentions the genus *Notacanthus* only once, in conjunction with the phenomenon now known as *submergence: Notacanthus & *Macrourus [are] two very remarkable Greenland genera, which inhabit deep water, [and] have recently been discovered on the coast of New Zealand & S. Australia. (*Big Species Book*, p. 555).

Here, CD was probably referring to the Spiny eel *Notacanthus chemnitzii* Bloch, 1788, which ranges in the North Atlantic from Northwestern Greenland to South Africa, and in the Pacific from Chile to New Zealand and Australia. This species, incidentally, is reported to feed heavily on CD's *Zoophites, i.e. sea anemones (Sulak 1990), certainly an interesting adaptation.

Descent Short title of *The Descent of Man, and Selection in Relation to Sex* (Darwin 1877b), probably CD's most daring work, as it deals with a species – *Homo sapiens* – he avoided in *Origin*.

Descent provided six *figures and text material for numerous entries in this book, on, for example, *seasquirts, Professor Möbius' *Pike, and *sexual selection, here illustrated by one of its results, *sexual dimorphism. The many cases illustrating this, and pertaining to fishes (in Chapter XII, pp. 330–47) are presented here on a per-species basis, much of it based on information supplied by *Günther. Indeed, CD thanked him for having helped transform this chapter from much the worst into one of the best (July 15, 1870, *Calendar*, no. 7276; see the *Calendar* for CD's other, so far unpublished, letters on this).

Development Also known as *ontogeny, the development of an individual from a fertilized egg to the adult form involves a series of stages which many biologists, foremost *Haeckel, have likened to a 'recapitulation' of their ancestors' different morphologies. CD deals with this and related issues of *embryology repeatedly, with fish being frequently invoked, e.g. In early stage, the wing of *bat, hoof, hand, paddle are not to be distinguished. At a still earlier <stage> there is no difference between fish, bird, etc. and mammal. It is not that they cannot be distinguished, but the arteries <illegible>. It is not true that one passes through the form of a lower group, though no doubt fish more nearly related to foetal state.[55]

This similarity of the earliest stage is remarkably shown in the course of the arteries which become greatly altered, as foetus advances in life and assumes the widely different course and number which characterize full-grown fish and mammals. How wonderful that in egg, in water or air, or in womb of mother, artery should run in same course. (*Foundations*, p. 42; n. 55, originally written across the page: They pass through the same phases, but some, generally called the higher groups, are further metamorphosed. [line break]? Degradation and complication? No tendency to perfection. [line break]? Justly argued against *Lamarck?). And: There is no object gained in varying form, etc. of foetus (beyond certain adaptations to mother's womb) and therefore selection will not further act on it, than in giving to its changing tissues a tendency to certain parts afterwards to assume certain forms.

Thus there is no power to change the course of the arteries, as long as they nourish the foetus; it is the selection of slight changes which supervene at any time during <illegible> of life.

The less differences of foetus – this has obvious meaning on this view: otherwise how strange that a [monkey] horse, a man, a bat should at one time of life have arteries, running in a manner, which is only intelligibly useful in fish! The natural system being on theory genealogical, we can at once see, why foetus, retaining traces of the ancestral form, is of the highest value in classification. (*Foundations*, pp. 43–5). Further, arguing against the likelihood of separate creation for three species of rhinoceros in Java, Sumatra and Peninsular Malaysia, respectively, CD rhetorically asks what we are to make of the fact That in possessing [...] useless abortive teeth, and in other characters, these three rhinoceroses in their embryonic state should much more closely resemble other mammalia than they do when mature. And lastly, that in a still earlier period of life, their arteries should run and branch as in a fish, to carry the blood to gills which do not exist. (*Foundations*, p. 250).

Devonian A series of Paleozoic rocks, including the Old Red Sandstone (*Origin VI*, p. 423). Named after Devonshire, in England, and pertaining to times between 418 and 362 million years ago, right after the *Silurian, and before the Carboniferous.

The Devonian is often referred to as the 'Age of Fishes' because of the flourishing, during that period, of many now extinct fish groups (see Moy-Thomas 1971).

Diacope marginata *See* **Snappers**.

Diadromy(-ous) Referring to regular, physiologically mediated migrations of certain stages of fish between fresh waters and the sea, and involving the majority of the members of a population (Myers 1949; McDowall 1988, 1997).

Diadromy has two basic forms, anadromy (as in *salmon), where spawning occurs in rivers, while the adults feed and grow in the sea, and catadromy (as in *eels), where spawning occurs in marine waters, and the adult stages feed and grow in rivers, lakes, and other freshwater bodies.

A third form of diadromy, 'amphidromy', has been proposed by Myers (1949), wherein larval fish migrate to the sea after hatching and return to fresh waters as juveniles, with spawning and post-juvenile growth thus occurring in fresh water. However, clear-cut cases seem hard to find (some *Galaxiidae?). McDowall (1997) feels that amphidromous fish are not truly diadromous because they spawn and perform most of their growth in the same medium.

One important aspect of diadromy is that it tends to be connected with *homing, especially so in *salmon. So far, however, only one author, McDowall (2001), has publicly wondered why it should be so. Perhaps the answer is that all fish 'home' (*see* **Obstinate Nature**), in which case the occurrence of diadromy is all there is to explain. And this may not be too difficult given the latitudinal comparisons of Gross *et al.* (1988), who showed that marine waters, at high latitudes, tend to be more productive than rivers and lakes, and conversely at low latitudes. Hence the preponderance of salmon-type anadromy at high latitudes, and of eel-type catadromy at low latitudes.

Diary Short title of *Charles Darwin's Diary of the Voyage of H.M.S. Beagle* edited by CD's granddaughter Nora Barlow (1933).

There is also a version of the *Diary* edited by Richard Darwin Keynes (1988), another of CD's descendants.

The *Diary* provided several of the entries in this book. However, its references to fishes were ignored if they overlapped substantially with the corresponding accounts in CD's *Journal, which was largely adapted from the *Diary* and is more widely available.

Difficulties CD compiled, in a courageous chapter titled Difficulties on theory (*Origin I*, pp. 171–206) or Difficulties of the theory (*Origin VI*, pp. 133–67), those features of organisms which he himself (or others) had identified as hard to explain through his theory of *natural selection. Most of these 'difficulties' were more apparent than real; for example, the alleged 'perfection' of the vertebrate *eye, and the alleged lack of a selection advantage for animals possessing 'imperfect eyes' over animals lacking eyes altogether, have been conclusively shown to be red herrings (Dawkins 1996, pp. 126–79), though they continue to be bandied about (Behe 1996).

In *Foundations, the manuscripts that anticipated *Origin*, CD included a section on Difficulties in the acquirement by Selection of complex corporeal structures, dealing, as may be expected, with the existence, in various organisms, of intermediate behaviours or structures, not yet 'perfect' but already conferring a selective advantage on their possessors. Examples are the shocking habits of *electric fishes, the aquatic habits of the *jaguar, and the *gasbladder, or the eyes of fishes: In considering the eye of a quadruped, for instance, though we may look at the eye of a molluscous animal or of an insect, as a proof how simple an organ will serve some of the ends of vision; and at the eye of a fish as a nearer guide of the manner of simplification (*Foundations*, p. 130).

Another set of perhaps more legitimate difficulties arises in conjunction with what is now called *altruism – though the examples discussed by CD would not now be seen as cases of altruism: The nature or condition of certain structures has been thought by some naturalists to be of no use to the possessor, but to have been formed wholly for the good of other species; thus certain fruit and seeds have been thought to have been made nutritious for certain animals – numbers of insects, especially in their larval state, to exist for the same end – certain fish to be bright coloured to aid certain birds of prey in catching them, etc. Now could this be proven (which I am far from admitting) the theory of natural selection would be quite overthrown; for it is evident that selection depending on the advantage over others of one individual with some slight deviation would never produce a structure or quality profitable only to another species [. . .]: the bright colours of a fish may be of some advantage to it, or more probably may result from exposure to certain conditions in favourable haunts for food, notwithstanding it becomes subject to be caught easily by certain birds. (*Foundations*, pp. 130–1).

The potential 'falsifiability' of natural selection, here stressed by CD, is a key reason why *Popper was wrong in considering this theory 'metaphysical' (as *creationism is). But then again, Popper is one of the many philosophers of science who derived most if not all their case studies from physics, and who therefore missed what Ghiselin (1969) called "the triumph of the Darwinian method".

***Diodon* spp.** *See* **Burrfishes**.

Distribution CD used the distribution of animals and plants for much of his argument against special creation, and he was quite frustrated about the state of knowledge on the distribution of fishes, then insufficient for the type of inference he had in mind (see, for example, what we now know of *eel reproductive migrations).

This frustration shines through in the quotes below, taken from *Notebook C* (p. 243): many fish of Taiti found at <New> Isle de France:[2] xx instance of wide range, where means of wide range[3] work this out – L. *Jenyns, about my fish. New Zealand & New Holland fish very similar. –[5] (nn. 2, 3, and 5 refer to Lesson (1826), pp. 27–8).

All the discussions <after> about affinity & and how one order first becomes developed & then another – (according as parent types are present) must follow after there is proof of the non creation of animals. – then argumen May be. – subterranean lakes, hot springs &c &c inhabited therefore mud wood be inhabited, then how is this effected by – for instance fish being excessively abundant (*Notebook C*, pp. 258–9).

Also: Mr Cuming informs me, that he has upwards of a hundred species of shells from the eastern Coast of Africa identical with those collected by himself at the Philippines and at the eastern coral-islands of the Pacific Ocean: now the distance from these islands to Eastern Africa is equal to that from pole to pole. Under similar circumstances Dr. Richardson has found that fishes have immense ranges. (Darwin 1846, p. 204 in *Collected Papers I*; Richardson 1846, pp. 190, 191; the naturalist Hughes Cuming [1791–1865] returned to England in 1839 after 20 years collecting living and fossil shells in the Pacific; Garber 1994).

Gradually, these large ranges, however, turned from a liability into an asset for his theory, and led CD to generate a bold prediction: Geologists finding in the most remote period with which we are acquainted, namely in the *Silurian period, that the shells and other marine *productions[59] in North and South America, in Europe, Southern Africa, and Western Asia, are much more similar than they now are at these distant points, appear to have imagined that in these ancient times the laws of geographical distribution were quite different than what they now are: but we have only to suppose that great continents were extended east and west, and thus did not divide the inhabitants of the temperate and tropical seas, as the continents now do; and it would then become probable that the inhabitants of the seas would be much more similar than they now are. [. . .] Many fish, I may add, are also common to the Pacific and Indian Oceans. (*Foundations*, p. 179; n. 59 reads: D'Orbigny shows that this is not so.; Orbigny 1834–47, see also Geoffroy-Saint Hilaire and Blainville 1834).

This bold prediction nicely describes the Tethys Sea, which ranged in time from the lower Cambrian to the later Tertiary, and in space from what are now the Eastern Central Pacific and the Caribbean (the Isthmus of *Panama did not exist then) in the west, via the Indian Ocean, to Southeast Asia in the east. The Tethys Sea was identified and named by Suess (1885), over 40 years after CD had suggested its existence (see Wegener 1966; Ekman 1967, Chapter iv; *see also* **Bony fishes**).

Divine intervention(s) CD's views on these were straightforward: I have reflected a good deal on what you say on necessity of continued intervention of creative power. I cannot see this necessity; & its admission, I think would make theory of nat. select. valueless. Grant a simple archetypal creature, like the *Mud-fish or Lepidosiren, with the five senses & some vestige of mind, & I believe Natural selection will account for production of every Vertebrate animal. (*Correspondence* to C. Lyell, Oct. 20, 1859).

The astute reader will note that this barb, directed against *creationism, also questions the 'contingent' evolution so strongly advocated by S. J. Gould in *Wonderful Life* (1989) and many of his other writings (including Gould 2002), even if CD himself noted that ! fish can never become a man (*Notebook B*, p. 227). The point is that fish, over the eons, did 'become' men and women.

Dogfish Common name used for small sharks, notably the widespread *Squalus acanthias* Linnaeus, 1758.

CD discussed their low fecundity, in contrast to that of *Cod: The Picked Dog-fish (Squalus acanthias) actually swarms on many coasts & yet is said to lay only six *eggs; whereas the Cod-fish sometimes lays above three million & a half.[1] (*Big Species Book*, pp. 206–7; n. 1 refers to Yarrell (1836), Vol. 2, p. 401, for the high abundance, and to Fleming (1822), Vol. 2, p. 356, who cites Rondelet (1554); *see also* **Eggs of fishes**).

CD described as follows two specimens of 'dogfish' – probably *S. acanthias* – from Tierra del Fuego Colour pale, 'Lavender purple,' with cupreous gloss, sides silvery; above with irregular quadruple chain of circular and oblong snow white spots; tip of dorsal and *caudal blackish, under part of caudal reddish, iris pearly white. Length of old specimen tip to tip 2 feet 3 inches. breadth from tip of pectoral to tip of other 8 inches. Young specimen out of belly; with it is posterior spine of old specimen. (*Fish in Spirits*, no. 840; also no. 882).

Dohrn, Anton German naturalist (1840–1909), whose correspondence with CD (Groeben 1982) shows that he failed to get across his *idée fixe* that vertebrates evolved from flipped-over annelid worms (Dohrn 1875) (Fig. 8), a notion

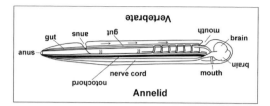

Fig. 8. Diagram to illustrate the transformation of an annelid into a vertebrate, as envisaged by Anton Dohrn and others. Redrawn with modifications from Fig. 182 in Wilder (1923).

originally proposed by Geoffroy-St. Hilaire (1822, 1830). This idea is well put by Semper (1875): "If the embryo of the annelide be turned so that its ventral surface lies upward, its section presents exactly the same arrangement of the organs as the Selachian embryo. Consequently, by the discovery of the segmental organs, the belly of the annulose animal is identified with the back of the vertebrate" (details in Wilder (1923), pp. 570–80).

CD had earlier encountered a version of this: There is no scale, according to importance of divisions in arrangement, of the perfection of their separation. – thus Vertebrate blend with Annelidæ by some fish (*Notebook D*, p. 370). However, CD, who in a letter to Dohrn expressed that the work astonished him (see *Calendar*, no. 9991) did not even mention it when later discussing *vertebrate origins. Having identified the larvae of ascidians (or *seasquirts) as progenitors, he did not bother much with *their* ancestry.

And yet, the recent discovery that similar homoeotic or *Hox* genes, in a wide range of vertebrates and invertebrates, determine their basic body plans and the relative location of their organ systems, is compatible with the suggestion that the vertebrates are derived from flipped-over, bilaterally symmetrical animals also ancestral to annelids, the *Ur-bilateria* of De Robertis and Sasay (1996). [I added the hyphen, lest some read the new name as *Urbi-lateria*, i.e. urban latecomers in Very Low Latin].

Gould (1998a, p. 334), when presenting the related story of Gaskell's (1908) derivation of vertebrates from arthropods, suggested that, while supporting Geoffroy-St. Hilaire's "old theory of inversion", the presently available data "do not support the silly notion that at some defining point in the march of evolutionary *progress, an arthropod literally flipped over to become the first vertebrate". This is rhetorical overkill: yes, no arthropod ever turned into a vertebrate, by flipping over or otherwise, and whether marching for progress or not. However, the idea that animals may develop the habit of 'flipping over', then, over evolutionary time, acquire anatomical features reflecting this flip is not "silly" at all: *flatfishes do it all the time, and CD used unflappable logic to explain how this could happen. (*See also* **Flatfish controversy**.)

And while it can be argued that flatfishes have flipped by only 90°, I can think of several fish species that have flipped by 180°, and do not seem the worse for it. One example is the Blotched upside-down catfish *Synodontis nigriventris* David, 1936, whose common and scientific names (*nigriventris* = blackbelly) say it all. [There is also my small but instructive contribution to the *Annals of Improbable Research* (Pauly 1995a), which deals with upside-down fishes, and which *is* silly.]

The neat part about this whole thing, as noted by Kinsbourne (1978), is that it provides a testable hypothesis to explain why vertebrates have the left side of their *brain connected to the right side of the body and *vice versa*. In our ur-bilaterian ancestors, the *eyes (they did have eyes; De Robertis and Sasay 1996) and their associated image processing cells (the original brain) may have experienced a first torsion when their bodies became reoriented by 90°, from vertical to horizontal (as in the *ontogeny of *flatfishes). Then, some populations may have returned to a vertical posture, but not by reversing the previous torsion (as seem to be presently occurring with *Greenland halibut), but by completing the 180° turn (see also Pinker (1994), pp. 303–5). This would allow interpreting the partial asymmetry of the *lancelet (Gee 1994), one of the simplest chordates, as a vestige of a rotation, with *Natural selection not having come around to re-establishing the original, full bilateral symmetry of the unflipped progenitors.

Dragonets Members of the family Callionymidae, one of the few taxa of colourful fishes in temperate waters: There is reason to suspect that many tropical fishes differ sexually in colour and structure; and there are some striking cases with our British fishes. The male *Callionymus lyra* has been called the *gemmeous dragonet* 'from its brilliant gem-like colours.' When fresh caught from the sea the body is yellow of various shades, striped and spotted with vivid blue on the head; the dorsal fins are pale brown with dark longitudinal bands; the ventral, *caudal, and anal fins being bluish-black. The female, or sordid dragonet, was considered by *Linnaeus, and by many subsequent naturalists, as a distinct species; it is of a dingy reddish-brown, with the dorsal fin brown and the other fins white. The sexes differ also in the proportional size of the head and mouth, and in the position of the eyes;[12] but the most striking difference is the extraordinary elongation in the male ([Fig. 9]) of the dorsal fin. Mr. W. Saville Kent remarks that this 'singular appendage appears from my observations of the species in confinement, to be subservient to the same end as the wattles, crests, and other *abnormal adjuncts of the male in gallinaceous birds, for the purpose of fascinating their mates.'[13] The young males resemble the adult females in structure and colour. Throughout the genus Callionymus,[14] the males is generally much more brightly spotted than the females, and in several species, not only the dorsal, but the anal fin is much elongated in the males. (*Descent*

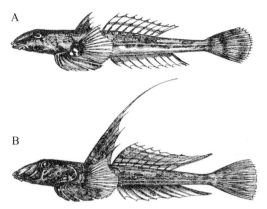

Fig. 9. Dragonet *Callionymus lyra*, Family Callionymidae, illustrating sexual dimorphism, a result of sexual selection. From p. 337 in *Descent II*, based on a woodcut by Mr G. Ford, prepared under the guidance of Dr. Günther. **A** Female: **B** Male.

II, pp. 335–6; n. 12 reads: I have drawn up this description from Yarrell's British Fishes, vol. I, 1836, pp. 261 and 266.; n. 13 cites Saville-Kent (1873b); n. 14 cites Günther (1861), pp. 138–51).

Linnaeus (1758, p. 249) indeed described two species of dragonet from Europe (in addition to one from India, now assigned to another family). One was *Callionymus lyra*, the other *C. dracunculus* (i.e. 'small dragon', or 'dragonet'). *C. dracunculus* is now a 'suppressed' name, synonymous with *C. lyra*.

Dredging An important research method, adapted from a commercial fishing gear, and used to obtain sampling marine sediments, and the *zoobenthos (Rehbock 1979). CD reports in his **Autobiography* on dredging trips he undertook when he was a student in *Edinburgh: I also became friends with some Newhaven fishermen, and sometimes accompanied them when they trawled for oysters, and thus got many specimens. (p. 50).

The samples obtained by systematic dredging surveys, notably by Edward Forbes and Karl Möbius, led to numerous, and much debated, generalizations about the distribution of *zoobenthos along depth gradients, and its tendency to form recurrent communities, respectively (Rumohr 1990). Rehbock (1979) provides an entry point into this literature, which does not concern fish directly, as dredges, contrary to trawls, do not usually capture fish and other agile animals.

Driftfishes Members of the Family Nomeidae. Two specimens taken on March 23, 1832 (*Zoology Notes*, p. 325) by CD at 17°12′S and 36°33′W (i.e. northeast of the *Abrolhos, Brazil) were tentatively attributed by Jenyns to *Psenes leucurus* (*Fish*, pp. 73–4), i.e. to what is now the Freckled driftfish *Psenes cyanophrys* Valenciennes, 1833.

Duckbills Members of the flatheaded Family Percophidae, which includes the Brazilian flathead *Percophis brasilianus* Quoy & Gaimard, 1825, described in *Fish* (p. 23), based on a specimen collected by CD in northern *Patagonia.

This fish is known in Brazil to be "de excelente gosto" (Carvalho-Filho 1994, p. 196), in line with CD, who noted: When cooked, was good eating. (*Fish in Spirits*, nos. 347, 692; see also **Oxford University Museum**.)

***Dules* spp.** *See* **Groupers** (I).

Ecology A term coined by *Haeckel to refer to the scientific discipline devoted to studying the relationships of organisms to their environment. More precisely, autecology refers to single species and their abiotic and biotic environments, while synecology deals with interrelationships among species.

Ecology is a relatively new discipline, at least compared with other branches of biology. This is understandable, as it was first necessary to describe and name a representative part of the Earth's species (see **Taxonomy**), and to understand their basic functional anatomy and physiology before their ecology could be understood.

This is well illustrated by CD's failure to apprehend the structure of pelagic *food webs, owing to his misunderstanding the nature of *plankton.

Still, CD was a better ecologist than most biologists of his time (see Stauffer 1960): the theory of *natural selection required plausible scenarios for the modification of all features of animals and plants (morphological, physiological, behavioural . . .) and this required knowledge of how they relate to their environment, and to each other.

Edinburgh A city in Scotland, home of the University at which young CD started the study of medicine in October 1825, abandoned in April 1827 (Ashworth 1935).

The many connections of the Darwin family to Edinburgh are reviewed in Shepperson (1961), along with the intellectual history that led to Scotland's 'Golden Age', crucial to the development of CD's ideas.

CD's first scientific writing – an account of his dissection of a *Lumpfish, performed under the guidance of his mentor, Prof. R. Grant – was written in Edinburgh, as was his subsequent, better-known discovery that the so-called ova of Flustra had the power of independent movement by means of cilia, and were in fact larvæ. (*Autobiography*, p. 50; *Flustra* is a Bryozoan, one of CD's Zoophites.)

Edmonston, John A taxidermist based in Aberdeen, from whom, in 1826, young CD learnt the basics of preparing animal specimens (*Corresp.* to sister Susan Darwin, Jan. 29, 1826).

What little information is available on Edmonston is summarized in a patronizingly titled contribution by Freeman (1978/79). Edmonston appears to have contributed much to CD's positive perception of people of African ancestry, and to his rejection of slavery, widely accepted by his English contemporaries.

Taxidermy was important in CD's time because lack of suitable, tight containers made it difficult to keep specimens in alcohol and other liquids. Hence, not only bird and mammal specimens were reduced to dried skins, but also fishes (see also **Collection; Galápagos**).

Eel The 'Common' or European eel *Anguilla anguilla* (Linnaeus, 1758), which enters river mouths as small transparent 'glass eels', grows, then returns to the sea to spawn (see also **Eels**).

CD wisely abstained from suggesting a precise spawning location for adult eel, then still a popular guessing game, and one which brought shame to virtually all practitioners, starting with Aristotle (see Aristotle 1962), who had them spontaneously generated by "earth guts that grow spontaneously in mud and in humid ground" (*Historia Animalium*, Book VI, 15). Indeed, this issue was resolved only through the 1922 expedition of the Danish scientist Johannes Schmidt to the Sargasso Sea, in the mid-Atlantic (Nikol'skii 1961, pp. 315–16; Muus and Dahlstrøm 1974, pp. 82–3), where eel now spawn. When the Atlantic was young and narrow, both European and American eel probably spawned on the shelf edge. As Europe and America drifted apart, and the Atlantic widened (Wegener 1966), these eel found themselves having to swim every year a few centimetres further offshore to reach the same spawning grounds. This, however, is the very mix of change and obstinacy leading, via millennia of *natural selection, to adaptations that look miraculous to those not firmly rooted in the natural sciences (*see also* **Distribution; Obstinate nature**).

On the other hand, CD, whose work is so neatly vindicated by the reproduction of Eel, mentions them in one of his many discussions of the evolution of *eyes. Here, he cites from an account in some French Transactions (*Correspondence* to J. M. Rodwell, Nov. 5, 1860), as follows: it has been 'remarked that fishes which habitually descend to great depths in the ocean have large eyes'.[1] And one most remarkable fact is on record, <which is worth giving, though of a most perplexing nature.> M. Eudes-Deslongchamps gives with great detail two cases[2] of eels taken from wells about 100 feet in depth, which had their eyes of immense size, so that their upper jaw in consequence projected over the lower. (*Big Species Book*, pp. 296–7; n. 1 refers to Richardson (1856), p. 219; n. 2 to Eudes-Deslongchamps (1835, 1842)).

But here comes the remarkable fact the first specimen was shown to *Agassiz, & he thought it was specifically identical with the common Eel. One of the wells was within the precincts of a prison; & it seems impossible to conjecture how the eel got in; & it seems, moreover, quite incredible that such an alteration could have supervened during one generation: it is, also, most improbable that there should be a *race of subterranean eels, for, I believe it is well established that the eel invariably breeds in the sea.

Surrounded with difficulty as this case is, we apparently have in the large eyes of these eels, & in the blind Gadus from the deep parts of the lake Leman, a parallel case to the opposite condition of the eyes of the Kentucky *cave-fish . . . (*Big Species Book*, p. 297; *see* **Burbot** for the Gadus of Lake Léman).

The context of this story is provided in a letter to C. *Lyell: very important characters may be modified by correlation of growth. But I doubt whether they throw light on abrupt origin of new forms. [. . .]. With respect to animals, besides the case of monstrous *Gold-fish with analogous fish in state of nature alluded to, I have wondrous case of monstrous eels, (examined by Agassiz) & apparently produced by darkness, but I cannot satisfy myself on case; nor does it appear certain that they breed. (*Correspondence*, Feb. 23, 1860; 'correlation' is here meant as the normal coincidence of one phenomenon, character, etc., with another; *Origin* VI, p. 432).

In Northern Europe, two *varieties of eel are often alleged to exist, one pointy-, the other broad-headed. Thurow (1953) demonstrated that these represent the extremes of a wide range of head shapes, mostly due to different diets: pointy-headed eels tend to feed on worms and small crustaceans, broad-headed ones on larger crustaceans and fishes.

This plasticity of head shape makes it seem quite possible for our imprisoned eels to have developed large eyes in the course of their *ontogeny. Eudes-Deslongchamps (1835) stresses how "impossible" it was for these macrophthalmic eels to have reached their well from any other water body, and that nobody ever gave eels to the prisoners.

But then, nobody gave birds to the Birdman of Alcatraz, and still he had a lot of them.

Eels A group of elongated fishes belonging to the Family Anguillidae, and allied groups, such as the Congridae (conger eels).

When referring to 'eel' CD generally means the European, or 'Common' *eel; e.g. in the quote Professor Ercolani has recently shown (Accad. Delle Scienze, Bologna, 28 December, 1871) that eels are *androgynous (*Descent* I, p. 161, n. 28; incidentally, Professor Ercolani (1871a, b) was quite wrong).

However, other eels show up, for example in *Tahiti, where CD, describing an improvised meal taken after the ascent of a steep, narrow valley, mentioned that a little stream, besides its cool water, produced eels and cray-fish. (*Journal*, Nov. 18, 1835).

There are three species of eel in Tahitian inland waters (Marquet 1992; Marquet and Galzin 1991, 1992): *Anguilla marmorata* Quoy & Gaimard, 1824, occurring mainly in running waters, below waterfalls; *A. megastoma* Kaup, 1856, in running waters, above waterfalls, and *A. obscura* Günther, 1872, in estuaries and shallow stagnant waters. Our best candidate for the species that contributed to CD's meal is thus *A. megastoma*.

Another species is the Shortfin eel *Anguilla australis* Richardson, 1841, sampled by CD in late December 1835 at the *Bay of Islands, New Zealand (*Fish in Spirits*, no. 1337), and whose colour *Jenyns (1842, p. 142) described as "similar to that of the common *eel".

(*See also* **Heron, Otters, Morays**; *Fish in Spirits*, no. 431).

Eelpouts Fishes of the Family Zoarcidae, not related to the eel(s), and represented here by three species.

The first is *Phucocoetes latitans* Jenyns, 1842, caught amongst kelp in the *Falkland Islands (*Fish in Spirits*, nos. 598–599; Fig. 10A).

The second, *Iluocoetes fimbriatus* Jenyns, 1842, is described in *Fish*, based on a specimen from *Chiloé, Chile (pp. 166–7). CD noted on this *Blennius* under stones (*Fish in Spirits*, no. 1080; Fig. 10B), mistaking his find for one of the *blennies.

The third, based on a small specimen with Sides with transverse bars of chocolate and brownish-red, separated by narrow grey spaces was originally assigned to *Conger*, a genus of *eels (*Fish*, p. 143; *Fish in Spirits*, nos. 866, 870). It is now called *Maynea puncta* (Jenyns, 1842).

Another member of the Zoarcidae is the European species *Zoarces viviparus* (Jenyns, 1758), whose German common *name (*Aalmutter*, meaning 'Mother of eels' and referring to their giving birth to complete, eely-looking young) shows that one must be careful with folk etymologies (*see* **Names, common**), which can be as misleading as scientific names.

Eggs of fish CD clearly felt that the large number of eggs in fish supported his theory of *natural selection, as it suggested – given habitat limitations – extremely high mortalities for both the eggs and the larvae hatching out of those eggs.

Thus, struggle & destruction follows inevitably in accordance with the law of increase so philosophically enunciated by Malthus[1] (*Big Species Book*, p. 176; Malthus 1826; n. 1 states Franklin & many others have clearly seen & exemplified the great tendency to increase in all the lower animals and plants; Franklin 1755).

CD was right (whatever one might personally feel about Malthus), and here is one of his few quantitative comments on this: Everyone must have seen statements of the number of eggs & seeds produced by many of the lower animals and plants.[2] To illustrate geometric progression one meets in works on arithmetic calculations such as, that a Herring in eight generations, each fish laying 2000 eggs, would cover like a sheet the whole globe, land & water (*Big Species Book*, pp. 176–7; n. 2 reads as follows: I will copy out a few instances of numbers of eggs & seed. Mr. Harmer in Phil.

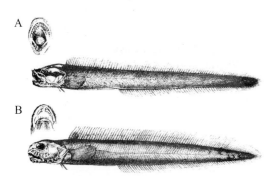

Fig. 10. Eelpouts, Family Zoarcidae, with inserts showing teeth (*Fish*, Plate XXIX). **A** *Phucocoetes latitans*, from kelp beds in the Falkland Islands. **B** *Iluocoetes fimbriatus* from Chiloé Island, Chile.

Transact. 1767, p. 280 weighed the whole & portions of roe and counted in this portion the number of eggs. The numbers differed considerably in different individuals

*Carp	203,109	and 101,200	lowest number
*Cod	3,681,760		
*Flounder	1,357,400	and 133,407	do
*Herring	36,960	and 21,285	do
*Smelt	38,272	and 14,411	do

[...]

(N.B. These observations [...] are confirmed by independent calculations by C.F. Lund in Acts of Swedish Academy Vol. 4). [For Smelt, the lower number in Harmer (1767), p. 285 is 38,27<u>8</u>; somebody made an error transcribing that number, but it was not necessarily CD.]

Clearly, high fecundity (and the high mortality that consequently ensues) implies an enormous role for predation, as recently restated for fishes by Christensen (1996).

Another twist to the issue of high fecundity is that it does not preclude rarity, or in CD's words How many rare fish there are existing in very scanty numbers, yet annually producing thousands of ova! (*Big Species Book*, p. 206; see also **Water-beetles**).

Egg development CD expressed much interest in the relationship between the development and survival of seeds and ova (*see* **Experiments**), notably fish eggs, as this topic closely relates to the issues of the geographic distribution of plants and animals.

One item on this, based on Stark (1838), is roe of Asterias in stomach of sammon remain after rest of animal digested (*Notebook C*, p. 267; and yes, the *spelling of *salmon is not correct, but what a great way to show how this English word should be pronounced!)

The most explicit comments on this issue are found, however, in CD's response to a letter of March 21, 1855 from John Davy, in which were described results of experiments later published in Davy (1856). It was shown therein that the survival of fertilized salmon eggs depends not only on their being kept moist, but also on temperature, and on their developmental stage.

To this CD commented: The case of the ovum exposed on the moss for three days, & the wonderful retention of vitality of the ova in very moist air seem of particular value in regard to the power of dispersal. Surely these results will, also, be of practical value.

Hardly anything has surprised me more than the non-developed ova having less tenacity of life than those much more fully developed. – I almost hope that sh^d you ever have another opportunity, it may be worth while to test this one point again.

I have been much struck by your experiments on the effects of rather high temperature; I have often speculated whether the ova accidentally introduced into the stomach of an herbivorous bird could escape the action of the gastric juices, but I had not at all calculated on the very injurious action of the mere temperature.

With many such experiments as yours, Geographical *Distribution would become in my opinion, a very different subject to what it is now ... (*Correspondence*, March 26, 1855).

Later, CD let Hooker know how pleased he was about Davy's results: My notions sometimes bring good; D^r. Davy has been experimenting at my request, (in order to see how fishes' ova might get transported) on the retention of vitality; & he found that salmon's ova, exposed for 3 whole days to open air, & even some sun-shine, & they produced fine young fish. D^r. Davy has sent a paper to Royal Soc. on the subject. – N.B. Remember to ask about my distinct case of 'a lady in N. America' who saw fishes' spawn adhering to a Ditiscus (*Correspondence*, April 7, 1855; for the lady in N. America see **Morris, Margaretta Hare**; for Ditiscus, *see* **Water-Beetles**).

Electric eel Common name of the fish which Linnaeus (1758) described as *Gymnotus electricus*, but later renamed *Electrophorus electricus* (Linnaeus, 1766). This fish, which can reach lengths of well above 2 m, and occurs in deep rivers of the Orinoco and Amazon, is feared for its ability to inflict powerful electric jolts. But scientists have loved it since 1729, when Richer, an astronomer, first reported to the French Academy on its electric abilities (Moller 1995, pp. 15 and following). For CD, the *Gymnotus*, as he called it, was a big challenge, as it was difficult, in his time, to conceive of transitory stages for the *analogous organs of this and other *electric fishes.

CD mentions a possible case of 'inherited memory' in electric eel in a letter (not seen) sent to G. J. Romanes on Jan. 24, 1877 or 78 (*Calendar*, no. 10813).

Electric fishes Common name for a taxonomically heterogeneous group, all capable of generating electric fields used to detect, and/or stun, potential prey and predators.

As suggested in Moller (1995), fishes can be arranged, in terms of their electrical abilities into four groups: (1) *No special ability* beyond that of generating a weak electric field when their muscles contract, as all animals do. (2) *Electrosensing only*, i.e. the ability to sense the electric fields generated by other animals. Electrosensing is the rule in *sharks, *rays, and *chimaeras, i.e. 470 species in 42 families and 14 orders. (3) *Weakly discharging*: the ability to generate a relatively weak electric field, used mainly for orientation when visibility is low, and for prey detection. The *elephant-fishes are an example of this group, consisting of 407 species in 11 families and 5 orders. (Note that this ability implies electrosensing, as well.) (4) *Strongly discharging*: the ability to generate strong electric fields, and to stun potential prey and predators, and the only ability of which CD was fully aware. This ability, which involves 38 species in 5 families and 4 orders, implies electrosensing as well (e.g. in *Gymnotus*, or *Torpedo*), except in the stargazers, Family Uranoscopidae, which stun without sense.

CD was worried by the implications, for his theory of *natural selection, of the existence of *electric organs in fish, because he thought they were [c]ases of organs in which there is no apparent passage or transition from other organ: or still better, if such transition can be shown in an unexpected manner. E.G. Electrical organs in Fish, seem to be really new organ & not any other changed. (*Correspondence* to *Huxley, Dec. 13, 1856). At the time this letter was written, CD was working on his *Big Species Book*, in which he elaborated upon this theme: The electric organs of Fish, – those wonderful organs which, as *Owen says, 'wield at will the artillery of the skies' – offer a special difficulty. Their intimate structure is closely similar to that of muscle;[3] but it is most difficult to imagine by what grades they could have arrived at their present state.[4] Nevertheless the fact, <recently discovered> that Rays[5] which have never been observed to discharge the feeblest shock, yet have organs closely similar to those of true electric fishes, shows that we are at present too ignorant to speculate on the stages by which these organs, now affording such a powerful means of defence to the *Torpedo & *Gymnotus, may have been acquired. But the special difficulty in this case lies in the fact that the Electric fishes, only about a dozen in number, belong to two or three of the most distinct orders or better sub-classes of Fish.[6] (*Big Species Book*, p. 363; n. 3 cites Owen (1846), p. 217; n. 4 reads Dr. Carpenter [1854] in his Principles of Comparative Physiology (iv. Edit.) has an interesting discussion on the Electric organs of fishes: compare p. 465–470, & 471; n. 5 cites Stark (1844) and adds on Jan 6, 1845 Mr. Goodsir read paper on same subject, & shows that the organ in the Ray is the middle & posterior of the *caudal muscle, greatly modified; see Goodsir (1855); n. 6 cites Valenciennes

(1841), p. 44; CD underestimated the number of electric fish species).

CD was so charged up with electric fishes that he discussed them with George Henry Lewes, "a man of letters" known mainly for having been the common-law husband of Mary Ann Evans, a.k.a. George Eliot (Freeman 1978, p. 188).

Thus, in a letter titled Against organs having been formed by direct action of medium CD writes to him that If you mean that in distinct animals, parts or organs, such for instance as the luminous organs of insects or the electric organs of fishes, are wholly the result of the external and internal conditions to which the organs have been subjected, in so inevitable a manner that they could be developed whether of use or not to their possessor, I cannot admit [your view.] Moreover, CD notes the strange implications of Lewes's notions: [if] I should apply the same doctrine to the electric organ of fishes [I would] have to make, in my own mind, the violent assumption that some ancient fish was slightly electrical without having any special organ for the purpose (Darwin and Seward 1903, Vol. I, pp. 306–7).

In any case, CD need not have worried. Roberts (2000), in his review of the Malapteruridae, a family of strongly discharging catfishes, points out that CD "failed to realize that electricity could modify the behaviour of small prey, either by immobilizing it or causing it to react in such a way that it would be more susceptible to predation. The initial behaviour leading to *natural selection favouring weak electric organs in fishes, is surely more efficient feeding on small organisms, such as tiny vermiform aquatic insect larvae. From this, it would be only a short step to weakly electrogenic conspecifics joining each other in foraging activities and the simplest form of electrocommunication (detecting weak electrical discharges signalling that conspecifics are engaged in feeding behaviour). And from this it seems a short step to the development of electrical signalling related to reproductive behaviour, and perhaps the most derived function of all, electro-location. There would be many opportunities for occurrence of the phenomenon known as 'intensification of function' leading to the gradual improvement of the electrogenic and electrosensory organs by means of natural selection and perhaps also *sexual selection."

Electric organs (I) The electric organs of fishes are given much emphasis in the first edition of Origin (1859) where they are discussed in the chapter titled *Difficulties on theory, and to which CD attached much importance, for example advising his doubting friend C. *Lyell: Please read p. 193 beginning 'The Electric organs' & trust me that the sentence 'In all these cases of two very distinct species &c &c' was not put in rashly; for I went carefully into every case (Correspondence, Sept. 23, 1860).

We shall follow the advice: The electric organs of fishes offer another case of special difficulty; it is impossible to conceive by what steps these wondrous organs have been produced; but, as *Owen and others have remarked, their intimate structure closely resembles that of common muscle; and as it has lately been shown that *Rays have an organ closely analogous to the electric apparatus, and yet do not, as Matteuchi asserts, discharge any electricity, we must own that we are far too ignorant to argue that no transition of any kind is possible.

The electric organs offer another and even more serious difficulty; for they occur in only about a dozen fishes, of which several are widely remote in their affinities. Generally when the same organ appears in several members of the same class, especially if in members having very different habits of life, we may attribute its presence to inheritance from a common ancestor; and its absence in some of the members to its loss through disuse or Natural selection. But if the electric organs have been inherited from one ancient progenitor thus provided, we might have expected that all electric fishes would have been specially related to each other. Nor does *geology at all lead to the belief that formerly most fishes had electric organs, which most of their modified descendants have lost. (*Origin I*, p. 193; see also Matteucci (1843, 1847)).

Electric organs (II) In the sixth, and last edition of *Origin* revised by CD, some of the *difficulties for the theory due to electric organs were resolved, while others had emerged. Thus, we can use the revised text to provide a typical example of how CD incorporated new information and stylistic improvements into subsequent editions of his books (added words are *underlined*, deleted words are crossed through). This is messy, so bear with us:

The electric organs of fishes offer another case of special difficulty; <u>for</u> it is impossible to conceive by what steps these wondrous organs have been produced; but, as Owen and others have remarked, their intimate structure closely resembles that of common muscle; and as it has lately been shown that Rays have an organ closely analogous to the electric apparatus, and yet do not, as Matteuchi asserts, discharge any electricity, we must own that we are far too ignorant to argue that no transition of any kind is possible. <u>But this is not surprising, for we do not even know of what use they are. In the *Gymnotus and *Torpedo they no doubt serve as powerful means of defence, and perhaps for securing prey; yet in the Ray, as observed by Matteucci, an analogous organ in the tail manifests but little electricity; even when the animal is greatly irritated; so little, that it can hardly be of any use for the above purposes. Moreover, in the Ray, besides the organ just referred to, there is, as Dr. R. M'Donnell has shown, another organ near the head, not known to be electrical, but which appears to be the real homologue of the electric battery in the Torpedo. It is generally admitted that there exists between these organs and ordinary muscle a close analogy, in intimate structure, in the distribution of the nerves, and in the manner in which they are acted on by various reagents. It should, also, be especially observed that muscular contraction is accompanied by an electrical discharge; and, as Dr Radcliffe insists, 'in the electrical apparatus of the torpedo during rest, there would seem to be a change in every respect like that which is met with in muscle and nerve during rest, and the discharge of the torpedo, instead of being peculiar, may be only another form of the discharge which attends upon the action of muscle and motor nerve.' Beyond this we cannot at present go in the way of explanation; but as we know so little about the uses of these organs, and as we know nothing about the habits and structure of the progenitors of the existing electric fishes, it would be extremely bold to maintain that no serviceable transitions are possible by which these organs might have been gradually developed.</u>

The electric <u>These</u> organs <u>appear at first to offer another and far more serious difficulty; for they occur in only about a dozen kinds of</u> fishes, of which several are widely remote in their affinities. Generally <u>When</u> the same organ appears <u>is found</u> in several members of the same class, especially if in members having very different habits of life, we may <u>generally</u> attribute its presence to inheritance from a common ancestor; and its absence in some of the members to strikethrough loss through disuse or Natural Selection. But <u>So that</u>, if the electric organs hadbe inherited from <u>some</u> one ancient progenitor thus provided, we might have expected that all electric fishes would have been specially related to each other; <u>but this is far from the case.</u> Nor does geology at all lead to the belief that formerly most fishes <u>formerly</u> had possessed electric organs, which most of their modified descendants have <u>now</u> lost. <u>But</u>

when we look at the subject more closely, we find in the several fishes provided with electric organs, that these are situated in different parts of the body – that they differ in construction, as in the arrangement of the plates, and, according to Pacini, in the process or means by which the electricity is excited – and lastly, in being supplied with nerves proceeding from different sources, and this is perhaps the most important of all the differences. Hence in the several fishes furnished with electric organs, these cannot be considered as homologous, but only as analogous in function. Consequently there is no reason to suppose that they have been inherited from a common progenitor; for had this been the case they would have closely resembled each other in all respects. Thus the difficulty of an organ, apparently the same, arising in several remotely allied species, disappears, leaving only the lesser yet still great difficulty; namely, by what graduated steps these organs have been developed in each separate group of fishes. (*Origin* VI, pp. 150–1; Radcliffe 1871; Pacini 1853).

Readers interested in more detailed changes from the first to the sixth editions, and in the intermediate versions of *Origin,* should consult Peckham's (1959) *variorum* edition.

Eleotris gobioides *See* **Sleepers.**

Elephantfishes Members of the Family Mormyridae, based on the genus *Mormyrus*. A group of about 20 genera and 200 species of freshwater fishes in tropical Africa and the Nile (Nelson 1994, p. 96). The common name is due to the shape of the head, whose elongated snout, in some species, resembles the trunk of an elephant (*see* **Monstrosities**). The elephantfishes are weakly discharging *electric fishes, and they use the fields they generate to navigate in the murky waters that are their home, and to locate their invertebrate prey. The integration of the electric signals thus received requires a tremendous amount of processing power, and therefore, elephantfishes have the heaviest *brain (per unit body mass) recorded in fish. In one species, the brain is indeed so large and active that it uses up to 60% of its owner's metabolic energy, a value three times higher than in humans (Nilsson 1996). This contrasts with *Pike, which is small-brained and, as CD did not tire of telling us, rather stupid.

Embryology The study of the development of the embryo, i.e. [t]he young animal undergoing development within the egg or womb (*Origin* VI, p. 433).

CD was well aware of the enormous support provided by embryology for a theory of common descent of organisms (Oppenheimer 1968), and this is expressed, among other things, in *Foundations*, which has a section titled Embryology.

This starts as follows: The unity of type in the great classes is shown in another and very striking manner, namely, in the stages through which the embryo passes in coming to maturity. Thus, for instance, at one period of the embryo, the wings of the *bat, the hand, hoof or foot of the quadruped, and the fin of the porpoise do not differ, but consist of a simple undivided bone.

At a still earlier period the embryo of the fish, bird, reptile and mammal all strikingly resemble each other. Let it not be supposed this resemblance is only external; for on dissection, the arteries are found to branch out and run in a peculiar course, wholly unlike that in the full-grown mammal and bird, but much less unlike that in the full-grown fish, for they run as if to aerate blood by branchiae on the neck, of which even the slit-like orifices can be discerned. How wonderful it is that this structure should be present in the embryos of animals about to be developed into such different forms, and of which two great classes respire only in the air. Moreover, as the embryo of the mammal is matured in the parent's body, and that of the bird in an egg in the air, and that of the fish in an egg in the water, we cannot believe that this course of the arteries is related to any external conditions. (p. 218).

However, *Haeckel's *dictum* does not obtain: it has often been asserted that the higher animal in each class passes through the state of a lower animal; for instance, that the mammal amongst the vertebrata passes through the state of a fish: but Müller denies this, and affirms that the young mammal is at no time a fish, as does *Owen assert that the embryonic jelly-fish is at no time a polype, but that mammal and fish, jelly-fish and polype pass through the same state; the mammal and jelly-fish being only further developed or changed. (*Foundations*, p. 219; Müller 1838–42; *see also* **Ontogeny recapiculates phylogeny**).

Also, Whatever may have been the form or habits of the parent-stock of the Vertebrata, in whatever course the arteries ran and branched, the selection of *variations, supervening after the first formation of the arteries in the embryo, would not tend from variations supervening at corresponding periods to alter their course at that period: hence, the similar course of the arteries in the mammal, bird, reptile and fish, must be looked at as a most ancient record of the embryonic structure of the common parent-stock of the four great classes. (*Foundations*, p. 226; *see also* **Vertebrate origins**).

Endemic Peculiar to a given locality (*Origin* VI, p. 433).

Engraulis ringens *See* **Peruvian anchoveta**.

Entomostraca A division of the class *Crustacea, having all the segments of the body usually distinct, gills attached to their feet or organs of the mouth, and their feet fringed with fine hair. They are generally of small size (*Origin* VI, p. 433). Thus defined, this group excludes the prawns, lobsters, crabs, etc., which make up the 'Malacostraca'. The term Entomostraca is generally used by CD for what we would today call *zooplankton. (*See also* **Food webs**.)

Eponym A place, process or thing named after a person; for example, the temperature and light (i.e. *latitude) threshold for atoll formation, now called 'Darwin Point' (Grigg 1982), or the unit of relative change in the structure of organisms called *'darwin,' and used in palaeontology.

Several fish species have been named eponymously after CD, notably the *wrasse *Pimelometopon darwini* (Jenyns, 1842), listed as *Cossyphus darwini* in the compendium of animals, institutions, places and plants named after CD that was published by Freeman (1978, pp. 81–6). However, the following fish species were described early enough to have been on that list: *Trachichthys darwinii* Johnson, 1866 – now *Gephyroberyx darwinii*; *Tetrodon darwinii* Castelnau, 1873 ("Dedicated to the greatest naturalist of the age") – now *Marilyna darwinii*; *Oncopterus darwinii* Steindachner, 1874 (described as *Rhombus in Fish); *Sebastes darwini* (Cramer in Jordan, 1896); *Graviceps darwini* (Whitley, 1958) and *Ogcocephalus darwini* (Hubbs, 1958). Many of these scientific *names have been shown, in the meantime, to be *synonyms of other species, based on rules too intricate to be documented here, and detailed in the *International *Code of Zoological Nomenclature*.

Moreover, a number of species were named after the *city* of Darwin (formerly 'Port Darwin'), northern Australia. These are *Lates darwinensis* Macleay, 1878; *Ophiocara darwiniensis* Macleay, 1878; *Agonostoma darwiniense* Macleay, 1878; *Holacanthus darwinensis* Saville-Kent, 1889; *Pomacentrus darwiniensis* Whitley, 1928; and *Epinephelus darwinensis* Randall and Heemstra, 1991 (see Eschmeyer (1998), pp. 456–7, and *FishBase for current names).

These species may be called 'second-order eponyms', with Darwin the city being a 'first-order eponym'. The common name *'Darwin's roughy' is also a second-order eponym of sorts, being based on the scientific name, itself a first-order eponym. [*See* **Retronym** for the concept of a 'first-order eponym'].

Other eponyms in this book are *guppy(-ies), named after R. J. L. *Guppy, and Mr *Fish, a reverse eponym, and apparently a man of little imagination.

Esox lucius See **Pike**.

Evolution The process by which the Earth's diverse plants, animals and other living organisms came into being. The key features of evolution, besides its incontestable factuality (notwithstanding *creationism), are descent with modification (*Origin* VI, p. 404), and the relatedness of increasing numbers of taxa as one moves back in time, eventually encompassing all living and extinct forms, i.e. shared ancestry.

Evolution was not a new idea when, upon his return from the voyage of the *Beagle*, CD started his search for the driving force of evolution (*see* **Lamarck; Non-Darwinian evolution; Notebooks; Spontaneous generation**). However, it is the mechanism that CD proposed – *natural selection – and the many examples in *Origin* and other major works which eventually turned evolution from an interesting concept into a 'scientific research programme' (Lakatos 1970), providing guidance and standards for the day-to-day research of successive generations of biologists (*see also* **Paradigm**).

Perhaps surprisingly, CD did not use the word 'evolution' in any work published prior to 1870, the first instance being on p. 2 of *Descent* (the next was in the sixth edition of *Origin*; see Freeman (1977), pp. 79–80, 129).

Exocoetus spp. *See* **Flying fishes**.

Experiments (I) Contrary to a widely held belief, CD did perform numerous experiments to test his various hypotheses (Ghiselin 1969; *see* **Lizards** for one of CD's most incisive experiments, on whether *Amblyrhynchus cristatus* consumes plants or fishes).

Other (also exceedingly simple) experiments he conducted tested the ability of fishes to withstand rapid changes of salinity (*see* **Burrfishes (II), Freshwater fishes (II)** and **Galaxiidae**), presumably to check on their ability to alternate between marine and freshwater habitats, and thereby increase their *range.

Most of the experiments documented in CD's *Experimental notebook* (*DAR 157a) dealt with issues related to the distribution of plants, which he had identified as a strong source of arguments against the doctrine of the immutability of species. Thus, this series of five entries on 'Experiments' mainly deals with the approach CD devised to tell whether (and how) fish could impact on the distribution of higher plants. Overall, these five entries match the standard structure of a scientific paper, with this entry corresponding to the 'Introduction,' II–IV corresponding to the 'Materials and Methods' sections of three successive series of experiments, and V corresponding to a joint 'Results and Discussion' section.

Experiments (II) CD's first series of experiments with fish, documented in this entry, was meant to test (a) whether fish would eat seeds; and (b) whether such seeds would germinate once they had passed through the intestinal tract of the fish.

These experiments are anticipated in notes that CD penned in 1840 (see annotations to Lyell 1838), but really started in the mid 1850s, as indicated in a letter to CD's cousin W. D. Fox: I am rather low today about all my experiments. - everything has been going wrong - the fan tails have picked the feathers out of the Pouters in their journey home - the fish at the Zoological Gardens after eating seeds would spit them all out again - Seeds will sink in salt water - all nature is perverse & will not do as I wish it, & just at present I wish I had the old *Barnacles to work at & nothing new (*Correspondence*, May 7, 1855).

More details on these experiments are available: Gave Gold Fish at Zoolog. Gardens. Canary Millet, Lettuce, Cabbage, Linseed Barley onion. They took them in mouth & kept them for some seconds [..] & then rejected them with force. The seeds had soaked 28 hours. I c^d. not get other fish to try, I noticed they cracked the Onion seed. (DAR 205.2.115, dated May 5, 1855).

And here is how we know what CD hoped these experiments to show: Everything has been going wrong with me lately; the fish at the Zoolog. Soc. ate up lots of soaked seeds, & in imagination they had in my mind been swallowed, fish & al, by a heron, had been carried a hundred miles, been voided on the banks of some other lake & germinated splendidly, – when lo & behold, the fish ejected vehemently, & with disgust equal to my own, all the seeds from their mouths. (*Correspondence* to J. D. Hooker, May 15, 1855).

Experiments (III) CD's second series of fish experiments tested a hypothesis more sophisticated than the first: whether the seeds consumed by fish would germinate if voided by a bird, after it had consumed the fish (*see also* **Dace**).

CD wrote on this: I am trying many *little* experiments, but they are hardly worth telling, though some I am sure will bear on distribution (*Correspondence* to C. J. F. Bunbury, April 21, 1856). But he did tell, if first to a close friend: I find Fish will greedily eat seeds of aquatic grasses, & that millet seed put into Fish & given to Stork & then voided will germinate. So this is the nursery rhyme of this is the stick that beat the pig &c &c. (*Correspondence* to J. D. Hooker, Jan. 20, 1857; Conner 1813). This experiment is recorded in CD's *Experimental notebook*, p. 19 (*DAR 157a).

While conducting his experiments, CD went on reading related material and asking his correspondents. Thus the question he wrote at the bottom of a letter received from Richard Hill, which mentioned the distribution of seeds by sea currents around Jamaica: Tropical Fish eat seed? (*Correspondence* from, Jan. 10, 1857).

The letter in which he posed this question to Hill is now lost, but we have the answer: "We have two distinctive River Mullets the Mountain Mullet, and the Hog-nose Mullet that both feed on the seeds of a Laurel called the Timber Sweet-wood, and wild fig. They take the fruit as they float down the stream, and can be caught with a line, if it be baited with these fruits" (*Correspondence*, March 12, 1857). Satisfied, CD annotated this letter Fish eating seeds.

Hill was a competent, well-published naturalist, with a good knowledge of Jamaican and other fishes (see, for example, Hill 1881). Hence, it is not surprising that the two mullets he mentions can be readily identified as *Agonostomus monticola* (Bancroft, 1834), and *Joturus pilchardi* Poey, 1860, respectively. Both live in rivers (but spawn in the sea, and hence are *catadromous), and both are reported to feed on a variety of items, including benthic plants (Cruz 1987).

Experiments (IV) Given the largely negative results of his own experiments (see above), CD resorted to an approach that had worked for many other problems: enlisting hired hands, friends, colleagues and family (Browne 1995, 2002).

Thus, CD talked one of his nephews, Edmund Langton, into performing some of his experiments. Langton reported on this to CD's son Francis as follows: "Will you tell your papa that I have tried the experiments with all the seeds but the *minnows took only a very little Dutch clover and spit it out again, and the Prussian *carp took one anthoxanthus seed and spit it out again but it was a rather cold day so I will try again" (*Correspondence*, Vol. 6, p. 365).

As well, papa hired James Tenant, the keeper of the aquarium at the gardens of the Zoological Society in London, who reported as follows:

"Dear Sir, I now send the results of the experiments of the Seeds which I have been trying ever since you was at the Gardens.

I have succeeded in getting Several of them to take them by letting them go a day or two without food the Minnow take the millet very and so does the Gold fish one minnow took 5 Seeds this day and the *Tench and Common Carp and *Barbel take the Wheat after it is Soaked well in Water for 12 Hours

As an Illustration of the fact that Barbel will take Wheat Dr Crisp – a Fellow of the Society who is a great Angler told me, that he has taken Barbel and dissected them And found a quantity of Wheat in them and he says he has caught them near the water mills where the wheat has been spilt into the river but never could get them to take it as a bait. I should be most happy to make any Experiment you might suggest and I remain your obedient Servt" (*Correspondence*, March 31, 1857).

CD then followed up with Edward Crisp, who responded that, indeed, he "caught the Barbel (several of them) last autumn near a milldam and was surprised to find wheat in their stomach." etc. (see *Correspondence*, April 4, 1857).

Finally, Tenant reported on the last experiments: "I have tried the Fish with Oats and the two sorts of Seeds that you sent The Tench and Cm Carp and Gold fish take Both Sorts equally well but I cannot get them to take Oats but if I ever succeed will let you know the result" (*Correspondence*, April 23, 1857).

We also know the nature of the seeds that CD sent, as he wrote in his *Experimental book* (*DAR 157a, p. 18): Fish ate Yellow Water Lilly & Potamogeton seeds – 3 Water Lily seeds passed through a Stork, but did not germinate.

Experiments (V) The result of all this experimenting and corresponding is that CD felt comfortable enough about the impact of fish on the distribution of plants to write about it in *Origin*, viz.:

Freshwater fish, I find, eat seeds of many land and water plants: fish are frequently devoured by birds, and thus the seeds might be transported from place to place. I forced many kinds of seeds into the stomachs of dead fish, and then gave their bodies to fishing-eagles, storks, and pelicans; these birds, after an interval of many hours, either rejected the seeds in pellets or passed them in their excrement; and several of these seeds retained the power of germination. (*Origin VI*. p. 327).

CD later developed this theme: I have stated that freshwater fish eat some kinds of seeds, though they reject many other kinds after having swallowed them; even small fish swallow seeds of moderate size, as of the yellow waterlily and Potamogeton. *Herons and other birds, century after century, have gone on daily devouring fish; they then take flight and go to other waters, or are blown across the sea; and we have seen that seeds retain their power of germination, when rejected many hours afterwards in pellets or in the excrement. When I saw the great size of the seeds of that fine waterlily, the Nelumbium, and remembered Alph. de Candolle's remarks on the distribution of this plant, I thought that the means of its dispersal must remain inexplicable; but Audubon states that he found the seeds of the great southern water-lily (probably, according to Dr. Hooker, the *Nelumbium luteum*) in a heron's stomach. Now this bird must often have flown with its stomach thus well stocked to distant ponds, and then getting a hearty meal of fish, analogy makes me believe that it would have rejected the seeds in a pellet in a fit state for germination. (*Origin VI*, p. 346; Candolle 1855).

Expression Short title of the book titled The Expression of the Emotions in Man and Animals (Darwin 1890). The second, posthumous edition of *Expression* is used here, as it elaborates on one of the only two references to fish in the entire work (*see* **Kingfisher** for the other).

In the first edition, CD wrote that: Hardly any expressive movement is so general as the involuntary erection of the hairs, feathers and other dermal appendages; for it is common throughout three of the great vertebrate classes.[11] These appendages are erected under the excitement of anger or terror; more especially when these emotions are combined, or quickly succeed each other. The action serves to make the animal appear larger and more frightful to its enemies or rivals, and is generally accompanied by various voluntary movements adapted for the same purpose, and by the utterance of savage *sounds. (p. 100).

Note 11, added to the second edition, edited by Francis Darwin based on CD's annotations to the first, reads as follows: Rev. S. J. Whitmee (Proc. Zool. Soc., 1878, pt. i, p. 132) describes the erection of the dorsal and anal fins if fishes in anger and fear. He suggests that the erection of the spines gives protection against carnivorous fish, and, if this is so, it is not hard to understand the association of such movements with these emotions. Mr F. Day (Proc. Zool. Soc., 1878, p. i, p. 219) criticizes Mr Whitmee's conclusions, but the description by Mr Whitmee of a spiny fish sticking in the throat of a bigger fish and being finally ejected, seems to prove that the spines are useful.

The Reverend Mr Whitmee, then a missionary in Samoa, thus appears to have been very observant. But (or: Moreover,) he clearly did not believe in the naïve adaptionism implicit in *creationism. Thus, he suggests in a note that the new edition of the *Encyclopedia Britannica* should be purged of the notion that the elegant and diversified colors of fishes reflect their *altruism, i.e. that they were created for the special gratification of 'man'. As he wryly concludes, "unfortunately, these quickly disappear when man gets possession of the fish".

Extinction The ultimate fate of all species – though CD differed from the 'catastrophists' of his days, notably Cuvier, as to the most common mode of species extinction.

Among his first, rather opaque musings on the topic of extinction, we have: In a decreasing population at any one moment fewer closely related [. . .] ultimately few genera (for otherwise the relationship would converge sooner) & lastly perhaps one single one. –

Will not this account for the odd genera with few species which stand between great groups, which are bound to consider the increasing ones.–

NB As Illustrations are many anomalous *lizards living; or the tribe fish extinct. or of Pachydermata, or of coniferous trees; or in certain shell cephalopoda. – <<Read Buckland>> (*Notebook B*, p. 206; Buckland 1836).

Then CD thought he understood extinctions: The view entertained by many geologists, that each *fauna of each Secondary epoch has been suddenly destroyed over the whole world, so that no succession could be left for the production of new forms, is subversive to my theory, but I see no grounds whatever to admit such a view. On the contrary, the law, which has been made out, with reference to distinct epochs, by independent observers, namely, that the wider the geographical range of a species the longer is its duration in time, seems entirely opposed to any universal extermination. The fact of species of mammiferous animals and fish being renewed at a quicker rate than mollusca, though both aquatic; and of these the terrestrial *genera being renewed quicker than the marine; and the marine mollusca being again renewed quicker than the Infusorial animalcula, all seem to show that the extinction and renewal of species does not depend on general catastrophes, but on the particular relations of the several classes to the conditions to which they are exposed. (*Foundations*, pp. 145–6).

Here, CD was very wrong: catastrophes – notably the impacts of large meteorites – are now recognized as being a major factor in species extinctions (see, for example, Raup 1986; Alvarez 1998). However, few contemporary writers believe that this reduces the role of *natural selection as a major evolutionary force (Dawkins 1996, 1998). One of these is the late S. J. Gould, as suggested in his many works cited here (I must tread carefully, as he insisted that that he didn't; see Wright 1999).

We now turn to specifics: I explain the fact of so many anomalous or what may be called 'living fossils' inhabiting now only fresh-water, having been beaten out & exterminated in the sea by more improved forms; thus all existing *Ganoid fishes are fresh-water as is *Lepidosiren & Ornithorhynchus, &c. (*Correspondence* to J. D. Hooker, Dec. 24, 1858). Or, put differently, we occasionally see an animal like the Ornithorhynchus or Lepidosiren, which in some small degree connects by its affinities two large branches of life, and which has apparently been saved from fatal competition by having inhabited a protected station. (*Origin, VI*, p. 105). Indeed, some few Ganoid fishes, have been preserved from utter *extinction by inhabiting rivers, which are harbours of refuge, and are related to the great waters of the ocean in the same way that islands are to continents. (*Descent I*, p. 159).

Or With respect to the Lepidosiren, Ganoid fishes, perhaps Ornithorhynchus, I suspect, as stated in the Origin, that they have been preserved, from inhabitiing fresh-water and isolated parts of the world, in which there has been less competition and less rapid progress in Natural Selection, owing to the fewness of the individuals which can inhabit small areas; and where there are few individual variations at most must be slower. There are several allusion to this notion in the Origin, as under *Amblyopsis, the blind *cave fish (Darwin and Seward 1903, Vol. I, p. 143). And again The preservations of ancient forms in islands appears to me like the preservation of ganoid fishes in our present freshwaters. (Darwin and Seward 1903, Vol. I, p. 481).

Yet, many of the aquatic 'living fossils' discovered in the twentieth century were marine forms, for example the coelacanth *Latimeria chalumnae* Smith, 1939, from South-eastern Africa (Balon *et al.* 1988), of which a close relative has been recently found in Sulawesi, Indonesia (Erdman *et al.* 1998; Pouyaud *et al.* 1999).

I suggest the score is Extinctions 2 : CD 0.

Eyes For some reason, the evolution of the eye is seen by many as particularly difficult for *natural selection to explain. The adherents of *creationism, indeed, invariably include some scientific-looking drawing of an eye in their pamphlets. Even Emma Darwin – CD's cousin and wife – had ophthalmic concerns (Dawkins 1996, p. 127), as did Karl *Popper.

This is strange, as the eyes are the organs for which intermediate steps that increase the fitness of their owners are not only easy to conceive, but occur in a wide range of animals (Wallace 1889, pp. 90–91; Dawkins 1996, pp. 126–79). Indeed, straightforward computer simulations show that, if the fitness of organisms is increased by spatial information, it will take only a few hundred thousand years for continuous small improvements to turn light-sensitive patches into eyes capable of focus (Nilsson and Pelger 1994). Thus, these authors conclude, "it is obvious that the eye was never a threat to Darwin's theory of evolution".

CD did not know this, and he devoted a section of the *Difficulties on Theory chapter of the first edition of *Origin* to this topic, but without references to fishes. However, he intended for its first US edition to include the following . . . (though in the fish Amphioxus), the eye is an extremely simple condition without a lens) (Burkhardt *et al.* 1993, p. 580; *see also* **Dohrn**).

And we find this bit in the sixth edition of *Origin*, with further elaboration: Within the highest division of the animal kingdom, namely, the Vertebrata, we can start from an eye so simple, that it consists, as in the *lancelet, of a little sack of transparent skin, furnished with a nerve and lined with pigment, but destitute of any other apparatus. In fishes and reptiles, as *Owen has remarked, 'the range of dioptic structure is very great.' (p. 145).

Why, in God's name, don't the creationists challenge evolutionists about ***BRAINS**? Wouldn't that be more interesting?

Falkland Islands A small archipelago in the southwestern Atlantic (about 500 km from Argentina, on the South American mainland), in the eighteenth century much visited by ships from the port city of Saint Malo, in France, thus called 'Malouines' in French (Verne 1997, p. 111), and thence 'Malvinas' in Spanish. Taken over by Britain in the early nineteenth century.

CD visited the Falkland Islands twice, from March to April 1833 and in March–April 1834, and these visits form the basis of a neat little book on *Darwin's Desolate Islands* (Armstrong 1992a), from which the following, dealing with CD's "diligence as a collector", is adapted: "One note shows that on occasion Darwin carefully noted the habitat of what he collected: Shells, kelp; excepting small thin bivalve on beach, March, East Falkland (DAR 29.1/6, specimen 1029). [. . .] Darwin was often careful to take notes of the *colours of specimens, or part of specimens which were likely to change or be lost by his preservation methods. The colours of the eyes of birds and the scales of fish were recorded [. . .].

Many of the specimens collected by Darwin and others aboard HMS *Beagle* still exist [. . .]. Fish specimens preserved in alcohol at the British Museum (Natural History) in South Kensington include the eelpout *Phucocoetes latitans*, caught by Darwin in the kelp-beds of Berkeley Sound, East Falkland, and specimens of *Aplochiton zebra*, i.e., syntypes of the Falkland trout, the well-known *endemic freshwater fish in the islands" [*see also Fish in Spirits*, nos. 553–555]. "Darwin had a certain amount of specialised equipment, including separate insect-nets for sweeping through grass on land and for use in aquatic environments [. . .]. Seine nets were sometimes employed for *fishing." (Armstrong 1992a, pp. 13–14). Indeed CD left no stone unturned, and *Fish in Spirits* reports one specimen (no. 558) from under stones.

Human impacts on the Falkland Islands began in the mid eighteenth century, with the 'secret' French settlement of the 'Isles Malouines'; Armstrong (1994) describes the devastations they and their successors brought about.

On freshwater environments, Armstrong writes: "Even now the freshwater *ecology of the Falklands has been little documented, so the effect of the introduction of the brown *trout [*Salmo trutta*] on the Falkland trout (*Aplochiton zebra*) and the several species of *Galaxiidae (*G. attenuatus*, *G. maculatus* and *G. smithii*) has not been much discussed. It has however been stated that the introduced trout preys on the smaller fish, and that there has been some reduction of the indigenous species."

Three species of Darwin's Fishes occur in the Falkland Islands: *Phucocoetes latitans* see **Eelpouts**; *Aplochiton zebra* (*see* **Galaxiidae**); and the Falkland sprat *Sprattus fuegensis* (*see* **Herrings**).

Fauna The totality of the animals naturally inhabiting a certain country or region, or which have lived during a geological period (*Origin* VI, p. 433).

Figures CD could not draw well, and this is one of the few self-critical observations in *Autobiography* that appear to be true. There, he notes his incapacity to draw, an irremediable evil in his life (p. 47). Hence the clumsiness of the few sketches in his letters and notes (with some brilliant exceptions in his *Zoological Notes*), and of the lone phylogenetic tree in *Origin*. This, however, did not prevent most of his books from featuring decent drawings: the *Zoology of the Beagle Voyage* was illustrated by hired artists. Indeed, as attested by his letters to Jenyns, the figures illustrating *Fish consumed much of CD's attention (and funding): You say there are about 37 new species of *Acanth[opterygians] & according to this proportion, there should be about 21 in the other orders. Now do you think it very desirable that all should be engraved

75

[?...]. I find money has gone rather quickly than I anticipated. I must therefore be a little stingy, as I am fully convinced, as before said, that your part will be most valuable, you may rely on it. I have no wish to carry my stinginess to any great extent.

You guess there will be 6 or 7 numbers each with 5 or 6 plates, giving about 36 plates, & therefore I presume you imagine about every other plate will contain two fishes. – I ask these questions, which I know you cannot answer, except most vaguely, that I may be able to come to some definite terms with the artist. Will you also tell me to how great a degree you would like the artist to come and lodge at Cambridge. – Richardson formerly had him in same manner at Chatham (*Correspondence*, Oct. 14, 1839).

CD also had woodcuts of fishes illustrating some of his work, executed by the well-known artist Mr G. Ford, from specimens in the British Museum, under the kind superintendence of Dr. Günther (*Descent II*, p. 335).

All the fish figures in this book were adapted from the lithographs prepared for *Fish* by Mr Benjamin Waterhouse *Hawkins, or from the woodcuts prepared for *Descent* by Mr Ford. This work, kindly performed by Ms Aque Atanacio, of the *FishBase project, involved scanning the originals, electronic enhancement (as far as possible) of the resulting pictures, reorientation of the fishes, if required (they now all point leftward, as all FishBase fishes do), and regrouping into panels with related species.

Filefishes Members of the Family Monacanthidae, related to the *triggerfishes, with which they share the mechanism that enables them to lock the first spine of their dorsal fin in an erect position. CD collected specimens of two filefish species in *King George's Sound, Australia. The first was the Chinese leatherjacket *Nelusetta ayraudi* (Quoy & Gaimard, 1824), described as *Aleuteres velutinus* in *Fish* (pp. 157–8; Very pale brown: fins pale orange.). The second was the Bridled leatherjacket *Acanthaluteres spilomelanurus* (Quoy & Gaimard, 1824), originally described as *Aleuteres maculosus* (*Fish*, pp. 156–7; Mottled with pale blackish green, leaving white spots.). The filefish species CD later wrote about are, however, the Broom filefish *Amanses scopas* (Cuvier, 1829), and the Pot-bellied leatherjacket *Pseudomonacanthus peroni* (Hollard, 1854): [in the..] *Monacanthus scopas* [the] male, as Dr. Günther informs me, has a cluster of stiff, straight spines, like those of a comb, on the sides of the tail; and these in a specimen six inches long were nearly one and a half inches in length; the female has in the same place a cluster of bristles, which may be compared with those of a tooth-brush. In another species, *M. peronii*, the male has a brush like that possessed by the female of the last species, whilst the sides of the tail in the female are smooth. In some other species of the same genus the tail can be perceived to be a little roughened in the male and perfectly smooth in the female; and lastly in others, both sexes have smooth sides. (*Descent II*, p. 331).

Günther expressed doubt on *Monacanthus* brushes (Calendar, no. 9316), but later (1880, pp. 685–6), he did describe sexual differences in the *caudal armature of filefishes.

Fish Short title of Volume IV of the *Zoology of the Beagle*, authored by L. *Jenyns. The first edition of *Fish* was published in four parts, over a period of 27 months (Sherborn 1897), a fact of great importance to taxonomists. Thus, although *Fish* is usually cited as 'Jenyns (1842),' it should be cited as 'Jenyns (1840–42).'

CD wrote to Jenyns on April 10, 1837, inviting him to describe the fishes collected during the voyage of the *Beagle* as part of its 'Zoology': The whole scheme is at present, merely floating in the air; but I was determined to let you know, as I should very much like to know what you think about it, & whether you would object to supply descriptions of the fish in such a work, instead of to transactions.

Jenyns accepted, and the two subsequently exchanged many letters regarding the contents and form of *Fish*; excerpts from these are presented under various headings (*see* **Jenyns (II); Hawkins; Figures**). Readers interested in a complete coverage of this exchange should consult Vol. 2 of CD's *Correspondence* (Burkhardt *et al*. 1986).

One interesting result of this exchange, and of Jenyns having access to CD's notes (i.e. **Fish in Spirits of Wine*) is that *Fish* differs from most taxonomic works of its time in that it includes reproducible description of *colours, and some ecological and behavioural observations.

Fish(es) A group of animals which includes "all those free-living, aquatic, cold blooded, gill-breathing craniates in which fins and not pentadactyl limbs are developed" (Moy-Thomas 1971, p. 1; see also Lecointre 2000). Thus defined, 'fishes' are a paraphyletic group, i.e., they do not include all descendants of their most recent common ancestor (De Queiroz 1988).

It is often wondered just what the correct use is of 'fish' vs. 'fishes.' The Glossary in Nelson and Paetz (1992) gives the rule that has gained widest acceptance: "the singular 'fish' is appropriately used when talking about any number of individuals of the *same* species, while the plural 'fishes' is used when discussing individuals of two or more species."

Be that as it may, CD had to turn fishes into tetrapods if he were to convince his colleagues of his theory being capable of explaining large evolutionary changes. The following quotes refer to this constant preoccupation: *M]ammalia in age long gone past. & still more so <<known>> with fishes and reptiles [. . .] We have not the slightest right to say there never was common progenitor to Mammalia & fish when there now exists such form as ornithorhyncus. [. . .] With unknown limits, every tribe appears fitted for as many situations as possible. For instance take *birds, animals, reptiles fish – Conditions will not explain status (Perhaps consideration of range of capabilities past & present might tell something) (*Notebook B*, pp. 192, 194, 198; see Zimmer (1998) for a captivating account of the transition from fishes to terrestrial vertebrates, true to CD's spirit).

Fish in Spirits of Wine Title of a copy of that part of CD's Zoological **Catalogue* dealing with fishes, kept as DAR 29.1 in the Darwin Archive in the library of Cambridge University, and now available as part of the **Zoology Notes*, edited by Keynes (2000). *Fish in Spirits of Wine*, originally extracted from CD's catalogue by Syms *Covington, and subsequently annotated by CD and *Jenyns, was transcribed from the original by Dr Jacqueline McGlade and is included in this book as Appendix I.

Fish in Spirits was used extensively by Jenyns when he wrote **Fish*, especially as far as live *colours and habitats are concerned. In most cases, CD's original words are simply transferred from *Fish in Spirits* to *Fish*.

Comparison of *Fish in Spirits* with the *Zoology Notes* shows *Fish in Spirits* to be enriched with explicit locations and dates, such as to enable Jenyns to better identify the specimens' sampling localities. Only few details were lost, notably the description of a *flyingfish colour, and of an unidentified fish sporting Sides transverse bars of 'chocolate & brownish red' separated by narrow grey spaces (compare no. 866 in *Zoology Notes* with same number in *Fish in Spirits*). More serious were Covington's errors in transcribing geographic coordinates: he tends to misrepresent the minutes, and recorded '3' instead of '38'°S in at least two instances (see specimens nos. 347 and 359).

Overall, however, *Fish in Spirits* is a more complete source of information on *Darwin's Fishes* than Keynes (2000). Hence our decision to include *Fish in Spirits* as Appendix I of this book, in spite of a large overlap with the fish entries of the *Zoology Notes*.

Fish, Mr Here is how CD introduces this gentleman: In the year 1869, Mr Fish[5] rejected my conclusions with respects to the part which *worms have played in the formation of mould, merely on account of their assumed incapacity to do so much work. He remarks that 'considering their weakness and their size, the work they are represented to have accomplished is stupendous'. Here we have an instance of that inability to sum up the effects of a continually recurrent cause, which has so often retarded the *progress of science, as formerly in case of geology, and more recently in that of the principle of *evolution. (*Worms*, p. 6; n. 5 refers to *Gardener's Chronicle*, 17 April 1869, p. 418). Clearly, this excellent gardener (*Variation II*, p. 129) should have read the work of *Lyell. There goes Mr Fish, in any case not really one of *Darwin's Fishes, only a reverse *eponym.

FishBase: A large database on the fishes of the world (over 28 000 valid species), available on CD-ROM (Froese and Pauly 2000) and on the Internet (www.fishbase.org), and used as a key source of information during the writing of *Darwin's Fishes*, notably on their taxonomic status and distribution.

Interested readers may use the Internet version (which also includes the full text of Froese and Pauly 2000) to access more information on these fishes and their relatives. FishBase also provides an extended coverage of many of the concepts introduced in this book, especially on nomenclature and classification. This is because, among other things, FishBase includes the full contents of William Eschmeyer's magisterial *Genera of Fishes* (over 10 000 genera reviewed; Eschmeyer 1990), and his *Catalog of Fishes*, three huge volumes evaluating the status of over 50 000 nominal species (Eschmeyer 1998). The numerous facts, e.g. on the feeding of fishes, and their *distribution, presented by FishBase in easily accessible form (as interactive graphs and maps) would have pleased CD, who used printed questionnaires (as we initially did with FishBase) to elicit precise answers to questions about the breeding of animals and other topics from his numerous correspondents (Freeman 1977, pp. 54, 111, 142), many of them parsons (Armstrong 2000).

CD would also have been pleased with the 'Darwinian' nature of FishBase, whose 'facts' are all subjected to the harsh environment created and maintained by thousands of critical users. We (the members of the FishBase team, led by Rainer Froese) hope that the surviving facts will contribute to generalizations worthy of the founder of evolutionary biology. Readers interested in working with FishBase, or in helping complete its coverage, will find it, and our e-mail and postal addresses, at www.fishbase.org.

Fishing The activity of catching fish, by *angling, or the deployment of other gear by humans, or as part of the foraging habits of *birds, or mammals such as otters and *jaguars. CD's writings contain numerous references to fishing, hence his being included in an anthology of fishing lore (Profumo and Swift 1985).

The quote below, from CD's *Journal* (p. 10), which concludes an entry on *St Paul's Rock, has no obvious context: The smallest rock in the tropical seas, by giving a foundation, for the growth of innumerable kinds of sea-weed and compound animals, supports likewise a large number of fish. The *sharks and the seamen in the boats maintained a constant struggle, who should secure the greater share of the prey caught by the lines. I have heard, that a rock near the Bermudas, lying many miles out at sea, and covered by a considerable depth of water, was first discovered by the circumstance of fishing having been observed in the neighbourhood.

This discovery is not explicitly mentioned in the history of Bermuda fisheries in Smith-Vaniz *et al.* (1999, pp. 53–9), but their account makes that story likely, given the abundance of fishes that the first colonists encountered in the early seventeenth century, the rapid build-up of a large local fishing fleet, and the ensuing decline of the nearshore resources.

FitzRoy, Robert Captain of the *Beagle from 1831 to 1836, aged only 26 years when he assumed command. His reports, notably his *Narrative* (FitzRoy 1839), and CD's own writing prove him to have possessed great technical skills, and to have been, on the whole, a rather genial host to the even younger CD.

However, History was not kind to FitzRoy. The contrast between this self-conscious, bigoted aristocrat, who fervently believed in slavery and *creationism, and his guest, the liberal and now immortal CD, is just too great. And neither did it help that, in 1865, FitzRoy ended up cutting his own throat – literally.

Flatfishes It may seem strange today, given the issues now warmed over to boost the sagging fortunes of contemporary *creationism, that the *asymmetry of flatfishes was, in CD's time, at the centre of the scientific debate about *natural selection and its implications. Hence the attention given to this group by CD: The Pleuronectidae, or flat-fish, are remarkable for their asymmetrical bodies. They rest on one side – in the greater number of species on the left, but in some on the right side; and occasionally reversed adult specimens occur. The lower, or resting surface, resembles at first sight the ventral surface of an ordinary fish: it is of a white colour, less developed in many ways than the upper side, with the lateral fins often of smaller size. But the eyes offer the most remarkable peculiarity; for they are both placed on the upper side of the head. During early youth, however, they stand opposite to each other, and the whole body is then symmetrical, with both sides equally coloured. Soon the eye proper to the lower side begins to glide slowly round the head to the upper side; but does not pass right through the skull, as was formerly thought to be the case. It is obvious that unless the lower eye did thus travel round, it could not be used by the fish whilst lying in its habitual position on one side. The lower eye would, also, have been liable to be abraded by the sandy bottom. [. . .] The chief advantages thus gained seem to be protection from their enemies, and facility for feeding on the ground. (*Origin VI*, p. 186). Straightforward enough, but it still led to a storm of controversy, some of which is captured in the next two entries.

We conclude this entry with *Platessa orbignyana*, a flatfish species represented in *Fish* (p. 137) by a specimen of 8 inches 9 lines collected by CD in *Bahia Blanca, Argentina, where it is said to be plentiful, and whose colours were Above dirty reddish brown: beneath faint blue: iris yellow. (*See also Fish in Spirits*, no. 394.)

Eschmeyer (1998, p. 1246) suggests that "*Platessa orbignyana* Valenciennes [is] probably not an available name. It appears that the plate from Valenciennes was published first and the name was made available from then . . ."

Jenyns also assigned another flatfish specimen, of "6 inches 6 lines," collected by CD at King George's Sound, Australia, to *P. orbignyana*. However, the latter specimen is now considered to have belonged to *Pseudorhombus jenynsii* (Bleeker, 1855), the Small-toothed, or 'Jenyns' flounder'. This species (or complex of species), which is *endemic to southern Australia (Gomon *et al.* 1994, p. 850), reaches 35 cm in total length, 1.5 kg, and is considered an excellent food fish.

Flatfish controversy (I) The first 'Flatfish' controversy covered here is that between CD and Mivart (1871), one of his most articulate critics, but still one whose position is more interesting when presented by CD than in the original: Mr Mivart has taken up this case, and remarks that a sudden spontaneous transformation in the position of the *eyes is hardly conceivable, in which I quite agree with him. He then adds: 'if the transit was gradual, then how such transit of one eye a minute fraction of the journey towards the other side of the head could benefit the individual is, indeed, far from clear. It seems, even, that such an incipient transformation must rather have been injurious.' But he might have found an answer to this objection in the excellent observations published in 1867 by Malm. The Pleuronectidae, whilst very young and still symmetrical, with their eyes standing on opposite sides of the head, cannot long retain a vertical position, owing to the excessive depth of their bodies, the small size of their lateral fins, and to their being destitute of a swim bladder. Hence soon growing tired, they fall to the bottom on one side. Whilst thus at rest they often twist, as Malm observed, the lower eye upwards, to see above them; and they do this so vigorously that the eye is pressed hard against the upper part of the orbit. The forehead between the eyes consequently becomes, as could be plainly seen, temporarily contracted in breadth. On one occasion Malm saw a young fish raise and depress the lower eye through an angular distance of about seventy degrees.

We should remember that the skull at this early age is cartilaginous and flexible, so that it readily yields to muscular action. It is also known with the higher animals, even after early youth, that the skull yields and is altered in shape, if the skin or muscles be permanently contracted through disease or some accident. With long-eared rabbits, if one ear lops forwards and downwards, its weight drags forward all the bones of the skull on the same side, of which I have given a figure. Malm states that the newly hatched young of *perches, *salmon, and several other symmetrical fishes, have the habit of occasionally resting on one side at the bottom; and he has observed that they often then strain their lower eyes so as to look upwards; and their skulls are thus rendered rather crooked. These fishes, however, are soon able to hold themselves in a vertical position, and no permanent effect is thus produced. With the Pleuronectidae, on the other hand, the older they grow the more habitually they rest on one side, owing to the increasing flatness of their bodies, and a permanent effect is thus produced on the form of the head, and on the position of the eyes. Judging from analogy, the tendency to distortion would no doubt be increased through the principle of inheritance (*Origin VI*, pp. 186–7; *Wallace 1889, p. 90 deals in similar fashion with Mivart's objections; Malm 186<u>8</u>, abstracted in Günther 1868b).

For all his exertion against natural selection (*see also* **Seahorses (II)**), the Roman Catholic Mivart was excommunicated just before he died, because he did believe in some from of evolution (Mivart 1871; Browne 2002, p. 356). As the Pope has since changed his mind about the whole thing (John Paul II 1996), one can only wonder if measures were taken to retrieve Mivart from Hell, where he has been needlessly burning for over a century.

Flatfish controversy (II) The second Flatfish controversy (1863–1871), as described in agonizing detail by Vorzimmer (1969), opposed two Scandinavian naturalists and their allies, August Wilhelm Malm (backed by J. C. Schiödte & R. H. Traquair) and Johannes Japetus Steenstrup (backed by C. W. Thompson). At issue were the mechanisms of the eye migration in 'flatfishes', with each side in this controversy working on what they thought were canonical species (Malm 1854, 1868; Steenstrup 1863; Thompson 1865; Traquair 1865; Schiödte 1868). CD described this controversy as follows: Schiödte believes, in opposition to some other naturalists, that the Pleuronectidae are not quite symmetrical even in the embryo; and if this be so, we could understand how it is that certain species, whilst young, habitually fall over and rest on the left side, and other species on the right side. Malm adds, in confirmation of the above view, that the adult *Trachypterus* *arcticus* [*deal fish], which is not a member of the Pleuronectidae, rests on its left side at the bottom, and swims diagonally through the water; and in this fish, the two sides of the head are said to be somewhat dissimilar. Our great authority on Fishes, Dr Günther, concludes his abstract of Malm's paper, by remarking that 'the author gives a very simple explanation of the *abnormal condition of the Pleuronectoids.'

We thus see that the first stages of the transit of the eye from one side of the head to the other, which Mr Mivart considers would be injurious, may be attributed to the habit, no doubt beneficial to the individual and to the species, of endeavouring to look upwards with both eyes, whilst resting on one side at the bottom. We may also attribute to the inherited effects of use the fact of the mouth in several kinds of flat-fish being bent towards the lower surface, with the jaw bones stronger and more effective on this, the eyeless side of the head, than on the other, for the sake, as Dr Traquair supposes, of feeding with ease on the ground. Disuse, on the other hand, will account for the less developed condition of the whole inferior half of the body, including the lateral fins; though *Yarrell thinks that the reduced size of these fins is advantageous to the fish, as 'there is so much less room for their action, than with the larger fins above.' [. . .] From the colourless state of the ventral surface of most fishes and of many other animals, we may reasonably suppose that the absence of colour in flat-fish on the side, whether it be the right or left, which is undermost, is due to the exclusion of light. But it cannot be supposed that the peculiar speckled appearance of the upper side of the sole, so like the sandy bed of the sea, or the power in some species, as recently shown by Pouchet, of changing their colour in accordance with the surrounding surface, or the presence of bony tubercles on the upper side of the *turbot, are due to the action of the light. Here *Natural selection has probably come into play, as well as in adapting the general shape of the body of these fishes, and many other peculiarities, to their habits of life. We should keep in mind, as I have before insisted, that the inherited effects of the increased use of parts, and perhaps of their disuse, will be strengthened by Natural selection. For all spontaneous *variations in the right direction will thus be preserved; as will those individuals which inherit in the highest degree the effects of the increased and beneficial use of any part. How much to attribute in each particular case to the effects of use, and how much to Natural selection, it seems impossible to decide. (*Origin VI*, pp. 186–8; Mivart, 1871, pp. 49–50, 180; Pouchet 1871, 1872; Traquair 1865; Yarrell 1836, Vol. 2, p. 211; *see* Greenland halibut for the exclusion of light being the cause for paleness of the underside of most flatfishes).

Flatheads Fishes of the Family Platycephalidae. Here represented by the Longhead flathead *Platycephalus inops*, a species *endemic to southern Australia, and originally described based on a specimen with a *length of 16 "unc." (i.e. 16 inches) collected by CD in *King George's Sound (*Fish*, pp. 33–5). Though Jenyns feared it might be a "mere *variety" of *P. laevigatus* Cuvier, 1829, subsequent authors (e.g. Gomon *et al*. 1994) have accepted this species, now referred to as *Leviprora inops* (Jenyns, 1840).

Flounder Here is CD carping on the issue of Design: As I before said I flounder hopelessly in the mud. (*Correspondence* to Asa Gray, Feb. 17, 1861).

Flounder is also the common name of *Platichthys flesus* (Linnaeus, 1758), a *flatfish which spawns numerous *eggs, and which is important to European fisheries. CD introduces the flounder as follows: Many animals have the right and left sides of their body unequally developed: this is well known to be the case with *flat-fish, in which the one side differs in thickness and colour and in the shape of the fins, from the other, and during the *growth of the young fish one eye is gradually twisted from the lower to the upper surface.⁶⁰ In most flat-fishes the left is the blind side, but in some it is the right; though in both cases reversed or 'wrong fishes,' are occasionally developed; and in *Platessa flesus* the right or left side is indifferently the upper one. (*Variations* II, pp. 27–8; n. 60 reads: *See Steenstrup* on the *Obliquity of Flounders*: in *Annals and Mag. of Nat. Hist.*, May, 1865, p. 361. I have given an abstract of Malm's explanation of this wonderful phenomenon in the *Origin of Species* 6ᵗʰ edit., p. 186.).

We have now to consider whether, when the male differs in a marked manner from the female in colour or in other ornaments, he alone has been modified, the *variations being inherited by his male offspring alone; or whether the female has been specially modified and rendered inconspicuous for the sake of protection, such modifications being inherited only by the females. It is impossible to doubt that colour has been gained by many fishes as a protection: no one can examine the speckled upper surface of a flounder, and overlook its resemblance to the sandy bed of the sea on which it lives. Certain fishes, moreover, can through the action of the nervous system, change their colours in adaptation to surrounding objects, and that within a short time.³² (*Descent* II, p. 344; n. 32 cites Pouchet 1871).

Flounder, Fine A species of large flatfish (up to 70 cm) ranging from Ecuador to Chile, living on soft bottom down to about 70 m, and described as a new species, *Hippoglossus kingii*, by Jenyns (*Fish*, p. 138). This species was based on a drawing (Fig. 11) made for Captain *FitzRoy, by Philip G. *King, based on a specimen sampled by CD in *Valparaiso, Chile, and later lost.

The valid scientific name of the Fine flounder is now *Paralichthys adspersus* (Steindachner, 1867), Family Paralichthyidae, with *Hippoglossus kingii* Jenyns (1842) being considered a synonym, although it was published earlier. The

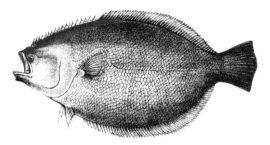

Fig. 11. The Fine flounder *Paralichthys adspersus*, Family Paralichthyidae (*Hippoglossus kingii* in *Fish*, Plate XXVI). Based on a lithograph by B. Waterhouse Hawkins, itself from a coloured drawing (now lost) by Mr. Philip G. King, midshipman of the *Beagle*.

arcana of the *Code suggest, in cases such as this, that a *holotype may be assigned to the species initially described ('type fixation'), and the subsequent name (here that proposed by Steindachner 1867) then declared a junior synonym. This is not encouraged, however, in the interest of nomenclatural stability. Still, King loses his *eponym.

Fly-catchers Birds of the Old world Family Muscicapidae, specialized in catching flying insects; similar to the New world's Tyrant fly-catcher, Family Tyrannidae.

CD reported as follows on the South American tyrant fly-catcher: I have frequently observed it, hunting a field, hovering over one spot like a *hawk [. . .]. At other times the Saurophagus haunts the neighbourhood of water, and there, like a *kingfisher, remaining stationary, it catches any small fish which may come near the margin. (*Journal*, July 5, 1832). And: I have often watched a tyrant flycatcher (*Saurophagus sulphuratus*) in South America, hovering over one spot and then processing to another, like a kestrel, and at other times standing stationary on the margin of the water, and then dashing into it like a kingfisher at a fish. (*Origin* VI, p. 141).

Flying The evolution of flight organs in various animals was perceived by many – including

CD – as a challenge to his nascent theory of *natural selection, and he devoted considerable efforts to identifying intermediate stages.

Thus, he mused: The Creator has made tribes of animals adapted preeminently for each element, but it seems law that such tribes, as far as compatible with such structure are in minor degree adapted for other elements. [. . .] every part would probably not be complete, if birds fitted solely for air & fishes for water (*Notebook B*, p. 181).

*Flying fish look like suitable intermediates, although, in fact, they do not fly. Rather, usually when attempting to escape from predators such as CD's *Albicore, flying fish accelerate, through a phase of rapid swimming right under, then on, the water surface, until they reach a speed sufficiently high for their modified pectoral fins to provide enough lift for take-off, then to glide, sometimes over long distances, before they fall back into the water.

CD contrasted this gliding with the true flight of *bats: Sailing *lizards. squirrels & Opossums <<& fish>>: *flying* lizards. – Mammalia. (*Notebook J*, p. 641). Also Could anyone. have foreseen, sailing, *climbing and *mud-walking fish? (*Notebook T*, p. 463).

The <<economy>> of world could have gone on without *Bats or *ostriches. – It can only be following out some great principle [. . .] It is clear birds made preeminently for air, yet if no birds, Mammalia would have taken place. There limit to this Adaptation. Fish could hardly have lived out of water. Though Crabs. – spider under water. (*Notebook B*, p. 202; *Marginalia*, bottom of pp. 148–51 in Lamarck 1830).

Discussing the gradations leading to perfect organs such as *eyes, ears or wings, CD mentions Suppose we had flying fish[51] and not one of our now called flying fish preserved, who would have guessed intermediate habits.

(*Foundations*, p. 16; n. 51 reads: That is truly winged fish.).

This theme of transition is further explored in his *Big Species Book*, which builds on *Foundations*: Seeing that we have flying Birds Mammals, & formerly flying reptiles, & seeing that so eminently an aquatic animal as a crab can by the contrivance to keep its branchiae moist exist on dry land, it is conceivable, that the so-called flying Fish, – which can glide to such great distances through the air, turning & rising slightly in its rapid course by the aid of its fluttering fin[1] – could have been converted into a perfectly flying animal: had this been the case, & our present flying-fish unknown, who would have ventured to have even conjectured that the sole use of an early transitional stage of the pectoral fin was to escape danger in the open ocean. (p. 341; n. 1 cites Owen 1846).

This last thought then went into *Origin* (VI, p. 140): Seeing that a few members of such water-breathing classes as the Crustacea and Mollusca as adapted to live on the land; and seeing that we have flying birds and mammal, flying insects of the most diversified types, and formerly had flying reptiles, it is conceivable that flying fishes, which now glide far through the air, slightly rising and turning by the aid of their fluttering fins, might have been modified into perfectly winged animals. If this had been affected, who would ever have imagined that in an early transitional state they had been the inhabitants of the open ocean, and had used their incipient organs if flight exclusively, as far as we know, to escape being devoured by other fish?

This then led to a generalization: we may conclude that transitional states between structures fitted for different habits of life will rarely have been developed at an early period in great number under many subordinate forms. Thus, to return to our imaginary illustration of the flying-fish, it does not seem probable that fishes capable of true flight would have been developed under many subordinate forms, for taking prey of many kinds in many ways, on the land and in the water, until their organ of flight had come to a high stage of perfection, so as to have given them a decided advantage over other animals in the battle for life. Hence the chance of discovering species with transitional grades of structure in a fossil condition will always be less, for their having existed in lesser numbers, than in the case of species with fully developed structures. (*Origin VI*, pp. 140–1; note that the last argument deals with a feature of the *fossil record which *punctuated equilibrium was meant to explain, and which is supposed to be incompatible with CD's gradualism).

Flying fish(es) Members of the Family Exocoetidae, named after their gliding behaviour, which superficially resembles *flying.

Shortly before he left for the voyage of the *Beagle*, CD received from his mentor, J. S. *Henslow, a letter at the end of which he left CD to his "better meditations on *Mermaids & Flying fish.-" (*Correspondence*, Oct. 25, 1831). It took four years for CD to start seeing mermaids, but flying fish took far less time: I have felt a little sea-sickness to day: which is too bad, as objects of interest are continually occurring. There were plenty of flying fish round the vessel but no large ones. (*Diary*, Feb. 10, 1832, halfway between the *Cape Verde Islands and the Brazilian coast).

A flying fish fell on the deck this morning; it struck the mast high up, near the main yard: sticking to the fish was a crab, the pain of which caused perhaps this unusual degree of action. (*Diary*, April 2, 1832, 130 miles east of *Rio de Janeiro).

One question this quote poses is whether fish feel pain – an issue much discussed, e.g., among anglers. While it seems that sharks and rays "lack the neural apparatus for the sensation of pain" (Snow *et al.* 1993; which would explain anecdotal accounts of eviscerated, frenzied sharks eating their own entrails), teleosts do appear to possess a nociceptive system (Chervova 1997; Erdmann 1999). Hence, yes, CD was probably right to attribute the behaviour of that flying fish to pain. Thus, we continue...

In the night anchored in the outer part of the harbor of Gallao [...t]he ocean teems with life, no one can watch the Flying-fish, Dolphin and Porpoises without pleasure. (*Diary*, July 19, 1835, in the harbor of Callao, near Lima, *Peru).

Yet, with all this, CD sampled only one flying fish specimen, near the end of his voyage, off the coast of Peru (18°S; *Fish in Spirits*, no. 1232). CD described it in his *Zoology Notes*, noting that the Whole upper body & fins were beautiful dark violet-blue. beneath snow white, but Jenyns wrote that "The colours were not noticed in the recent state" because *Covington, in a rare omission, failed to transfer this description into *Fish in Spirits*. In any case, Jenyns was unsure of his attribution of this specimen to *Exocoetus exsiliens*, noting that the taxonomy of this group was in flux (*Fish*, pp. 122–3). It still is, and so the attribution of this specimen to the Sparrow (or Whitetip) flying fish *Cheilopogon xenopterus* (Gilbert, 1890), is tentative as well. This is especially so since a record from Peru would represent an extension of the reported range of *C. xenopterus*, now stretching, according to Grove and Lavenberg (1997) from Cabo San Lucas, Baja California to Panama and the *Galápagos.

Food-fish The crew of the *Beagle*, and CD in particular, consumed fish regularly, and much of the fresh food supply was assured by *angling and other forms of *fishing. Also, it is evident that CD enjoyed meals with fish as their main course. Emma Darwin's recipe book in DAR attests to this, as does the fact that CD's household spent, at least in 1876, about as much on fish and game as on stationary, stamps & newspapers (Browne 2002, p. 388).

The following quotes, of which the first documents CD's only attempt at writing a humorous dialogue, illustrate these culinary aspects of fishes: The scene is the hinterland of *Rio de Janeiro, Brazil: On first arrival we unsaddle our horses & give them their Indian corn. Then with a low bow ask the Signor to do us the favor to give us something to eat. 'Anything you choose, Sir' is his answer. For the few first times vainly I thanked providence for guiding us to so good a man. The conversation proceeding, the case usually became deplorable: 'Any fish can you do us the favor of giving?. 'Oh no Sir.' 'Any soup.' 'No Sir.' 'Any bread.' 'Oh no Sir.' 'Any dried meat.' 'Oh no Sir.' If we were lucky, by waiting 2 hours we obtained fowls, rice & farinha.- It not unfrequently happens that the guest is obliged to kill with stones the poultry for his own dinner. [. . .]. At Campos Novos, we fared sumptuously; having rice & fowls, biscuit & wine & spirits for dinner; coffee in the evening, & with it for breakfast fish. (*Diary*, April 9, 1832).

CD's other accounts of how the food-fishes required for the *Beagle* were procured are equally vivid: Torrents of *rain & the atmosphere was so thick that it was impossible to continue the survey. We remained therefore at anchor. The bottom was rocky & in consequence plenty of fish: almost every man in the ship had a line overboard & in a short time a surprising number of fine fish were caught. I also got some Corralines which were preeminently curious in their structure. (*Diary*, August 26, 1832; and yes, rocky bottoms, which include coral reefs, usually have higher biomasses of large fishes than sandy or muddy bottoms: see the ecosystem models in Christensen and Pauly 1993).

The party who went out to shoot fresh provisions brought home 2 deer, 3 Cavies & an *Ostrich. With the net also a most wonderful number of fish were caught; in one drag more than a tun weight were hauled up, including ten distinct species. (*Diary*, September 16, 1832).

Again I walked to Punta Alta to look for fossil bones: on the road I crossed the track of a large herd of the Guanaco or American Camel: the marks were as large as a cow, but more cloven. We laid in a good stock of fresh provisions for sea, as 6 deer were shot & great numbers of fish caught. (*Diary*, October 16, 1832).

According to his *Diary* (Nov. 18, 1835), CD's experience in Tahiti was bucolic: The Tahitians having made a small fire of sticks, placed a score of stones about the size of a cricket ball on the burning wood. In about ten minutes' time, the sticks were consumed & the stones hot. They had previously folded up in small parcels made of leaves, pieces of beef, fish, ripe & unripe Bananas, & the tops of the wild Arum. These green parcels were laid in a layer between two of the hot stones & the whole then covered up by earth, so that no smoke or steam escaped. In about a quarter of an hour, the whole was most deliciously cooked; the choice green parcels were laid on a cloth of Banana leaves; with a Cocoa nut shell we drank the cool water of the running stream & thus enjoyed our rustic meal . . .

Food webs CD frequently considered issues which we would now view as relating to food web *ecology.

Thus, with regard to coral reefs, he noted that It has been shown [. . .] that there are some species of large fish, and the whole tribe of the Holothuriae, which prey on the tenderer parts of the corals. On the other hand, the polypifers [. . .] must prey on some other organic beings; the decrease of which from any cause, would cause a proportionate destruction of the living coral. The relation therefore, which determines the formation of reefs on any shore, by the vigorous growth of the efficient kind of coral must be very complex, and with our imperfect knowledge quite inexplicable (*Coral Reefs*, p. 84; Andrew and Gentien (1982) think that upwelled, nutrient-rich water explains much of the high productivity of coral reefs).

He wondered, similarly, about pelagic food webs: In deep water, far from the land, the number of living creatures is extremely small: south of the *latitude 35°, I never succeeded in catching any thing besides some *beroe, and a few species of minute *crustacea belonging to the *Entomostraca. In *shoaler water, at the distance of a few miles from the coast, very many kinds of crustacea and some other animals were numerous, but only during the night. Between latitudes 56° and 57° south of Cape Horn the net was put astern several times; it never, however, brought up any thing besides a few of two extremely minute species of Entomostraca. Yet *whales and seals, *petrels and albatross, are exceedingly abundant throughout this part of the ocean. It has always been a source of mystery to me, on what the latter, which live far from the shore, can subsist. I presume the albatross, like the condor, is able to fast long; and that one good feast on the carcass of a putrid whale lasts for a long siege of hunger. It does not lessen the difficulty to say, they feed on fish; for on what can the fish feed?

It often occurred to me, when observing how the waters of the central and intertropical parts of the Atlantic,[5] swarmed with *Pteropoda, Crustacea, and Radiata, and with their devourers the *flying-fish, and again with their devourers the *bonitos and *albicores, that the lowest of these pelagic animals perhaps possesses the power of decomposing carbonic acid gas, like the members of the vegetable kingdom. (*Journal,* December 6, 1833; n. 5 reads: From my experience, which has been but little, I should say that the Atlantic was far more prolific than the Pacific, at least, than in that immense open area, between the west coast of America and the extreme eastern isles of Polynesia.).

CD guessed wrongly regarding the basis of food webs: usually, *zooplankton do not function as primary producers (though flagellates can function as either photo- or heterotrophs, while the hydrozoan *Velella velella* Linnaeus, 1758 hosts zooxanthellae, as do corals and giant clams of the genus *Tridacna*). In aquatic ecosystems, *phytoplankton does the bulk of the decomposing [of] carbonic acid gas, and hence provide the basis for the food webs which CD describes. His question on what can the fish feed is thus answered: they feed largely on herbivorous *zooplankton (see food webs in Christensen and Pauly 1993, and www.ecopath.org). On the other hand, CD hit the bullseye with his guess about seabirds feeding on floating carcasses (Johnstone 1977).

Finally, we should mention a source of seabird food which CD missed, and which many contemporary ecologists (except Kraus and Stone 1995) also overlook, owing to the minor role large cetaceans now play in marine ecosystems: whale faeces.

Fossil Here is one of CD's statements on the importance of fossil(s), however incomplete the *fossil record may be: On the theory of descent, the full meaning of the fossil remains from closely consecutive formations being closely related, though ranked as distinct species, is obvious. As the accumulation of each formation has often been interrupted, and as long blank intervals have intervened between successive formations, we ought not to expect to find [. . .], in any one or in any two formations, all the intermediate varieties between the species which appeared at the commencement and close of these periods: but we ought to find after intervals, very long as measured by years, but only moderately long as measured geologically, closely allied forms, or, as they have been called by some authors, representative species; and these assuredly we do find. We find, in short, such evidence of the slow and scarcely sensible mutations of specific forms, as we have the right to expect (*Origin* VI, p. 307).

Moreover, He who believes that each [. . .] species was independently created, will, I presume, assert that each species has been created with a tendency to vary [...]. To admit this view is, as it seems to me, to reject a real for an unreal, or at least for an unknown, cause. It makes the works of God a mere mockery and deception; I would almost as soon believe with the old and ignorant cosmogonists, that fossil shells had never lived, but had been created in stone so as to mock the shells living on the sea-shore. (*Origin VI*, pp. 130–1; CD probably refers to Gosse (1857), whom various authors, notably Lankaster (1995), perhaps unjustly, accused of stating that God created fossils to test our faith in Creation).

And indeed, the mockery would be quite elaborate, given the similarity between fossils and living forms, here attested from Sidon (= Saïda), along the coast of what is now Syria, from a source dating from the Crusades: "A certain marvellous stone was brought to the king, in appearance like a quantity of scales, of the which when one was raised you saw beneath, between the two stones, the shape of a fish of the sea. And the fish was of stone, but nothing of its form was wanting, neither eyes, nor fins, nor colour, any more than if it had been living. The king asked for one of these stones, and found a *tench in it, of a brown colour like any other tench" (De Joinville's '*Histoire de St Louis*,' as cited and translated in Pictet and Humbert (1866a, b)).

However, Tench are not listed in the comprehensive paleontological survey of Müller (1985). Neither are recent populations of Tench reported from Syria or Lebanon, though they occur in nearby Turkey (Blanc *et al.* 1971), while their introduction in 1947 from Europe into the Jordan River appears to have been unsuccessful (Krupp and Schneider 1989). Perhaps this fossil Tench was a *wrasse: at least one source suggests that French speakers use 'tanche' for wrasses (Péronnet *et al.* 1998, p. 327). The divine mockery is here quite elaborate.

Fossil fishes CD was of two minds regarding the value of fish fossils: The bones of vertebrated animals are much more rarely found than the remains of the lower marine animals, and they are almost in proportion more valuable. A person not acquainted with the science will hardly be able to imagine the deep interest which the discovery of a skeleton, if of higher organization than a fish, in any of the oldest formations would most justly create. [...].

Fishes' bones are found occasionally in all sedimentary strata, and are highly interesting. (*Collected Papers I*, 1849, p. 234).

The books by Moy-Thomas (1971) and Müller (1985), and especially that of Long (1995) on the 'Rise of Fishes' suggest that it is the second of CD's statements, on fossil fishes being highly interesting, which holds.

Fossil record The ensemble of known *fossils, including *fossil fishes, available at any one time for drawing inferences on the *evolution of life.

CD was well aware of the 'imperfect' nature of the fossil record. Indeed, by discussing the frequency with which different organisms, living in different environments, become fossilized, he became one of the founders of the discipline now known as *taphonomy: So have fish and reptiles been at one time more closely connected in some points than they now are. Generally in those groups in which there has been most change, the more ancient the fossil, if not identical with recent, the more often it falls between existing groups, or into small existing groups which now lie between other large existing groups. (*Foundations*, p. 137).

If the Palaeozoic system is really contemporaneous with the first appearance of life, my theory must be abandoned, both inasmuch as it limits *from shortness of time* the total number of forms which can have existed on this world, and because the organisms, as fish, mollusca[11] and star-fish found in its lower beds, cannot be considered as the parent forms

Freshwater fishes (II)

of all the successive species in these classes. But no one has yet overturned the arguments of Hutton and *Lyell, that the lowest formations known to us are only those which have escaped being metamorphosed (*Foundations*, p. 138; n. 11, a pencil insertion by the author, reads: The parent-forms of Mollusca would probably differ greatly from all recent – it is not directly that any one division of Mollusca would descend from first time unaltered, whilst others had become metamorphosed from it. Hutton 1795).

Also: Some of the Secondary formations which contain most marine remains appear to have been formed in a wide and not deep sea, and therefore only those marine animals which live in such situations would be preserved.[16] (*Foundations*, p. 140; n. 16 reads: Neither highest or lowest fish (i.e. Myxina <?> or *Lepidosiren) could be preserved in intelligible condition in fossils; for '*Myxina*' see **Hagfishes**).

Foundations Short title of a small book, *The Foundations of the Origin of Species*, edited by Francis Darwin, based on a 1842 sketch and a 1844 essay by CD, outlining his theory of *natural selection, and published posthumously (Darwin 1909).

Foundations provided material integrated in several entries of this book, notably on *vertebrate origins and *development.

Freshwater fishes (I) Generally, when CD refers to 'fish(es),' marine species are implied. CD's explicit references to freshwater (usually abbreviated F.W.) fishes are not rare, however. Here are examples: Yarrell remarks he has somewhere met conjecture that all salt-water fish were once salt water (as they almost must have been on elevation of continents) but Ogleby well answers that nearly all F.W. Fish are Abdominals.·. that order first converted – is it an old order Geologically? (*Notebook D*, p. 342; discussion between William Ogilby, *Blyth, *Yarrell and CD is documented on pp. 340–1; note that quote should probably have read: ... that all fresh-water fish were once salt water. Abdominals are members of a group of fishes, now known to be paraphyletic, defined by the position of their pectoral fins).

Also: It has been found possible to accustom marine fish to live in fresh water; but as such changes in fish and other marine animals have been chiefly observed in a state of nature, they do not properly belong to our present subject. [i.e. the effects of domestication] (*Variations II*, p. 294). Other examples may be found in the entries referring to various freshwater fish species.

Freshwater fishes (II) An important point CD wanted to make concerning freshwater fishes is that many of them are relicts of earlier ages, surviving in fresh water because they do not have to compete against advanced marine forms. Thus, [w]ith respect to *Lepidosiren *Ganoid fishes, perhaps Ornithorhynchus I suspect, as stated in Origin, that they have been preserved from inhabiting F. Water & isolated parts of the world, in which there has been less competition & less rapid *progress in Nat. Selection, owing to the fewness of individuals which can inhabit small areas. &c &c. – & where there are few individuals, variations almost must be slower (*Correspondence* to *Lyell, 18/19 Feb. 1869). Correspondingly, we find in *Origin*: All freshwater basins, taken together, make a small area compared with that of the sea or of the land. Consequently, the competition between fresh-water productions will have been less severe than elsewhere; new forms will have been then more slowly produced, and old forms more slowly exterminated. And it is in freshwater basins that we find seven genera of Ganoid fishes, remnants of a once preponderant order: and in fresh water we find some of the most anomalous forms now known in the world, as the Ornithorhynchus and Lepidosiren, which, like *fossils, connect to a certain extent orders at present widely sundered in the natural *scale. These anomalous forms may be called living *fossils; they have endured to the present day, from having inhabited a confined area, and from having been exposed to less varied, and therefore less severe, competition. (*Origin* VI, pp. 83–4; interestingly, Wallace underlined the competition between fresh-water productions will have been less severe than elsewhere in his copy of *Origin*, and added "survival of the fittest"; see Beddall 1988b).

CD concluded these considerations (in *Origin* VI, p. 296) by stating that a few members of the great and almost extinct group of Ganoid fishes still inhabit our fresh waters. Therefore the utter extinction of a group is generally, as we have seen, a slower process than its production.

Freshwater fishes (III) Another aspect of the biology of freshwater fishes that interested CD was their distributional *range. Hence: [T]he wide ranging power of freshwater productions can, I think, in most cases be explained by their having become fitted, in a manner highly useful to them, for short and frequent *migrations from pond to pond, or from stream to stream within their own countries; and liability to wide dispersal would follow from this capacity as an almost necessary consequence. We can here consider only a few cases; of these, some of the most difficult to explain are presented by fish. It was formerly believed that the same freshwater species never existed on two continents distant from each other. But Dr Günther has lately shown that the *Galaxias attenuatus* inhabits Tasmania, New Zealand, the Falkland Islands, and the mainland of South America. This is a wonderful case, and probably indicates dispersal from an Antarctic centre during a former warm period. This case, however, is rendered in some degree less surprising by the species of this genus having the power of crossing by some unknown means considerable spaces of open ocean: thus there is one species common to New Zealand and to the Auckland Islands, though separated by a distance of about 230 miles. On the same continent freshwater fish often range widely, and as if capriciously; for in two adjoining river systems some of the species may be the same, and some wholly different. It is probable that they are occasionally transported by what may be called accidental means. Thus fishes still alive are not very rarely dropped at distant points by whirlwinds; and it is known that the ova retain their vitality for a considerable time after removal from the water. Their dispersal may, however, be mainly attributed to changes in the level of the land within the recent period, causing rivers to flow into each other. Instances, also, could be given of this having occurred during floods, without any change of level. The wide difference of the fish on the opposite sides of most mountain-ranges, which are continuous, and which consequently must from an early period have completely prevented the inosculation of the river systems on the two sides, leads to the same conclusion. Some freshwater fish belong to very ancient forms, and in such cases there will have been ample time for great geographical changes, and consequently time and means for much *migration. Moreover Dr Günther has recently been led by several considerations to infer that with fishes the same forms have a long endurance. Salt-water fish can with care be slowly accustomed to live in freshwater; and, according to Valenciennes, there is hardly a single group of which all the members are confined to fresh water, so that a marine species belonging to a freshwater group might travel far along the shores of the sea, and could, it is probable, become adapted without much difficulty to the fresh waters of a distant land. (*Origin* VI, pp. 343–4; fig. 22 of McDowall (1970) shows that the distribution of '*Galaxias attenuatus*' suggested by Günther (1866) still holds; however the name is a synonym of *G. maculatus*, which happens to be based on a *type specimen collected by CD; *see* **Galaxiidae**).

The claim that freshwater groups often have members that can travel far along the shores of the sea is completely justified. A good example is the Mozambique tilapia *Oreochromis mossambicus* (Peters, 1852), belonging to a decidedly freshwater group, the *Cichlidae, and which, after its *introduction into freshwater ponds in numerous tropical and subtropical countries, documented in *FishBase, has further expanded its range by travelling along coastlines, and becoming, in the process, one of the worst piscine pests on Earth.

Frogs *See* **Amphibians.**

Fry Young, and by extension small, fishes. The latter meaning is implied in the quote One evening near Rozario, as it was going dark, we were anchored in a Narrow Riacho or arm; here there were many smaller fry, & I saw one of these birds rapidly flying up & down ploughing the water as described in Maldonado (*Zoology Notes*, p. 160, pertaining to *scissor-beak as the predator and *characins as the prey; *see also Fish in Spirits,* no. 747).

It has long been known that the young of fish that live in deeper waters tend to have inshore nurseries, and this was used by *Jenyns (*Fish*, pp. 79–80) to infer that *silversides caught by CD off the Brazilian coast some miles from the land (*Fish in Spirits* no. 367) "were not so very young, as the fry of most fish keep close in shore".

Fucus *See* **Kelp.**

Gadus The genus of which *Gadus morhua*, the Atlantic *cod, is the *type species.

Also used by CD to refer to cod-like fishes in general.

Galápagos A small archipelago in the Central eastern Pacific, 1000 km from the Ecuadorian mainland, named (by Spaniards) after its *endemic giant tortoises, or more precisely after their shells, assimilated to *saddles*.

The Galápagos Islands, visited by CD from September 15 to October 20, 1835, are often, if falsely, believed to be the place where CD 'hit' on his theory of *evolution (whatever this is thought to be), when he first saw the birds now called *Darwin's finches. This myth has been refuted by numerous authors.

Indeed, upon his return to England, CD was particularly interested in the identity of the *fish* he sampled in the Galápagos. Thus, he wrote to *Henslow: Will you ask Leonard if he will look at the few Galapagos fish first, that is if it does not quite break through the order, in which the whole will be examined. The dried fish, I believe nearly all come from those islands. (*Correspondence*, March 28, 1837).

CD did learn a lot from these and his other specimens, especially about what we would now call 'biogeography,' and this is reflected in all his subsequent writings. Here is one of the generalizations he derived: If we now look to the character of the inhabitants of small islands, we shall find that those situated close to other land have a similar *fauna with that land, whilst those at a considerable distance from other land often possess an almost entirely peculiar fauna. The Galapagos Archipelago is a remarkable instance of this latter fact; here almost every bird, its one mammifer, its reptiles and sea shells, and even fish, are almost all peculiar and distinct species, not found in any other quarter of the world (*Foundations*, p. 159).

Grove and Lavenberg (1997, p. 8) suggest that Valenciennes (1855), based on fish sampled in 1838 by a French expedition, "was the first to suggest the uniqueness of the Galápagos ichthyofauna". As we can see from the above quote, Valenciennes was not (at least if we ignore the fact that *Foundations* was published long after it was written). Moreover, while Grove and Lavenberg (1997, p. 1) point out that 9.2% (41 of 444) of the fish species of the Galápagos are *endemics, they did not provide comparative data from other islands, and thus, based on their account, one cannot tell whether 9.2% is a large or a small number, CD and Valenciennes notwithstanding. [Comparative data, albeit pertaining only to benthic fishes, are now available; *see* **Ascension Island; Saint Helena Island**.]

What is established, on the other hand, is that CD was the first naturalist to collect fish in the Galápagos (in Chatham Island, now San Christobal, in the southeastern part of the archipelago), and that of the fifteen species he collected there, five happened to have been endemics.

The species from the Galápagos listed in Jenyns' *Fish* are: *Serranus albomaculatus* (p. 3); *Serranus labriformis* (p. 8); *Serranus olfax* (p. 9); *Prionotus miles* (p. 29); *Scorpaena histrio* (p. 35); *Prionodes fasciatus* (pp. 47, 164); *Pristipoma cantharinum* (p. 49); *Latilus princeps* (p. 52); *Chrysophrys taurina* (p. 56); *Gobius lineatus* (p. 95); *Cossyphus darwini* (p. 100); *Gobiesox poecilophthalmos* (p. 141); *Muraena lentiginosa* (p. 143); *Tetrodon annulatus* (p. 153); and *Tetrodon angusticeps* (p. 154).

The Galápagos Islands are presently under much threat from human activities (Jackson 1993; Powell and Gibbs 1995), with valiant attempts to halt the damage (Reck 1999). Thus, these islands continue to serve as a laboratory, this time with regard to tests of our ability to preserve the biodiversity handed over to us by long evolutionary processes.

Galápagos shark(s) When in the singular, 'Galápagos shark' refers to a species widespread in tropical seas, *Carcharhinus galapagensis*

Galaxiidae

(Snodgrass & Heller, 1905), and which reaches 2–3 m. In the plural, 'Galápagos sharks' refers to sharks around the *Galápagos Islands. As it turns out, CD has something to say about both. Let's start with the singular. Galápagos shark are common around *Saint Paul's Rock (Lubbock and Edwards 1981). Based on its behaviour and abundance, Edwards and Lubbock (1982) suggest that this species was the one in CD's account that the men of the *Beagle caught a large number of fine large fish & would have succeeded much better had not the sharks broken so many of their hooks and lines: they contrived to land three of these latter fish, & during our absence 2 large ones were caught from the ship (*Diary*, February 16, 1832).

Lubbock and Edwards (1981) suggest that, around Saint Paul's, *C. galapagensis* feeds on *Exocoetus volitans*, the Tropical two-wing *flying fish, which congregates in enormous numbers close to the surface, and which also appears to support other inhabitants of Saint Paul's Rock.

Now we turn to the plural: At the Galapagos Islands the great land-*lizards (Amblyrhynchus) were extremely tame so that I could pull them by the tail whereas in other parts of the world *large* lizards are wary enough. The aquatic lizard of this same genus, lives on the coast-rocks, is adapted to swim & dive perfectly, & feeds on submerged *algae: no doubt it must be exposed to danger from the sharks; & consequently, though quite tame on the land, yet I could not drive them into the water & when I threw them in, they always swam directly back to the shore . . . (*Big Species Book*, p. 496). Note that the sharks of which the Galápagos lizards were afraid (as established by CD's *experiment) may not have been Galápagos shark.

Galaxias spp. *See* **Galaxiidae**.

Galaxiidae A family consisting of over 50 species of freshwater or diadromous fishes, and "major elements of the relatively depauperate Southern Hemisphere freshwater fish fauna" (Waters *et al*. 2000).

CD collected four of their species, and *Jenyns created a new genus ('*Mesites*') for three nominal species, *Mesites maculatus* from a fresh water brook, Hardy Peninsula, Tierra del Fuego (*Fish in Spirits*, no. 543) (Fig. 12A), and from the Santa Cruz River, Patagonia (*Fish*, p. 120, *Fish in Spirits*, no. 952), *M. alpinus*, also from the Hardy Peninsula (i.e. from Alpine fresh water lake; *Fish in Spirits*, no. 536), and *M. attenuatus*, from the *Bay of Islands, New Zealand (*Fish*, pp. 118–22) (Fig. 12B). However, *Mesites* had already been assigned to a genus of beetles (Eschmeyer 1990, p. 242), while *M. maculatus* turned out, like *M. attenuatus*, to be a synonym of the Inanga, *Galaxias maculatus* (Jenyns, 1842), one of the most widely dispersed species of freshwater fishes (McDowall 1970; *see* **Freshwater fishes (III)**). This leaves *Galaxias alpinus* (Jenyns, 1842) as the second species to be accounted for, plus two species of *Aplochiton*, an uncontested genus created by Jenyns.

Aplochiton taeniatus Jenyns, 1842 was collected by CD in Goree Sound, Patagonia, in the freshwater part of a river mouth, consistent with the wide occurrence of this species in Chilean lakes (McDowall and Nakaya 1987; McDowall 1988). Moreover, upon being placed in salt water: they immediately died (*Fish in Spirits*, no. 526; *Fish*, p. 133) (Fig. 12C). Still, based on their distribution, the case can be made that some of their populations may be *diadromous, or better *amphidromous (McDowall 1988, 1997), as is *Galaxias maculatus* (Fischer, 1963).

The Falkland trout *Aplochiton zebra* Jenyns, 1842, finally, was based on three specimens collected by CD in a freshwater lake of the Eastern Falklands connected to the sea by a short brook, again consistent with amphidromy. CD described its dull leaden colour, but not the zebra-like stripes after which the species is named (Fig. 12D). CD, however, noted that the species was common [. . .] that it is good eating, and grows to be about half as large again

Ganoid fishes

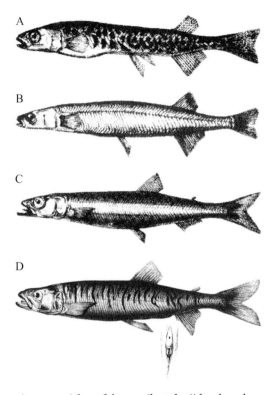

Fig. 12. Fishes of the Family Galaxiidae, based on lithographs by B. Waterhouse Hawkins. **A** *Galaxias maculatus* (*Mesites maculatus* in *Fish*, Plate XXII), based on a specimen from the Hardy Peninsula, Tierra del Fuego, Argentina. **B** Also *G. maculatus*, but based on a specimen from the Bay of Islands, New Zealand, originally assigned to another species (*Mesites attenuatus*; Plate XXII). **C** *Aplochiton taeniatus,* based on a specimen from Goree Sound, Tierra del Fuego, Argentina (*Fish*, Plate XXIV). **D** *Aplochiton zebra*, taken from a freshwater lake in the Falklands (Plate XXIV); insert: "anal and generative orifices".

as the individuals procured, which measured about 25 cm (*Fish*, pp. 131–2).

The Galaxiidae belong to the few fishes which CD sampled during the voyage of the *Beagle* and to which he later returned. In this case, it is when he was discussing the distribution of *Freshwater fishes that he returned to the Galaxiidae. Many of the biological facts required for this were derived from his cor-respondence with *Günther (see, for example, *Calendar*, no. 7983).

Ganoid(s) A type of heavy *scale observed in fishes such as the *gar-pikes, sturgeons and other ancient fishes. This observation was reified into a taxon, the 'Ganoïdes' in the classifications of Agasssiz (1833–44; see Fig. 1 in Patterson 1981). However, it was shown soon thereafter that the '*ganoid fishes' do not form a single, monophyletic group (Müller 1844; see also Nelson 1994). This may not be much of a problem, as CD used the term 'Ganoids' in a broad sense, to refer to fishes intermediate between the selaceans and the teleosteans (Burkhardt *et al.* 1993, p. 582).

CD used the fact that the few species assigned (in his day) to the 'Ganoids' were largely confined to fresh water to infer that the previously ubiquitous and diverse groups to which they belong were, in the marine realm, outcompeted by modern forms. This is most probably correct (*see* **Extinction; Freshwater fishes (II); Lungfishes**).

Ganoid fishes Fish covered with peculiar enamelled bony scales. Most of them are extinct (Glossary of *Origin* VI, p. 434; see above for the status of Ganoids, now fully extinct, at least in modern fish classifications).

CD had this thing about caves harbouring primitive survivors of past ages. Thus, responding to a report of a fossil mammalian skull sporting both primitive (marsupial) and advanced (rodent) features, CD marvelled What an extraordinary creature this must be. If it turn out, as is too probable, a comparatively modern cave-fossil, the fact will not be altogether too pleasant for me, seeing how intermediate it is. I must coolly assume it is a very old form & like one of the Ganoid fishes, which are so comparatively common in S. America, that it has lived almost to the present day. (*Correspondence* to Hugh Falconer, April 22, 1863; *see* **Cavefishes; Lungfishes; Lungfish, South American**).

Gar-pike One of the common names of the Longnose gar *Lepisosteus osseus* Linnaeus, 1758, one of the seven species of Lepisosteidae, a family of primitive freshwater fishes, previously counted among the *Ganoids.

A *heterocercal caudal fin (i.e. with the upper lobe more developed than the lower, as in sharks) is still formed in the larvae of gar-pike (Agassiz 1857a), and only later does the tail become homocercal (symmetrical), the change providing an illustration of how *ontogeny 'recapitulates' phylogeny (Haeckel 1866, Vol. 2, p. 300).

J. D. Dana wrote to CD about this as follows: "One of the most interesting of our peculiar tribes, as you undoubtedly know, is that of the Gar-pikes, of which there are several genera & near two dozen known species occurring over the Continent between Cuba & the northern Lakes – and not represented elsewhere on the globe. – It is not the point in view, yet I may mention here a fact of geological interest brought out by *Agassiz at our Assoc. meeting a fortnight since. There were some young individuals, alive, shown, which had the tail of the Ancient *Ganoids – That is, the vertebræ were actually continued to the extremity of the upper lobe –[6] This upper lobe [...] drops off as the animal grows & the fish then is of the modern type of form" (*Correspondence*, Sept. 8, 1856; n. 6 reads: This fact was eventually published in Agassiz 1857[a]).

CD was pleased: What a striking case of vertebræ in tail of young Gar-Pike; I wish with all my heart that *Agassiz would publish in detail on his theory of parallelism of geological & embryological development; I *wish* to believe, but have not seen nearly enough as yet to make me a disciple. (*Correspondence*, Sept. 29, 1856).

However, CD is slightly disingenuous here: his mind was very much set, in 1856, against Agassiz's *threefold parallelism', i.e. his creationist interpretation of parallels between the fossil record and the ontogeny of recent fish (see, for example, Agassiz 1849, 1857a; see also Gould 1977, pp. 63–8, and *Calendar*, no. 1588).

Gasbladder *See* **Swimbladder**.

***Gasterosteus* spp.** *See* **Sticklebacks**.

Gastrobranchus A synonym of *Myxine*, a genus of *hagfishes. On these, CD jotted in *Notebook C* (p. 247): Gastrobranchus <<only>> 2 species one in Northern Hemisphere 2^d in southern. The species from the Northern Hemisphere must have been *Myxine glutinosa* Linnaeus 1758, the first hagfish to be described. The other species, *Gastrobranchus coecus* Bloch, 1795, was later shown to be a synonym of *M. glutinosa* (Lloris and Rucabado 1991).

The species from the Southern Hemisphere is probably the very one sampled by CD in Tierra del Fuego, *Myxine australis* Jenyns, 1842 (Fernholm 1998), referred to as *M. glutinosa* in Lloris and Rucabado (1991).

However, nearly 60 species of hagfish have been described since, most in the genera *Myxine* and *Eptatretus* (Martini 1998; Wisner 1999). Thus, whatever CD had in mind here, it would not have worked out.

Gay, Claude French naturalist (1800–73). Gay conducted most of his field work in Chile, where, in 1834, he met CD and gave him specimens of the *Marblefish *Aplodactylus punctatus* and the Patagonian redfish *Sebastes oculatus* (*S. oculata* in *Fish*, p. 37; *Fish in Spirits*, no. 1014).

In return, CD cites Gay several times, notably in *Geology*. Details of Gay's life and work are given in Bauchot *et al.* (1990, p. 87).

Gear selectivity The process by which a given sampling (or fishing) gear, e.g. a net, retains specimens with certain attributes (e.g. a given size range), while letting others escape.

CD was well aware of the potentially biasing effects of gear selectivity on the composition of samples. He wrote on fish sampling: With many species the males are of much smaller size than the females, so that a large number of males would escape from the same net by which the females were caught. (*Descent II*, p. 249).

Gear selectivity is thus as important a topic for ichthyologists as it is for fisheries scientists (see accounts in Beverton and Holt 1957; Hilborn and Walters 1992).

Gemmules Particles hypothesized by CD to explain the transmission of attributes from one generation of organisms to the next. He defines them as minute granules which are dispersed throughout the whole system, [and which], when supplied with proper nutriment, multiply by self-division and are ultimately developed into units like those from which they were originally derived. [...] they are collected from all parts of the system and constitute the sexual elements, and their development in the next generation form a new being. (*Variation II*, p. 370).

The existence of gemmules was not among CD's best guesses. *Mendel, who had just published work demonstrating the existence of what later came to be called 'genes' felt that CD, in this case, had "succumbed to an impression without giving the matter proper thought" (cited in Orel (1996), p. 194).

The score: Genes 7 : Gemmules 1.

Genus According to the *Code, a "group of phylogenetically related *species or a single species forming a taxonomic category ranking between the *family and the species. Also a group term used in classifying organisms; contains one or more related species; the rank within the genus group next below the family group and above subgenus; a taxon at the rank of genus. In scientific *classifications, the genus is an assemblage of species possessing certain characters in common, by which they are distinguished. The genus is subordinate to phylum, class, order and family and superordinate to species. Zoological genera have unique names." Perhaps one should add to this that the genera are what people generally identify as 'kinds' when they identify plants and animals, the common names thus given referring to species only in cases of economically or culturally important organisms (Berlin 1992). Linnaeus (1758) tapped into this when he created the binomial system of scientific nomenclature.

Geology The science devoted to the study of rocks, and of the formation of major features of the Earth. It is not widely appreciated that CD was a famous geologist (and palaeontologist) long before he became a famous biologist, owing to the informative letters he sent to *Henslow during the voyage of the *Beagle* (1831–6) and which the latter published on CD's behalf.

CD's contributions to geology were later consolidated into three books (*see* **Geology**). Of these, *Coral Reefs is the one that contributed most to establishing CD's reputation as one of the two leading geologists of his time. (The other, of course, was Charles *Lyell.)

Geology Short title of *The Geology of the Voyage of H.M.S. Beagle. Part III. Geological Observations on South America* (Darwin 1846), with *Parts I* and *II*, it may be recalled, dealing with *Coral Reefs, and *Volcanic Islands*, respectively.

Fish show up only once in *Geology*, when CD describes sedimentary formations in Chile: The sandstone contains fragments of wood, either in the state of lignite or partially silicified, shark's teeth, and shells in great abundance, both high up and low down the sea-cliffs. (p. 127, and referring to a formation near Navidad) and (pp. 128–129): These concretions are remarkable from the great number of large silicified bones, apparently of cetaceous animals, which they contain; and likewise of a shark's teeth, closely resembling those of the *Carcharias megalodon.

Geophagus A genus of the family *Cichlidae, named after their detritivorous feeding habits ('earth eaters'; Barlow 2000, p. 32) consisting, according to Kullander and Nijssen (1989, pp. 30–1), of about ten species widely distributed in the Amazon and Orinoco basins, and in the Guianas. Many species practise *mouth-brooding, although there is variation

Gobies

from immediate uptake of the eggs to keeping them on the substrate for some time.

Gerres spp. *See* **Mojarras.**

Ghoti A manner of writing 'fish' that may have been proposed by George Bernard Shaw (source not traced) to illustrate what may be perceived as the irrationality of the rules for pronunciation in English, with *gh* as in rou*gh*, *o* as in w*o*men, and *ti* as in pala*ti*al. Thus, this book could have been named *Darwin's Ghoties*. No wonder CD had problems with his *spelling.

Goatfishes Members of the Family Mullidae, so named because their chin has two long barbels with chemosensory organs, used to probe sand flats or holes in reefs for their prey, usually small invertebrates.

Three species of goatfish were collected by CD: *Upeneus flavolineatus* Cuvier, 1835, reported as Dull silvery, with a yellow stripe on the side, in the *Cocos Islands (*Fish*, pp. 24–5); *Parupeneus trifasciatus* (Lacepède, 1801) in *Tahiti (assigned to *Upeneus* in *Fish in Spirits*, no. 1320 and *Fish*, pp. 25–6); and the West African goatfish *Pseudupeneus prayensis* (Cuvier, 1829). *Jenyns described this species (also assigned to *Upeneus*) based on a specimen sampled by CD in Porto Praya, St Jago Island, *Cape Verde Islands – hence the name. The specimen's colour was Vermilion, with streaks of iridescent blue. (*Fish*, p. 26; *Fish in Spirits*, no. 18).

Gobies Members of the Gobiidae, a very speciose family of small fishes. In the tiniest forms, reaching full maturity at 10 mm, many organs are reduced to a fixed number of cells (Miller 1979), as in some invertebrates, e.g. rotifers. Also, relative gill area (i.e. gill area per unit of body mass) is not only low to start with, but rapidly decreases with increasing body size, as predicted by a theory linking the *oxygen supply to fish (via gills) with their growth (Pauly 1982).

Most gobies are marine, but some are catadromous, while a few live entirely in freshwater (Bell 1999). Also, the Gobiidae include

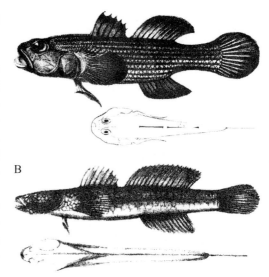

Fig. 13. Gobies, Family Gobiidae, with dorsal views as inserts. **A** *Bathygobius lineatus* (*Gobius lineatus* in *Fish*, Plate XIX), based on a specimen from Chatham Island, Galápagos. **B** *Ophiogobius ophicephalus* (*Gobius ophicephalus* in *Fish*, Plate XIX), from the Chonos Archipelago (South of Chiloe), Chile.

the *mudskippers, which have become largely terrestrial, and drown if kept under water for too long.

The Gobiidae are represented here by *Bathygobius lineatus* (Jenyns, 1842) and *Ophiogobius ophicephalus* (Jenyns, 1842), collected by CD in the *Galápagos (Fig. 13A) and *Chonos Archipelago (Fig. 13B), respectively, and initially assigned to the genus *Gobius* (*Fish*, pp. 95–8; *Fish in Spirits*, no. 1139, 1287).

CD later wrote that Certain fishes, belonging to several families, make nests, and [in] some of them [both sexes] take care of their young when hatched. [. . .] But the males of certain fishes do all the work, and afterwards take exclusive charge of the young. This is the case with the dull-coloured gobies,[36] in which the sexes are not known to differ in colour (*Descent II*, pp. 344–5; n. 36 refers to Cuvier 1829, p. 242.). However, Cuvier's page 242 contains nothing on the colour of gobies, while p. 243 mentions their 'brown-blackish bodies', 'white lines', 'red lines', 'whitish fins', 'brown spots with grey and red', 'varied colours', etc. Definitely not a good source for dull colouring.

Gobiesox spp. *See* **Clingfishes**.

Gobius spp. *See* **Gobies**.

Goldfish (I) Collective name for strongly modified forms of *Carassius auratus* (Linnaeus, 1758), Family Cyprinidae, resulting from over eight hundred years of selection by Chinese breeders (Li Zhen 1988). [Note that several other cyprinid species – e.g. the *Tench and the Crucian *carp – occasionally produce individuals with 'gold' colour, and some have been bred as ornamental fish.]

As it pertains to the fish most strongly modified by artificial selection, the literature on the 'gold' varieties of *C. auratus* received lots of CD's attention. However, CD's first mention of goldfish is based on a field observation; Octob. 25th. [1838, in Windsor Park . . .] Saw what was said to be hybrid between silver & gold fish (*Notebook E*, p. 405). Only much later did CD present a comprehensive account: Besides mammals and birds, only a few animals belonging to the other great classes have been domesticated; but to show that it is an almost universal law that animals, when removed from their natural conditions of life, vary, and that *races can be formed when selection is applied, it is necessary to say a few words on goldfish [. . .]. Goldfish (*Cyprinus auratus*) were introduced into Europe only two or three centuries ago; but they have been kept in confinement from an ancient period in China. Mr *Blyth[50] suspects, from the analogous *variation of other fishes, that goldencoloured fish do not occur in a state of nature.

These fishes frequently live under the most unnatural conditions, and their variability in colour, size, and in some important points of structure is very great. M. Sauvigny has described and given coloured drawings of no less than eighty-nine *varieties.[51] Many of the varieties, however, such as triple tail-fins, etc., ought to be called *monstrosities; but it is difficult to draw any distinct line between a variation and a monstrosity.

As goldfish are kept for ornament or curiosity, and as 'the Chinese are just the people to have secluded a chance variety of any kind, and to have matched and paired from it',[52] it might have been predicted that selection would have been largely practised in the formation of new breeds; and this is the case. In an old Chinese work it is said that fish with vermilion *scales were first raised in confinement during the Sung dynasty (which commenced A. D. 960), 'and now they are cultivated in families everywhere for the sake of ornament.'

In another and more ancient work, it is said that 'there is not a household where the goldfish is not cultivated, in rivalry as to its colour, and as a source of profit,' etc.[53] Although many breeds exist, it is a singular fact that the variations are often not inherited. Sir R. Heron[54] kept many of these fishes, and placed all the deformed ones, namely, those destitute of dorsal fins and those furnished with a double anal fin, or triple tail, in a pond by themselves; but they did 'not produce a greater proportion of deformed offspring than the perfect fishes.'

Passing over an almost infinite diversity of colour, we meet with the most extraordinary modifications of structure. Thus, out of about two dozen specimens bought in London, Mr Yarrell observed some with the dorsal fin extending along more than half the length of the back: others with this fin reduced to only five or six rays: and one with no dorsal fin. The anal fins are sometimes double, and the tail is often triple. This latter deviation of structure seems generally to occur 'at the expense of the whole or part of some other fin';[55] but Bory de Saint-Vincent[56] saw at Madrid goldfish furnished with a dorsal fin and a triple tail. One variety is characterized by a hump on its back near the head; and the Rev. L. Jenyns[57] has described a most singular variety, imported from China, almost globular in form like a *Diodon, with 'the fleshy part of the tail as if entirely cut away, the *caudal fin being set on a little behind the dorsal and immediately above the anal.'

In this fish the anal and caudal fins were double; the anal fin being attached to the body in a vertical line: the eyes also were enormously large and protuberant. (*Variations* I, pp. 312–13; with notes as follows: 50: The Indian Field, 1858, p. 255 [Also in *Correspondence* from Blyth, Jan. 23, 1856]; 51: Yarrell's British Fishes, vol. i, p. 319 [referring to Sauvigny 1780]; 52: Mr Blyth, in the Indian Field, 1858, p. 255; 53: W. F. Mayers, Chinese Notes and Queries, August 1868, p. 123; 54: Proc. Zoolog. Soc., 25 May, 1842 [see Heron 1842]; 55: Yarrell's British Fishes, vol. i, p. 319; 56: [Bory de Saint-Vincent] Dict. Class. d'Hist. Nat., vol. v, p. 276; 57 [Jenyns'] Observations in Nat. Hist., 1846, p. 211. Dr. Gray has described, in Annals and Mag. of Nat. Hist., 1860, p. 151, a nearly similar variety, but destitute of a dorsal fin.)

Goldfish (II) Following *Variations*, CD's next major work, *Descent*, further developed the goldfish theme: The gold-fish (*Cyprinus auratus*), judging from the analogy of the golden variety of the common *carp [. . .] may owe its splendid colours to a single abrupt variation, due to the conditions to which this fish has been subjected under confinement.

It is, however, more probable that these colours have been intensified through artificial selection, as this species has been carefully bred in China from a remote period.[29] Under natural conditions it does not seem probable that beings so highly organized as fishes, and which live under such complex relations, should become brilliantly coloured without suffering some evil or receiving some benefit from so great a change, and consequently without the intervention of *Natural selection. (*Descent* II, pp. 342–3; n. 29 reads: Owing to some remarks on this subject, made in my work On the Variation of Animals under Domestication, Mr. W. F. Mayers (Chinese Notes and Queries, August, 1868, p. 123) has searched the ancient Chinese encyclopedias. He finds that gold-fish were first reared in confinement during the Sung dynasty, which commenced A.D. 960. In the year 1129 these fishes abounded. In another place it is said that since the year 1548 there has been 'produced at Hangchow a variety called the fire-fish, from its intensely red colour. It is universally admired, and there is not a household where it is not cultivated, in rivalry as to its colour, and as a source of profit'; see also **Cyprinidae; Prussian carp; Race**).

Gonopodium The anal fin (or its anterior portion, or its rays) of a male fish when modified to serve as a copulatory organ, as, for example, in *livebearers. (*See also* **Jenynsiinae**.)

Gouramy(-ies) Members of the Family Belontiidae, which also includes several species of pugnacious 'fighting fishes,' such as the well-named *Betta pugnax* (Cantor, 1849).

The courtship behaviour of a species of this family, *Macropodus opercularis* (Linnaeus, 1758) was described by CD, based on work conducted in Paris with some of the first specimens introduced in Europe: A more striking case of courtship, as well as of display, by the males of a Chinese Macropus has been given by M. Carbonnier, who carefully observed these fishes under confinement.[27]

The males are most beautifully coloured, more so than the females. During the breeding-season they contend for the possession of the females; and, in the act of courtship, expand their fins, which are spotted and ornamented with brightly coloured rays, in the same manner, according to M. Carbonnier, as the peacock. They then also bound about the females with much vivacity, and appear by 'l'étalage de leurs vives couleurs chercher à attirer l'attention des femelles, lesquelles ne paraissaient indifférentes à ce manège, elles nageaient avec une molle lenteur vers les mâles et semblaient se complaire dans leur voisinage'. [. . . . the display of their vivacious colours, attempted to attract the attention of the females, which did not appear indifferent to this show; they swam slowly toward the males, and seemed to enjoy being close to them.]

After the male has won his bride, he makes a little disc of froth by blowing air and mucus out of his mouth. He then collects the fertilized ova, dropped by the female, in his mouth; and this caused M. Carbonnier much alarm, as he thought that they were going to be devoured. But the male soon deposits them in the disc of froth, afterwards guarding them, repairing the froth, and taking care of the young when hatched. I mention these particulars because, as we shall presently see, there are fishes, the males of which hatch their *eggs in their mouths; and those who do not believe in the principle of gradual *evolution might ask how could such a habit have originated; but the difficulty is much diminished when we know that there are fishes which thus collect and carry the eggs; for if delayed by any cause in depositing them, the habit of hatching them in their mouths might have been acquired. (*Descent II*, pp. 341–432; n. 27 refers to Carbonnier (1869b, 1870); the fishes whose males hatch their eggs in their mouth are *catfishes, and *Cichlidae).

Grant, Robert Zoology professor at the University of *Edinburgh, and CD's mentor from 1825 to 1827 (Ashworth 1935).

Grant (1793–1874), who specialized in marine invertebrates, was an ardent evolutionist and disciple of *Lamarck, the author of an evolutionary theory then widely perceived as inimical to the established order (*see* **Spontaneous generation**), beliefs which, in fact, appear to have impacted quite negatively on Grant's career (Desmond 1984).

CD, in his *Autobiography*, gives very little credit to Grant, and this has led to the notion that the teacher had little influence on his student, a notion restated, for example, in Tort (1996, p. 2019). The best refutation may be CD's own account of the dissection of a *Lumpfish, which is far too rich in ecological and evolutionary content not to have been influenced by a mature researcher. Or alternatively, CD was not, then, the disconnected teenager that his *Autobiography* portrays.

Great Chain of Being An old concept according to which all living organisms are interlinked more or less linearly by a directed evolutionary process, wherein all existing groups represent various ascending stages, like the steps of an escalator.

The great zoologist *Lamarck elaborated an evolutionary scheme which built on the *Great Chain of Being*, and which greatly influenced CD's grandfather, Erasmus Darwin, and CD's zoology professor in *Edinburgh, Robert *Grant.

The metaphor of a Chain of Being implies the concept of 'missing links' (transitional forms that should, but do not, occur in the *fossil record) more strongly than does the evolutionary 'tree' metaphor that CD proposed as an alternative, or the 'bushes' of *punctuated equilibrium. Hence it is not surprising that presumed missing links were being exhibited by P. T. Barnum (yes, he of the Circus!) as early as 1842. These included the "preserved body of a Feejee *Mermaid [. . .], *Ornithorhincus*, or the connecting link between the seal and the duck; two distinct species of *flying fish, which undoubtedly connect the bird and the fish; the Siren, or Mud Iguana, a connecting link between the reptiles and fish" (cited in Lovejoy (1936), p. 236).

That CD was influenced by this thinking is evidenced by the casualness with which, in works such as *Foundations*, *Origin* or *Descent*, he places the 'fish' in an intermediate position between *seasquirts and *vertebrates, and his frequent use of *Lepidosiren* (Barnum's 'Siren, or Mud Iguana') as transitory form between fishes and reptiles. (*See also* **Vertebrate origins**).

Great white shark The circum-global Great white shark *Carcharodon carcharias* (Linnaeus, 1758), Family Lamnidae, is most certainly one of the 'large' unidentified *sharks whose behaviour is briefly discussed by CD in his accounts of the voyage of the *Beagle*.

Information on the biology of the Great white shark is available in Klimley and Ainley (1996), and we discuss here only the maximum size reached by *C. carcharias*, which has been greatly overestimated by earlier authors. This tradition apparently started with an ichthyologist well known to CD: *Günther (1870), extrapolating from the size of the teeth of *C. rondeletti* (a junior *synonym), estimated a maximum length of 36.5 ft (11.1 m), a figure later rounded up to 40 ft, or 12.2 m (Günther 1880, pp. 320–1; Randall 1987b; Helfman *et al.* 1997, pp. 182–3). Bigelow and Schroeder (1948) gave 21 ft, or 6.4 m, as the largest size for great white shark, and thus Randall (1973) asked: "[w]hy have no white sharks been recorded by actual measurement between 21 and 36.5 feet in length?"

Based on re-measurements of the teeth stored at what was then still known as the 'British Museum (Natural History)', examination of *whale carcasses with teeth marks, and a close reading of Günther (1870), Randall (1973) concluded that the maximum length of 36.5 ft was a printer's error, and should have read 16.5 ft (5m). He conceded, however, that "it seems likely that white sharks more than 21 feet long swim in our seas today and remain to be captured". Randall (1973) further cautions that, as the maximum sizes inferred for fossil *Megatooth shark were based on Günther's estimate of 36.5 ft, these values will tend to be overestimates as well.

Greenland halibut Common name of *Reinhardtius hippoglossoides* (Walbaum, 1792), one of the largest *flatfish in the North Atlantic, the females reaching 120 cm and 45 kg. CD uses what appears to be this species to illustrate one end of the range of adaptations prevailing among flatfishes: different members, however, of the family present, as Schiödte remarks, 'a long series of forms exhibiting a gradual transition from Hippoglossus pinguis, which does not in any considerable degree alter the shape in which it leaves the ovum, to the soles, which are entirely thrown to one side.' (*Origin* VI, p. 186; Schiödte 1868).

The inference that *Hippoglossus pinguis* corresponds to Greenland halibut is indirect, as there is no type specimen for *Pleuronectes pinguis* Fabricius, 1824 that taxonomists could use to establish the synonymies linking these names (Eschmeyer 1998, p. 1341). However, Greenland halibut is the flatfish that behaves least like other pleuronectiforms. Thus, it often swims in an 'upright' position (i.e. like most other fish), and its dark colouring is similar on both sides, a fact compatible with CD's suggestion that the paleness of the underside in other flatfish is due to the exclusion of light; *Origin* VI, p. 188). Indeed, one has the impression that this fish is evolving back to symmetry, from a more asymmetrical ancestor. (*See* **Dohrn**, for an elaboration of this.)

We conclude this entry by noting that Greenland halibut is (confusingly) also known as *turbot, and that it provided a reason (or pretext, depending on one's degree of cynicism; see Harris 1998) for the March 1995 conflict between Canada and Spain.

Groupers (I) A group of carnivorous, *hermaphroditic, perch-like fishes, of which the largest can attain 3 m and 400 kg. CD sampled many (smaller!) groupers, and they are presented here in the same sequence as used by *Jenyns in *Fish*: The Whitespotted sandbass *Serranus albo-maculatus*, now *Paralabrax albomaculatus* (Jenyns, 1840), is now known to be a *Galápagos endemic (Grove and Lavenberg 1997). CD, based on Syme (1821), described its *colour rather extensively: Varies much. Above pale blackish-green; belly white; fins, gill-covers, and part of the sides, dirty reddish orange: on the side of the back, six or seven good-sized snow-white spots, with not a very regular outline. In some specimens, the blackish-green above becomes dark, and is separated by a straight line from the paler under parts. Again, other specimens are coloured dirty 'reddish-orange,' and 'gallstone yellow,' the upper parts only rather darker. But in all, the white spots are clear; five or six in one row, and one placed above. Sometimes the fins are banded longitudinally with orange and black-green. (*Fish in Spirits*, no. 1304, *Fish*, pp. 3–5; Fig. 14A).

Next, Jenyns presents the Dengat grouper *Serranus goreensis*, now referred to as *Epinephelus goreensis* (Valenciennes, 1830). This occurs along the West African coast and offshore islands, from Senegal to southern Angola, over different bottom types (rock, sand, mud), but its biology is little known. Jenyns described this species based on a single specimen which CD caught by hook at Porto Praya, St Jago Island, Cape Verde Islands (*Fish*, p. 6; *Fish in Spirits*, no. 17).

Serranus aspersus is based on a "small, and probably not full-sized" specimen collected by CD at Quail Island, and is now considered a synonym of *Epinephelus marginatus* (Lowe, 1834), the Dusky grouper, which ranges from the Mediterranean along the West African coast, around South Africa to Mozambique in the Indian Ocean. CD described the coloration of his specimen as follows: Dark greenish, black above, beneath lighter; sides marked with light emerald green: tips of the anal, *caudal, and hinder part of the dorsal, saffron yellow; tips of the pectorals orpiment orange, with Jenyns adding that "these colours have been much altered by the action of the spirit" (*Fish*, p. 7; 'orpiment' is a soft yellow mineral, arsenic trisulfide, used as a pigment; see also *Fish in Spirits*, no. 43).

Back to the Galápagos with the Starry grouper, *Serranus labriformis*, now *Epinephelus labriformis* (Jenyns, 1840), which ranges from Mexico to *Peru. CD, described his specimen as Mottled brown-yellow, black and white upper and lower edges of tail, edges of ventral and dorsal, (Art[erial] And purplish red) (*Fish in Spirits*, no. 1273; Fig. 14B).

Serranus olfax, which Jenyns initially planned to call *Serranus 'galapagensis'*, is indeed a Galápagos endemic (Rosenblatt and Zahuranec 1967), now referred as *Mycteroperca olfax* (Jenyns, 1840). A common large mottled brown fish (*Fish in Spirits*, no. 1275), locally known as *Bacalao, it supports an important fishery, documented by the former director of the Charles Darwin Research Station, Galápagos (Reck 1979, 1983; Fig. 14C).

Specimens of Barred rockfish *Plectropoma patachonica* (Jenyns, 1840) were sampled by CD off Northern Patagonia (38°20′S), and in the mouth of the Plata River, at a depth of 40 fathoms (*Fish*, pp. 11–12). Specimens from the latter area were described as above salmon coloured (*Fish in Spirits*, no. 358), and as closely allied with the Brazilianum (*Zoology Notes*, p. 342, no. 710). More details are available on the fish from higher up the Plata River: Many specimens exceeded a foot in length. Above aureous -coppery, with wave like lines of dark brown, these often collect into 4 or 5 transverse bands, fins leaden Colour; beneath obscure; pupils dark blue. CD also noted that One specimen, when caught vomited up small fish and a *Pilumnus*. Mʳ Earle states these fish are plentiful at Tristan da Cunha, where it is called the Devils fish; from the bands being supposed the marks of the Devils fingers. Was tough for eating, but good. This sort was taken in very great number. Coast of Patagonia, August (*Fish in Spirits*, no. 348; *Pilumnus* spp. are hairy crabs, Family Xanthidae).

Some authors (e.g. Menni *et al.* 1984) view this species as a *synonym of *Acanthistius*

Groupers (II)

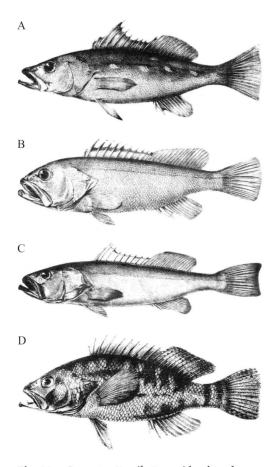

Fig. 14. Groupers, Family Serranidae, based on lithographs by B. Waterhouse Hawkins of specimens taken in the Galápagos by CD.
A *Paralabrax albomaculatus* (*Serranus albo-maculatus* in *Fish*, Plate II), based on a specimen from Charles Island. **B** *Epinephelus labriformis* (*Serranus labriformis*), from Chatham (San Christobal) Island (Plate III). **C** *Mycteroperca olfax* (*Serranus olfax*), Chatham Island (Plate IV). **D** *Serranus psittacinus* (*S. fasciatus* and *Prionodes fasciatus* in *Fish*, Plate IX), Chatham Island.

The next species in *Fish*, **Diacope marginata*, **Arripis georgianus* and **Aplodactylus punctatus*, are not groupers. Thus, we move to *Dules auriga* Cuvier, 1829 from **Maldonado Bay, which CD described as follows: Sides with numerous waving longitudinal lines of brownish red; intermediate spaces greenish-silvery, so figured as to look mottled. Head marked with lines of dull red and green. Ventral and anal fins dark greenish blue (*Fish*, p. 16; *Fish in Spirits*, no. 696). Next is *Dules leuciscus* Jenyns, 1840, collected by CD in the Matavai River, *Tahiti, and thus noted as a fresh water Fish (*Fish in Spirits*, nos. 1330, 1331), though it is a marine fish (*see* **Lagoon**). And not only this: it is not even a grouper (*see* Index).

Finally we have *Prionodes fasciatus*, collected by CD at Chatham (San Christobal) Island, Galápagos, in the form of a specimen that was Pale yellowish brown, with numerous transverse bars, of which the upper part is reddish black, the lower vermilion red; gill covers, head, and fins, tinted with the same (Fig. 14D). Jenyns described it as a *croaker on pp. 47–9 of *Fish* (published in 1840), but recanted on p. 164 (published in 1842), and assigned it to *Serranus*. And indeed, it is now identified as *Serranus psittacinus* Valenciennes, 1846, a species ranging from the Southern Gulf of California to Peru. Hastings and Peterson (1986) believe that this species uniquely differs from the protogynous hermaphroditism common in groupers in that some populations include both simultaneous *hermaphrodites and secondary males (derived from hermaphrodites, and 'sneaking' as do *jacks). However, it is likely that this occurs in other serranids as well.

Groupers (II) *Serranus* is the type genus of the Family Serranidae (Subfamily Epinephelinae), comprising the groupers *sensu stricto*. *Serranus* is presently defined much more narrowly than in CD's time, when it included most groupers then known. On the other hand, many nominal species have been shown to be synonyms of

brasilianus (Cuvier, 1828), while others (e.g. Carvalho-Filho 1994) consider these two species to be separate, with *A. brasilianus* defined as a Brazilian endemic, ranging from *Rio de Janeiro to Sao Paulo, and *P. patachonica* from Sao Paulo to Comodoro Rivadavia, Argentina (about 46°S).

other, previously known species, and hence the many name changes listed above.

CD commented on grouper sex as follows: [a]ll the vertebrata are bisexual, except as it would appear some fish of the genus Serranus[2] but from what we know of the habits of fish, an occasional cross seems far from improbable. (*Big Species Book*, p. 44; n. 2 refers to Quatrefages (1856), p. 80 (note 2), who cites "Dufossé, médecin à Marseille", i.e. Dufossé (1854), p. 300. Johannes (1981) discusses the spawning aggregations that groupers form when spawning – yes, group sex, for which they often swim a rather long way; Zeller 1997, 1998).

Later, CD returned to this feature of groupers, noting that: *Hermaphroditism has been observed in several species of Serranus, as well as in some other fishes, where it is either normal and symmetrical, or *abnormal and unilateral. Dr Zouteveen has given me references on this subject, more especially to a paper by Professor Halbertsma, in the *Transact. of the Dutch Acad. of Sciences*, vol xvi. Dr *Günther doubts the fact, but it has now been recorded by too many good observers to be any longer disputed. Dr M. Lessona writes to me, that he has verified the observations made by Cavolini on Serranus. (*Descent I*, p. 162, n. 28; Halbertsma 1864; Cavolini 1787; for information on Zouteveen, see annotation to Darwin (1872); an account of Michele Lessona, an Italian zoologist and translator of CD's work, is given in Tort (1996), p. 2623; notably, he showed that gastropod molluscs are very liable to have their heads bitten off by fishes.; *Variation I*, p. 359).

Günther's summary of Halbertsma (1864; see annotation) does not suggest disagreement. Thus, his doubts must have been expressed elsewhere. And indeed doubts can be perceived in Günther (1880, p. 157), where he tells us that "All fishes are *dioecious*, or of distinct sex. Instances of so-called *hermaphroditism* are, with the exception of *Serranus,* abnormal individual peculiarities, and have been observed in the Cod-fish, some Pleuronectidae, and in the Herring."

As it turns out, all groupers, not only *Serranus*, are protogynous *hermaphrodites (Nelson 1994, p. 335), and so are the *wrasses, and other, predominantly tropical fishes. These add up, however, to only 2% of all extant fish species (see *FishBase), and hence Günther may have been correct in emphasizing the dioecism of fishes. CD, of course, emphasized the 2%, given his vision of the ancestral vertebrate as a hermaphrodite. (*See* **Vertebrate origins**.)

Growth The process, linking size and time, by which living organisms increase their length and mass as they become older (*see* **Oxygen**).

However, CD anticipated Thompson (1917) and Huxley (1932) in usually referring to 'growth' as a process involving *size*-related changes in body shape and relations between body parts, a process known as allometry.

One exception to this pertains to *barnacles, whose *time*-related growth was described by CD, based on an extract from a letter by William Thompson: I examined a great number of Balani, in reference to the growth made by them during the present season, and found it to average three lines in diameter, and at most four lines. I saw a few minute specimens, only one line in diameter, showing that the species continued to breed until lately: these latter were probably not more than four weeks old. The young of the present year are plainly distinguished from the older ones, by their pure white colour and fresh appearance. Judging from the size of this year's specimens, and of the older ones on the same stones, I am of opinion that the term of life of the species is two years. Of the older shells, which I examined and found living in the spring, nine tenths are now dead, the walls only remaining, the opercular valves having been washed away. (Darwin 1851, Vol. II, pp. 272–3, Part I; *Corresp.*

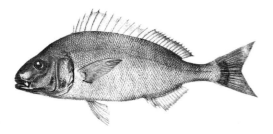

Fig. 15. Sheephead grunt *Orthopristis cantharinus*, Family Haemulidae (*Pristipoma cantharinum* in *Fish*, Plate X), based on a specimen from the Galápagos Archipelago.

Sept. 28, 1848; Balani refers to *Balanus*, a common genus of barnacle; *see* **Length** for 'line').

To this, CD added: Mr. Thompson goes-on to say, that the individuals which had, on July 3, a basal diameter of from two and a half to three lines, had attained, by the 30th of September, a diameter of four and a half lines, this being here the maximum size of the species.

A real growth curve, predicting size from age, could be computed from these observations, but we shall abstain from this, leaving it to anyone wanting to follow up on Crisp's (1983) extension of CD's barnacle work.

Grunts Members of the Family Haemulidae, common on and around coral reefs. Represented here by the Sheephead grunt *Orthopristis cantharinus* (Jenyns, 1840), collected by CD in the *Galápagos, and originally described as *Pristipoma cantharinum* (*Fish in Spirits*, no. 1266 (Older specimen): Blueish silvery; no. 1302 (Young): silvery above, shaded with brown and iridescent with blue; fins and iris sometimes edged with blackish brown. Flap of the gill cover edged with black. Fig. 15).

Every grunt grunts, i.e. grunts grunt (hear it at www.fishbase.org; click 'Fish sounds' under 'Information by Topics'). A description of these *sounds may be found in Fish and Mowbray (1970).

Günther, Albert An extremely productive fish taxonomist (1830–1914), born in Esslingen, Germany, who became, in 1857, on Richard *Owen's recommendation (Browne 2002, p. 224) assistant naturalist at the (then) British Museum (Natural History), and, in 1875, Keeper of its Zoological Department. While compiling his huge *Catalogue of Acanthopterygian Fishes in the collection of the British Museum* (Günther 1859–70) and other works, he maintained an active correspondence with CD, and responded to his numerous questions on various aspects of the biology and distribution of fishes. Günther is therefore cited numerous times, in several of CD's books. On the other hand, the volume that summarizes Günther's ichthyological work during this period of intellectual exchange, his *Introduction to the Study of Fishes* (Günther 1880) does not once mention CD, or *evolution. Was he caught between Owen and CD?

Guppy We shall not deal here with Dr H. B. Guppy, who corresponded with CD on various topics related to coral reefs, and who has nothing to do with *guppies. Rather, the person to mention here is the Reverend John Lechmere Guppy. Around 1870, he sent fish specimens from Trinidad to A. *Günther, who described one of these as *Girardinus guppii* Günther, 1866. However, this turned out to be a junior synonym of *Poecilia reticulata* Peters, 1860, 'the' guppy. Other authors also described fish sent by J. L. Guppy, whose many *eponyms may be found in *FishBase. Further details may be found in P. L. Guppy (1922), whose name assures us of a certain continuity, though it is not clear of what.

Guppy(-ies) *See* **Livebearers**.

Gurnards Fishes of the Family Triglidae, also known as searobins. CD sampled *Galápagos gurnard (*Prionotus miles* Jenyns, 1840) just where its common name suggests it was taken.

The specimen (Fig. 16) was [a]bove mottled brilliant tile red; beneath silvery white. (*Fish*, pp. 29–30; note indication in *Fish in Spirits*, no. 1267, that Jenyns considered naming the species *Prionotus* 'ruber', i.e. 'red'). CD

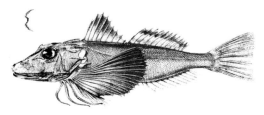

Fig. 16. Galápagos gurnard *Prionotus miles*, Family Triglidae, based on a specimen from Chatham (San Christobal) Island (*Fish*, Plate VI). The insert shows a jaw with a central notch and "six short, but well developed teeth on each side, followed by some minute denticles".

also sampled a Bluefin gurnard *Chelidonichthys kumu* (Cuvier, 1829) in the *Bay of Islands, New Zealand (*Trigla kumu* in *Fish in Spirits,* no. 1341; *Fish*, p. 27; Whole body bright red). In the Bay of *Rio de Janeiro, Brazil, CD sampled a specimen of the Blue searobin *Prionotus punctatus* (Bloch, 1793), and noted Fish, swimming on surface. Rio bay. above and sides olive brown, with red spots and marks; beneath silvery white; edges of pectoral fin Prussian blue, emitted a sound like a croak (*Fish in Spirits,* no. 269; *Fish* pp. 28–9).

Based on a letter from *Yarrell, answering his queries on the distribution of various genera (*Correspondence*, July 29, 1845), CD noted on p. 390 of the second Edition of his *Journal* that the *Galápagos fishes belong to twelve genera, all widely distributed, with the exception of *Prionotus*, of which the four previously known species live on the eastern side of America (Burkhardt *et al*. 1987, p. 232). As currently defined, the genus *Prionotus* contains 22 species: 7 on the eastern, and 15 on the western seaboard of the Americas (*see* FishBase). The score here: *Prionotus* 1 : CD 0.

Later, based on Dufossé (1862), CD discussed another aspect of gurnard biology: the Trigla produces pure and long-drawn *sounds which range over nearly an octave. (*Descent II*, p. 347). While Dufossé was referring to "*Trigla lineata* Lin.", i.e. to the Streaked gurnard *Chelidonichthys lastoviza* (Bonnaterre, 1788), recordings of fishes now or earlier assigned to '*Trigla*' exist (Tavolga 1968; Fish and Mowbray 1970) which do range well over an octave – as do many other fishes.

Gymnotus *See* **Electric eel.**

Haeckel, Ernst German biologist (1834–1919), exuberant supporter of *evolution and *natural selection, though with a twist later seen to have contributed to *social Darwinism. In his heyday, however, Haeckel was a magnificent artist (see Hanken 1999), wordsmith (*see* for example, **Ecology; Ontogeny**) and grower of evolutionary trees, and many of the transitory forms he postulated have since been discovered to actually exist.

His public disputes (notably with Hensen, on the nature and proper study of *plankton) are the stuff of legend (at least in Germany), as was his loud advocacy of the 'biogenic law' (Haeckel 1866), according to which "ontogeny recapitulates phylogeny".

Hagfishes Members of the Class Myxini, Superclass Agnatha, or jawless fish-like vertebrates, referred to as lowest fishes by CD (*Foundations*, p. 140; n. 16).

This ancient group (Fernholm 1998; Martini 1998), of which *Myxine glutinosa* Linnaeus, 1758 was the first to be described, is represented here by a single species, *Myxine australis* Jenyns, 1842, sampled by CD among *kelp, at Goree Sound, *Tierra del Fuego, but also occurring in the South Shetland Islands (Fernholm 1990; *see also* **Gastrobranchus**).

CD noted that, when alive, the specimen was Above coloured like an earth worm, but more leaden; beneath yellowish and head purplish and that it was very vivacious, And retained its life for a long time. Had great powers of twisting itself and could swim tail first; when irritated it struck at any object with its teeth; and by protruding them, in its manner, much resembled an adder striking with its fangs. Head most curiously ornamented with tentacles; vomited up a sipunculus. When caught. This fish is abundant amongst the rocky islets, having found one on the beach nearly dead, I observed a milky fluid transcending through the row of internal pores or orifices. It would appear to be Mixinidus (*Fish in Spirits*, no. 515; *Fish*, p. 159; sipunculids are worm-like eaters of bottom muck; small wonder the hagfish gagged).

The fisheries of the world, in their relentless search for new 'resources' to exploit have recently discovered hagfishes, both as food and as raw material for a particularly fine leather (ill-named 'eelskin'). Needless to say, several populations, e.g. off the Korean Peninsula, have been devastated, while 'nuevas pesquerías' (Arancibia *et al.* 2000) are being announced from other locations. Unfortunately, these are not 'new fisheries' – they are the same old fisheries, only elsewhere, and for yet another group that we would have never thought of exploiting before. If experience is any guide (see Ludwig *et al.* 1993), this 'new resource' will be gone before we have time to set up an effective management system.

Hamilton Smith, Charles English officer and zoologist (1776–1859), whom the mature CD read (see annotation to Smith 1852) and frequently cited (Tort 1996, pp. 4036–7). We know when CD first met him: while waiting for the Beagle to sail from Plymouth, he dined with Mr Harris, (the author of several papers on Electricity) & met there several very pleasant people. Colonel Hamilton Smith, who is writing on fishes with Cuvier. . . . (*Diary*, Nov. 4, 1831; William Snow Harris, also known as 'Thunder and Lightning Harris', invented a method for protecting ships from lightning, first adopted by the Russian Navy).

Bauchot *et al.* (1990) compiled short biographies for all the persons they could find to have contributed to Cuvier and *Valenciennes' monumental *Histoire Naturelle des Poissons*. Colonel Hamilton-Smith is not listed.

On the other hand, Hamilton Smith co-translated and edited the fish volume of Cuvier's *Règne animal* (Griffith and Smith 1834), and so he did work with Cuvier. As he writes (in Griffith and Smith 1834, n. 1, p. 16), he "had the honour repeatedly to witness

[Cuvier] with the dissecting knife in his left, and the pencil in his right hand, laying open the parts required, and sketching in a manner so spirited and bold, as to be only surpassed by its fidelity".

Handicap principle A recent theory (Zahavi and Zahavi 1997) aiming to interpret acoustic, visual or chemical signals emitted by animals to conspecifics, or would-be predators.

The 'signals' in question, which range from the display of brightly coloured body appendages to various 'risky' behaviours, have the common attribute that they are costly and/or risky to produce or to possess, and thus send to competitors, members of the other sex, or to potential predators, true information about the sender's fitness.

The handicap principle elegantly explains many aspects of *sexual selection, notably the extreme expression of apparently useless (or even deleterious) male attributes (such as extremely long fins, which, while perfectly symmetrical, are unsuited for predator evasion), previously attributed (Fisher 1930) to 'runaway selection' (*see also* **Non-Darwinian evolution; Asymmetry**).

Hare, Margaretta *See* **Morris, Margaretta Hare.**

Hawk(s) Raptorial birds, members of the Family Falconidae.

CD lists two species of hawk as consuming fish, viz. *Craxirex galapagoensis*, which feeds on all kinds of offal thrown from the houses, and dead fish and marine *productions cast up by the sea (*Birds*, p. 25) and *Milvago chimango*, which is found in *Chiloe & on the coast of *Patagonia, & I have seen it T. del Fuego. [. . .] The Chimango frequently frequents the sea coast, & the border of lakes & swamps, where it picks up small fish (*Birds*, p. 14; *Ornithological Notes*, p. 236).

Hawkins, Benjamin Waterhouse English artist (1807–89) who drew all but one of the *figures (plates) in *Fish*, and whom Jenyns refers to as '*Waterhouse*' in n. 19, p. 40.

Here is CD's account of how this came about, in a letter to Jenyns: You must be surprised at not having heard from me before, but owing to a succession of headaches I have been prevented until today seeing *Yarrell. – I think he has hit upon the right artist, namely W. Hawkins, who engraved the fish for Richardsons volume – I have written to him to talk to me over terms, &c. Richardson's plate are done on Zinc, but it seems now generally considered that stone is preferable (*Correspondence*, Oct. 14, 1839; Richardson 1836).

His manner of drawing fishes was described by CD as follows: Mr Hawkins first draws a careful outline, which the Describer inspects & approves & then he put this on stone & fill up details from the fish itself, – so that he might have a dozen outlines ready for your inspection [. . .] & afterward completes them & then you inspect them at a second visit (*Correspondence* to Jenyns, Oct. 14, 1839).

Hawkins is also famous for his collaboration with Richard *Owen, for whom he made life-size models of *Iguanodon, Megalosaurus*, etc., for the Crystal Palace of the Great Exhibition of 1851, in London.

Heckel *See* **Haeckel** (*see also* **Spelling!**).

Heliases crusma *See* **Damselfishes.**

Helotes octolineatus *See* **Tigerperches.**

Henslow, John Stevens British naturalist (1796–1861), botany professor at Cambridge, CD's friend and mentor in natural history. He was appointed in 1825 to the Chair of Botany at Cambridge University, and became one of the most influential professors at that institution (Jenyns 1862), though he never accepted evolutionist ideas.

During the voyage of the *Beagle*, Henslow managed that part of CD's *collection of specimens shipped ahead of the return of the *Beagle*. He also turned several of his letters into papers presented at meetings of learned societies, thus establishing CD's reputation even before his return. Barlow (1967) edited the correspondence between CD and Henslow, now integrated into CD's *Correspondence*.

Hermaphrodite(-ism) Possessing the organs of both sexes (*Origin VI*, p. 434), a condition which CD also called androgynous.

Three forms of hermaphroditism are recognized: protogyny, in which female gonads are formed first, which later turn into male gonads; protandry, where the converse occurs; and true hermaphroditism, in which each individual is simultaneously male and female. Although individual cases of hermaphroditism occur in individuals of most animal species (including humans), animal species in which hermaphroditism is the rule tend to be rare.

Hermaphroditism is, however, common in sessile organisms such as *barnacles, and *seasquirts, for reasons discussed, for example, in Maynard Smith (1978). In fish, about 500, i.e. 2% of all extant species, appear to be hermaphroditic (*see* **FishBase**). Of these, many are *groupers (i.e., belonging or related to the genus *Serranus*), *parrotfishes and *wrasses (protogyny), *damselfish (protandry), and *porgies (mode varying among and within species). Some degree of sexual plasticity is reported from other groups (e.g. the *Cichlidae, see Barlow 2000, pp. 55–9). However, hermaphroditism occurs only rarely in *eels, notwithstanding CD accepting an account to the contrary.

Only one fish species has been so far documented as a true hermaphrodite: the Mangrove rivulus *Rivulus marmoratus* Poey, 1880 (Soto *et al.* 1992), which produces one egg at a time, and fertilizes it. Thus, CD was almost right when he wrote Sexes are never combined in same individual. Quatrefages seems to believe in Serranus being hermaphrodite. I forget author's name. (*Correspondence* to *Huxley, Dec. 16, 1857; *see* **Groupers (II)** for more on Quatrefages and the forgotten author's name).

CD considered the existence of hermaphroditism in fishes to bear on *vertebrate origins. This is summarized in the argument below: It has long been known that in the vertebrate kingdom one sex bears rudiments of various accessory parts, appertaining to the reproductive system, which properly belong to the opposite sex; and it has now been ascertained that at a very embryonic period both sexes possess true male and female glands. Hence some remote progenitor of the whole vertebrate kingdom appears to have been hermaphrodite or androgynous[28]. [... However] we have to look to fishes, the lowest of all the classes, to find any still existent androgynous forms (*Descent I*, p. 161; n. 28 refers to *Serranus*).

Also: It may be suggested, as another view, that long after the progenitors of the whole mammalian class had ceased to be androgynous, both sexes yielded milk, and thus nourished their young; and in the case of the Marsupials, that both sexes carried their young in marsupial sacks. This will not appear altogether improbable, if we reflect that the males of existing syngnathous fishes receive the *eggs of the females in their abdominal pouches, hatch them, and afterwards, as some believe, nourish the young[30]; that certain other male fishes hatch the eggs within their mouths or branchial cavities; and that certain male *toads take the chaplet of eggs from the females, and wind them round their own thighs, keeping them there until the *tadpoles are born; (*Descent I*, p. 163; n. 30 refers to *Hippocampus*).

Herons Long-legged, fish-eating birds of the Family Ardeidae. CD notes that Mr <*Lyell> <<*Waterhouse>> has frequently heard that Herons bring *eels alive to their nests; & then they may be picked up beneath the trees – Are there any Fish seed-eaters. This important in Transport of Fish. Let a *Hawk fly at Heron – (*Notebook A*, p. 88; *see also* **Experiments**).

Herring Common name of *Clupea harengus* Linnaeus, 1758, the *type species of the genus *Clupea*, itself the type genus of the Family Clupeidae, including all *herrings.

Herring were crucial to the development of European marine fisheries (Cushing 1988; Sarhage and Lundbeck 1991), and the study of their biology largely shaped fisheries science, when it emerged as a discipline of its own, at the turn from the nineteenth to the twentieth century (Went 1972). Notably, studies of meristic features of Herring (number of scales, vertebrae and other countable features), showed that they present a vast range of variation (*Big Species Book*, p. 117) and quickly led to the emergence of the concept of a 'stock', roughly similar to that of a 'population.'

Stocks are populations (or more precisely the exploited part thereof) isolated enough for their exploitation not to impact on the status of similar adjacent entities. (Note that this definition allows stocks to be distinguished even in cases when the genetic flow between adjacent stocks precludes differentiation by standard genetic methods, such as the study of allele frequencies through electrophoresis.)

The identification of numerous stock and coastal stocklets, e.g. those in the North Sea, did not prevent the emerging industrial fisheries of the late nineteenth and early twentieth centuries from exploiting this species as if it consisted of a single large, panmixing population (Cushing 1975, 1988). Consequently, given the *obstinate nature of their spawning habits, these coastal stocklets were depleted, and never recovered (see Pitcher 2001 and references therein). They are now often alleged never to have existed, another sad case of shifting baseline (Pauly 1995b).

For additional information on Herring *see* **Eggs of fishes; Grouper (II); Owen; Variety.**

Herrings Members of the genus *Clupea*, and/or of the Family Clupeidae, the latter also including

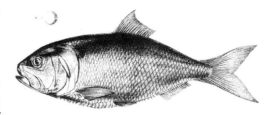

Fig. 17. Argentine menhaden *Brevoortia pectinata*, Family Clupeidae (*Alosa pectinata* in *Fish*, Plate XXV), from Bahia Blanca, Argentina. Insert shows a magnified scale from the nape.

sardines (pilchards), sprats, menhaden, and other small schooling fishes.

The family includes several species for which Jenyns, based on specimens collected by CD, provided the original description. Thus, *Ramnogaster arcuata* (Jenyns, 1842) was originally described as *Clupea arcuata* in *Fish* (*Fish*, p. 134), based on specimens from *Bahia Blanca (*Fish in Spirits,* no. 371, 416: back blue, belly silvery). The Falkland sprat *Sprattus fuegensis* (Jenyns, 1842) was similarly described as *Clupea fuegensis* (*Fish*, pp. 133–4), based on a specimen Caught at night. 2 miles from shore. Cape Ines, *Tierra del Fuego (*Fish*, pp. 133–4; *Fish in Spirits*, no. 838).

This is completed by the South American pilchard *Sardinops sagax* (Jenyns, 1842), described as *Clupea sagax* (*Fish*, pp. 134–5; *Fish in Spirits*, no. 1251), with specimens from San Lorenzo Island, near Callao, the harbour of Lima, *Peru, and by the Argentine menhaden *Alosa pectinata* (pp. 135–6), from Bahia Blanca (silvery, dorsal scales iridescent with green and copper: head greenish, tail yellow; *see also Fish in Spirits*, nos. 390, 391; Fig. 17), now *Brevoortia pectinata* (Jenyns, 1842).

Heterocercal Describes a fish tail with one lobe larger than the other. This condition is very common in ancient fishes and in *sharks, whose tail has an upper lobe that is usually much larger than the lower lobe.

Hippopotamus(-i)

Here is CD's version: the embryos of the existing vertebrata will shadow forth the full-grown structure of some of those forms of this great class which existed at the earlier periods of the earth's history: and accordingly, animals with a fish-like structure ought to have preceded birds and mammals; and of fish, that higher organized division with the vertebrae extending into one division of the tail ought to have preceded the equal tailed, because the embryos of the latter have an unequal tail (*Foundations*, p. 230).

In modern *bony fishes, or 'teleosteans,' the tail is usually homocercal (or equal-tailed as CD put it; see, for example, Fig. 17), or diphycercal (i.e., lacking lobes; see Fig. 10). However, larval teleosts, such as the *Gar-pike, often sport heterocercal tails, and this may be seen as another instance where *ontogeny recapitulates phylogeny.

Hippocampus spp. *See* **Seahorses.**
Hippoglossus kingii *See* **Flounder, Fine.**
Hippoglossus pinguis *See* **Greenland halibut.**
Hippopotamus(-i) Large *mammals of Africa, related to the *whales, and whose two species spend most of the daytime in rivers, shallow lakes and other water bodies, while grazing on land during the night. The common hippo (*Hippopotamus amphibius*) is the larger and more aquatic of the two, and is generally considered to have several subspecies. Hippos enter this book though their illustration of the ample geographic distribution of large pelagic *sharks, a topic that intrigued CD, as illustrated by a cryptic comment on the back cover of *Notebook R* (p. 81): Range of Sharks – Nothing For Any Purpose.

We now turn to the hippos in that same *Notebook R*: Dampier's last voyage to New Holland P 127 – Caught a shark 11 ft long. 'Its maw was like a leathern sack, very thick & so tough that a sharp knife could not cut it: in which we found the Head & Boans of a Hippotomus; the hairy lips of which were still sound and not petrified, and the jaw was also firm, out of which we pluckt a great many teeth, 2 of them, 8 inches long, & as big as a mans thumb, the rest not above half so long; The maw was full of jelly which stank extreamly.' – This shark was caught in Shark's Bay. Lat. 25°. the nearest of the E Indian Islands, namely Java is 1000 miles distant! Where are Hippotami found in that Archipelago? Such have never been observed in Australia. (pp. 22–3; quotes are from Dampier 1703).

There are no hippos is Southeast Asia (i.e. in the E Indian Islands), and we will first examine the possibility that the shark in question swam all the way from the nearest possible location (East Africa) with a half-digested hippo head in its stomach. Here are the required elements:

(1) Large sharks are known to be able to keep ingested preys in their stomach for up to a month without digesting them (McCormick *et al.* 1963, p. 233; Budker 1971, pp. 101–3; see also the "Shark arm mystery" in Anon. 1986, pp. 114–19, which highlights the tattoo on a man's chopped arm, regurgitated 18 days after the murder and its ingestion by a tiger shark);

(2) The great circle distance between Shark Bay, Western Australia, and the nearest coast off the African mainland, i.e. Mozambique, is about 8000 km if care is taken to avoid a collision with the coast of Madagascar;

(3) A pelagic shark 11 ft long (3.35 m) may be assumed able to maintain a swimming speed of about two body lengths per second (i.e. a conservative two thirds of the sustained speed of pelagic fishes predictable for such sizes; see 'Swimming Table' in *FishBase). This would correspond to about 24 km h^{-1}, or 580 km d^{-1}. Let's assume 500 km d^{-1} to be on the conservative side, noting that this is still a very large figure, well in excess

of the "minimum velocity" of 71 km d^{-1} reported by Boustany *et al.* (2002), based on satellite tagging data for a *Great white shark;

(4) Hence it would require only 16 days for our shark, after having ripped the head off a carcass of *H. amphibius kiboko* (the east African subspecies) that had floated down the mouth of a Mozambican river, to reach Shark Bay in Australia, there to be caught and autopsied by Capt. Woode Roger's men. This is well within the known periods of delayed digestions reported from large sharks. And the trip would have remained under a month, as well, if the shark's speed had approached 1 body length per second.

But then, what Woodes Roger's men found need not have been the head of a hippo. Indeed the description above – hairs, teeth, etc. – is entirely compatible with the head of a dugong (*Dugon dugong*), a *sirenian that was extremely abundant in western Australia (Jackson *et al.* 2001), where it is still common. And it is more *parsimonious to assume that a shark attacked a dugong in Shark Bay, Australia, then relaxed near the scene of the crime, rather than consuming a hippo in Mozambique, then engaging in a one-month post-prandial race straight across the Indian Ocean.

Histoire naturelle des poissons Magisterial review of most of the knowledge on fish up to the early nineteenth century, the "golden age of ichthyology", authored by Georges Cuvier and Achille Valenciennes, published in 22 volumes from 1828 to 1850, and covering 4055 species, of which 2311 were new to science (Bauchot *et al*. 1990).

In spite of its heft (11 253 pages, 650 plates), the *Histoire naturelle des poissons* remained incomplete, for reasons not established (Monod 1963). It nevertheless represents a major achievement, comparable in scope and in terms of the international collaboration it required (documented in Bauchot *et al.* 1990) to the effort that has gone in the past decade into mapping the human genome. Indeed, the painstaking documentation of the world's biodiversity, as exemplified by the work of taxonomists such as Cuvier and Valenciennes, may be viewed as 'mapping the Earth genome' – a task still incomplete, though we now can use powerful relational databases to assist us (*see* **FishBase**).

Holocentrus See **Squirrelfishes**.

Holotype The *Code* defines a holotype as "the single specimen selected by the original describer of a species to be the standard-bearer for the new name".

Many of the fish specimens collected by CD became holotypes, and most of these are listed in Appendixes II and III.

Homing CD here made two fundamental observations concerning homing. The first was that [t]he *migration of young birds across broad tracts of the sea, & the migration of young salmon from fresh into salt-water, & the return of both to their birth-places, have often been justly advanced as surprising instincts (*Big Species Book*, p. 490; *see* **Diadromy**).

The second observation was that the power in migratory animals of keeping their course is not unerring, as may be inferred from the numbers of lost Swallows often met with by ships in the Atlantic:[1] the migratory *salmon, also, often fails in returning to its own river, 'many Tweed salmon being caught in the Forth'. (*Big Species Book*, p. 493; n. 1 includes In regard to the Salmon, see Scropes Days of Salmon Fishing p. 47).

And indeed, some salmon end up spawning in the 'wrong' stream. However, the fraction of such erring salmon is low, in line with the low (but non-zero) probability that a given run (i.e. population), in a given year, loses access to its spawning ground (e.g. because of a landslide).

The concept of *obstinate nature proposed by Cury (1994) generalizes these ideas to other fishes and animal groups.

Homologous organs *See* **Analogous organs**.

Hox A class of genes regulating, during the embryonic growth of various animals, the sequential arrangement of body parts (including limbs) along their longitudinal axis.

Recently, to the amazement of many biologists, *hox* genes were found to occur in a number of phyla, ranging from arthropods to chordates (see Zimmer 1998). Besides supporting the notion of a common ancestry for all extant animals, this discovery also implied an underlying commonality of body construction among metazoans. Also, this discovery provides support to earlier naturalists, such as Anton *Dohrn, who argued that vertebrates can be derived from annelid-like ancestors, once they are flipped over dorso-ventrally.

Huxley, Thomas Henry English zoologist and evolutionist (1825–85), and a friend to CD. Desmond (1994), who studied Huxley's extensive publications and correspondence (5000 letters) in great detail, shows that many of his often quoted repartees and pronouncements were invented by hagiographers, and hence none of this material will be reproduced in this entry.

However, being concerned with fishes, we shall mention Huxley's service as Inspector of Fisheries, from 1881 to 1884, and refer the interested reader to Desmond's biography for Huxley misreading signs of overfishing (already then), dissecting *fishes* and being asked what this had to do with regulating *fisheries* (already then), and ending up fighting a bureaucracy devoted to maintaining itself, rather than the natural resources it was mandated to sustain (already then).

Hybrids The offspring of the union of two different species (*Origin VI*, p. 434), e.g. the *mule. The infertility (or low fertility) of hybrids is part of the modern definition of *species, though it had long been known to breeders. The term 'mule fish' was common in CD's time (see, for example, Knight 1828, referring to hybrid of Trout and Salmon).

Hybrids are relatively common among fishes, particularly between freshwater species (Schwartz 1972; *but see* **Surgeonfishes**). Thus, CD noted Octob. 25th . [. . .in Windsor Park . . .] Saw what was said to be hybrid between silver & *gold fish (*Notebook E*; p. 405).

Later CD also asks: Whether <Yar> knows whether Shaws hybrids between *Trout & *Salmon were fertile & whether homogeneous (*Notebook Q*, p. 511; here Yar is William *Yarrell).

Shaw's contributions of 1836, 1838 and 1840 do not mention hybrid of Trout and Salmon, but point out that 'parr' are the young of Salmon. Also, Shaw (1840, p. 558) compares the young of *Salmo salar* with those of Trout *Salmo trutta*.

The issue of mule fish came up once more with regard to a fragment kept in DAR 205.7 (letters), in which CD asks Are the statements on hybrid fishes in the Phil. Trans. For 1771 to be depended on? (*Correspondence*, early Sept. 1851; p. 58; Forster 1771, pp. 318–19).

Two notes are attached to this fragment. The first reads: Believes he has seen many mule fish in fish – between trout & salmon. – Mr Knight. Phil. Trans. 1828. p. 319. (Knight 1828). The second note mentions Walker (probably Walker 1803a and/or 1803b), and states: Scropes Day of Salmon Fishing p. 34 M^r. Shaw crossed Trout and Salmon (*Correspondence*, early Sept. 1851). It seems that CD confused the works of Shaw and Knight – or at least confused me.

Hybrid vigour The increase of size and performance (other than fertility) that often characterizes the offspring of parents belonging to different *species.

This topic is discussed by CD in chapter XVII of *Variations*, titled: On the good effect of crossing, and on the evil effect of close interbreeding.

Therein, on p. 112, among the evidence cited by CD, is a quote by Mr. Abraham Dee Bartlett, Superintendent of the London Zoological Garden, stating that Among all hybrids of vertebrated animals there is a marked increase in size (pers. comm. to Murie 1870, p. 40).

CD had carefully read Murie's rather long-winded paper, which, among other things, reaffirms the earlier observation (Murie 1868) that the growth of adult *Salmon (*Salmo salar*) raised entirely in fresh water is "retarded" relative to the growth of sea-going individuals, as also previously reported by Yarrell (1839). This observation was doubted by some ichthyologists, notably *Günther (cited in Murie 1868 and 1870). Günther believed that salmon cannot complete their life cycle entirely in freshwater (which they can), and that therefore, the fish in question had been hybrids of *S. salar* and another, smaller salmonid. We can thus see how Murie's argument is strengthened by Bartlett's quotation.

Hydrocyon hepsetus *See* **Characins**.
Hygrogonus *See* **Cichlidae**.

Ichthyology The scientific discipline devoted to the study of fish. Until recently, 25 000 extant species of fish were thought to exist. However, *FishBase has added numerous valid species to this estimate (see www.fishbase.org), and 5000 to 10 000 new species may be described in the next decades.

This book presents some 320 species of fish, most of them Darwin's Fishes as defined earlier. Hence, it may be argued that this book covers about 1% of ichthyology. This is puny compared with Eschmeyer (1998), the first since Linnaeus (1758) to review all fish species known at his time (though *Günther came close). Other comprehensive accounts of ichthyology are Cuvier (1828/1995); Nelson (1994), and Helfman *et al.* (1997).

The word 'Ichthyology' was also used by CD to mean 'ichthyofauna', referring to the fish species occurring in a given area. Thus, in *Notebook C* (p. 295), CD noted the view of Geoffroy-Saint Hilaire and Blainville (1834, p. 95) that the Icthiology of S. America. more peculiar than its ornithology X p. 12 do. excepting *salmons.

And here, finally, is CD's challenge to ichthyologists: it is useless to speculate <<not only>> about beginning of animal life.: generally, but even about great division, our <only> question is not, how there come to be fishes & quadrupeds, but how the come to be, many genera of fish &c &c at present day (*Notebook C*, p. 257).

Ichthyosaurs Extinct reptiles formerly occupying the niche now filled by dolphins and other small and medium-sized marine *mammals (Motani 2000), with some having very large *eyes, indicative of deep diving (*see* **Lumpfish**, and Motani *et al.* 1999).

CD notes on these that: according to Professor *Owen,[22] the Ichthyosaurians – great sea-*lizards furnished with paddles – present many affinities with fishes, or rather, according to *Huxley, with *amphibians; a class which, including in its highest division *frogs and toads, is plainly allied to the *Ganoid fishes. (*Descent I*, pp. 158–9; n. 22 cites Owen (1860), p. 199, but fish affinities are discussed on p. 200).

Iguana *See* **Lizards**.

Iluocoetes fimbriatus *See* **Eelpouts**.

Introduction The results of a transfer, e.g. of fish, from one country (or ecosystem) to another (Lever 1996; Welcomme 1988). Introductions have expanded the original distribution *ranges of many fish species, notably of *salmon and *sunfishes.

CD did not write much about fish introductions. The few instances pertain to *Wels catfish, and – implicitly – to the different impacts of fish introductions in *Mauritius and North America. Most introductions have adverse impacts, and quantitative information on this, and a list of villainous species, are available from *FishBase.

Iquique City in northern Chile, described by CD in his *Journal* (entry of July 12, 1835). The city (then of about a thousand inhabitants) and its surroundings were, in 1836, still part of *Peru, but were annexed by Chile in the aftermath of the War of the Pacific (1870–1; Nyrop 1981).

The type specimen of *Engraulis ringens* (Jenyns, 1842), the *Peruvian anchoveta, was collected by CD in or near the harbour of Iquique, in mid-July, 1836 and detailed in *Fish* (pp. 136–7).

The War of the Pacific was fought over natural resources, including guano, the accumulated droppings of millions of seabirds feeding on anchoveta, the most abundant fish in the Humboldt Current system (*see* **Punctuated equilibrium**).

Iquique now is a large, busy city, and a major landing port for the purse seine fisheries off Northern Chile, where millions of tonnes of Peruvian anchoveta are caught, all reduced to fishmeal, fed to pigs and farmed *salmon (as is *Capelin), thus precluding another war over guano. Why? Because the bird populations along the Peru/Chile coast, unable to compete with the industrial fishery, crashed in 1965, seven years before the anchoveta did (details in Pauly and Tsukayama 1987), and never recovered (while the fishery did recover, at least partly). Thus, so little guano is now produced that nobody would want to go to war over it. But then, you never know.

Islands The peculiar *fauna and flora CD encountered on the nearly 40 islands he visited during the voyage of the *Beagle* were crucial to his eventual rejection of *creationism, and his development of an alternative theory.

This is nicely illustrated in the following quote, which also explains why fishes, whose distributions were then not well known, did not contribute much, if at all, to CD's crucial insight: With respect to whether *Galapagos beings are *species, it should be remembered that Naturalists are prone, fortunately, to take their ideas, which are arbitrary & empirical, from their own Fauna, which in this case is only true criterion. – Hence is highly unphilosophical to assert, that they are not species, until their breeding together has been tried. [...]

From these views we can deduce why small islands should possess many peculiar species. for as long as physical change is present with respect to new arrivers, the small body of species would far more easily be changed. – Hence the Galapagos Isld^s are explained. On distinct Creation, how anomalous, that the smallest newest & most wretched isl^d should possess specics to themselves. – Probably no case in world like Galapagos. no hurricanes. – isl^{ds} never joined, nature & climate very different from adjoining coast. Admirable explanation is thus offered. – From these views, one should infer that Mollusca would offer few species, or rather be slowly changed & vertebrata much so. – so far true, but do not fish offer a most striking anomaly to this. Have they wide *ranges? Agassiz has shewn that they most widely differ (*Notebook J*, p. 640; Agassiz 1833–44).

Jacks A word with several meanings, of which three are relevant to this book:

(1) Diminutive male salmon, often staying in freshwater or within coastal waters, which return to the spawning grounds earlier than the full-sized males, and compete with them by 'sneaking' onto the spawning females (see Groot and Margolis (1991) for details on several species of Pacific salmon). This is the fish CD probably had in mind when he wrote: animals, belonging to quite distinct classes, are either habitually or occasionally capable of breeding before they have fully acquired their adult characters. This is the case with the young males of the salmon. (*Descent* II, p. 485, n. 39);

(2) Another common name for *Esox reticulatus*, the North American *pike, or Chain pickerel (Scott and Crossman 1973, pp. 363, 374);

(3) Fishes of the Family Carangidae.

The following carangid jacks (or scads) were collected by CD: *Parona signata* (Jenyns, 1841), from Northern *Patagonia, originally *Paropsis signata* (*Fish*, pp. 66–8; *Fish in Spirits* no. 449; Fig. 18A); Bigeye scad *Caranx torvus*, now *Selar crumenophthalmus* (Bloch, 1793), from *Tahiti (*Fish in Spirits*, no. 1321; *Fish*, pp. 69–71; Fig. 18B); *Caranx georgianus* (*Fish*, pp. 71–2), now *Pseudocaranx dentex* (Bloch & Schneider, 1801), from *King George's Sound, Australia; and *Rainbow runner *Seriola bipinnulata* from the Cocos Islands (*Fish*, pp. 72–3), now *Elagatis bipinnulata* (Quoy & Gaimard, 1825).

The horse mackerels of the genus *Trachurus* also belong to the Carangidae. They are here represented by the Greenback horse mackerel *Trachurus declivis* (Jenyns, 1841), based on a specimen also collected at King George's Sound and originally assigned to *Caranx* (*Fish in Spirits*, no. 1374; *Fish*, pp. 68–69; Fig. 18C). According to Linholm and Maxwell (1988), distinct pop-

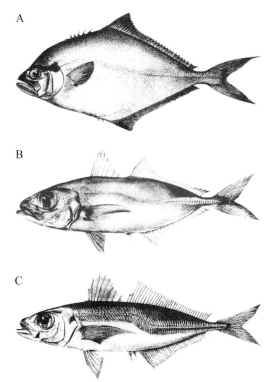

Fig. 18. Jacks, Family Carangidae. **A** *Parona signata* (*Paropsis signata* in *Fish*, Plate XIII), based on a specimen from Bahia Blanca, Argentina. **B** *Selar crumenophthalmus* (*Caranx torvus* in *Fish*, Plate XV) from Tahiti. **C** *Trachurus declivis* (*Caranx declivis*, Plate XIV), from King George's Sound, Australia.

ulations (or 'stocks' in the parlance of fisheries scientists) of *Trachurus declivis*, also known as 'jack mackerel' support important fisheries in southern Australia.

Jaguar A carnivore whose peculiar habits CD used to draw inferences on the evolution of *whales.

But first, let's see what CD says about the *ecology of jaguars: The wooded banks of the great rivers appear to be the favourite haunt of the jaguar; but south of the Plata, I was told, they frequented the reeds bordering lakes: wherever they are, they seem to require water. [...] their common prey is the carpincho, so that it is generally said, where the carpinchos are plentiful there is little danger of the jaguar. Falconer states, that near the mouth of the Plata, on the southern side, the jaguars are numerous, and that they chiefly live on fish; this account I have heard repeated. (*Journal*, October 12, 1833; 'carpinchos' refers to the large rodent *Hydrochoeris *capybara*; for 'Falconer' see Falkner (1774)).

Here is how CD envisaged the transition from a land-based to a fully aquatic carnivore: All the discussion about affinity & how one order first becomes developed & then another – (according as parent types are present) must follow after there is proof of the non creation of animals. – then argumen May be, – subterranean lakes, hot springs &tc &tc inhabited therefore mud wood be inhabited, then how is this effected by – for instance, fish being excessively abundant & tempting the Jaguar to use its feet much in swimming, & every development giving greater vigour to the parent so tending to produce effect on offspring – but WHOLE *race of that species must take to that particular habitat. – All structures either direct effect of habit, or heredetary (*Notebook C*, pp. 258–9; CD defines habitat as the locality where a plant or animal usually lives; *Origin* VI, p. 434).

For it has been maintained by several authors that one species, for instance of the carnivorous order, could not pass into another, for instance into an *otter, because in its transitional states its habits would not be adapted to any proper conditions of life; but the jaguar is a thoroughly terrestrial quadruped in its structure, yet it takes freely to the water and catches many fish; will it be said that it is *impossible* that the conditions of its country might become such that the jaguar should be driven to feed more on fish than they do now; and in that case is it impossible, is it not probable, that any the slightest deviation in its instincts, its form of body, in the width of its feet, and in the extension of the skin (which already unites the base of its toes) would give such individuals a better *chance* of surviving and propagating young with similar, barely perceptible (though thoroughly exercised), deviations? (*Foundations*, p. 132).

Later, in the first edition of *Origin*, the jaguar became a much criticized bear. In the sixth edition of *Origin* the bear had disappeared, although CD's jaguar/bear story still provides a plausible scenario for the evolution of *whales (Zimmer 1998), whatever their actual ancestry (Milinkovich *et al*. 1993; Wong 2002; Thewissen *et al*. 2002). Notably, the vertical oscillations of whale flukes could be interpreted as vestiges of the body undulations typical of most running carnivorous mammals, in contrast to *ichthyosaurs, whose caudal fins moved horizontally, just as did the bodies of the walking reptiles that were their ancestors.

Jenyns, Leonard (I) English naturalist and clergyman (1800–93), author of *Fish, i.e. volume IV of the *Zoology of the Beagle*.

CD's input into *Fish* was substantial: he sampled all the fish reported upon by Jenyns, who also had access to all of CD's field notes (notably on sampling sites, and on live *colours, based on Syme 1821; *see also* **Fish in Spirits of Wine**). Further, it is CD who 'superintendented' (i.e. guided, directed, edited) the publication of *Fish*, as amply documented in his *Correspondence.

Still, it is Jenyns who identified and/or named CD's specimens, and we shall briefly meet the man before discussing his work. Leonard Jenyns was born in London in 1800, the son of George Leonard Jenyns, vicar of Swaffham Prior, Cambridgeshire. In 1828, he became vicar of Swaffham Bulbeck, also in Cambridgeshire. By the mid 1830s, he had published works (Jenyns 1835, 1837) that had established his reputation as a naturalist, one of two reasons why CD invited him to document the *collection of fish he had amassed during the voyage of the *Beagle* (Jenyns 1840–42; Armstrong 2000).

The second reason was friendship: Jenyns' sister Harriet (1797–1857) had married, in

1823, J. S. *Henslow, CD's mentor and friend, and this had provided, long before CD went on the *Beagle, numerous opportunities for CD and Jenyns to meet and to gradually appreciate each other.

However, CD's relationship with Jenyns' was strained at first, CD finding him selfish & illiberal (*Correspondence*, Feb 26, 1829) – apparently because he had refused to exchange some of his specimens with the youthful Darwin, then engrossed in collecting insects. CD then competed with Jenyns: I think I beat Jenyns at *Colymbetes (May 18), and: I am glad of it if it is merely to spite Mr. Jenyns (July 15, 1829).

Gradually, things settled, and CD could report to his cousin William Darwin Fox: I have seen lots of him lately, & the more I see the more I like him (*Correspondence*, July 9, 1831). A more sedate account of the growth of CD's relationship with Jenyns may be found in *Autobiography* (pp. 66–7).

Upon receiving an inheritance, Jenyns changed his name to Blomefield, and, in 1849, moved to the Isle of Wright, then, shortly thereafter to Swainswick, near Bath, where he founded the Bath Natural History Society. In 1887, he published an autobiography (later reprinted, see Jenyns 1889), and passed away in 1893.

The specimens described in Jenyns' *Fish* represent the bulk of *Darwin's Fishes, and include many new species. The non-taxonomists Barrett and Freeman (1989, Vol. 6, p. 5), without reference to the standards of Jenyns' time (Pancaldi 1991; Winsor 1991), described him in their Foreword to *Fish* as "an excellent field naturalist" who, however, found "it difficult to know where to draw the line between different species, particularly where small differences were concerned [. . .] Darwin himself might have regarded such minor deviations as intra-specific, and relegated such to variations, or as we say today, *subspecies or *races. Such differences were accentuated by the distortions and colour changes due to preservation in spirits. They were also difficult to assess because of the small number of specimens of each form available to study."

However, as established through *FishBase, few (less than 20%) of Jenyns' new species are now considered synonyms of previously described species, although most are assigned to different *genera (see various entries in this book). This confirms Wheeler (1973), who describes Jenyns as "a methodical worker who sets out his information in well-ordered manner".

Jenyns appears to have committed only two big blunders in his scientific life. The second is not to have grasped the implications of CD's theory of *natural selection, which he thought could only explain transitions among related groups, but not bridge the chasms separating phyla, and humans from "brutes".

His first blunder, of course, was that he turned down the offer he had received – before CD – to accompany *FitzRoy on his planned voyage with the *Beagle* (Browne 1995, pp. 151–2).

Jenyns, Leonard (II) The writing of *Fish required, as did the other parts of the *Zoology of the Beagle*, numerous interventions from CD, its editor.

These interventions are well documented in Vol. 2 of CD's *Correspondence*, and abbreviated versions of two typical letters to Jenyns are presented here. The first includes: Henslow tells me he hears a groan occasionally escape from you, when you mention my fishes. [. . . .] I would sooner all the fish had rotted, than Mr Gray[1] described them & with this exception, if you had not taken pity on them, they most surely would have remained for many a long year unlooked at and unnamed. – I understand that you have made your mind that some of the fish are undescribed. – If D'Orbigny's Voyage[2] is not in Public Library[3], you ought to urge Mr Lodge[4] to get it; as it is most important work & will certainly contain some of my fish, at least some from the Patagonian coast. The Coquilles' Voyage[5] possibly may contain some from the Falkland Isl[ds]. – Henslow will deliver you a portfolio of coloured drawings of fish, made for Capt. FitzRoy by the artists on Board of the Beagle[6].

The locality of each is mentioned, it is just possible that where only one of a marked genus occurs at a place you might be able to identify it, in which case we might have a coloured drawings. –

I will have any or all the fish you think new lithographed. – You need not be in any hurry about the fish (without you want to finish it & say you have done all you can do) as the fish will be published last of the vertebrate animals (*Correspondence*, Dec. 3, 1837, with notes adapted from Burkhardt *et al.* (1986), p. 62: n. 1 refers to the then Assistant Zoology Keeper at the British Museum, a procrastinator; n. 2: see Orbigny (1834–47); 3: referring to the library of Cambridge University, not one of its constituent colleges; n. 4: Cambridge University Librarian; n. 5: Lesson (1830–31); n. 6: drawings have not been located).

The second letter includes: I really am sincerely grieved to hear the fish give you so much trouble. – I beg you to remember, that I should be very sorry that you should give up time to them, which could be better otherwise be spent. – Whatever you may choose to do I shall look at it as a clear gain, for otherwise all the specimens would in all probability have been entirely useless & my trouble in collecting them quite lost. [...].

For the credit of English zoologists, do not despair and give up; for if you do, then will it be said that there was not a person in Great Britain with knowledge sufficient to describe any specimen which may be brought here. [...].

You ask me how many species of fish would be published in each number; there would probably be from 12 to 16.- The numbers come out on alternate months, but as I have said, invertebrate animals might alternate with the fish if more than one number is produced (*Correspondence*, Dec. 4, 1837).

Fish ended up being published in four parts (Sherborn 1897), but the invertebrates were not included in the **Zoology of the Beagle*. More interesting here is the cajoling to which CD resorted to keep his authors working, and which any scientist who has edited a multi-authored work will easily recognize.

Here the cajoling includes appeals to professional standards (English zoologists) and national pride (not a person in Great Britain. . . .). The latter undoubtedly refers to the then dominant position of French naturalists, particularly in ichthyology.

Jenyns' questions As illustrated by the previous entry, the preparation of *Fish* involved an extensive correspondence between CD and Jenyns, some of which was devoted to specific questions by Jenyns on the specimens supplied by CD.

Here are answers CD provided: I will now answer your questions as far lies in my power. – (1) I can offer no opinion on probable identity of fish in Rio-Negro & Santa Cruz as I know nothing of the *ranges of fish, but there is a great space on Patagonian coast without any rivers. – 2. I found the fish (947) at Santa Cruz high up lying dead on the bank, & and I also found the little fish 952 in numerous streamlets entering the rivers high up. 3. I know no such isl[d] as Goree near Porto Praya; there is Guritti near Maldonado in La Plata; by reference to the number I shall be able to tell certainly. 4. The Bank Aiguille I am almost certain is the same with Lagulhas off the *C[ape] of Good Hope. 5. Fish 354 came from Lat 37° 26′ S. on East coast of Patagonia (south of the Plata). 6. Oualan is one of the Caroline Archipel in the North Pacific. (7) Fish (1331) came from the River of Matavai in Otaheite. 8. I might easily have confounded ventral and anal fin (*Correspondence*, Oct. 14, 1839; *Fish in Spirits*, nos. 354, 947, 952, 1331).

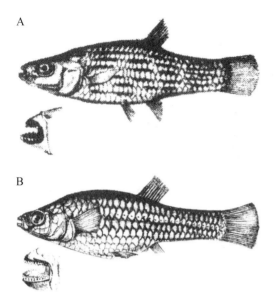

Fig. 19. One-sided livebearer *Jenynsia lineata*, Subfamily Jenynsiinae, Family Anablepidae, based on specimens collected by CD in Uruguay (*Fish*, Plate XXII). **A** Specimen originally assigned to *Lebias lineata*, from Maldonado, with insert showing teeth. **B** Specimen originally assigned to *Lebias multidentata*, from freshwaters near Montevideo, with insert showing teeth.

Comments may be added to a few of these answers: (2) the fish in question is the *Creole perch; (3) CD later wrote on this: St. Jago in the Cape de Verds. is about 360 miles from Goree, on the coast of Africa; hence any species very likely would be identical from the 2 places (*Correspondence* to Jenyns, Oct. 17, 1839); (4) 'Bank Aiguille' ('aiguille' is French for needle, similar to the Portuguese 'agulha') does indeed correspond to 'Agulhas Bank,' east of the *Cape of Good Hope, and a place at which, in the years near 1500, as noted by Portuguese navigators, the magnetic North (to which a compass needle points) coincided with the geographic North; (7) this refers to the *grouper *Dules leuciscus*, from *Tahiti.

Jenynsiinae A subfamily of the Anablepidae, allied to the killifishes, wherein the anal fin is turned into tube-like gonopodium, used by the males to fertilize the females.

In the Jenynsiinae, gonopodium and female genital aperture are either dextral or sinistral, the males and the (viviparous) females being equally divided into sinistral and dextral mating types (Wischnath 1993). Such *asymmetry is rather rare among bilateral animals, though male *Homo sapiens,* when having trousers made in upmarket shops, will be asked if they 'carry' (or 'dress') right or left.

The Jenynsiinae are represented here by the Onesided livebearer *Jenynsia lineata* (Jenyns, 1842), sampled by CD in fresh water near Montevideo, and in Maldonado, Uruguay. CD's specimens, which differed slightly from each other, were originally described as two species, *Lebias lineata* and *L. multidentata* (*Fish*, pp. 116–18; *Fish in Spirits*, no. 470; Fig. 19).

Journal Short title of *Journal of Researches into the Geology and Natural History of the Various Countries visited by H.M.S. Beagle, under the Command of Captain FitzRoy, R.N. from 1832 to 1836* (Darwin 1839). The *Journal* contributed many entries to this book.

Jurassic A geological period of the Mesozoic era, lasting from 180 to 135 million years ago, between the *Triassic and the *Cretaceous. Mentioned by CD when discussing the origins of *bony fishes.

Keeling Islands *See* **Cocos Islands**.

Kelp Brown *algae, forming, in some coastal areas, dense beds resembling underwater forests.

CD's description of kelp beds reads as follows: There is one marine *production, which from its importance is worthy of a particular history. It is the kelp or *Fucus giganteus* of Solander. This plant grows on every rock from low-water mark to a great depth, both on the outer coast and within the channels. I believe, during the voyages of the *Adventure* and *Beagle*, not one rock near the surface was discovered, which was not buoyed by this floating weed. The good service it thus affords to vessels navigating near this stormy land is evident; and it certainly has saved many a one from being wrecked.

I know few things more surprising than to see this plant growing and flourishing amidst those great breakers of the western ocean, which no mass of rock, let it be ever so hard, can long resist. The stem is round, slimy, and smooth, and seldom has a diameter of so much as an inch. A few taken together are sufficiently strong to support the weight of the large loose stones to which in the inland channel they grow attached; and some of these stones are so heavy, that when drawn to the surface they can scarcely be lifted into a boat by one person.

Captain Cook, in his second voyage, says, that at Kerguelen Land 'some of this weed is of a most enormous length, though the stem is not much thicker than a man's thumb. I have mentioned, that on some of the *shoals upon which it grows, we did not strike ground with a line of twenty-four fathoms. The depth of water, therefore, must have been greater. And as this weed does not grow in a perpendicular direction, but makes a very acute angle with the bottom, and much of it afterwards spreads many fathoms on the surface of the sea, I am well warranted to say that some of it grows to the length of sixty fathoms and upwards.'

Certainly at the Falkland Islands, and about *Tierra del Fuego, extensive beds frequently spring up from ten and fifteen fathom water. I do not suppose the stem of any other plant attains so great a length as 360 feet, as stated by Captain Cook. Its geographical *range is very considerable; it is found from the extreme southern islets near Cape Horn, as far north, on the eastern coast (according to information given me by Mr Stokes), as lat 43°, and on the western it was tolerably abundant, but far from luxuriant, at *Chiloe, in lat. 42°. It may possibly extend a little further northward, but is soon succeeded by a different species. We thus have a range of fifteen degrees in latitude; and as Cook, who must have been well acquainted with the species, found it at Kerguelen Land, no less than 140° in longitude.

The number of living creatures of all orders, whose existence intimately depends on the kelp, is wonderful. A great volume might be written, describing the inhabitants of one of these beds of sea-weed. Almost every leaf, excepting those that float on the surface, is so thickly incrusted with corallines, as to be of a white colour. We find exquisitely-delicate structures, some inhabited by simple hydra-like polypi, others by more organized kinds, and beautiful compound *Ascidiae.[48] On the flat surfaces of the leaves various patelliform shells, Trochi, uncovered molluscs, and some bivalves are attached. Innumerable crustacea frequent every part of the plant. On shaking the great entangled roots, a pile of small fish, shells, cuttle-fish, crabs of all orders, sea-eggs, star-fish, beautiful Holuthuriae (some taking the external form of the nudibranch molluscs), Planariae, and crawling nereidous animals of a multitude of forms, all fall out together. Often as I recurred to a branch of the kelp, I never failed to discover animals of new and curious structures.

In Chiloe, where, as I have said, the kelp did not thrive very well, the numerous shells, corallines, and crustacea were absent; but there yet remained a few of the flustraceae, and some compound Ascidiae; the latter, however, were of different species from those in Tierra del Fuego. We here see the fucus possessing a wider range than the animals which use it as an abode.

I can only compare these great aquatic forests of the southern hemisphere with the terrestrial ones in the intertropical regions. Yet if the latter should be destroyed in any country, I do not believe nearly so many species of animals would perish, as, under similar circumstances, would happen with the kelp. Amidst the leaves of this plant numerous species of fish live, which nowhere else would find food or shelter; with their destruction the many *cormorants, divers, and other fishing *birds, the *otters, seals, and porpoises, would soon perish also (*Journal*, June 1, 1834; n. 48 reads: I have reason to believe that many of these animals are exclusively confined to this station.).

What CD describes here is how kelp create and support a coastal ecosystem. Perhaps as a surprise to some, its enormous biomass does not make kelp a '*keystone species'. Rather this role is taken by the (low-biomass) species that has the largest impact on the biomass of kelp.

Note that CD almost correctly describes the range of the Fucus giganteus of Solander, i.e. the giant kelp *Macrocystis pyrifera* (Linnaeus, 1771, pp. 311–12), a member of the Laminariales (Lessoniaceae). This species, for which Gaudichaud (1826, p. 121) proposed the name *F. giganteus* (an illegal name according to the code of botanical nomenclature) before Harvey and Hooker (1845, p. 157) linked it with Daniel Solander (Captain Cook's *naturalist), is indeed widely distributed along rocky shores in the cold waters of the Southern Hemisphere. However, *M. pyrifera* also reaches into the northeastern Pacific, all the way to California, where it is used for alginate production. Thus, it extends far more than a little further northward of Chiloé (www.seaweed.ie/search/searchchoice.html).

The cast of characters is completed by Freeman (1978, p. 267), who informs us that John Lort Stokes (1811–85), later to become Admiral, was a mate and assistant surveyor during the second voyage of the *Beagle*.

Keystone species A living component of an ecosystem, whose indirect influence on the structure and biomass of that system is well above what would be expected, given its own biomass and food consumption (Power *et al.* 1996).

By this definition, *kelp are not keystone species in the kelp-dominated systems described by CD. Rather, their keystone species tend to be sea otters and/or fishes which, by consuming sea urchins, prevent these from overgrazing the kelp and thus replace richly structured ecosystems by 'urchin barrens' (Paine 1980; NRC 1999).

Power *et al.* (1996) give numerous examples from freshwater, coral reefs and shelf systems wherein fishes act as keystone species.

King, Philip Gidley Midshipman on the voyage of the *Beagle*; he drew the flatfish later described as *Hippoglossus kingii* in *Fish* (pp. 138–9). He also was the son of Philip Parker King, the Captain of the *Beagle* on its first voyage to South America, from 1826 to 1830 (see King *et al.* 1836), the source of much confusion.

King George's Sound The *Beagle* was anchored in this semi-enclosed water body in southwestern Australia from March 6 to 14, 1836. There, CD collected a number of fishes, many of them caught by net in Princess Royal Harbour, i.e. in the inner part of the Sound, about 35°, 03′ S; 117° 53′E (Armstrong 1985; Nicholas and Nicholas 1989; *Fish in Spirits* nos. 1365–94).

FitzRoy (1839, p. 628) confirmed the use of a net, and noted that "twenty different kinds" of fish were caught. However, *Fish* includes only ten species with King George's Sound as sampling location: *Aleuteres maculosus*;

Aleuteres velutinus; *Apistus –?*; *Arripis georgianus*; *Caranx declivis*; *Caranx georgianus*; *Dajaus diemensis*; *Helotes octolineatus*; *Platycephalus inops*; and *Platessa orbignyana*. Another species, the globefish *Diodon nicthemerus*, was probably also sampled at King George's Sound, although the evidence for this is indirect (*see* **Burrfishes**).

CD did not think much of the colonists around King George's Sound, writing that the inhabitants live on salted meat & of course have no fresh meat or vegetables to sell, they do not even take the trouble to catch the fish with which the bay abounds: indeed I cannot make out what they are or intend doing. (*Diary*, March 6, 1836).

Kingfishers Fish-eating birds, whose feeding behaviour was described by CD: Mr Bartlett informs me that the flamingo and the kagu (*Rhinochetus jubatus*) when anxious to be fed, beat the ground with their feet in the same odd manner. So again kingfishers, when they catch a fish, always beat it until it is killed; and in the Zoological Gardens they always beat the raw meat, with which they are sometimes fed, before devouring it. (*Expression*, pp. 49–50).

Note that the South American Kingfisher *Ceryle (Alcedo) americana* does not sit in the stiff & upright manner, of the Europaean species; when seated on a twig perpetually elevates & depresses its tail. – Note low, like the clicking of two small stones. Is said to build in trees: In stomach fish, internal coating of that organ bright orange. (*Ornithological Notes*, Barlow 1963, p. 215).

CD also studied another kingfisher species, *Ceryle (Alcedo) torquata*: Alcido. female. This bird is abundant in T. del Fuego, in la Plata (Brazil?) & Southern Chili. This specimen came from the island of *Chiloe [. . .]. In the stomach of this bird was a Cancer brachyurus & small fish. CD's *Ornithological Notes* (pp. 249–50) are the source of a similar remark on p. 42 of *Birds*.

Koi Coloured forms of *Carp *Cyprinus carpio*, initially bred in Japan, where they form the basis of a sizeable pet industry (Balon 1995a, b; Barrie, 1992).

This includes regular *beauty contests, at the end of which the losers are eaten. CD does not mention Koi, though he discusses several domesticated forms of Carp. A pity: Koi are really beautiful.

***Labrus* spp.** *See* **Wrasses**.

Lagoon Small body of water, covering either the interior, shallow part of atolls, or a depression behind a coastline.

CD generally uses the word 'lagoon' to refer to the interior of an atoll, i.e. the shallow, rubble-filled depression on top of a sunken volcano's tip (*see Coral Reefs*). The animal life in such lagoons is quite exuberant, although strong rains can reduce the salinity of the water therein sufficiently to induce mass mortalities, as CD reports from the *Cocos Islands.

Coastal lagoons also experience vast changes in salinity, both due to occasional rains, and seasonally, whether the sand banks that usually separate them from the sea break open during the high waters or not (Bird 1994; Pauly 1975). However, contrary to coral reef fishes, most coastal marine fishes (and invertebrates) can withstand relatively low salinities (especially so in the tropics). Thus, coastal lagoons often support highly productive fisheries for marine fish (and shrimps), enhanced in many cases by regulating the exchanges of water and fishes in and out of the lagoons (Pauly 1976; Pauly and Yáñez-Arancibia 1994). Hence CD's *Zoology Notes* (p. 58): In my geological notes, I have mentioned the lagoons on the coast which contain either salt or freshwater. – The Logoa near the Botanic Garden is one of this class. – the water is not so salt as the sea, for only once in the year a passage is cut for the sake of the fish.

Lamarck, Jean-Baptiste The great French zoologist and evolutionist (1744–1829), whose major contribution was to get organismic *evolution *as a process* accepted by a wide range of European scientists – including CD, who praised his lofty genius. (*Notebook C*, p. 275) – and by part of the public (Humphreys 1995; *see also* **Spontaneous generation**).

Unfortunately, the evolutionary mechanism Lamarck proposed was incorrect, or as CD put it the Lamarckian doctrine of all modifications of structure being acquired through habits, & being then propagated, is false (*Big Species Book*, p. 365), and this view is reflected in the first edition of *Origin*.

Yet, despite these brave words CD reverted to Lamarckian notions, if after intense critique by colleagues, as reflected in subsequent editions of *Origin* (Vorzimmer 1969). This, if anything, indicates the historic importance of the Lamarckian doctrine.

Lampreys Elongated, fish-like chordates lacking bones and jaws, whose approximately 40 species are usually grouped into one Family, the Petromyzontidae, the sole representative of their Order, the Petromyzontiformes and Class, the Cephalaspidomorphi, itself sister group of the Myxini, or *hagfishes (Nelson 1994, pp. 30–3). All occur in cold waters of the Northern Hemisphere, and many are *parasites.

Two representative species are the (*anadromous) Sea lamprey *Petromyzon marinus* Linnaeus 1758, and the European river lamprey *Lampetra fluviatilis* (Linnaeus 1758); like all lampreys, they die after spawning.

CD wrote on lampreys only once, when he mentioned their ammocoete larvae as example of a primitive vertebrate form: Considering the Kingdom of nature as it now is, it would not be possible to simplify the organization of the different beings, (all fishes to the state of the *Ammocœtus, Crustacea to – ? &c) without reducing the number of living beings . . . (*Notebook E*, p. 423; *see also* CD's annotation to Quatrefages De Bréau, 1854).

Both the common and the scientific name of river lampreys (*Lampetra* spp.) have their origin in the Latin *lambere*, to lick, and *petra*, stone, referring to the habit, common in riverine species, of using their mouths to attach themselves to rocks.

Lancelet Common name of fish-like lower Chordates, earlier assigned to the catch-all genus '*Amphioxus*'.

Lancelets, even though perhaps not rightfully included among 'fishes', are quite interesting in themselves. Thus, CD marvels: every naturalist who has dissected some of the beings now ranked as very low in the *scale, must often have been struck with their really wondrous and beautiful organisation.

Nearly the same remarks are applicable, if we look to the great existing differences in the grades of organisation within almost every class excepting birds; for instance [. . .] amongst fish, of the *shark and *Amphioxus, which latter fish in the extreme simplicity of its structure closely approaches the invertebrate classes. But mammals and fish hardly come into competition with each other; the advancement of certain mammals or of the whole class to the highest grade of organisation would not lead to their taking the place of and thus exterminating fishes. Physiologists believe that the *brain must be bathed by warm blood to be highly active, and this requires aërial respiration; so that warm-blooded mammals, when inhabiting the water, live under some disadvantages compared with fishes. In this latter class members of the shark family would not, it is probable, tend to supplant the Amphioxus; the struggle for existence in the case of the Amphioxus must lie with members of the invertebrate classes. (Notes for US edition of *Origin*, in Burkhardt *et al.* (1993), pp. 578–9).

[T]he lancelet or amphioxus, is so different from all other fishes, that Häckel maintains that it ought to form a distinct class in the vertebrate kingdom. This fish is remarkable for its negative characters; it can hardly be said to possess a brain, vertebral column, or heart, etc.; so that it was classed by the older naturalists among the worms. Many years ago Professor Goodsir perceived that the lancelet presented some affinities with the *Ascidians, which are invertebrate, *hermaphrodite, marine creatures permanently attached to a support. (*Descent I*, p. 159; Goodsir 1844; see **Asymmetry; Dohrn; Haeckel**).

***Latilus* spp.** *See* **Sandperches**.

Latitude The angular distance, north or south of the equator, of a point on the Earth's surface.

CD, after reviewing much literature on the *distribution of organisms concluded that latitude is more important element than longitude (*Notebook C*, p. 316).

Of course, this does not mean that other factors do not play a role. Thus, in the following quote, we see how latitude provides the key to understanding the composition of certain *fauna, but not of others: Sir J. Richardson says the Fish of the cooler temperate parts of the S. Hemisphere present a much stronger analogy to the fish of the same latitudes in the North, than do the strictly Arctic forms to the Antarctic. (*Correspondence* to J. D. Dana, April 5, 1857; the statement attributed to Richardson may be based on Richardson (1846), p. 191, or on his letter to CD in *Correspondence*, July 15, 1856, or both).

We conclude that, in any case, latitude is so important that once their evolutionary affinities are sorted out, most of the remaining aspects of the biology of fishes can be inferred from their latitudinal distribution (Pauly 1994d, 1998), and similarly for coral reefs (Grigg 1982) and other organisms (Ekman 1967). Note that this statement is qualified: sorting out 'evolutionary affinities' includes locating ancestors in the framework of plate tectonics (Le Grand 1988; Pielou 1979), which provides, among other things, an explanation as to why Arctic fishes differ so much from Antarctic fishes.

***Lebias* spp.** *See* **Jenynsiinae**.

Length Here a measure of body 'size', itself the major physiological and ecological attribute of organisms (Goolish 1991; Peters 1983; Pauly 1997; Schmidt-Nielsen 1984).

Fish length is measured in different ways, depending on circumstances, fish taxa, and professional traditions. Fish taxonomists usually mean 'standard length' (SL) when they write 'length', as this length defines 'body length' proper. SL is measured from the tip of the snout to the end of the caudal peduncle, and thus can be obtained even when the *caudal fin of preserved specimens is broken (or in species that do not have caudal fins). On the other hand, fisheries scientists generally use fork length (FL, to the tip of the shortest, central caudal rays, if any) or total length (TL, measured to the tip of the extended caudal fin lobes, if these can be bent backward and aligned with the main body axis), both of which can be measured more quickly in the field. Need we stress that anglers prefer to use the latter length to report on their exploits?

The lengths in Jenyns' *Fish* are expressed in 'inches' (1 inch = 2.54 cm) and 'lines' (12 lines = 1 inch). Note also that the abbreviation 'unc.', used by Jenyns for his short Latin descriptions of new species, refers to 'uncia', a twelfth.

Lepidosiren See **Lungfishes; Lungfish, South American.**

Ling Common name of *Molva molva* (Linnaeus, 1758), a member of the Family Lotidae, Order Gadiformes (cod-like fishes), which reaches up to 200 cm and lives over rocky bottoms in deeper waters of the North Atlantic.

CD mentions this fish when discussing variations in the anatomy of plants and animals: Mr. Couch[6] has seen the common ling Gadus molva with two cirri on the throat & G. mustela with five barbs. (*Big Species Book*, p. 110; n. 6 refers to Couch (1825), p. 73). There are no problems with CD's first case: Ling usually sport one barbel or 'cirrus,' and hence two cirri are noteworthy. However, as couched by CD, the second case can easily lead to misunderstandings: the species in question, now called *Ciliata mustela* (Linnaeus, 1758), a small, shallow-water fish also belonging to the Family Lotidae, is also known as Fivebearded rockling, because it has one barbel on the lower jaw plus four on the snout (Muus and Dahlstrøm 1974).

Couch originally wrote on this: "the variety of this fish which possesses five barbs has been supposed to be a distinct species; but from attentive examination I am convinced that this is a mistake: both varieties are common, frequenting the same places, and having no other marks of difference. Both of them are subject to great varieties of colour, from a light yellow with brown spots, to an uniform reddish-brown. Nor is the number of cirri an objection to this supposition, as I have seen the common Ling with two cirri at the throat."

Thus, Fivebearded rockling appear to exist that have five barbs plus one barbel, 'Sixbearded rockling' as it were.

Linnaeus, Carolus A great Swedish physician and naturalist (1707–78), to whom a grateful sovereign granted nobility in 1761, hence his new name Carl von Linné, a neat mix of German and French bits (with the central bit pronounced 'fonn', not 'vonn'). More importantly, Linnaeus proposed hierarchical *categories* into which *taxa* could be fitted, as well as binomens for the scientific names of species (Linnaeus 1758), thus ploughing and seeding the field which CD later harvested (Cain 1957; Blunt 1971).

With regard to fishes, it must be mentioned, however, that Linnaeus took his classification and species from Artedi (1738), whom Günther (1880, pp. 9–11) and many others see as the founder of modern ichthyology.

Livebearers Members of the Subfamily Poeciliinae (Family Poeciliidae), whose males, which sport a *gonopodium derived from anal fin rays, internally fertilize the females, which give birth to live young.

This very speciose group, now considered distinct from the Cyprinidontidae (pupfishes), includes the *Guppy and the mollies (*Poecilia*) of aquarium fame, and *Gambusia*, widely disseminated for mosquito control (Gerberinch

Lizards

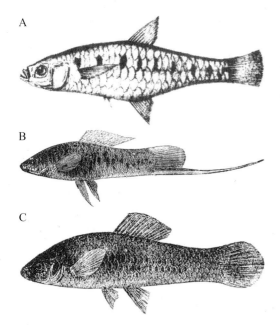

Fig. 20. Livebearers, Family Poeciliidae.
A *Cnesterodon decemmaculatus* (*Poecilia decem-maculata* in *Fish*, Plate XXII). Lithograph by B. Waterhouse Hawkins, based on a specimen from Maldonado, Uruguay. **B** Male *Xiphophorus hellerii*. **C** Female *Xiphophorus hellerii*. (B, C based on woodcuts by Mr. G. Ford, originally published in *Descent II*, p. 338.)

and Laird 1968; Welcomme 1988), but now, equally widely considered a pest (data in *FishBase).

During the voyage of the *Beagle*, CD sampled two species of livebearers. The first is *Poecilia unimaculata* Valenciennes, 1821, now *Poecilia vivipara* Bloch and Schneider 1801, sampled from a freshwater ditch in *Rio de Janeiro, where they occurred in great numbers (*Fish in Spirits*, no. 195), and from a salt *lagoon, in Rio as well, and also occurring in great numbers (*Fish in Spirits*, no. 210). The other species is the Ten-spotted livebearer *Cnesterodon decemmaculatus* (Jenyns, 1842), from *Maldonado, and described as *Poecilia decem-maculata* (*Fish*, pp. 114–16; *Fish in Spirits*, no. 669; Fig. 20A).

CD later wrote about two other species of livebearers: In the very distinct family of the Cyprinodontidae – inhabitants of the fresh waters of foreign lands – the sexes sometimes differ much in various characters. In the male of the *Mollienesia petenensis*,[16] the dorsal fin is greatly developed and is marked with a row of large, round, ocellated, bright-coloured spots; whilst the same fin in the female is smaller, of a different shape, and marked only with irregularly curved brown spots. In the male the basal margin of the anal fin is also a little produced and dark coloured.

In the male of an allied form, the *Xiphophorus Hellerii* (Fig. 20B), the inferior margin of the *caudal fin is developed into a long filament, which, as I hear from Dr *Günther, is striped with bright colours. This filament does not contain any muscles, and apparently cannot be of any direct use to the fish.

As in the case of the *Callionymus, the males whilst young resemble the adult females in colour and structure (Fig. 20C). Sexual differences such as these may be strictly compared with those which are so frequent with gallinaceous birds.[17] (*Descent II*, pp. 337–8; n. 16 reads: With respect to this and the following species I am indebted to Dr. Günther for information [from a letter to CD probably sent after Dec. 20, 1870; see Calendar no. 5734]; see also his paper on the 'Fishes of Central America,' in Transact. Zoolog. Soc., vol. vi, 1868, p. 485; n. 17 refers to Günther 1861, *Catalogue of Fishes in the British Museum*, vol. 3, p. 141, in which one can find, in addition to the comparison to gallinaceous birds, two generalizations important for ichthyology, i.e. that "in almost all fishes, larger individuals [will not be found in the shallow] localities where smaller ones abound", and that "young fishes are [. . .] more active in their habits than old ones", with the latter generalization resulting from the interactions between body size and *oxygen consumption).

Lizards A reptilian suborder (Lacertilia) that includes the geckos, iguanas, chameleons, Komodo dragons, Gila monsters, and other groups of uncertain affinities (Parker *et al.* 1964, pp. 496–7). Two species of iguanas, one aquatic (*Amblyrhynchus cristatus* or *A. subcristatus*), the other terrestrial (*A. demarlii* or *A. cristatus*) are described by Bell (1843), based on specimens from the *Galápagos Islands.

CD discusses in some detail the behaviour of the former species, the marine iguana, and an abbreviated version of his account is presented here: This lizard is extremely common on all the islands throughout the Archipelago. It lives exclusively on the rocky sea-beaches, and is never found, at least I never saw one, even ten yards inshore. [. . .] These lizards were occasionally seen some hundred yards from the shore swimming about; and Captain Collnett, in his *Voyage*, says 'they go out to sea in *shoals to fish.' With respect to the object, I believe he is mistaken; [. . .] I opened the stomach of several, and in each case, found it largely distended with minced sea-weed, of that kind which grows in thin foliaceous expansions of a bright green or dull red colour. I do not recollect having observed this sea-weed in any quantity on the tidal rocks; and I have reason to believe it grows at the bottom of the sea, at some little distance from the coast. If such is the case, the object of these animals occasionally going out to sea is explained. [. . .]

The nature of this lizard's food, as well as the structure of its tail, and the certain fact of its having been seen voluntarily swimming out at sea, absolutely prove its aquatic habits; yet there is in this respect one strange anomaly; namely that when frightened it will not enter the water. [. . .] I several times caught [the] same lizard, by driving it down to a point, and though possessed of such perfect powers of diving and swimming, nothing would induce it to enter the water; and as often as I threw it in, it [either tried to conceal itself in the tufts of sea-weed, or it entered some crevice. As soon as it thought the danger was past, it crawled out on the dry rocks . . .]

Perhaps this singular piece of apparent stupidity may be accounted for by the circumstance, that this reptile has no enemy whatever on shore, whereas at sea it must often fall a prey to the numerous *sharks. Hence, probably urged by a fixed and hereditary instinct that the shore is its place of safety, whatever the emergency may be, it there takes refuge. (*Journal*, Oct. 3, 1835; and not only sharks: Jackson (1993), p. 234 reports marine iguana remains from the stomach of Giant hawkfish *Cirrhitus rivulatus* Valenciennes, 1846).

The Baconian nature of the experiment conducted here cannot be denied. Indeed, when CD opened the stomach of several of these lizards, and reported on the *algae therein, he followed the rules Bacon (1620) had proposed for science, then a 'new tool' (*novum organum*). Notably, with his cutting knife, CD followed the second of the "Aphorisms concerning the interpretation of Nature and the Kingdom of Man", namely "Neither the bare hand nor the understanding left to itself are of much use. It is by instruments and other aids that the work gets done, and these are needed as much by the understanding as by the hand . . .".

Loaches Small bottom-dwelling fishes of the Families Cobitidae and Balitoridae, occurring in rivers, ponds and shallow lakes, and attaining their greatest diversity in South Asia. CD first mentions loaches in the *Big Species Book*, and refers to the combined digestive and respiratory functions of their gut: in the Loach (*Cobitis*) the whole alimentary canal acts of course for its proper end, but likewise in aid of the lungs, 'as this fish swallows air & voids carbonic acid'. (p. 355; CD's quote is from Owen (1846); note the gut 'aiding' the lungs, rather than the gills!).

Then, CD points out that their *swim-bladder is used only for hearing: [i]ndeed, in some fish (Owen, Hunterian Lectures. Fish p. 210) as the *Cobitis barbatula*, the swim-bladder apparently subserves no other function. (*Big Species Book*, p. 356, n. 4).

Finally, loaches are mentioned in *Origin VI*: We should be extremely cautious in concluding that an organ could not have been formed by transitional gradations of some kind. Numerous cases could be given among the lower animals of the same organ performing at the same time wholly distinct functions; thus in the larva of the dragonfly and in the fish Cobites the alimentary canal respires, digests, and excretes. In the Hydra, the animal may be turned inside out, and the exterior surface will then digest and the stomach respire. In such cases *Natural selection might specialize, if any advantage were thus gained, the whole or part of an organ, which had previously performed two functions, for one function alone, and thus by insensible steps greatly change its nature (p. 147).

To which of these three species CD last refers is unclear, as they were all originally assigned to the genus *Cobitis* (Linnaeus, 1758). It may have been the Weatherfish *Misgurnus fossilis* Linnaeus, 1758, now endangered in much of Europe (Kouril *et al.* 1996), which lives in muddy habitats, and survives low dissolved *oxygen concentrations by swallowing air at the surface. This air is passed to the intestine (of which parts are modified for respiratory purposes), and the exhalant gas is then vented.

This provides another, if tenuous link with CD: as noted by Browne (1995, p. 487), when he complained about his stomach, he often meant a part of his digestive tract that the conventions of his time – and of ours as well – made difficult to mention. In fact it was quite a confession when he once mentioned to his cousin Fox that all excitement & fatigue brings on such dreadful flatulence that in fact I can go nowhere (*Correspondence*, Dec. 18, 1847).

And since we are dealing with revolting topics, we might also mention that another loach, the Japanese weatherfish *Misgurnus anguillicaudatus* (Cantor, 1842), is used for a dish obtained by adding the live fish to a pot of tofu and water. The pot is then heated, whereupon the fish, mistaking the tofu for mud, bury themselves into the tofu. The tofu–fish aggregate is then cooked and served, probably with soy sauce. This dish is so popular that Japanese projects encouraged the culture of *M. anguillicaudatus* in areas, and its export from countries, where it was originally unknown, notably the terraced rice fields of the Philippine Cordillera (Bocek 1982).

Lophobranchii Old name for a Superfamily (now called 'Syngnathoidea') of 'tuft-gilled' fishes, comprising the *pipefishes, *seahorses, and ghost *pipefishes, and mentioned by CD when discussing the *reproduction of seahorses.

Lumped (-er; -ing) In taxonomy, lumping is the act of aggregating several distinct taxa under the same name, and is done by lumpers.

Thus, grouping the *Lumpfish, a *teleost, with the *Chondrichthyes is a serious case of lumping, while grouping all the Atlantic and Pacific *salmon together with Trout into the Genus *Salmo*, as done in Linnaeus (1758), was less excessive.

The converse of lumping is 'splitting,' and is done by splitters. *Jenyns, who so well described the fish collected by CD during the voyage of the *Beagle*, succinctly described these two trends in taxonomy in a letter to CD: "As you are well aware, I am no stickler for the multitude of so-called species created by so many naturalists of late years, & I always thought the time was not distant – when, after the brain-splitting process has been carried to it's utmost length – some at least could see the necessity of retracing their steps, & again uniting a large number of the forms they had so carefully separated." (*Correspondence*, Jan. 4, 1860).

These splitting–lumping oscillations are still going back and forth: see, for example, Nelson (1994, p. 328) on the split of the *Lumpfish from the snailfishes.

Lumpfish Common name of *Cyclopterus lumpus* Linnaeus, 1758, a member of the odd-shaped Family Cyclopteridae, or lumpsuckers, whose names derive from their pelvic fins, modified into a sucking disc (Nelson 1994, p. 328).

The following quote, taken from CD's Early notebook (entry of March 16, 1827), on a dissection he conducted with *Grant, his zoology teacher, is the oldest bit of scientific prose CD appears to have penned. And sure enough, it contains the characteristic mix of theoretical and empirical considerations which we shall encounter again in the more mature scientist: Procured from the black rocks at Lieth a large Cyclopterus Lumpus (common lump fish). Length from snout to tail 23. inches, girth 19.. It had evidently come to the rocks to spawn & was left there stranded by the tide; its ovaria contained a great mass of spawn of a rose colour. Dissected it with Dr. *Grant. It appeared very free from disease & had no intestinal *worms: its back however was covered with small crustaceous animals.

Eyes small. Hence probably does not inhabit deep seas? Stomach large. Liver without gallbladder. Kidneys situated some way from the Vertebrae: an unusual fact in *cartilaginous Fishes. Air bladder was not seen. *Brain very small; the optic nerves being nearly as large as the spinal cord, neither the brain or spinal matter nearly filling its cavity. The valves in the heart were very distinct; the peduncle strong. The body was not covered with <skin> scales, but slimy & remarkably thick. The sucker on its breast was of a white colour. I believe it is generally a reddish yellow? The plebs differ whether it is edible. (*Collected Papers* II, 1827, p. 285; *see also* **Swimbladder**).

A number of issues are raised by this account, notably:

1. is the inference eye small hence does not inhabit deep seas valid?
2. are kidneys situated some way from Vertebrae really an unusual fact?
3. is *Cyclopterus lumpus* edible? And, perhaps, most interestingly:
4. how could young CD 'see' all this during a simple dissection?

That evolving large *eyes improve the ability to see in deep water is a generally valid inference, recently used, for example, to postulate deep diving in *Ophthalmosaurus*, an *ichthyosaur with particularly large eyeballs (Motani *et al*. 1999).

Concerning (2), the solution is quite simple: in CD's time, *C. lumpus*, though a *bony fish, was still classified, because of its thick cartilaginous skin, among the *cartilaginous fishes, together with the sharks and rays (see, for example, Kirby 1835, p. 390). These, however, differ from the bony fishes in their kidneys lying very close to, or even "on each side of the vertebral column" (Daniel 1922, p. 305). Hence CD's wonderment, from which as good a case as any can be derived for the utility of a sound *classification.

Item (3) is quickly answered: not only is *C. lumpus* edible, and supports growing fisheries, now that traditional target species have been depleted (Neis *et al*. 1997), but the plebs (again!) do consume its roe, i.e. CD's great mass of spawn of a rose colour, as substitute for the sturgeon-based *caviar* which the less plebeian pretend to love.

As for item (4), the answer requires an entry of its own, on *Seeing.

Lungfishes Members of the Dipnoi, or Dipneusti ('two-breathers'), a formerly diverse group with only three recent genera (two monospecific) in three families, and whose African and South American forms, at the end of the wet season, dig themselves in the mud, and spend the dry summer months in a slimy cocoon, safely hibernating, or rather aestivating (Johnels and Svensson 1954).

The recent lungfishes, initially perceived as amphibians (see, for example, the title of Krefft 1870, and CD's discussions of *Lepidosiren), consist of the Australian lungfish *Neoceratodus forsteri* (Ceratodontidae), the South American lungfish *Lepidosiren paradoxa* (Lepidosirenidae; literally 'scaly *mermaids'), and the *African lungfishes (*Protopterus*, four species, Family Protopteridae), all presented below. The last genus was previously *lumped with *Lepidosiren*, which caused great confusion, notably with the editors of CD's *Correspondence*, who write that "*Owen described a new species of the South American lungfish *Lepidosiren annectens* in 1841" (Burkhardt *et al.* 1991, p. 479).

However, when CD mentions *Lepidosiren*, he refers to the West African lungfish (*Protopterus annectens*) of which Owen (1839) provided a detailed account, wherein he also reviewed the earlier description of *Lepidosiren paradoxa* Fitzinger, 1837.

Lepidosiren annectens is often used by CD to represent lungfishes in general, and a transitional form within the lower *tetrapods: The Lepidosiren – *Amblyrhyncus & Toxodon, <all> equally aberrant – the two former connecting classes like Toxodon <<In orders>> – Fish & reptiles in former case – Reptiles & Birds & Mamm. in ornityhyrhycus – is not this right? – (*Notebook E*, p. 448); Look at the mud-fish (Lepidosiren annectens), which is so intermediate in structure, that although the greatest living authority considers it to be certainly a fish, many highly competent judges class it as a reptile: if then there be any truth in our theory, it would not be ridiculous to suppose that the Lepidosiren could be modified by natural selection into an ordinary fish, or into a reptile. (*Big Species Book*, p. 384); The Lepidosiren is also so closely allied to *amphibians and fishes, that naturalists long disputed in which of these two classes to rank it (*Descent I*, p. 159; *Toxodon* are extinct, giant relatives of the *Capybara, first discovered by CD in Patagonia and described by Owen 1840).

CD considered the low diversity of lungfishes to be an indication of their inability to compete with other forms, a theme which reappears frequently in his writing. Thus: The more aberrant any form is, the greater must be the number of connecting forms which have been exterminated and utterly lost. And we have some evidence of aberrant groups having suffered severely from extinction, for they are almost always represented by extremely few species; and such species as do occur are generally very distinct from each other, which again implies extinction. The genera Ornithorhynchus and Lepidosiren, for example, would not have been less aberrant had each been represented by a dozen species, instead of as at present by a single one, or by two or three. We can, I think account for this fact only by looking at aberrant groups as forms which have been conquered by more successful competitors, with a few members still preserved under unusually favourable conditions. (*Origin VI*, p. 378).

Moreover, such aberrant forms as lungfishes appear to have lost parts of organs more fully developed in their ancestors: Owen considers the simple filamentary limbs of the Lepidosiren as the 'beginning of organs which attain full functional development in higher vertebrates'; but according to the view lately advocated by Dr Günther, they are probably remnants, consisting of the persistent axis of a fin, with the lateral rays or branches aborted. (*Origin VI*, p. 399; Owen 1849; 'aborted' refers here to an organ whose development has been arrested at a very early stage. *Origin VI*, p. 430).

Lungfishes, African Members of the genus *Protopterus*, Family Protopteridae. The African lungfishes consist of four species, of which *Protopterus annectens* (Owen, 1839), formerly *Lepidosiren annectens*, became quite important to the development of CD's view of phylogeny.

Lungfishes were earlier considered to be *amphibians (Owen 1839) and, therefore, evolutionary links between fishes and reptiles.

Lungfish, Australian Common name of *Neoceratodus forsteri* (Krefft, 1870), sole survivor of the nine species of a genus that is over 100 million years old, and thus one of the oldest living vertebrate genera.

N. forsteri differs from South American and African lungfishes in having a single, instead of a double, *swimbladder in the adults, and lacking the ability to aestivate during dry spells; and in having pectoral and pelvic *fins shaped like paddles.

These paddle-like fins are interesting here, as the number of their digits differs from the five in archetypal vertebrates (Owen 1849). Thus, CD states in *Descent* (I, p. 36, n. 38), I also attributed, though with much hesitation, the frequent cases of *polydactylism in men and various animals to reversion. I was partly led to this through Professor *Owen's statement, that some of the Ichtyopterygia possess more than five digits, and therefore, as I supposed, had retained a primordial condition; but Professor Gegenbaur (Jenaischen Zeitschrift, vol. v, part 3, p. 341), disputes Owen's conclusion. On the other hand, according to the opinion lately advanced by Dr *Günther, on the paddle of Ceratodus, which is provided with articulated bony rays on both sides of a central chain of bones, there seems no great difficulty in admitting that six or more digits on one side, or on both sides, might reappear through reversion. (Günther 1871, pp. 531–5; Günther 1872a; Gegenbaur 1870a).

Fossils described since CD wrote this show that the earliest tetrapods (*Ichthyostega, Acanthostega*) had a variable number of digits (up to eight) on their limbs, and hence polydactylism in men and various animals can indeed be perceived as a reversion. We would nowadays explain this, with Zimmer (1998, pp. 57–85), through the working of *Hox* genes that are not switched off at the 'right' time.

Lungfish, South American Common name of *Lepidosiren paradoxa* Fitzinger, 1837, Family Lepidosirenidae, the only lungfish species in the New World.

Lyell, Charles English geologist (1795–1875), author of best-selling geological texts (Lyell 1830, 1838), whose main thesis is often reduced to the concept of 'uniformitarianism,' i.e. that rates of geological processes occurring *at present* can be used to infer the rates at which the same processes occurred *in the past*. As these rates (e.g. erosion, mountain uplift), both present and past, tend to be slow, Lyell provided for CD's *natural selection the 'deep time' that this theory required (see also Lewis 2000). On the other hand, this position could be (and was) understood as implying a minor role for catastrophes, invoked by, for example, Cuvier, to explain various features of the fossil record. However, we now know that major catastrophes, leading to the *extinction of major parts of the world fauna and flora, occurred (Raup 1986; Alvarez 1998) as well as the continuous processes emphasized by Lyell, just as we now know that light consists of *both* waves and particles.

CD had Volume 1 of Lyell's *Principles of Geology* with him on the *Beagle* (the copy was given to him by *FitzRoy, which is quite ironical in view of his hatred of the theory of evolution CD later developed), while Volume 2 reached him on October 26, 1832, these having an immense impact on the development of CD's views, both with regards to *coral reefs, and later *natural selection.

The quote below documents some of CD's jotting while reading Lyell (1838): p. 426 Sauroid fish in coal. true fish & not intermediate between fish & reptiles – yet osteology closely resembles reptiles [. . .] p. 428 Upper *Silurian, fishes oldest formation highly organized (*Notebook D*, p. 374).

Macropus According to Eschmeyer (1990, p. 233), an invalid emendation of the generic name *Macropodus*, referring to the *Gouramy and related species.

Macrourus See **Rattails**.

Madeira A group of five small islands off the coast of Morocco, belonging to Portugal, and named after 'wood', i.e. once abundant forests.

CD, based on Richardson (1846, pp. 189, 191), and referring to *submergence, compared its ichthyofauna with that of Japan: I may mention that I was myself much struck by finding a very rare genus of parasitic *cirripedes on crabs, from Madeira & Japan. Some of the fish, also, from Madeira, as I am informed by the Rev: R. B. Lowe represent those of Japan. (*Big Species Book*, p. 543; the personal communication implied here refers to work published as Lowe 1843).

Testing the species overlap between Madeira and Japan is straightforward: according to *FishBase (Nov. 2000), there are 490 species of marine fish in Madeira, and 3297 in Japan. Of these, 159 are shared between Madeira and Japan. Put differently, 33% of Madeiran species occur in Japan, and 4.5% of the Japanese species occur in Madeira. The Reverend Mr Lowe was right, though mainly because large pelagic fishes have extremely broad distributions.

We note, finally, that Madeira is the type locality of *Darwin's roughy *Gephyroberix darwinii* (Johnson, 1866), an *eponym.

Maldonado City about 75 miles east of *Montevideo, the capital of what is now Uruguay. Maldonado was then a most quiet, forlorn little town [...] there is scarcely a house, an enclosed piece of ground, or even a tree to give it an air of cheerfulness. (*Journal*, July 26, 1832; Falconer 1937, who reports on much improvement).

Maldonado Bay is near the mouth of the estuary of La Plata, but it is "bathed with pure sea water" (Parodiz 1981, p. 53). The fish sampled from May to July 1833 by CD in the area of Maldonado, and described in *Fish*, are, however, a mixture of freshwater, brackish and marine species: *Dules auriga* (see **Groupers (I)**); *Otolithus guatucupa* and *Corvina adusta* (see **Croakers**); *Atherina argentinensis* (see **Silversides**); *Chromis facetus* (see **Cichlidae (I)**); *Poecilia decem-maculata* (see **Livebearers**); *Lebias lineata* (see **Jenynsiinae**); *Tetragonopterus interruptus* and *Hydrocyon hepsetus* (see **Characins**); and *Diodon rivulatus* (see **Burrfishes**).

Mammals(-ia) A class of vertebrates, characterized by milk glands used to feed the young, and by hair, providing the insulation that helps them maintain their body temperature within a narrow range.

The 5000 species of mammal, apparently all derived from early Insectivora, are mainly terrestrial, but about 120 species have become secondarily adapted to live in aquatic environments, and are collectively called *marine mammals, even though a few species (notably dolphins and *sirenians) live in freshwater bodies, notably the lower courses of large rivers (Jefferson *et al.* 1993).

CD sugggested that the freshwater dolphins of the Family Platanistidae are an an extremely isolated and intermediate, very small family. Hence to us they are clearly remnants of a large group; and I cannot doubt that we have here a good instance of precisely like that of the *ganoid fishes, of a large ancient marine group, preserved in fresh – water, where there has been less competition, and consequently little modification (Darwin and Seward 1903, Vol. II, p. 8; *see also* **Extinctions**).

The adaptations to aquatic life are most elaborate in the cetaceans, or *whales, and in the *sirenians, which cannot survive on land. Still, CD believed they were maladapted to aquatic life, at least when compared to fishes: Physiologists believe that the *brain must be bathed by warm blood to be highly active, and this requires aërial respiration; so that warm-blooded mammals, when inhabiting the water, live under some disadvantages compared with fishes. (Burkhardt *et al.* 1993, p. 579; *see* **Complexity** for a refutation).

CD thought extensively about the origins of mammals: My theory drives me to say that there can be no animal at present time having an intermediate affinity between two classes – there may be some descendent of some intermediate link. – the only connection between two such classes will be those of analogy, which when sufficiently Multiplied becomes affinity yet often retaining a family likeness, & this I believe the case. = any animal really connecting the fish & Mammalia, must be sprung from some source anterior to giving off these two families, but we see analogies between fish. – Birds same remarks. (*Notebook C*, p. 302). And: I cannot conceive any existing reptile being converted into a Mammal. From *homologies I sh.^{d} look at it as certain that all Mammals have descended from some single progenitor. What its nature was, it is impossible to speculate. More like, probably, the Ornithorhynchus or Echnida than any known form as these animals combine Reptilian characters (& in lesser degree Bird character) with Mammalian. We must imagine some form as intermediate as is *Lepidosiren now in between Reptiles & Fish, between Mammals & Birds on the one hand (for they retain longer the same embryological character) & Reptiles on the other hand.

With respect to a mammal not being developed on any island, besides want of time for so prodigious a development, there must have arrived on the island, the necessary & peculiar progenitor having characters like the embryo of a mammal, & not an *already developed* reptile, Bird or Fish. – We might give to a Bird the habits of a mammal, but inheritance would retain almost for eternity some of the bird-like structure, & prevent our new creation ranking as a true mammal. (*Correspondence* to C. Lyell, Sept. 1, 1860).

Marblefishes Fishes of the Family Aplodactylidae, occurring off southern Australia, New Zealand, and *Peru/Chile.

These coastal, predominantly herbivorous fishes are represented here by *Aplodactylus punctatus* Valenciennes, 1832. Its description was based on a specimen "first sent from Valparaiso, by M. D'Orbigny, where it was also observed by M. *Gay. Mr Darwin's *collection contains a specimen, which has unfortunately lost the number attached to it; but as he made a collection on that coast, it was probably obtained in the same locality." (*Fish*, p. 15).

Marginalia Notes handwritten at the margins of books or other texts.

One of the most famous marginalia in history is a note on number theory by Pierre de Fermat (1601–65), claiming: "I have discovered a truly remarkable proof which this margin is too small to contain", which spawned a quest of over three centuries for what became known as 'Fermat's Last Theorem' (Singh 1997).

CD's marginalia, on the other hand, have so far remained largely unexamined. Only the marginalia of the books in his personal library were published (di Gregorio and Gill 1990); those on his 2500 reprints – about a quarter of a million *words (Vorzimmer 1963, p. 374, n. 11) – are still unpublished.

CD's book marginalia pertaining to fishes are all included in this volume, either in the bodies of entries, or [in square brackets] following the reference in question. Here, the numbers refer to the sequence of columns (two per page) used by di Gregorio and Gill (1990) for their *Marginalia*.

The *Marginalia* confirm the impression one gets from the *Notebooks* that CD, after he had discovered the principle of *natural selection (in late 1838), actively searched the literature for supporting examples, i.e. that his mind was no longer 'open' to a range of alternative explanations.

Also, the *Marginalia*, written for his benefit only, are interesting in that they prove CD to be far more judgmental of the work of other naturalists – e.g. *Agassiz – than may be inferred from his publications, or even from his letters (Sheets-Pyenson 1981). This provides for a 'layered' interpretation of CD's writing – and indeed of the evolution of his ideas – wherein we can observe the progression of an idea from a comment on a book he read, to an entry in a notebook and/or to a letter asking field naturalists for more facts, then finally to a sequence of very similar paragraphs in several of his books. I leave it to the reader to identify such cases from the CD quotes reproduced here.

Marine mammals *See* **Mammals.**

Mauritius Island state in the Indian Ocean, previously called 'Isle de France', whose capital, Port Louis, was a port of call to H. M. S. *Beagle* from April 29 to May 9, 1836.

Armstrong (1990) described the natural history, landscape, economy and society of Mauritius in 1839, as seen and reported by CD, while Keynes (2000, p. 367) lists the few specimens he collected there, including an unidentified fish (catalog no. 1449) .More interestingly, CD noted that [i]t would be superfluous to give the cases amongst my notes of the enormous increases of Birds, fish, frogs and insects, when turned out into new countries: the one island of Mauritius would afford striking instances of all these classes except fishes; & for fish we may turn to N. America (*Big Species Book*, p. 178; the pointer to North America may refer to Richardson, 1836, though this work does not mention introduced fishes, and only the common *carp had been introduced at the time into the continental USA; *see* **FishBase**).

The previous citation expanded upon earlier jottings in *Notebook B* (p. 234): Frogs attempted to be introduced I isle de France p. 170, <<Fish introduced>> (Saint-Pierre 1773, p. 169: "On a fait venir jusqu'à des poissons étrangers," i.e. "even foreign fishes were introduced").

Actually, for information on the fishes introduced into Mauritius, we can turn to *FishBase, which lists 17 freshwater fish species as being introduced until the end of the twentieth century into Mauritius, with details on the reasons for, and some effects of, these introductions.

Megatooth shark One of the few *sharks for which CD gives a scientific name (*Carcharias megalodon*), which Keyes (1972) and others now consider a synonym of *Carcharodon megalodon* (Agassiz, 1835).

The context for CD's report is the stepformed terraces in Herradura Bay, near *Coquimbo, Chile, of which Mr *Lyell concluded [...] that they must have been formed by the sea during the gradual raising of the land. Such is the case: on some of the steps which sweep around from within the valley, so as to front the coast, shells of existing species both lie at the surface, and are embedded in a soft calcareous stone. This bed of the most modern tertiary epoch passes downward into another, containing some living species associated with others now lost. Amongst the latter may be mentioned shells of an enormous perna and an oyster, and the teeth of a gigantic shark, closely allied to, or identical with the Carcharias Megalodon of ancient Europe; the bones of which, or of some cetaceous animal, are also present, in a silicified state, in great numbers. (*Journal*, May 14, 1835; *Perna* is a genus of mussels. Essentially the same observations are presented on pp. 127–9 of *Geology, and on p. 43, Vol. I of CD's *Collected Papers; also note that, as shown by Keyes (1972), *C. megalodon* did not occur only in ancient Europe).

CD collected one of the shark teeth in question, and later donated it, along with other items, to the Royal College of Surgeons (*Correspondence* to Caroline Darwin, August 9–12, 1834). Thus, the *Descriptive Catalogue of the Fossil Organic Remains of Reptilia and Pisces Contained in the Museum of the Royal College of Surgeons of England* lists, as its item 433: "A tooth, with part of one side broken away, of the great extinct Shark (*Carcharodon megalodon*, Ag.). It is of almost equal size with No. 431 [from the Miocene tertiary formation of Malta]. From the older tertiary deposits at Coquimbo, Chile, South America. Presented by Charles Darwin, Esq., F.R.S." (Anon. 1854).

CD recalled this donation in a letter to Richard *Owen: A year or two ago, some one (I think Mʳ. Searles Wood) applied to me for teeth of Carcharias; & I then looked carefully & could not find a noble specimen, which I found in the older Tertiary beds at Coquimbo in Chile; & which I am almost certain I showed once to you (hence your vague memory of the fact) & which I *rather* think I left with you. If not at the College it is gone the way of all flesh – (a very inappropriate remark for a silicified fossil) – You would know my specimen by having a small number attached to it, if not removed. (*Correspondence*, Dec. 23, year uncertain: 1847–1854).

All that is left is to mention that *C. megalodon* may have reached 16 m, i.e. twice the size of the *Great white shark (Randall 1973; Gottfried *et al*. 1996; Helfman *et al*. 1997, p. 184), but not 25 m, 30 m or even longer, as suggested in some accounts (Lacepède 1803, pp. 555–6; McCormick *et al*. 1963, p. 211; Janvier and Welcomme 1969; Müller 1985, p. 155; Keyes 1972).

Mendel, Gregor The Austrian monk (1822–84) who, based on experiments conducted in his garden, invented genetics, and whose key work (Mendel 1866), published in an obscure provincial journal, is supposed to have been completely ignored, for several decades, before it was 'rediscovered' at the beginning of the twentieth century by H. De Vries and others.

This is an urban myth: Mendel's work – mostly conducted in a greenhouse – was cited by at least four different scientists, including the author of an entry on 'hybridism' in the ninth edition of the *Encyclopedia Britannica* (1881–95), before it became fully appreciated (Zirkle 1964). Neither does the treatment of Mendel's work by the scientific community of the late nineteenth century illustrate well the concept of 'premature discovery' (Garfield 1980), since CD at least could probably have absorbed this work, had he known of it, and used it to fill the need met by his *ad hoc* invention of *gemmules. Zirkle (1964) also points out that CD, in the second edition of *Variations I* (1875; p. 345), cited a paper by H. C. H. Hoffman which itself cited Mendel (1869) five times. The only lesson to be drawn from this is that things that can go wrong do, as the proverbial Murphy knew (Matthews 1997).

Another urban myth that may be mentioned here is that Mendel was a cheat, his ratios of F_1 peas being too good to be true (Fisher 1965). Yet several authors have shown that the ratios Mendel published could be achieved without malfeasance (Hewlett 1975; Orel 1996, pp. 199–209).

The point here is that it is too easy to gratuitously accuse dead scientists of unethical behaviour. If at all, let's pick fights with live ones! (*See also* **Blyth; Wallace**.)

Merlu French common name for a number of cod-like fishes, notably European hake *Merluccius merluccius* (Linnaeus 1758). [Note 'mer-lucius' = 'sea pike,' which is another name for hake.]

This name shows up in a strange account of what appears to be an even stranger specimen collected by CD in Tierra del Fuego: Merlus; caught in Good Success Bay, by hook and line; colours reddish brown and white variously marked. Eye with singular fleshy appendage; to the fish are sewn another pair from another specimen. March (*Fish in Spirits*, no. 546).

Two species of hake occur around Tierra del Fuego (Lloris and Rubacado 1991), viz. *Merluccius gayi* Guichenot, 1848 and *Merluccius hubbsi* Marini, 1933. However, neither of these species has fleshy appendages near the eyes. Suggestions anyone?

Mermaids CD's friend and mentor, J. S. Henslow, wrote to him, shortly before he embarked on the *Beagle, that he left him to his "better meditations on Mermaids and Flying fish." (*Correspondence*, Oct. 30, 1831).

And indeed, CD soon wrote about *flying fishes. On the other hand, it took him four years, spent mainly at sea, to finally become interested in mermaids. Here is his sole account of these creatures: There are two things in Lima which all Travellers have discussed; the ladies 'tapadas,' or concealed in the saya y manta, & a fruit called Chilimoya. To my mind the former is as beautiful as the latter is delicious. The close elastic gown fits the figure closely & obliges the ladies to walk with small steps, which they do very elegantly & display very white silk stockings & very pretty feet. They wear a black silk veil, which is fixed round the waist behind, is brought over the head & held by the hands before the face, allowing only one eye to remain uncovered. But then, that one eye is so black & brilliant & has such powers of motion & expression, that its effect is very powerful. Altogether the ladies are so metamorphised, that I at first felt as much surprised as if I had been introduced among a number of nice round mermaids, or any other such beautiful animal. (*Diary*, p. 332; CD forgets here that we do not speak Spanish as well as he did then. Thus: a 'tapada' is a woman concealing her face under a 'mantilla', i.e. a silk or lace scarf covering head and shoulders, while 'chilimoya' is *Anana cherimola*, the Peruvian custard apple).

Banse (1990) gives an update on the biology of real mermaids. (*See also* **Sirenians; Lungfishes**.)

Mesites attenuatus *See* **Galaxiidae**.

Migrations The directed movements performed seasonally, or at least twice in their lifetime, by many fishes and other animals, and linking different parts of their *range (see, for example, Harden-Jones 1968). *Diadromy, i.e. a life cycle involving spawning migrations from fresh to marine waters, or conversely, is in fishes usually combined with precise *homing.

The word 'migration' is also used, by CD and some other authors, to mean other forms of movements, including range extensions, but this is misleading.

Minnows A name applied throughout the world to a number of small fishes, though usually with a modifier (Bronze minnow, Chub minnow, etc.).

Being simply a 'minnow', on the other hand, is the privilege of the 'Eurasian,' or 'Common' minnow *Phoxinus phoxinus* (Linnaeus, 1758), which, together with *Carp, *Goldfish and others, belong to the Family Cyprinidae.

CD mentions this species of minnow frequently, notably in his account of the intellectual *tour de force* exhibited by Möbius' *Pike. Minnows are not particularly bright, either. Yet various aspects of their behaviour pose interesting questions to evolutionary biologists (see, for example, Murphy and Pitcher 1997). The most interesting of these behaviours is their release, when attacked, of a substance called 𝔖𝔠𝔥𝔯𝔢𝔠𝔨𝔰𝔱𝔬𝔣𝔣 (literally: 'fright-stuff'; German is like that).

The evolutionary question is, obviously, that there is little point, in the context of *natural selection, in an animal devoting considerable energy (Wisenden and Smith 1997) to synthesizing a substance which, when released, warns unrelated conspecifics that a predator is around. *Altruism should not go that far, especially if remote predators are attracted by the scent of such an 'alarm substance', as seems to be the case.

One explanation is, possibly, that if a secondary predator is included among those attracted, this may cause the primary predator to release its uneaten prey in order to flee faster (Mathis *et al*. 1995; Smith 1992). As fish with scars attributable to unsuccessful attacks frequently occur in nature, this scenario does not seem too far-fetched. A similar explanation has been proposed for the bioluminescent flashes emitted by planktonic *Noctiluca* (Burkenroad

1943), which had so captivated CD (*see* **Bioluminescence**).

Missing links *See* **Great Chain of Being.**

Möbius, Karl-August German zoologist (1825–1908); performed some of the earliest quantitative studies in *ecology (here: *zoobenthos). Möbius proposed the concept of 'biocoenoses' as living superorganisms, using oyster beds as his example (Möbius 1877; Rumohr 1990).

The compilation by Junker and Richmond (1996) of the correspondence between CD and German scientists includes only one letter from Möbius to CD, sent in 1881(no. 13431). However, it can be assumed that Möbius contacted CD earlier: it is unlikely that CD accidentally stumbled on the obscure publication in which Möbius presented the story, much cited by CD, of the *Pike which painfully learned not to consume the *Minnow with which it shared an aquarium.

Given the importance of this story, at least to CD, it would have been nice to get details on its background. Unfortunately, it was not based on Möbius's own work. Rather, it was communicated to him by "Oekonomierath Amtsberg, of Stralsund", i.e. by an official from a small city on the Baltic coast. In any case, the study does not appear to have been replicated, although pike predation on minnows, and the latter's evasive actions, have been studied in great detail (see, for example, Magurran and Pitcher 1987).

There is, of course, a better-known Möbius: August Ferdinand Möbius (1790–1868), he of the strip. However, as is often the case with such things, it appears that A. F. Möbius was not the first to discover or to describe the topological oddity now known as the 'Möbius strip.' Another German mathematician, Johann Benedict Listing, appears to have done it first . . . Oh, how easy to get hold of gossip on the World Wide Web!

Mojarras Members of the tropical Family Gerreidae, noted for their strongly protrusile mouths (used for feeding on small bottom invertebrates), and shiny grey bodies; hence their other name of 'silver biddies'.

The mojarras are represented here by two species, the Common silver biddy *Gerres oyena* (Forsskål, 1775), via a specimen collected by CD in the *Cocos Islands (*Fish*, p. 59–60; White, silvery), and the Jenny mojarra *Eucinostomus gula* (Quoy & Gaimard, 1824), sampled by CD in *Rio de Janeiro, and described as *Gerres gula* in *Fish* (pp. 58–9).

Monacanthus scopas *See* **Filefishes.**

Monsters A term used by CD for sharks (*see* **Burrfishes**), and for fishes with two heads (also called *monstrosities): Whenever two bodies or two heads are united, each bone, muscle, vessel, and nerve on the line of junction appears as if it had sought out its fellow, and had become completely fused with it. Lereboullet[2], who carefully studied the development of double monsters in fishes, observed in fifteen instances the steps by which two heads gradually became united into one. (*Variations II*, p. 333; n. 2 refers to Lereboullet (1855a), p. 855, and (1855b), p. 1029).

Monstrosity(-ities) In CD's view, organisms that suffered *abnormal development. Thus, monsters are generally sterile & not often inheritable. (*Correspondence* to C. *Lyell, Feb. 18/19, 1860).

CD covered monstrosities in *Origin*, but felt his point had been misunderstood, owing to *Origin* being only an abstract of the *Big Species Book* he had wanted to publish, and which he called an M.S. Thus, he writes: Here again comes in the mischief of my *Abstract*: in fuller M.S. I have discussed parallel case of a normal fish like a monstrous *Gold-fish: I end my discussion by *doubting*, because all cases of monstrosities which resemble normal structures, which I could find were not in allied groups. (*Correspondence* to C. *Lyell, Feb. 18/19, 1860).

Also: Here is the evil of an abstract; in my fuller M.S. I have discussed a very analogous case of a normal fish like an extremely monstrous Gold Fish (*Correspondence* to J. D. Hooker, Feb. 20, 1860).

And here is the account in the M.S.: As monstrosities can not be clearly distinguished from *variations, I must say a few words on some of the conclusions arrived at by those who have studied the subject. Geoffroy St. Hilaire & his son Isidore[1] repeatedly insist on the law that monstrosities in one animal resemble normal structures in another. [...]. To give two or three of the best instances from Mr. Isidore Geoffroy; – in the pig, – which has the snout much developed & which is allied, but as *Owen has shown, not so closely as we formerly thought to the Tapir & Elephant, a monstrous trunk is developed oftener than in any other animal: the frequent monstrosity of three, four or even a greater number of breasts in woman seems to stand in relation to the fact of most mammals having more than two mammae: *Carps are very subject to a curious monstrosity causing their heads to appear as if truncated, & an almost exactly similar but normal structure is met with in the species of *Mormyrus, a genus of fish belonging to the same Order with the carp[2]. (*Big Species Book*, pp. 318–19; n. 1 refers to Geoffroy-Saint-Hilaire (1832–37, 1841); n. 2 to Geoffroy-Saint-Hilaire (1841), Vol. 1, pp. 353 & 436).

CD's argument is probably correct. However, the last example (in n. 2) is misleading: *Mormyrus* and other *elephantfishes belong to the Order Osteoglossiformes (bony tongues), which is not closely related to the Cypriniformes (carp and their allies).

Montevideo Capital of *Uruguay, a frequent and challenging port of call of the *Beagle* in 1833 (details in *Diary* and Parodiz 1981).

CD sampled only two fish species in Montevideo: the Rio de la Plata onesided livebearer *Lebias multidentata* (*Jensyniinae); and a *mullet.

Morays Members of the Family Muraenidae, consisting of tropical and subtropical *eel-like fishes with pore-like gill openings and lacking pectoral fins.

Several species of moray were collected by CD during the voyage of the *Beagle* (*see*, for example, *Fish in Spirits*, after no. 157). However, only two were properly described in *Fish*. The first of these is the Jewel (Or lentil) moray, *Muraena lentiginosa* Jenyns, 1842, "known to have bitten people who have tempted them" (Jackson 1993, p. 236), and whose description was based on two specimens, one larger from Charles (= Floreana) Island (fine dark purplish brown with yellow circular spots; *Fish in Spirits*, no. 1299), the other (Dark reddish-purple brown, with pale, or whitish-brown spots: eyes bluish) from a tidal pool at Chatham (= San Christobal) Island, Galápagos (*Fish*, pp. 143–4; *Fish in Spirits*, no. 1286). Interestingly, Grove and Lavenberg (1997, p. 156) also note that the colour markings change during ontogeny.

The second species is the Ocellated moray *Gymnothorax ocellatus* (Agassiz 1831), ranging from the Greater Antilles to the South of Brazil, collected by CD in the harbour of Rio de Janeiro and described as *Muraena ocellata* (*Fish*, p. 145; *Fish in Spirits*, no. 301).

The other moray specimens collected by CD, in Porto Praya, Cape Verde Islands, and Tahiti, were both too small to have developed the adult attributes used for species identification (*Fish*, pp. 145–6). The former specimen, recorded as no. 46 in the *Zoology Notes* (p. 322), does not reappear in *Fish in Spirits*, whereas the latter does, as no. 1327, suggesting that *Covington erroneously skipped the record in question.

Finally, we note that the former British Museum (Natural History) has in its register, for April 8, 1853, the entry "*Muraena* In spirit W. Indies". The adjacent page, at the same level, reads: "Presented by C. Darwin Esq. Dep." (see Appendix II); I am not aware of this specimen having been identified.

Mormyrus See **Elephantfishes.**

Morris, Margaretta Hare (I) Sometime in mid 1855, Richard Chandler Alexander, one of CD's many correspondents, received a letter from Margaretta Hare Morris, answering one of CD's queries on the manner in which fishes colonize water bodies such as mountain lakes.

Her solution was that *water-beetles, especially *Dytiscus marginalis*, could easily have done the job. Thus, she noted: "In 1846 I had the pleasure of studying the history of this insect, on the shore of several of our mountain lakes in north Pennsylvania, and found that it fed on fish and on fish roe that it found near the margin of these Lakes, destroying numbers of the lake trout which are found only in these inland seas, in *deep water*, and never in the outlets, most of these lakes are supplied with water by springs at the bottom, only, and have no communication with other lakes, emptying into rivers, where lake trout are never seen.

While on a visit in Montrose, Susquehanna Co Pennsylvania, I was presented with a *Dyticus marginalis*, which flew into a window, attracted by the light of a lamp, he must have flown at least three miles, as there are no lakes nearer that town, but several of some miles in extent about that distance from Montrose – this specimen had no roe on it – but those feeding near the margin of the lake were covered with it – leaving no doubt in my mind as to the fact that they thus carried the *eggs, from lake to lake and peopled them with fish that had no other means of being transported." (*Correspondence*, June 15, 1855).

Alexander sent this letter to CD, who annotated it, writing with regard to the last sentence: this is the most important statement in Letter, then rather pompously corrected himself: this is the only important statement in Letter (*Correspondence*, Vol. 5, p. 356.).

Somehow, Alexander must have written about this to CD before he sent him Morris' letter, since CD dismissed her even before he got her letter: if Dr. Alexander can give me no more information, the case even for *credulous* me is worthless; I have constantly observed eggs of some parasite adhering to the body of Nepa, a *water-bug*, if the Lady uses the term correctly, it is perhaps these ova, & not of Fish. (*Corresp.* to Hooker, April 13, 1855).

The editors of the *Correspondence* completed this sad tale with the statement that "Margaretta Hare Morris has not been identified" (Vol. 5, p. 356); this is rectified below.

Morris, Margaretta Hare (II) An entomologist whose published work was "instrumental in controlling two devastating agricultural pests – the seventeen year locust and the Hessian fly" (Kass-Simon *et al*. 1990, p. 255).

Margaretta Hare Morris joined the American Association for the Advancement of Science in 1850 (Rossiter 1982, p. 76), and became, in 1859, the first female member of the Philadelphia Academy of Natural Sciences. Eckhardt (1950), who provides details on her life and work, notes that she would have been "one of the early greats" in the Academy, had she been a man.

I suspect that CD (and Alexander?) would not have dismissed her so easily, had she, indeed, been a man.

That *Dytiscus marginalis* larvae (also known as 'water tigers') prey on fish *eggs, larvae and juveniles is well established (McCormick and Polis 1982). Further, there is no reason to assume that an entomologist as accomplished as Morris would have been unable to distinguish a parasitized adult beetle from one with fish eggs attached to its body or legs. Indeed, in the male *Dytiscus*, the front legs have 'suckers' through which a sticky fluid is excreted (Miall 1895). Although these suckers are used primarily to hold the female during copulation, it is likely that they also retain small objects, including fish eggs.

Note, finally, that water-beetles are commonly involved in "cross-predation" with fish, wherein the insects, both larvae and adults, prey on fish eggs (see CD's own note on this in '*Showers of fishes'), larvae and juveniles, while the larger fish prey on the larval and adult beetles (McCormick and Polis 1982). The existence of such feeding loops may actually confer evolutionary benefits on water beetles that bring fish eggs along when they colonize newly formed water bodies.

Margaretta 1 : Charles 0.

Mother Carey's chickens Sailors' name for Storm *petrels, probably referring to the Virgin Mary ('*Madre cara*'; Evans 1981, p. 759). Carl Safina (pers. comm.) suggests that the name 'Mother Carey's chicken', now fallen into disuse, referred particularly to Wilson's storm petrel, *Oceanites oceanicus*, a bird that is easily identifiable while feeding because it will 'walk on water' more than other storm petrels (Harrison 1987, p. 215).

CD mentioned this bird only once in relation to fish, while sailing near the Abrolhos, Brazil: since leaving *Bahia, the only things we have seen here were a few sharks & Mother Carey's chickens (*Journal*, March 24–6, 1831).

Motion sickness A condition due to sensory conflicts, wherein the inputs from the organ sensing the effects of gravity (the inner ear in humans, or the otoliths in fishes) fail to match, within the *brain's centre for sensory–motor integration, with inputs from other sensory systems, notably vision and proprioception.

It is well attested that CD suffered extensively from motion sickness during the voyage of the *Beagle*. In fact, he complained about it regularly in his letters to his family and friends. Here is his first mention, to his chief Lord of Admiralty, his friend and mentor J. S. *Henslow, of what was to become a constant lament: The two little peeps at sea-sick misery gave me but a faint idea of what I was going to undergo (*Correspondence* 18 May–16 June 1832).

That fishes, on the other hand, can also become 'seasick' may come as a surprise to some. But the evidence is there: Cod that vomited while being transported in the holding tank of a small craft (McKenzie 1935). This was subsequently confirmed by *Carp experiencing the simulated weightlessness of parabolic aircraft flights (Clément and Berthoz 1994), and by longer term exposure of the Killifish *Fundulus heteroclitus* (Hoffman et al. 1977), and normal and 'labyrinthectomized' *Koi (Mori et al. 1996) to the weightlessness of space.

Mouth-brooding The way certain fishes protect their eggs and/or larvae from predation and parasitism; common in *Cichlidae, *catfishes, amblyopsid *cavefishes and other groups. Note that it is not only the females that practise mouth-brooding. In the Blackchin tilapia *Sarotherodon melanotheron* Rüppel, 1852 (Cichlidae), the males incubate the eggs and guard the hatchlings in their mouth, losing much weight, and acquiring *parasites in the process (Pauly 1974, 1976). This should make the males a valuable resource for females to compete for, and this is indeed what seems to occur (Barlow 2000, p. 77), in a reversal of traditional sex roles similar to that occurring in *sea horses and *pipefishes.

Catfishes also exhibit mouth-brooding, which intrigued CD: The males of certain other fishes inhabiting South America and Ceylon, belonging to two distinct Orders, have the extraordinary habit of hatching within their mouths or branchial cavities, the *eggs laid by the females.[38] I am informed by Professor *Agassiz that the males of the Amazonian species which follow this habit, 'not only are generally brighter than the females, but the difference is greater at the spawning-season than at any other time.' The species of *Geophagus act in the same manner; and in this genus, a conspicuous protuberance becomes developed on the forehead of the males during the breeding-season. (*Descent II*, p. 345; n. 38 reads: Professor Wyman, in *Proc. Boston Soc. of Nat. Hist.*, 15 September, 1857. Also Professor Turner, in

Journal of Anatomy and Phys., 1 November 1866, p. 78. Dr. Günther has likewise described other cases, referring to Günther 1868a; Agassiz wrote a letter on this on July 22, 1868; *Calendar*, no. 6286, though it had been published three years earlier, in Agassiz 1865).

Mudfish *See* **Lungfishes.**

Mudskippers Members of the tropical and subtropical genus *Periophthalmus*, and allied genera of *gobies (Family Gobiidae), characterized by their adaptations to life on mudflats (Gibson 1993). CD refers to their behaviour as mud-walking (*Notebook T*, p. 463). We encounter the mudskippers again in the *Big Species Book*: Certain fish use their pectoral fins for splattering over the mud, for jumping & even for climbing trees; if fish had become, like land-crabs & onisci, terrestrial animals, how easily ancient transitional uses of the pectoral fins, might have baffled all conjecture. (p. 341; onisci is CD's plural for *Oniscus*, a woodlouse). (*See also* **Climbing fish**.)

Mud-walking fish *See* **Mudskippers.**

Mugil *See* **Mullets.**

Mule The sterile offspring of a mare and a male donkey; or generally, any *hybrid. Here is an example of CD's use of the term: Lord Moreton's law cannot hold with fishes, <<& there are mule fishes>> & reptiles & those which <lay> <<have>> their *eggs, <inter>, impregnated externally; nor can it be a *necessary* concomitant, with moths, which can be impregnated externally- (*Notebook E*, p. 417).

The fanciful 'law' in question, based on breeding experiments between a quagga (the dam) and an Arabian chestnut (the sire), is the "extraordinary fact of so many striking features, which do not belong to the dam, being in two successive instances, communicated through her to the progeny, not only of another sire, who has them not" (Morton 1821, p. 22).

It is in such cases that we can see how the absence of a correct theory of inheritance inhibited everybody's thinking, including CD's. As an interesting aside, we may note that it is precisely the availability of such theory which made it possible to recreate the extinct quagga (at least in its external features). This animal, as DNA analysis established, was not a separate equine species, but a subspecies of Burchell's or Plains zebra. This made it possible, through *selection involving zebras from several countries in southern Africa, to breed zebras increasingly similar to quaggas. (My source here is the in-flight magazine of South African Airways, very hard to trace; see instead www.museums.org.za/sam/quagga/quagga.htm.)

Mullets Fishes of the Family Mugilidae, characterized by special adaptations (narrowly spaced gill rakers, long guts) enabling them to derive their nutrition from the fine detritus that accumulates on top of shallow sediments (Odum 1970).

CD collected mugilids in *Bahia Blanca and *Montevideo, South America (*Mugil liza* Cuv. et Val.? *See Fish*, pp. 80–1; *Fish in Spirits*, no. 393: Back coloured like Labrador feldspar; iris coppery; plentiful; no. 458), in the *Cocos Islands (*Mugil?* pp. 81–2), and in *King George's Sound, Australia (*Dajaus diemensis* Richardson, 1840; *Fish*, pp. 82–3).

Updating Jenyns' names is straightforward. The first refers to the Liza, *Mugil liza* Valenciennes, 1836. '*Mugil?*' refers to a specimen so badly preserved that Jenyns preferred not to formally identify it. This was probably the Squaretail mullet *Liza vaigiensis* (Quoy & Gaimard, 1825), given that it is the only mullet species reported from the Cocos Islands, despite a century of vigorous sampling (*see* FishBase). Finally, the fish from King George's Sound was a specimen of *Aldrichetta forsteri* (Valenciennes, 1836), the Yellow-eye mullet (*see also* **Names, updated**).

Muraena spp. *See* **Morays.**

Myxine(-a) *See* **Hagfishes.**

Names, common Used in this book as label for entries on those of 'Darwin's Fishes' that have such names in English, and capitalized (as for bird names), following a recent trend among ichthyologists.

Common names are what most people know about most fishes (and similarly for other groups of organisms). Thus, listing common names, and their etymologies is a way to start documenting the (traditional) knowledge held about organisms by the speakers of a given language. *FishBase contains over 150 000 common names, in over 200 languages, which can be used for this purpose, and for the folk taxonomy they imply (*see*, for example, **Eelpout**). Moreover, common names allow in some cases inferences on language universals, notably the tendency for speakers of various language families to use common names with high-frequency sounds ('i'; 'ee') for small fishes and low-frequency sounds ('a'; 'o') for large fishes (Berlin 1992; Palomares *et al.* 1999).

CD would have liked the idea that common names (of fish) could throw their own (if feeble) light on our origins, as he believed the *evolution of languages, like that of our bodies, to support a relatively recent origin for all humankind. This recent, common origin is now well established (Ruhlen 1994; Stringer and McKie 1996), but was much contested earlier.

Names, scientific Since the publication of Linnaeus (1758), newly described living organisms are almost always attributed 'binomina,' i.e. scientific (not 'Latin') names consisting of a unique genus name, always capitalized (e.g. *Salmo*) and a species epithet, never capitalized (e.g. *salar*), both jointly defining a species. Further, the *International *Code of Zoological Nomenclature* recommends that the name(s) of the author(s) of the original description be appended to the binomen, e.g. *Salmo salar* Linnaeus, 1758, usually with a comma between the author and the year, as shown here. However, if a subsequent analysis of the *taxonomy of a group (a 'revision') moved a species from its original genus to another one, the name of the original author(s) is (are) put in bracket, hence *Salvelinus fontinalis* (Linnaeus, 1758). [The authors' names do not usually call for references; however, in this book they are treated as if they did, thus providing further links to the taxonomic literature of CD's time.] The *Code* contains a huge number of rules, in addition to the few presented here, and many were derived from a report by Strickland *et al.* (1843), of which CD was a co-author. Further details are presented in the context of various entries presenting scientific names (*see also* **Eponym; Occam's Razor; Species; Synonym**).

Names, updated An anonymous historian who reviewed the draft of this book felt that updating the scientific names of fish used by CD (and Jenyns) is "unacceptable in a historical reference book". Careful readers of *Darwin's Fishes* will note that the original version of the names used by CD, Jenyns and other Victorian scientists (another modern term deemed inappropriate) are all presented in their original form. However, where required, these names were also 'updated,' i.e., the fish in question were given, as well, the scientific names now considered valid (see **Index to the Fishes**). This enables readers to link CD's and his colleagues' accounts to the living, growing body of knowledge that is still being assembled (some of it in electronic databases such as *FishBase), on these very organisms, by contemporary biologists. What better method is there to bring to light the contents of the musty tomes to which such historians have relegated the founders of our biological disciplines?

Natural selection This concept is at the very core of CD's work. While various notions of evolution, usually structured around the *Great

Chain of Being, had become widespread among naturalists in the early nineteenth century (e.g. *Grant), an explanatory mechanism was still required, as *Lamarck's effort had stalled. CD's solution was based on three, easily verifiable observations: (a) That all plants and animals produce more offspring (seeds or other propagules, or eggs, or young) than required to replace themselves, and that these would saturate their environments within a few generations, were they all to survive, grow and also reproduce; (b) that these offspring differ in all their measurable attributes (morphological, physiological, behavioural), and that these difference are inherited; and (c), that given the set of environmental conditions these offspring happen to encounter, those that happen to possess appropriate attributes will tend to survive, grow and reproduce (i.e. will be 'fitter,' or better adapted) than their siblings. This will cause a shift in the distribution of attributes in the population in question, which, given continued environmental pressure, will tend to increasingly differ from other populations of the same species, until a new species has emerged (hence the '**Origin of Species by Means of Natural Selection;*' see also* **Punctuated equilibrium; Speciation**). It took CD lots of thinking to get there, and to follow on some of the implications of his discovery, as illustrated in these early musings on this topic: many of every species are destroyed either in egg or [young or mature (the former state the more common)]. In the course of thousand generations infinitesimally small differences must inevitably tell;[24] when unusually cold winter, or hot or dry summer comes, then out of the whole body of individuals of any species, if there be the smallest differences in their structure, habits, instincts [senses], health, etc., <it> will on an average tell; as conditions change a rather larger proportion will be preserved: so if the chief check to increase falls on seeds or *eggs, so will, in the course of 1,000 generations, or ten thousand, those seeds (like one with down to fly) which fly furthest and get scattered most ultimately rear most plants, and such small differences tend to be hereditary like shades of expression in human countenance.

So if one parent <?> fish deposits its egg in infinitesimally different circumstances, as in rather shallower or deeper water, etc., it will then <?> tell. (*Foundations*, p. 8; n. 24 states that In a rough summary at the close of the Essay, occur the words: "Every creature lives by a struggle, smallest grain in balance must tell.).

Our theory requires that the first form which existed of each of the great divisions would present points intermediate between existing ones, but immensely different. Most geologists believe *Silurian *fossils are those which first existed in the whole world [. . .]. Not so Hutton or *Lyell: if first reptile of Red Sandstone <?> really was first which existed: if Pachyderm of Paris was first which existed: fish of *Devonian: dragon fly of Lyas: for we cannot suppose them to be progenitors: they agree too closely with existing divisions. (*Foundations*, pp. 26–7; Hutton 1795).

These ideas have been much refined since they were originally proposed, and here is the most concise version of natural selection so far presented, which Casti (1989, p. 148) calls "Darwin's Formula":

Adaptation = Variation + Heredity + Selection.

This simple formula provides a key to interpreting most of CD's writings, including much of what he wrote about fishes, as presented in this book.

Naturalist Used formerly for what is now a 'biologist'; also a position, in the seventeenth and eighteenth centuries, on many of the vessels of the British and other European navies.

Naturalists – often also performing as medical officers – were expected to report on the *fauna and flora of the regions visited, and to collect specimens for their national museums. These specimens, sampled in all major regions of the world are what led, for example, to monumental works such as Cuvier and Valenciennes' *Histoire Naturelle des Poissons*, for which Bauchot *et al.* (1990) listed all the contributing naturalists.

Strangely enough, CD was not really the naturalist of the *Beagle* during her voyage of 1831-6, though he acted as if he were (save the medical bits). The naturalist of the *Beagle* was the ill humoured and petulant Robert McCormick, a surgeon who, perhaps unsurprisingly, became so jealous of the growing role assumed by CD – a private guest of Captain *FitzRoy – that he left during a landing in Brazil, in April 1832, and returned to England (Burstyn 1975). Gruber (1969) cites CD as having commented He is no loss. McCormick was succeeded by Benjamin Bynoe as an 'acting surgeon' whom CD thanked for his very kind attention to me when I was ill at Valparaiso (*Journal*, viii, n. 1), and who never attempted to compete with him for the naturalist spot.

Nictitating membrane A feature defined by CD in *Descent* (I, p. 17): The nictitating membrane, or third eyelid, with its accessory muscles and other structures, is especially well developed in *birds, and is of much functional importance to them, as it can be rapidly drawn across the whole eye-ball. It is found in some reptiles and *amphibians, and in certain fishes, as in *sharks.

The nictitating membrane is also defined in ecological terms in the glossary of *Origin's* sixth edition: a semi-transparent membrane, which can be drawn across the eye in birds and reptiles, either to moderate the effect of strong light or to sweep particles of dust, etc. from the surface of the eye.

Non-Darwinian evolution Modes of *evolution not explicitly covered by CD's theory of *natural selection, nor his theory of *sexual selection.

Non-Darwinian evolution is implied, evidently, in theories proposed prior to CD's *Origin* (1859), notably that of *Lamarck. CD reluctantly accepted this theory as complementary to *natural selection, as evidenced in the sixth Edition of *Origin* (Vorzimmer 1969). The non-Darwinian theories of evolution proposed since *Origin* may be roughly grouped into two sets:

(a) Those proposed as alternative to (and generally incompatible with) CD's theories; and
(b) Those meant (or shown) to complement CD's theories.

Most notable in the former group are various attempts to revive Lamarckism, as exemplified by Arthur Koestler's book on *The Case of the Midwife Toad* (1971), or frank appeals to ignorance, which may be paraphrased as: "I don't know the biochemical pathway leading to light-sensitive pigments, and hence *eyes cannot have evolved" (Behe 1996). Group (a) also includes attempts to cast evolution as a process guided by a higher entity, e.g. God (John Paul II 1996; which brings us back to good old *creationism), or 'morphogenic fields' (Sheldrake 1988). Frontal attacks of this sort have proven quite sterile in scientific terms, though they are quite interesting as expressions of various cultural fringes.

The group in (b), on the other hand, has considerably enriched our understanding of evolution. One example is the theory of 'neutral selection', or "mutation-random drift theory" (Kimura 1983), which attempts to explain some aspects of genetic variability not well covered by standard accounts (but see critical reviews in Gillois 1991, 1996).

Another example is endosymbiosis (Margulis 1981), wherein organisms originally swallowed by, or which had invaded the body of, another organism, establish themselves within that host organism, and provide some service to it, 'room service' as it were. In such cases, evolution of new (common) body plans, and the conquest of new niches, is much accelerated over the evolutionary rate generated by *natural selection acting alone (Margulis and Sagan 1995; Sapp 1994).

Such acceleration also occurs through 'larval transfer', wherein, during the evolution of some animals, genes coded for a body form in one lineage are transferred, via some form of hybridisation, to another, unrelated form. This mechanism appears in several phyla, and has been proposed to explain the origin of chordates (Williamson 1998; *see also* **Dohrn; Vertebrate origins**).

Another example of a non-Darwinian theory that complements CD's work is *handicap theory, recently proposed to account for previously puzzling features of signals sent to conspecifics, or to potential predators, by the members of a given population.

The theory of *punctuated equilibrium proposed by Eldredge and Gould (1972) probably also belongs here, though there is an ongoing debate as to its standing *vis-à-vis* selectionist explanations (see, for example, Dawkins 1996, 1998). Gould (1997a, b) called such explanations "Ultra-Darwinian", and those proposing them "Darwinian Fundamentalists".

The latest exchanges on this have been rather undignified, and likely to turn off non-specialists interested in learning more about evolution (Wright 1999). Indeed, it is only the creationists who gain from food fights around what Eldredge (1995) calls the "High Table of Darwinism".

Notacanthus *See* **Deep-sea spiny eels.**

Notebooks Devices used by CD for brainstorming, and for recording questions to be followed up later. Several of CD's notebooks cover the period after his return from the *Beagle voyage, in 1836, to 1844, when he completed a first comprehensive account of *natural selection, in form of the manuscript later referred to as *Foundations*. These documents, here referred to as '*Notebooks*', are witness to the most creative period of CD's life, and are characterized by a free flow of ideas and associations. Unfortunately, this is largely obliterated, in this book, by my selection of quotes dealing with fishes.

The *Notebooks*, originally not meant to be read by others, also document CD's personal views and value judgments (Barrett and Gruber 1980), as do the *Marginalia*. Both the *Notebooks* and the *Marginalia* show that in the late 1830s, his mind gradually closed, with facts being sought mainly to articulate, and later to support, his nascent theory.

There are fifteen *Notebooks*, whose sequence and pagination roughly follow the dates at which CD started them (adapted mainly from Figure 1 in Barrett *et al*. 1987): R: May 1836 – June 1837 (*Red Notebook*) [pp. 17–82]; A: June 1837 – December 1839 [pp. 83–140]; G: 1838 (*Glen Roy*) [pp. 141–66]; B: July 1837 – March 1838 [pp. 167–236]; C: March–July 1838 [pp. 237–328]; D: July–September 1838 [pp. 329–94]; E: October 1838 – July 1839 [pp. 395–456]; T: July 1839–1841 (*Torn Apart*) [pp. 457–71]; S: Summer 1842 [pp. 472–4]; Z: 1837–1839 (*Zoology Notes, Edinburgh Notebook*) [pp. 475–86]; Q: 1839–44 (*Questions and Experiments*) [pp. 487–516]; M: July–September 1838 [pp. 517–60]; N: October 1838 – December 1839 [pp. 561–96]; O: 1838–40 (*Old & Useless Notes*) [pp. 597–630]; and J: 1838 (*Abstract of J. Macculloch*) [pp. 631–42].

Barrett *et al*. (1987) group these notebooks as follows: '*Geology*' (A, G); '*Transmutation of Species*' (B, C, D, E, Q, S, T, and Z); and '*Metaphysical Enquiries*' (J, M, N, and O), with J referring mainly to Macculloch (1837).

Some of these had been published, either as complete entities (e.g. by Herbert 1980b) or in part (e.g. by Barrett and Gruber 1980) before they were assembled in the complete edition used here as a reference.

Weinshank *et al*. (1990) subsequently used the texts recovered by Barrett *et al*. (1987) for a *Concordance to CD's *Notebooks*, and this was used to verify that all of CD's entries with FISH and related terms are included here.

Some of CD's other notebooks are mentioned in the annotation to Davy (1828), and the entries on ***Catalogue***, **Cormorants**, **Experiments (I)**, and **Lumpfish.**

Obstinate nature A concept developed by Philippe Cury, meant to emphasize the tendency of successive generations of organisms to repeat what their parents did.

For fishes, this involves spawning at the very same place and time of the year at which they were hatched (Cury 1994). This phenomenon, which implies some form of imprinting, is well known in *homing *salmon. More recently, it has been demonstrated in the *Cichlidae (Barlow 2000, pp. 239–40). Here is a quote from a book CD read, documenting its occurrence in *Shad (not 'herring'): "Mr. Franklin told me that in that part of New England where his father lived, two rivers flowed into the sea, in one of which they caught great numbers of herring, and in the other not one. Yet the places where these rivers discharged themselves into the sea were not far apart. They had observed that when the herrings came in spring to deposit their spawn, they always swam up one river, where they used to catch them, but never came into the other. This circumstance led Mr. Franklin's father, who had settled between the two rivers, to try whether it was not possible to make the herring also live in the other river.

For that purpose, he put out his nets, as they were coming up for spawning and he caught some. He took the spawn out of them, and carefully carried it across the land to the other river. It was hatched, and the consequence was that every year afterward they caught more herring in that river, and it is still the case.

This leads one to believe that the fish always like to spawn in the same place where they were hatched, and from which they first put out to sea, being as it were accustomed to it." (Kalm 1753; Vol. 1, pp. 154–5 in 1937 edition).

Obstinate nature, which may be seen as a form of stabilizing selection (Schmalhausen 1949), leads, when applied to fishes, to very *parsimonious reinterpretations of much of what was previously known of their life-history strategies and *ecology, and to striking new inferences, still being tested. While generally compatible with the slow changes that CD preferred, the concept of obstinate nature is also eminently compatible with *punctuated equilibrium.

Occam's Razor A concept much used in Science, related to the nature of scientific explanations.

The concept goes back to theologian-cum-philosopher William of Ockham (1280–1349?), to whom the following two phrases are attributed: "*Frustra fit per plura, quod potest fieri per pauciora*" (It is vain to do with much what can be done with less) and "*Entia non sunt multiplicanda praeter necissitatem*" (Entities are not to be multiplied beyond necessity). Ockham, named after his birthplace in Surrey, but whose name is usually latinized, studied at Oxford, became a Franciscan (a revolutionary idea at the time) and then practised in Avignon, where he made a strong case for his order's ideal of poverty, ultimately concluding that the Pope (who certainly was not 'poor') was a Heretic.

Unsurprisingly, he was excommunicated, and his teaching banned, and to add injury to insult, he died of the plague.

Over the centuries, Occam's two phrases, increasingly separated from their original context, were gradually distilled into an injunction to scientists to keep their explanations of phenomena as simple, or *parsimonious, as possible (Pauly 1994c), notwithstanding Dunbar (1980). The incisive formula, however, appears to have been the idea of the French philosopher Etienne de Condillac, who in 1746 proposed a '*rasoir des nominaux*' as a device to cut through the tangles of nominalism (Safire 1999).

Occam's Razor thus implies cutting from one's argument all matters not supported by strong evidence, or irrelevant to making one's point. This now appears obvious – yet the consequences of this dictum for all sciences, including biology, are enormous, whether they are left implicit (as is commonly the case) or made explicit. Indeed, Occam's dictum largely defines good science, as well as its antithesis, 'adhockery'. The latter is the generation of *ad hoc* hypotheses for every new realization of the same phenomenon, similar to the sausages that emerge each time the crank of a sausage machine is turned.

Adhockery is easy to detect: it is usually prefaced by sentences such as: "This may be so in theory; however, in the case of the [taxonomic group, process, place or period] I work on, things are more complicated. . . ." This is then followed by a complex hypothesis, backed by data for only the very taxonomic group, process or place in question. (*See also* **Difficulties of theory**.)

Yet, as the accounts of major scientific new ideas indicate, the key step in their discovery has been simplification, i.e. the elimination of all 'entities' not related to the problem at hand. *Linnaeus, for example, forced all of the Earth's organisms into a system of unique scientific *names, thus cutting away the often page-long chunks of Latin prose (i.e. the descriptions) previously used for species identification.

Similarly, *Lyell managed to cut (for a while) ancient catastrophes out of the geological discourse, thus giving CD the deep time he needed for *evolution by means of *natural selection.

CD himself cut the long established link between 'Design' and 'Function', and thus between Creator and Creation.

Ontogeny recapitulates phylogeny The observation that during its development, "an individual repeats the most important changes in form evolved by its ancestors during their long and slow paleontological development" (Haeckel 1866, Vol. 2, p. 300; Gould 1977, p. 77).

There is an enormous literature on this observation (reviewed in Gould (1977) and Arthur (2002)), and many variants. One of these was proposed by Milne-Edwards (1844) and endorsed by CD, namely that the more widely two animals differ from each other, the earlier does their embryonic resemblance cease; thus a fish on the one hand, & mammals together with birds on the other hand branch off from the common embryonic form at a very early period, whereas mammals & birds being more closely related to each other than to fish, diverge from each other at a later period. This seems to be in accord with Mr. Brullé's principle that the more each part is changed from the common archetype, the earlier it is developed; for as a fish differs in nearly all its organization from a mammal more than a bird differs from the mammal, the fish as a whole would have to be differentiated at an earlier period than the bird. (*Big Species Book*, pp. 303–4; CD later agreed with *Huxley when the latter criticized Mr. Brullé's principle as applied to the development of fish *brains).

Whatever our view about Haeckel and what he called his "phylogenetic law", it is true that at the age of six weeks, human embryos sport four round paddles at the ends of their limbs, with webbing between the fingers and toes (Smith 1999), as had our fish-like ancestors.

Ophidium(-on) *See* **Cusk eels**.

Origin Short title of *On the Origin of Species by means of Natural selection, or the Preservations of Favoured Races in the Struggle for Life*, CD's masterwork of 1859.

So much has been written about *Origin* that there is no point dealing with this here (but see entries on **Foundations** and **Big Species Book**).

The fish-related quotes extracted from *Origin* and included here stem from the first and/or the sixth edition (1876; the last revised by CD). Readers interested in the intervening texts may consult the variorum edition (Peckham 1959).

Ornithological Notes Short title of a contribution by CD's granddaughter, Nora Barlow (1963), who edited the notes on *birds that CD penned during the voyage of the *Beagle*. These notes, extracted from CD's larger set of *Zoology Notes* (Keynes 2000), were scanned for references to fishes, incorporated into the entries referring to various bird species.

Ostracion punctatus *See* **Boxfishes.**

Ostriches Large flightless birds of Africa, of which two relatives, *Rhea americana* and *R. darwinii*, occur in the vast 'pampa' of Argentina.

The former species, also known as 'nandu', gained entry into this book through the following quote: The ordinary habits of the ostrich are familiar to every one. They feed on vegetable matter; such as roots and grass; but at *Bahia Blanca, I have repeatedly seen three or four come down at low water to the extensive mud-banks which are then dry, for the sake, as the Gauchos say, of catching small fish. (*Journal*, August 1833; the same information, still unverified, may also be found in *Birds*, p. 121).

Otolithus spp. *See* **Croakers.**

Otters Aquatic mammals, belonging to the Order Carnivora, and inhabiting a wide range of mainly freshwater systems, where they consume fish, and invertebrates such as sea urchins and bivalves in the case of the marine otters.

CD reported on freshwater otters from Tahiti, where: Shaded by a ledge of rock, beneath a façade of columnar Lava, we ate our dinner. My guide before this had procured a dish of small fish & fresh-water prawns. They carried with them a small net stretched on a hoop; where the water was deep in eddies, they dived and like otters by their eyesight followed the fish into holes & corners & thus secured them. (*Diary*, Nov. 18, 1835; the small fish in question appear to have been *eels).

CD also discusses marine otters, if implicitly, in his comments on *kelp, but the most prescient comments on the ecological role of otters was passed on to CD by Francis Galton, who wrote him, with reference to "[i]ndirect effects" that "M^r. Young of Invershin told me years ago that he did not approve wholly of killing otters in order to preserve salmon. Otters killed a few salmon but they killed many trout & the Salmon *fry had no greater enemy than trout. Therefore he actually preserved the otters in more than one instance with the view to the advantage of the Salmon." (*Correspondence*, Dec. 9, 1859).

One wishes this were known to those who ask for 'culls' of *marine mammals to help rebuild fish populations that have been overfished. (*See also* **Keystone species**.)

Ova Used by CD for what may also be called *eggs. Thus the note, relevant to the *distribution of fishes and other animals: Fish & drift sea weed – may transport ova of shells. – Conchifera. hermaphrodites – *eggs in groups. (*Notebook C*; p. 316), and the question: Will ova of fishes & Mollusca <<& Frogs>> pass through bird stomachs & live? (*Notebook D*, p. 392).

CD tells us, incidentally, how the above question could be answered: Make Duck eat Spawn, eggs of snails, row of fish & kill them in hour or two (*Notebook Q*, p. 496; *see also* **Experiments I–VI; Fecundity**).

Owen, Richard English anatomist (1804–92), now famous for his feuds with evolutionists, especially *Huxley (Rupke 1994).

His relationship with CD started rather well: he authored that part of *Zoology* devoted to *Fossil Mammalia* (Owen 1840). Also, he is cited in much of CD's writing, by virtue of the broad scope of his work, which included the original description of one of the six extant species of *lungfish, and his development of criteria for distinguishing *analogous organs from their homologous counterparts.

However, his text on the anatomy of the vertebrates, including fishes (Owen 1846) very nicely illustrates the 'old school,' wherein lots of unconnected facts are presented, or where the theory consists mainly of attributing to animals a 'striving toward Man'.

Wilder (1923, p. 568) comments that "[n]ever was there a more stupendous result of the labor of a single human life than this great work of Owen, and yet of the entire structure reared by his incessant toil all that remains is the large amount of accurate description and a great enrichment of osteological nomenclature. It was a house built upon the sand."

That Owen could have believed he developed a theory comparable to CD's is hard to believe, but so it was. One effect is that Owen gradually became an outspoken critic of emergent Darwinism, often scoring minor hits against some of its unsubstantiated claims (Gould 1998a, pp. 119–40), while more often shooting himself in the foot, e.g. by attempting to use the dinosaurs, which he was first to describe, to refute CD (Gould 1998b).

But it is his *Palaeontology* (Owen 1860, p. 151), and specifically his review of fossil teleosts, that gives the game away: "those species such as the nutritious *cod, the savoury *herring, the rich-flavoured *salmon, and the succulent *turbot, have greatly predominated in the period immediately preceding and accompanying the advent of man, and that they have superseded species which, to judge by the bony *Garpikes (*Lepidosteus*), were very much less fitted to afford mankind a sapid and wholesome food". Patterson (1981) is right in dismissing him as a "pious victualler".

Oxford University Museum A place of many wonders, including some parts of CD's *collections.

Chancellor *et al.* (1988) present a detailed account of the state of CD's collection of zoological specimens (consisting mainly of crustaceans) in this museum. This includes the following items pertaining to fish, citing numbers and texts as written by CD on the specimens' metal and paper labels, and in his *Catalogue* and *Zoology Notes*:

145 Shells. Crustacea & Fish (Caught at Bahia from Feb 29th to March 17th -);

347 Fish. Coast of Patagonia Latitude 38° 20′ August 26th – Soundings 14 fathoms. Caught by hook & line V.77. (Monte Video: August 1832). Chancellor *et al.* (1988) remark here that: "the fish referred to is *Percophis brasilianus* described by Jenyns (1840–2: 23). This fish was caught by hook and line in fourteen fathoms water on the coast of Patagonia, in lat. 38° 20′ and was noted by Darwin as when cooked, was good eating. This specimen is no longer in the Zoological Museum Cambridge and is presumably not extant." (*see also* Fish in Spirits, no. 347).

551 Sphaeromida from stones & a Crust. Macrouri. from stomach of a *Gadus.- (*1833 March Tierra del Fuego [. . .] G[ood] S[uccess] Bay; see also Fish in Spirits*, note before no. 538: on Gadus being common about Cape Fairweather; it leaves the coast in March);

1269 Fish [. . .], which refers to the *wrasse described as *Cossyphus darwini* (*Fish* pp. 100–2), now *Pimelometopon darwini* (Jenyns), and the specimen is kept at the Museum of Natural History, in London. (*See* **Collections**.)

Now this was admittedly a tad boring; but so are some museums, as well.

Oxygen The element required by animals (along with food) for metabolism and *growth. Most atmospheric O_2 originates from plants (notably *phytoplankton) as a by-product of a reaction wherein carbon dioxide (CO_2) and water (H_2O) are used, along with the energy in sunlight to synthesize carbohydrates (Schrödinger 1967; *see also* **Food webs**).

To be metabolically useful, oxygen must be *inside* an organism, and this is a big problem for water-breathers such as fish, which must deal with a medium containing 30 times less oxygen than air, which is 55 times more viscous than air, and in which diffusion is 300 000 times slower than in air (Pauly 1981). These characteristics imply that a significant fraction of the routine metabolism of fish must be devoted to breathing itself, with values of up to 40% having been estimated in *Tench.

Moreover, oxygen cannot be stored by animals in more than minute amounts – at least compared with food, which can be stored for relatively long periods as stomach contents (as in *sharks) or as lipids (as *cod liver oil). Thus, in fish and other water-breathers, oxygen supply through the gills largely determines activity levels and growth rates.

Moreover, gill size (i.e. surface areas) determines the ultimate size fish can reach, with large gills implying large ultimate size and vice versa. Consequently, ultimate size is also influenced by environmental factors that affect oxygen consumption (i.e. metabolic rate), notably water temperature. Thus, warm-water fishes, other things being equal, tend to remain smaller than their cold-water congeners, and not only because there is less dissolved oxygen in warm water.

A *parsimonious theory articulated around this limiting effect of oxygen has been presented by Pauly (1979, 1981, 1984, 1994a, 1997, 1998, with more to come). It is generally ignored, though it nicely explains various aspects of growth mentioned, but not resolved by CD, or by others who touched upon this subject (*See* **Growth; Reproductive drain hypothesis**).

Panama, Isthmus of The narrow stretch of land separating the Atlantic and Pacific Oceans at latitudes from 8 to 9° North, and 80° West.

CD used the isthmus of Panama to illustrate a 'law,' i.e. that barriers of any kind, or obstacles to free *migration, are related in a close and important manner to the differences between the *productions of various regions. (*Origin I*, p. 347). Also, in that same first edition, CD states that: No two marine *faunas are more distinct, with hardly a fish, shell, or crab in common, than those of the eastern and western shores of South and Central America; yet these great faunas are separated only by the narrow, but impassable, isthmus of Panama. (p. 348).

In the sixth edition, however, the isthmus of Panama had become quite passable: The marine inhabitants of the eastern and western shores of South America are very distinct, with extremely few shells, crustacea or echinodermata in common; but Dr *Günther has recently shown that about thirty percent of the fishes are the same on the opposite sides of the isthmus of Panama; and this fact has led naturalists to believe that the isthmus was formerly open. (p. 317).

Günther's (1869) demonstration was based on counts of 193 marine and brackish-water species on either the Atlantic side (Belize to Panama) and/or the Pacific side (Guatemala to Panama) of the isthmus, of which 59 occurred on both sides, i.e. 30.6%. The corresponding figures, based on analysis of contemporary records and the modern taxonomy in *FishBase, are 1461 (689 on the Atlantic side and 801 on the Pacific side), with only 60 species (3.7%; mainly large pelagics) shared between the two sides of the isthmus. This is one order of magnitude less than Günther calculated and, incidentally, an indication of the need to keep abreast of taxonomy when discussing issues of biodiversity (see also Collins 1996).

Much to his credit, CD changed his mind in the face of Günther's then current data, from the Isthmus of Panama being impassable, and the faunas on its both sides being very distinct, to being formerly open, with both sides sharing many species. However, the above quotes illustrate an instance where CD's recycling of 'facts' may have gone too far, especially since he does not highlight his flip.

This is particularly true since Günther (1869, p. 398) made the need for a correction of the relevant part of *Origin* quite explicit: "Mr Darwin (Origin of Species, 3rd edit, p. 378) was not acquainted with [the fact that so many species were shared], which by no means militates against his argument, but merely modifies it."

Paradigm Originally a 'pattern' (e.g. Latin: *piscis, piscem, piscis, pisci, pisce*), used in the much quoted, but less read classic by Kuhn (1970) to refer to the complex of methods, and of key experiments, results, personalities and texts that articulate a scientific discipline and its research programme (see also Lakatos 1970).

Paradigms and paradigm shifts, while necessary to enable progress within a discipline – indeed, to define what it is that is worth *seeing – are not perceived as such by practitioners (Kuhn speaks of their "invisibility"). Rather, they define, at any given period, what is considered 'good science' within a discipline. Hence, practising scientists usually do not, and most probably need not, reflect much on 'their' paradigm, and should even less brag about 'shifting' it. It is enough that they should attempt to push the limits of their disciplines, while striving for *parsimony. Later, historians can always attempt to distinguish the paradigms further articulated from those that shifted, etc.

And yes, CD did create a new paradigm, though one could not tell if one were to read only conventional philosophers of science, whose lazy authors (excepting Ruse (1986),

Wilson (1991), and a few others) do not seem ever to be able to look beyond Newton and Einstein.

Parasites An organism living upon or in, and living at the expanse of, another organism (*Origin VI*, p. 438). CD wrote very little about the parasites of fishes, and indeed [t]he incalculable host of parasites which pass their whole lives on or in the bodies of other animals do not here especially concern us. (*Big Species Book*, p. 506; note the pun involving 'host' = an army, or many, and host = what parasites need).

However, CD's *Catalogue* and *Zoology Notes* indicated that he collected some fish parasites during the voyage of the *Beagle*: 351 Isopod (Bopyrus?) on fish (Coast of Patagonia [. . .] 14 fathoms.); 802 Crust. Isopoda. – I believe certainly was on the body of a large *dog fish. colour above. mottled greenish grey & tile red, edge dark brown. (Watchmen Cape L. 48°.18'., January 1834); 1269 Fish; 1270 Crust Parasit on Fish (1269) (1835 Septemb. Galapagos. Chatham Isd.; the parasite in question was identified as a copepod by Chancellor *et al.* (1988), who also noted that the specimen is "in fair condition"). Later, CD also wrote that: Many parasitic Crustaceans have their limbs atrophied when attached for life to fishes. (*Big Species Book*, p. 294).

Fish, while having lots of parasites (Möller and Anders 1989), generally do not function *as* parasites. Only the *lampreys are considered parasites as defined above, as are the diminutive males attached to females in many, but not all, species of ceratoid anglerfish (Pietsch 1976). One example of an anglerfish with parasitic males is the Triplewart seadevil *Cryptopsaras couesii* Gill, 1883. [As an aside, one might wonder why anyone would want to be attached to an aquatic triplewarted devil, female or not. Some males are like that, though.]

Other fish species, e.g. the Cookiecutter shark *Isistius brasiliensis* (Quoy & Gaimard, 1824), the Malawi eyebiter *Dimidiochromis compressiceps* (Boulenger, 1908), or the False cleaner fish *Aspidontus taeniatus* Quoy & Gaimard, 1834, take only chunks of larger fish, and may be considered 'predators', though picky ones.

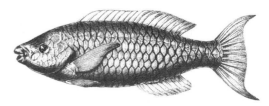

Fig. 21. Singapore parrotfish *Scarus prasiognathos*, Family Scaridae (*Scarus chlorodon* in *Fish*, Plate XXI), based on a specimen from the Cocos Islands.

The *Remora, usually attached to sharks, and the pearlfishes (Family Carapidae), which spend the daylight hours inside the guts of sea cucumbers, may represent cases of commensalism. Further, some fish are brood parasites, i.e. function as the *Cuckoo fish CD was wondering about.

Paropsis signata See **Jacks**.

Parrotfishes Fishes of the Family Scaridae, possessing teeth fused to form a beak capable of biting chunks of live corals and encrusting *algae, the latter usually forming the bulk of their diet.

CD collected his first scarid species in *Tahiti (*Fish in Spirits*, nos. 1318–19), i.e. the Globehead parrotfish *Scarus globiceps* Valenciennes, 1840, described by *Jenyns as *Scarus lepidus* (*Fish*, pp. 107–8). Later, in the *Cocos Islands, CD again collected a specimen of *S. globiceps* (this time identified as such; *Fish*, pp. 106–8). His collection there also included the Singapore parrotfish *Scarus prasiognathos* Valenciennes, 1840 ('*Scarus chlorodon*' on pp. 105–6 of *Fish*; Fig. 21), and '*Scarus* sp.' (p. 109), which could be any of the other eleven *Scarus* species in the Cocos Islands (Allen and Smith-Vaniz 1994).

'*Scarus*', the name then used for all parrotfishes, can be easily misread or misspelled *Sparus* (a genus of *porgies), and this appears to have happened in CD's journal entry for the Cocos Islands: there are here two species of fish, of the genus Sparus, which exclusively feed on coral. Both are coloured of a splendid bluish-green, one living invariably in the *lagoon, and the other amongst the outer breakers. Mr Liesk assured us that he had repeatedly seen whole *shoals grazing with their strong bony jaws on the tops of the coral branches. I opened the intestines of several, and found them distended with a yellowish calcareous matter. These fish, together with the lithophagous shells and nereidous animals, which perforate every block of dead coral, must be very efficient agents in producing the finest kind of mud, and this, when derived from such materials, appears to be the same with chalk. (*Journal*, April 6, 1836).

In a footnote referring to Quoy and Gaimard (1824–5), our ever-intrepid CD then added that It has sometimes been thought (*vide* Quoy in Freycinet's) *Voyage*, that coral-eating fish were poisonous; such certainly was not the case with these Spari. If CD tested parrotfish by consuming one, then he exposed himself to a real danger, as parrotfishes, in the Pacific, are often toxic, owing to their ingestion of *ciguatera (Auerbach 1991; Bagnis *et al*. 1974; *see also* **Burrfishes (I)**).

Observing parrotfishes (probably the species he also sampled) our lucky CD turned the *Spari* back into the *Scari* they always should have been: On the outside of the reef much sediment must be formed by the action of the surf on the rolled fragments of coral; but, in the calm waters of the *lagoon, this can take place only in a small degree. There are, however, other and unexpected agents at work here: large shoals of two species of Scarus, one inhabiting the surf outside the reef and the other the lagoon, subsist entirely [. . .] by browsing on the living polypifers. I opened several of these fish, which are very numerous and of considerable size, and I found their intestines distended by small pieces of coral, and finely ground calcareous matter. This must daily pass from them as the finest sediment (*Coral Reefs*, pp. 19–20).

Later CD returned to the feeding adaptation of parrotfish, asking himself [w]hat are the "palatal Tritores" found in the coraliferous mountain Limestone [?] are they allied to the jaws of the Cocos fish (*Notebook R*, p. 30).

Ancestral parrotfishes may have had triturating teeth on the roofs of their mouths. Still, CD was quite wrong in his belief, at the time this was written, that parrotfishes and other coral reef animals had produced the chalk of the *Cretaceous.

Parsimony(-ious) The simplicity (and brevity) of an explanation, relative to the complexity of a set of observed phenomena seen as requiring explanation.

In science, parsimonious explanations are preferred over intricate ones, which often only restate the phenomena they are supposed to explain. As put by *Popper (1980, p. 142): "[s]imple statements, if knowledge is our object, are to be prized more highly than the less simple ones because they tell us more; because their empirical content is greater; and because they are better testable".

*Natural selection and *sexual selection, as proposed by CD, are excellent examples of parsimonious theories, as they explain, with a few *words, a wide array of phenomena. (*See also* **Occam's Razor, Obstinate nature, Oxygen**).

Patagonia Southeastern part of the South American mainland, encompassing several provinces of the Republic of Argentina (Santa Cruz, Chubut, Rio Negro, etc.), all south of Buenos Aires, and north of *Tierra del Fuego.

The fish collected by CD along the Patagonian coast were assigned to specific segments of that coast when described in Jenyns' *Fish*. Thus, in addition to the five species from *Bahia Blanca, the following eight species were sampled in 'Northern Patagonia': *Plectropoma patachonica* (*see* **Groupers**); *Pinguipes fasciatus* (*see* **Sandperches**); *Percophis brasilianus* (*see*

Duckbills); *Umbrina arenata* (see **Croakers**); *Paropsis signata* (see **Jacks**); *Atherina incisa* (see **Silversides**); *Mugil liza* (see **Mullets**); and *Alosa pectinata* (see **Herrings**). Only one species, *Aphritis porosus* (see **Thornfishes**) originated from Central Patagonia, while two species were from Southern Patagonia: *Perca laevis* (see **Creole perch**) and *Mesites maculatus* (see **Galaxiidae**).

Penguins Flightless birds known for their swimming ability. Though most abundant around Antarctica, where they exhibit a remarkable tolerance to cold and feed on krill (*Euphausia superba*), a few species reach further north. The northernmost species (*Spheniscus mendiculus*) occurs in the *Galápagos.

CD performed a simple behavioural experiment (similar to the one he did with *lizards in the Galápagos) with the penguins in the *Falkland Islands: [o]ne day, having placed myself between a penguin (Aptenodytes demersa) and the water, I was much amused by watching its habits. It was a brave bird; and till reaching the sea, it regularly fought and drove me backwards. [. . .] When at sea and *fishing, it comes to the surface, for the purpose of breathing, with such a spring, and dives again so instantaneously, that I defy any one at first sight to be sure that it is not a fish leaping for sport. (*Journal,* March 19, 1834; at least for Antarctic penguins, not lingering at or near the surface makes a difference to the ability of Leopard seals to catch them, as these seals are capable of [un]dressing a penguin, their favourite prey, with only a few bites).

Yet, in spite of their aquatic prowess, penguins are not fishes. CD, even when still looking for his key to the evolutionary riddle, was well aware not only of the gulf that separates penguins from fishes, but also of the difficulties involved in connecting such widely separated groups: Thus, we have a species as soon as once formed by separation or change in part of country, repugnance to intermarriage <increases> – settles it.? We need not think that fish & penguins really pass into each other.– The tree of life should perhaps be called the coral of life, base of branches dead; so that passages cannot be seen. – this again offers contradiction to constant succession of germ in *progress. – << not only makes it excessively complictated>> Is it thus fish can be traced right down to simple organization – birds – not. (*Notebook* B, pp. 176–7; original shows here two simple branching trees; contradiction refers to *Lamarck). And: between Mammalia & fishes, one penguin, one tortoise shows hiatus – but not saltus – when Linnaeus put whale between cow & hawk a frolicsome saltus. (*Notebook C*, p. 287).

Perca laevis See **Creole perch**.

Perch Common name of *Perca fluviatilis* Linnaeus, 1758, the most common representative, in Europe, of the Family Percidae, i.e. the perches. 'Perch' is also used for other percids; thus CD wrote upon reading p. 292 of Agassiz (1850) case of variability in a *Perch, good as for Agassiz (*Marginalia* 11–12).

This refers to Yellow perch *Perca flavescens* (Mitchill, 1814), whose many local forms *Agassiz, in one of his many leaps from logic, saw as further proof of the fixity of species.

Percophis brasilianus See **Duckbills**.
Percopsis See **Trout-perch**.
Perilampus perseus A small species of the Family *Cyprinidae, occurring in India, described by McClelland (1839, p. 395), and emphasized in CD's annotation to this work.

McClelland thought that this and similar small species were created with the high fecundities required to feed predators (p. 458), whose search for their prey is further facilitated by the latter's bright colours, the whole arrangement implying a form of what may be called 'suicidal *altruism' (p. 230). CD, in his annotation of this work, points out that these notions, if true, would force him to say farewell to my thesis, i.e. to abandon *natural selection. The current identity of '*Perilampus perseus*' cannot be established, owing to the insufficient description in McClelland (1839), and the absence of a *holotype. It was probably as species of *Danio* or *Brachydanio*, genera well known amongst aquarists.

Peru The country of South America where CD began to see *mermaids, but still had enough control over himself to collect what turned out to be two new fish species (*Otolithus analis* and *Clupea sagax*), or even three if *Iquique, now in Chile, is counted as a Peruvian city (as it was in 1835). This third species was *Engraulis ringens*, the *Peruvian anchoveta, presented below.

Peruvian anchoveta The most abundant member of the Family Engraulidae, the Peruvian anchoveta is the fish that got this book started (*see* **Preface**, and Pauly 1992).

The species was described by Jenyns (1842, pp. 135–7) from a specimen (no. 1229 in *Fish in Spirits*) collected by CD in *Iquique, formerly in Peru, now in Chile (clearly a great source of misunderstandings). From the mid 1960s to the early 1970s, the Peruvian anchoveta supported the largest single-species fisheries in the world, with official landings of 12 million tonnes in 1970 (underestimating the true catch of 16–20 million tonnes; Castillo and Mendo 1987); before it quite predictably collapsed. This, and the fact that I have authored numerous papers on *Engraulis ringens*, and been the senior editor of two books on its *ecology and fisheries (Pauly and Tsukayama 1987; Pauly *et al*. 1989; *see* **Punctuated equilibrium**), almost made me write more than necessary about a species which, most people believe, contributes nothing but the nasty, salty bits on pizzas. Hence, as many do when ordering a pizza, we shall 'hold the anchovies'.

Petrels Seabirds of the Family Procellariidae, represented here by *Mother Carey's chickens, and Grey petrels, which have nothing to do with *bird's nest soup, and whose peculiar feeding habits were discussed by CD in the context of oceanic *food webs.

The description of the Grey petrel *Puffinus cinereus* in *Birds includes an account of their behaviour, based on observations made by CD near Port Famine, *Patagonia: Their flight was direct and vigorous, and they seldom glided with extended wings in graceful curves, like most other members of this family. Occasionally, they settled for a short time on the water; and they thus remained at rest nearly the whole of the middle of the day. When flying backwards and forwards, at a distance from the shore, they evidently were *fishing: but it was rare to see them seize any prey. They are very wary, and seldom approach within gun-shot of a boat or of a ship; a disposition strikingly different from that of most of the other species. The stomach of one, killed near Port Famine, was distended with seven prawn-like crabs, and a small fish. In another, killed off the Plata, there was the beak of a small cuttlefish. (p. 138).

The killing of specimens was not always as neat as the text above seems to imply. This is illustrated by the following quote from the *Ornithological Notes*: Puffinus. this bird is very abundant in the Sts. of Magellan, near Port Famine [. . . .]. One being slightly wound[ed], was quite unable to dive. Stomach much distended, with a small fish & seven or eight good size Crust. Macrouri (pp. 230–1).

On the other hand, Harrison (1987), describing the flight of the Grey petrel (now *Procellaria cinerea*), emphasizes their awkward, stiff wingbeats, as well as their tendency to fly high, i.e. away from ships (p. 207). Thus, CD's observations on the flight of Grey petrels are confirmed.

Phosphorescence *See* **Bioluminescence**.
Phucocoetes latitans *See* **Eelpouts**.
Phytoplankton The community of microscopic plants which inhabits the lighted zone of the oceans, and of freshwater systems. By fixing carbon dioxide, and releasing *oxygen,

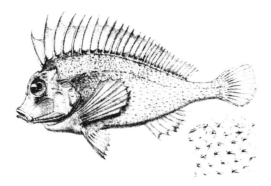

Fig. 22. Pigfish *Congiopodus peruvianus*, Family Congiopodidae (*Agriopus hispidus* in *Fish*, Plate VII), based on a juvenile from Tres Montes Peninsula, southern Chile. The insert shows magnified scales.

phytoplanktonic organisms, like other plants, contribute to closing the biological cycles within which animals consume oxygen and release carbon dioxide. CD was unaware of the existence of phytoplankton, and this affected his perception of marine *food webs. (*See* **Plankton**.)

Pigfishes Members of the Congiopodidae, small, predominantly benthic fishes from the Southern Hemisphere, with a strange appeare[ance, due to their] bony face (*Fish in Spirits*, no. 1123).

Jenyns described a new species of pigfish which he called *Agriopus hispidus* (*Fish*, p. 38), based on two specimens of less than 5 cm, caught in middle of cave near the Peninsula of Tres Montes, southern Chile (*Fish in Spirits*, no. 1123; Fig. 22).

These specimens (Pale reddish orange, with black spots on the fins, and a dusky shade on the back) differed from the previously described *Agriopus peruvianus* (which reaches about 30 cm) only in minor features of their dentition. Interestingly, *Jenyns was hesitant about the status of this species, noting that "[p]erhaps it may be the same as the species brought from the coast of Chili by Mr Cuming, and briefly noticed by Mr Bennett in the *Proceedings of the Zoological Society* (1832, p. 5), but which this last gentleman did not venture to describe as new" (*Fish*, p. 40; *see* **Distribution** for more on Cuming).

And indeed, in the Appendix of *Fish* (p. 163), *Jenyns backtracked, acknowledging that "[n]otwithstanding what I have advanced in regard to this species, further consideration has inclined me to suspect, that it may prove ultimately only the young of the *A. peruvianus*".

This example illustrates a widespread phenomenon in fish taxonomy, wherein 'new' species turn out to refer to juvenile forms, with the newly proposed scientific names then turning into *synonyms. As the generic name *Agriopus* itself has become a synonym as well (Eschmeyer 1990, p. 17), the valid name for our species is *Congiopodus peruvianus* (Cuvier, 1829).

Pike Common name of *Esox lucius* Linnaeus, 1758, also known as the 'European' pike, of which the North American pike *Esox reticulatus* Rafinesque, 1814, may be a *synonym.

CD describes the latter as follows: The colours of the pike (*Esox reticulatus*) of the United States, especially of the male, become, during the breeding-season, exceedingly intense, brilliant, and iridescent.[24] (*Descent II*, p. 340; n. 24 cites *The American Agriculturist* (1868), p. 100).

Though appreciating their *beauty, CD did not think much of the intellectual capacity of pike: What a contrast does the mind of an infant present to that of the pike, described by Professor *Möbius, who during three whole months dashed and stunned himself against a glass partition which separated him from some *minnows; and when, after at last learning that he could not attack them with impunity, he was placed in the aquarium with these same minnows, then in a persistent and senseless manner he would not attack them! (from *A biographical sketch of an infant*. *Collected Papers II*, pp. 191–200; Möbius 1873).

This story had already been used by CD in *Descent (I, pp. 75–6): A curious case has been given by Prof. Möbius[23] of a pike, separated by a plate of glass from an adjoining aquarium stocked with fish, and who often dashed himself with such violence against the glass in trying to catch the other fishes, that he was sometimes completely stunned. The pike went on thus for three months, but at last learnt caution, and ceased to do so. The plate of glass was then removed, but the pike would not attack these particular fishes, though he would devour others which were afterwards introduced; so strongly was the idea of a violent shock associated in his feeble mind with the attempt on his former neighbours. If a savage, who had never seen a large plate-glass window, were to dash himself even once against it, he would for a long time afterwards associate a shock with a window-frame; but very differently from the pike, he would probably reflect on the nature of the impediment, and be cautious under analogous circumstances. Now with monkeys, as we shall presently see, a painful or merely a disagreeable impression, from an action once performed, is sometimes sufficient to prevent the animal from repeating it. If we attribute this difference between the monkey and the pike solely to the association of ideas being so much stronger and more persistent in the one than the other, though the pike often received much the more severe injury, can we maintain in the case of man that a similar difference implies the possession of a fundamentally different mind? (n. 23 gives as source: Die Bewegungen der Thiere, etc. 1873, p. 11., i.e. Möbius 1873).

And why not recycle it again, to illustrate another aspect of life's grandeur? So here we go, this time in *Worms (pp. 93–4): It is surprising that an animal so low in the *scale should have the capacity [to plug their burrows by dragging in objects], as many higher animals have no such capacity. [. .] Even a pike continued during three months to dash and bruise itself against the glass sides of an aquarium, in the

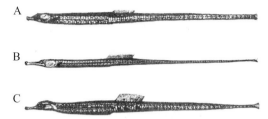

Fig. 23. Pipefishes, Family Syngnathidae (*Fish*, Plate XXVII). **A** Banded pipefish *Micrognathus crinitus* (*Syngnathus crinitus* in *Fish*), from Bahia Blanca, Argentina. **B** Deep-bodied pipefish *Leptonotus blainvilleanus* (*Syngnathus acicularis*), based on a female specimen from Valparaiso, Chile. **C** Network pipefish *Corythoichthys flavofasciatus* (*Syngnathus conspicillatus*), from Tahiti.

vain attempt to seize minnows on the opposite side.

In the first two quotes, CD faults the pike for *not* attacking the minnows, i.e. for having learnt something. In the third quote, the pike is faulted *for* attacking the minnows, i.e. for not learning fast enough. This doesn't seem fair to pike.

Pimelodus spp. *See* **Catfishes.**
Pinguipes spp. *See* **Sandperches.**
Pipefishes (I) Very thin, elongated fishes, members of the Family Syngnathidae ('fused jaws'). In this family (to which the *seahorses also belong), it is the males that incubate the *eggs, while the females compete for access to males (Berglund and Rosenqvist 1993; Kuiter 2000).

The three well-preserved pipefish specimens that CD collected were described by Jenyns as three new species of *Syngnathus*. *Fish in Spirits* (no. 1154) lists another specimen, "in very bad condition, & thrown away" by Jenyns. Moreover, he had been anticipated by several years for two of his presumed new species.

The single hit was the Banded pipefish *Micrognathus crinitus* (Jenyns, 1842), which ranges from Florida to Argentina, and which CD had sampled at *Bahia Blanca (*Fish*, pp. 148–9; *Fish in Spirits*, no. 138; Fig. 23A). Guimarães

(1999) recently discussed the status of this species, so far confused by the occurrence of two forms, one common, plain or inconspicuously banded, the other conspicuously banded (the 'ensenadae' form), and now shown to intergrade.

The second of CD's species, the Deep-bodied pipefish, was sampled in Valparaiso, Chile (*Fish in Spirits*, no. 1075; Fig. 23B), and described as *S. acicularis* by Jenyns. It is now recognized as *Leptonotus blainvilleanus* (Eydoux & Gervais, 1837), which ranges from Argentina to Northern Chile (De Buen 1966). The third species is the Network pipefish *Corythoichthys flavofasciatus* (Rüppel, 1838), occurring throughout the Indo-Pacific and whose description by Jenyns – as *S. conspicillatus* – was based on a specimen from *Tahiti (*Fish in Spirits*, no. 1328; Fig. 23B).

However, what is interesting about pipefishes is their reproductive habits, as recorded in *Notebook D* (p. 387): Notice the Syngnathus, or Pipe fish the male of which receives <young> <<eggs>> in belly. – analogous to men having mammæ; and: The Pipe-fish is instance of part of the *hermaphrodite structure being retained in the male. – <like> <<far>> more than marsupial bones, & even more than Mammae, which have given milk. (*Notebook E*, pp. 412; source is Fries (1839a), pp. 96–7).

Also: Annals of Natural History (p 225. 1838) account of metamorphosis in the young of Syngnathus.= curious as showing generality of law. even in fish (*Notebook E*, p. 421; Fries 1839b).

We also have from *Notebook T* (p. 461): Translation of P. Fries most curious paper on the Pipe-fish – which he divides into two divisions, one of which are marsupial & the other have young which undergo metamorphosis & are provided with fins, & hence do not require sac. – but the male in these hatch young – are there not some Marsup. Mammalia, which <do> have not sack, – Most curious fact & this paper deserves fresh study & whole order of the fish. – *Embryology. p. 97. for Man Chapt see Yarrell Syngnathus. (The most curious paper is Fries (1839a), p. 96 and (1839c), p. 451; Man Chapt is a manuscript chapter written in 1785 by John Walcott and cited by both Fries (1839a), p. 97 and Yarrell (1836), Vol. 2, pp. 327–8).

Then we have this neat quote, highlighting CD's interest in *colours: One of the most striking instances ever recorded of an animal being protected by its colour (as far as it can be judged of in preserved specimens), as well as by its form, is that given by Dr Günther[33] of a pipe-fish, which, with its reddish streaming filaments, is hardly distinguishable from the sea-weed to which it clings with its prehensile tail. (*Descent II*, p. 344; n. 33 refers to Günther (1863) [should be 1865b], p. 327, and to plates xiv and xv, displaying the Weedy sea dragon *Phyllopterix taeniolatus* (Lacepède, 1804) and the Leafy sea dragon *Phycodurus eques* (Günther, 1865) respectively. Both are *endemic to southern Australia (Coleman 1980, p. 88; Gomon *et al*. 1994, pp. 462–4). Their biology is discussed by Groves (1998), whose article also includes beautiful pictures.

Pipefishes (II) Our second entry on pipefishes illustrates what can happen – even when one is as careful as CD was – when citing dubious sources: In most of the *Lophobranchii (pipefish, Hippocampi, etc.) the males have either marsupial sacks or hemispherical depressions on the abdomen, in which the ova laid by the female are hatched. The males also show great attachment to their young.[39] (*Descent II*, p. 346; n. 39 cites Yarrell, *Hist. of British Fishes*, vol. ii, 1836, pp. 329 and 338).

Here are the Yarrell quotes from which CD took his line on the great attachment of Pipefish to their young: "M. Risso notices the great attachment of the adult Pipefish to their young, and this pouch probably serves as place of shelter to which the young ones retreat in case of danger. I have been assured by fishermen that if the young were shaken out of the pouch into the water over the side of the boat, they did not swim away, but when the parent fish was held in the water in a favourable position, the young would again enter the pouch" (Yarrell 1836, Vol. 2, p. 329; p. 338, also cited by CD, refers only to morphological features of *Syngnathus ophidion*).

Also, referring to a figure in "Rondeletius" (i.e. Rondelet 1554), Yarrell notes "below the figure, in that work, of [*Syngnathus* sp.], several of the young are represented as swimming near the abdomen of the parent fish. This figure of Rondeletius is copied in Willughby, plate I, 25, fig. 6." (Yarrell 1836, Vol. 2, pp. 329–30; Willughby 1686).

I translated as follows Risso's comments on pipefish reproduction: "a marked tenderness and limitless devotion for their young appear to characterize the feelings of the fishes in this genus. Thus, *Syngnathus* species have a very peculiar development: the *eggs do not hatch in the belly of the female. Rather, they flow into a membranous sack that forms below their tail, and which appears to open up longitudinally when the young, which appear fully formed to the light, are developed sufficiently to look after themselves."

Thus Risso (1810, p. 71), besides assigning the pregnancy to the wrong sex, does not suggest that the parent looks after the young once they are hatched; rather, he suggests the opposite.

Now to Rondelet's figure, as reproduced by Willughby (1686). This indeed shows seven small pipefishes obediently aligned under the body of their mother, while the eighth is pointing downward. This *mise-en-scène*, unfortunately, involves just one young too many for a perfect match with a syngnathous Snow White and six of her dwarfs, with the seventh, Dopey, wandering about, as he always does . . .

Clearly, what we have here is a phantasm, similar to the composite published by Bloch (1785), which sported the head and body of the Anglerfish *Histrio histrio* and the luring apparatus of *Antennarius striatus*, and which confused ichthyologists for 200 years, until Pietsch and Grobdecker (1987, pp. 69–70) squashed the thing.

We note, finally, that the anatomy of the brood pouch of male pipefish precludes the young getting back in, once they have been extruded. It does seem that Yarrell's fishers made fun of him, and indirectly, of CD.

Pipefishes, ghost Members of the Family Solenostomidae, from the tropical Indo-West Pacific. Members of the *Lophobranchii, which also includes the *pipefishes and *seahorses (Kuiter 2000).

After discussing the observation that male seahorses tend to be brighter than the females, CD notes that The genus Solenostoma, however, offers a curious exceptional case,[40] for the female is much more vividly-coloured and spotted than the male, and she alone has a marsupial sack and hatches the *eggs; so that the female of Solenostoma differs from all the other Lophobranchii in this latter respect, and from almost all other fishes, in being more brightly-coloured than the male. It is improbable that this remarkable double inversion of character in the female should be an accidental coincidence. As the males of several fishes, which take exclusive charge of the eggs and young, are more brightly coloured than the females, and as here the female Solenostoma takes the same charge and is brighter than the male, it might be argued that the conspicuous colours of that sex which is the more important of the two for the welfare of the offspring, must be in some manner protective. But from the large number of fishes, of which the males are either permanently or periodically brighter than the females, but whose life is not at all more important for the welfare of the species than that of the female, this view can hardly be maintained. (*Descent II*, p. 346; n. 40 reads: Dr. Günther, since publishing an account of this species in *The Fishes of Zanzibar*, by Colonel Playfair, 1866, p. 137, has re-examined the specimens, and has given me the above information.; see Playfair and Günther (1867); *see also* **Altruism**).

Plagiarism The use of somebody else's *words without proper attribution, with the aim of appearing to others as the author of these words.

CD, like most major authors, has been accused of plagiarism, notably by Eiseley (1959), who attempted to show that he had stolen key ideas from Edward *Blyth, of all people. Other accusations of plagiarism have been levelled against CD, notably concerning *Wallace, but the publication of *Foundations*, and especially of the *Notebooks*, which establish CD's priority by about two decades, should have finally settled this issue (as does the title of Wallace 1889).

Intellectually far more interesting is what may be called 'self-plagiarism', the re-publication of the same material by the same author (Garfield 1982), illustrated here by CD's recycling of Möbius' *Pike, his *Megatooth shark's tooth, and other stories.

I have made this even worse, by occasionally using part of the same story to document CD's interest in different fish species, e.g. the *flatfishes. One might call this 'allogenous self-plagiarism'.

Plagiostomous Earlier name for *cartilaginous fishes, especially *sharks and *rays.

Used by CD only once, when discussing reproduction: The males of Plagiostomous fishes (sharks, rays) and of Chimaeroid fishes are provided with claspers which serve to retain the female, like the various structures possessed by many of the lower animals. (*Descent II*, pp. 330–1; *see* **Chimaera**).

Plagusia *See* **Sole.**

Plaice Common name of *Pleuronectes platessa* Linnaeus, 1758, the *flatfish *par excellence* (or at least that after which the Family Pleuronectidae, and the Order Pleuronectiformes, including all flatfishes, are named).

CD mentions plaice only once: Perhaps the lesser number of teeth in the proportion of four to seven in the upper halves of the two jaws of the plaice, to twenty-five to thirty in the lower halves, may likewise be accounted for by disuse. (*Origin VI*, p. 188).

Plankton A term coined by Hensen (1887) and pertaining to the assemblage of microscopic *algae (*phytoplankton) and animals (*zooplankton) which drift in marine (and fresh) water, and form the basis of most aquatic *food webs (Smetacek 1999).

The absence of a clear concept of the nature and ecological role of plankton is evident in CD's writings, although he spent much time, especially in the first months of the voyage of the *Beagle* collecting and studying zooplankton.

Lack of a concept does not preclude enthusiasm, however: I proved to day the utility of a contrivance which will afford me many hours of amusement & work, it is a bag four feet deep, made of bunting, & attached to <a> semicircular bow: this by lines is kept upright, & dragged behind the vessel. This evening it brought up a mass of small animals & tomorrow I look forward to a greater harvest. (*Diary,* Jan. 10, 1832, near the Equator).

I am quite tired having worked all day at the produce of my net. The number of animals that the net collects is very great & fully explains the manner so many animals of large size live so far from land. Many of these creatures, so low in the *scale of nature, are most exquisite in their forms & rich in colours. It creates a feeling of wonder that so much *beauty should be apparently created for such little purpose. (*Diary*, Jan. 11, 1832).

Hensen, in 1887, proposed more than a concept for 'drifting organisms': he clearly identified *phytoplankton as the basis of aquatic *food webs (Smetacek 2001), and herbivorous *zooplankton as their first-level consumers. From this, he inferred the necessity for fisheries research, then a nascent discipline, to rigorously quantify the distribution of plankton, thus founding the 'Kiel School of Planktonologists' (Kölmel 1986; Schlee 1973, p. 229). In addition, he developed the vertically hauled 'Hensen net', field tested during Hensen's 'Plankton-Expedition' in the North Atlantic, in 1889. Its results established that plankton, away from coastal waters, has a predictable distribution over large areas and contrary to the expectations of the time, high biomass (and production) at high latitudes, and conversely at low latitudes (see, for example, the global map of primary production in Longhurst (1998), for both sets of patterns).

These results were hotly disputed by *Haeckel (1890, pp. 8–10; 90–103), who had earlier worked on the taxonomy of planktonic organisms, and who, blinded by a "variability" that he could not handle, aggressively opposed any application of mathematics (including descriptive statistics, a "dangerous science") to biological problems (Porep 1980; Smetacek 1999). He thus remained unaware of what has been called, in a neat phrase, the "unreasonable effectiveness of mathematics in the natural sciences" (Wigner 1960).

CD was different, and conceded in his *Autobiography*: I have deeply regretted that I did not proceed far enough at least to understand something of the great leading principles of mathematics; for men thus endowed seem to have an extra sense (p. 58). CD would have admired Hensen's work.

Platycephalus inops *See* **Flatheads.**
Plecostomus barbatus *See* **Catfishes.**
Plectropoma patachonica *See* **Groupers.**
Poachers Members of the Family Agonidae, small fishes covered with bony plates, represented here by *Agonopsis chiloensis* (Jenyns, 1840). This was described, based on a specimen collected by CD at *Chiloé, Chile, as *Aspidophorus chiloensis*

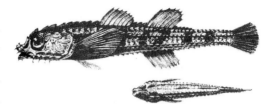

Fig. 24. Poacher *Agonopsis chiloensis*, Family Agonidae (*Aspidophorus chiloensis* in *Fish*, Plate VII), based on a specimen from Chiloé, Chile. Insert: dorsal view.

in *Fish* (pp. 30–3; *Fish in Spirits*, no. 1153; Fig. 24).

Poecilia spp. *See* **Livebearers.**
Pollan, Irish Common name of *Coregonus pollan* Thompson 1835, Family *Salmonidae, endemic to Lough Neagh (type locality) and Lough Erne, North Ireland.

CD refered to this fish, when after having read Thompson (1839), and other papers on the insular faunas, he noted: a freshwater fish peculiar to Ireland (*Notebook E*, p. 421).

Polydactylism The condition of limbed vertebrates having, on one or more limbs, more fingers and/or toes than the five that are expected. Occasionally, there are several supernumerary digit; but usually only one. (*Variation I*, p. 457).

CD was interested in polydactylism because he thought it could be used to draw inferences on the origin of fingers and toes, and ultimately on the origins of limbs (*see* **Lungfish, Australian**). We shall not follow him there, but rather use this to document another case of how CD used his network of correspondents to follow up on a hunch, verify it the best he could, then build it into his next book.

This story starts with a letter from CD to Francis Trevelyan Buckland, asking whether he had heard of an article in *'The Field'* about the fins of fishes growing again after being cut off, and inquiring whether he had heard of the re-growth of organs such as tail or finger or toe in the mammals or birds, and noting I am privately informed that regrowth occurs with monstrous additional fingers with men (*Correspondence*, Jan. 26, 1863; *see also* **Monstrosity**).

Buckland answered with the suggestion that he, CD, should write to naturalist J. J. Briggs, which CD promptly did: Dear Sir, [. . .]. I remember seeing in The Field an excellent article, written I believe by you, on the regrowth of the fin's of fishes when cut off. [. . .] I have not kept the copies, & if it not be asking too great a favour, I should be very much obliged if you would inform me (1) what kinds of fish were tried (2) what or whether pectoral, dorsal &c were cut (3) whether whole or half or quarter the fin was cut off (4) whether the bony rays were again formed, & whether the fin ultimately appeared perfect. I wish to quote the fact on your authority in a work which I am preparing for publication. . . . (*Correspondence*, Feb. 2, 1863; Briggs 1862).

CD also checked with a trusted colleague: My dear Huxley, Reflecting over the plate of the Ray fins, I suspect that I have been blundering; & that in six-fingered men the increase is generally confined to metacarpal and digits. If no Fish would do ?? (*Correspondence*, Feb. 8, 1863; plate of the Ray fins not identified). And: . . . . May I say that the digits, (divided by so many joints in the rays) are indefinite in numbers & very generally more than five in the pectoral fins of Fishes? How are Sharks &c in this respect? These being one of the oldest orders would help (*Correspondence*, Feb. 16, 1863).

The epistolary exchange did help, and CD felt confident enough to add to the work he was preparing for publication an account stating that he had been informed by Briggs and Buckland that when portion of the pectoral and tail fins of various freshwater fishes are cut off, they are perfectly reproduced in about six weeks time and that fishes sometimes have in their pectoral fins as many as twenty metacarpal and phalangeal bones, which, together with the bony filaments, apparently represent our digits with their nails (Burkhardt *et al.* 1999 give *Variations* II, pp. 15–17 as the source of this; the index of Vol. 20 of CD's *Collected Works* does indeed list "*Fishes, regeneration of portion of fins of, ii, 15*" but the text of p. 15 deals only with mammals, as does that of an 1896 edition published by D. Appleton, New York. Go figure.)

And yes, the fins of fish do regrow following amputation (see, for example, Marí-Beffa *et al.* 1996; Santamaría *et al.* 1996). As for the regrowth of human fingers: it does seem to happen as well, at least for the part "beyond the outermost crease of the of the outermost joint" of children's fingers. The account of Becker and Selden (1985, Chapter 7), which discusses this in some details, does not mention that monstrosities accompany such regeneration. But then, CD had abandoned this notion by the time *Variations* (I, p. 468) was published.

Pomotis *See* **Sunfishes.**

Popper, Karl One of the few modern philosophers of science, along with Thomas Kuhn (he of the *paradigm), with an impact on the thinking, if not always on the practice, of scientists. (We shall abstain from discussing here the impacts of work by earlier philosophers, e.g. *Occam, Bacon (1620), or Descartes (1637). This is another book.)

Popper's merit is to have realized the full implications of the fact that it is usually difficult to prove that something is (true), but straightforward to prove that it is not (true). From this, Popper (1980) derived a 'decision criterion' wherein a proposition can be considered 'scientific' if – and only if – a procedure exists (or can be imagined) that *can* show this proposition to be *wrong*. Or put differently: a proposition must be potentially falsifiable if it is to be scientific. This is why a hypothesis that passed a test is not 'verified' but only 'corroborated', and still may fail to pass the next test.

Having discovered this criterion in the late 1920s (Popper 1934), he applied it to a number of then emerging theories, notably Freud's psychoanalysis, and to CD's theory of *natural selection, as he understood it. He concluded that both of these theories were not scientific, but rather 'metaphysical', i.e. that, like religions, they could not possibly be falsified. Unfortunately for him, he had not bothered to learn much about the science he was evaluating, and hence the caricatural aspect of his evaluation of natural selection. [On the other hand, he was probably right about psychoanalysis; see, for example, Medawar (1982).]

Thus, we have Popper claiming that "The real difficulty of Darwinism is the well-known problem of explaining an *evolution which *prima facie* may look *goal-directed*, such as that of our *eyes, by an incredibly large number of very small steps; for according to Darwinism, each of these steps is the result of a purely accidental mutation. That all of these independent accidental mutations should have had survival value is difficult to explain" (Popper 1979, pp. 269–70). And it goes downhill from there, because Popper, at this point, had still not understood that natural selection requires more than 'mutations'.

Fortunately for his reputation, Popper (1985, pp. 242–3) later conceded: "I have changed my mind about the testability and the logical status of the theory of natural selection; and I am glad to have had an opportunity to make a recantation." So here we are.

Porgies Fishes of the Family Sparidae, named after the genus *Sparus*, with over 100 species in marine and brackish waters (Cuvier and Valenciennes 1831; *see also* FishBase). Two are represented here.

The first of these is the Galápagos porgy *Calamus taurinus* (Jenyns, 1840), sampled by CD where its common name suggests it was, and described by Jenyns (1842) as *Chrysophrys taurina*. CD describes its colours as White, with four dark brown much interrupted bands, giving a mottled appearance; head coloured with the same; top of the head, ridge of the back,

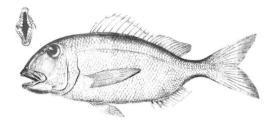

Fig. 25. Galápagos porgy *Calamus taurinus*, Family Sparidae (*Chrysophrys taurina* in *Fish*, Plate XII), from Chatham (San Christobal) Island, Galápagos; insert shows teeth.

edges of the dorsal, *caudal and ventral fins, tinted with fine azure blue. (*Fish*, pp. 56–7; *Fish in Spirits*, no. 1274; Fig. 25).

The other species of porgy CD discussed is the Black sea bream *Spondyliosama cantharus* (Linnaeus, 1758): The males of *Cantharus lineatus* become, during the breeding-season, of deep leaden-black; they then retire from the *shoal, and excavate a hollow as a nest. 'Each male now mounts vigilant guard over his respective hollow, and vigorously attacks and drives away any other fish of the same sex. Towards his companions of the opposite sex his conduct is far different; many of the latter are now distended with spawn, and these he endeavours by all the means in his power to lure singly to his prepared hollow, and there to deposit the myriad ova with which they are laden, which he then protects and guards with the greatest care.[26] (*Descent II*, p. 341; n. 26 cites Nature, May, 1873, p. 25, i.e. Saville-Kent 1873a).

Sparids, i.e. fish of the genus *Sparus*, were, doubly misidentified by CD as producers of *Cretaceous chalk: CD mistook them for coral-feeding *parrotfish (*Spari* instead of *Scari*). Moreover, the *Scari* and pre-*Scari* of eons past would not have sufficed to produce the layers of chalk accumulated during the Cretaceous.

Prionodes fasciatus *See* **Groupers.**
Prionotus spp. *See* **Gurnards.**
Pristipoma cantharinum *See* **Grunts.**

Production(s) The generation of tissue (both somatic and reproductive) by animals and plants (Winberg 1971).

When CD uses 'productions', however, he means the *organisms* that are produced. Thus, he writes, for example, the shells and other marine productions in North and South America (*Foundations*, p. 179), or barriers of any kind, or obstacles to free *migration, are related in a close and important manner to the differences between the productions of various regions. (*Origin I*, p. 347). This use of 'productions' is now obsolete.

Progress A slippery concept, of which CD was quite wary, even in the early musings that led to his development of *natural selection: There must be some law, that whatever organization an animal has, it tends to multiply & IMPROVE on it. – Articulate animals must articulate. in vertebrates tendency to improve in intellect, – if generation is condensation of changes, then animals must tend to improve. yet fish same as, or lower than in old days (*Notebook D*, p. 347).

Also: Why has organization of fishes & Mollusca (& plants ???) been so little progressive <<!Agassiz makes it wonderfully *changed*, since *Cretaceous period, whether progressive I know not>> (*Notebook E*, p. 414; Agassiz 1833–44).

CD later became even more explicit on this: If you *quote me about progressive development, I sh.^{d} wish it to be added, that such notion only applies to all organic beings taken in mass: for I do not doubt that many whole groups have retrograded in one sense. The whole mass of fish are now lower. (ie less closely allied to the *higher* reptiles) than they were formerly (*Correspondence* to J. D. Hooker, July 2, 1859).

This point reappears in material prepared for the first US edition of *Origin*: Even at the present day, looking to members of the same class, naturalists are not unanimous which forms are highest; thus some look at the Selaceans or *sharks from their approach in some important point of structure to reptiles as the highest fish; others look at the *teleosteans as the highest.

The *ganoids stand intermediate between the selaceans and the teleosteans; the latter, at the present day, are largely preponderant in number, but formerly selaceans and ganoids alone existed; and in this case, according to the standard of highness chosen, so will it be said that fishes have advanced or have retrograded in organisation. To attempt to compare in the scale of highness members of distinct types seems hopeless; who will decide whether cuttle-fish be higher than a bee? (Burkhardt *et al*. 1993, pp. 581–2).

The sixth edition of *Origin* amplifies this, and concludes the relevant section (p. 308) by adding to the bee of the previous sentence – that insect which the great Von Baer believed to be 'in fact more highly organized than a fish, although upon another type'? (Von Baer 1828 & 1837; Nägeli 1866).

Finally, we have from the same source (p. 97): But we shall see how obscure this subject [of progress] is if we look, for instance, to fishes, among which some naturalists rank as highest those which, like the sharks, approach nearest to *amphibians; whilst other naturalists rank the common bony or teleostean fishes as the highest, inasmuch as they are more strictly fish-like, and differ most from other vertebrates classes. (*See also* **Complexity**; **Spencer**.)

***Psenes* spp.** *See* **Driftfishes**.

Pteropoda A group of small snails, whose adaptation for life in the *zooplankton involves a foot modified into a pair of winglike lobes (hence the Greek name 'wingfooters'), and whose tiny calcareous shells, when sedimented, form a characteristic ooze. Some species, therefore, are used as index fossils.

CD mentions pteropods as the food of *flying fishes, while discussing pelagic *food webs.

Puffers Fishes of the Family Tetraodontidae, characterized by what military types would call a strong defensive posture: an ability to inflate themselves, thereby becoming too large to handle for many would-be predators; poisonous innards, of '*fugu*' fame; fused teeth forming a strong 'beak', and along with it, a habit of biting such that they give men a reason not to swim naked (Halstead 1978; Pauly 1991).

The largest among puffers is the Starry toadfish *Arothron stellatus* (Bloch & Schneider, 1801), which reaches up to 120 cm and lives on patch reefs and coral slopes throughout the Indo-Pacific. CD collected a small specimen of this species, which *Jenyns described as new (as *Tetrodon aerostaticus*), while noting its similarity to Bloch's species. However, the sampling locality is unknown, the label attached to the specimen having been lost (*Fish*, p. 152).

Another puffer collected by CD is *Tetrodon implutus* Jenyns, 1842 from the *Cocos Islands. This species was never again reported from there (see Allen and Smith-Vaniz 1994).

Also, CD collected two puffer species in San Christobal (= Chatham) island, *Galápagos, which Jenyns described as new, and assigned to the genus '*Tetrodon*' (*Fish*, pp. 153–4). The first of these was the Bullseye puffer, now *Sphoeroides annulatus* (Jenyns, 1842), the other the Narrow-headed puffer *Sphoeroides angusticeps* (Jenyns, 1842 (Fig. 26)). The former, the commonest member of its family in the Galápagos, is interesting in that it buries itself at night, following daytime feeding (Merlen 1988). Moreover, the specimen should have been easy to catch, given that the members of this species clearly like to swim under and close to the hull of boats (pers. obs., December 2000).

CD described the colour pattern Jenyns used to name this fish: Beneath snow white. Above dark brownish-black, this colour forming a series of broad oval rings, one within another; the outer and largest ring includes nearly the entire surface of the back and sides. The upper surface is, in addition, marked with round

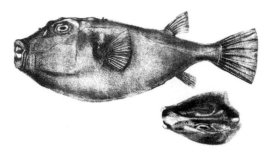

Fig. 26. Narrow-headed puffer *Sphoeroides angusticeps*, Family Tetraodontidae (*Tetrodon angusticeps* in *Fish*, Plate XXVIII), based on a specimen from Chatham (San Christobal) Island, in the Galápagos, now held at the University Museum of Zoology, Cambridge. The insert shows a dorsal view of the head.

spots of a darker shade. Pectoral and dorsal fins yellowish brown. Iris, inner edge clouded with orange; pupil dark green-blue. Jenyns trusted CD in naming the fish '*annulatus*', since he wrote about the specimen that "In its present state, there is no indication of the rings noticed above" (*Fish*, p. 154). He was right to trust CD in this, as the rings in question are indeed very visible in representations based on live colours (see, for example, Merlen 1988, Fig. 93).

CD also noted that this fish, which he called a Diodon, also made a loud grating noise (*Fish in Spirits*, no. 1293).

As for *S. angusticeps*, all we can say is that CD described it Above dull green, base of Pectorals and Dorsal black, a white patch beneath the Pectorals, Inflatable. (*Fish in Spirits*, no. 1265). These colours also had faded in the specimen Jenyns examined (*Fish*, p. 154). Jenyns had misidentified this specimen at first, as he wrote "Tetrodon aerostaticus" beside CD's description (*Fish in Spirits*, no. 1265).

Punctuated equilibrium CD, based on *Lyell, perceived *evolution by *natural selection as working slowly and steadily, at rates that we would now estimate as a few *darwins (Fig. 27A). The problem with this view of evolution is that it

Punctuated equilibrium

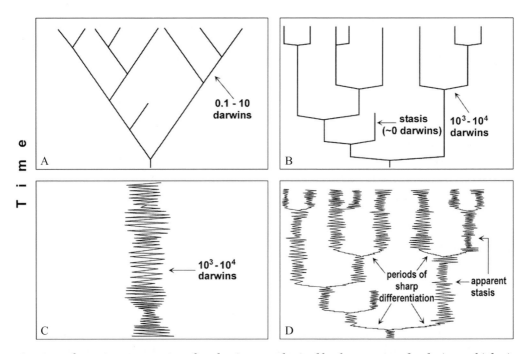

Fig. 27. Schematic representation of mechanisms emphasized by three groups of evolutionary biologists, and their synthesis. **A** Classical Darwinian view of evolution as the gradual differentiation of clades by natural selection for a long time *at the same low rate* (a few darwins). **B** Punctuated equilibrium, in which speciation and further differentiation occur *at extremely high rates* (*c*.10^3 to 10^4 darwins), but during periods so short as to appear instantaneous in the fossil record, these periods being then followed by long periods of 'stasis', during which natural selection is constrained (*c*. 0 darwin) **C** Evolution in real time implies high rates of change (*c*. 10^3 to 10^4 darwins), e.g. for species rapidly oscillating between two forms. **D** The rapid oscillations in **C** correspond to periods of apparent 'stasis' in the fossil record.

generates innumerable '*missing links'. CD's explanation for the scarcity of transitional form was the imperfection of the *fossil record, for which he gave a number of reasons, while creating, in the process, the discipline of *taphonomy (Eldredge 1995, pp. 95–6; *see also* **Flying**).

Eldredge and Gould (1972) proposed punctuated equilibrium as a solution to this and related problems. Herein, the emergence of a new taxon is viewed as involving a small population, and occurring during a very short period (e.g. after a catastrophe, or the invasion of a new habitat), following which the newly evolved taxon remains unchanged, or in 'stasis' for long periods of time, i.e. until it becomes extinct. This 'rectangular' vision of evolution (Douglas and Avise 1982; Fig. 27B), neatly explains why transitional forms are rare in the fossil record (because they were rare when alive, as well) and has been accepted by many practitioners.

Punctuated equilibrium implies extremely strong variability in evolution rates, from thousands of *darwins during periods of differentiation to essentially zero darwin during periods of stasis, and is often presented in a manner that de-emphasizes the role of *natural selection, CD's key discovery. However, students of what, in absence of a better term, may be called 'evolution in real time' (Wiener 1995; Grant 1999) have shown that, as argued by CD, natural selection works *all the time*, and generates evolution rates that are more or less constant, though much higher than CD would have thought possible (Fig. 27C).

This contradicts at least the naïve version of punctuated equilibrium, wherein 'stasis' is perceived as a period where species 'do not change'. Indeed, species have to change, even when they live in ecosystems that remain self-similar over long periods, if these systems oscillate more or less regularly between alternative states. However, in such cases, the changes are not cumulative, i.e. do not lead to new behaviours or forms (Fig. 27D).

An example of an oscillating ecosystem is the Humboldt Current system, off *Peru and Chile, which, along with the adjacent ecosystem around the Galápagos Islands, is impacted every 3–5 years by El Niño events (Quinn et al. 1978; Grove 1985; Bakun 1996), superimposed on the overall self-similarity of an upwelling that has lasted for millions of years (De Vries and Pearcy 1982). In such cases, colonists such as the ancestors of *Darwin's finches, or of the *Peruvian anchoveta, would, upon arrival from their ecosystem of origin, experience an initial period of steady, directed changes toward the phenotype required for their new habitat, which would result in the quasi-horizontal lines in evolutionary trees drawn by punctuated equilibrists. The newly adapted species would then be forced to track short-term climate oscillations ('evolution in real time'). In the Humboldt Current, this requires rapidly evolving from adaptations to El Niño conditions to those adaptations appropriate for the intervening La Niña years, and vice versa. (Note that punctuated equilibrists have no device to represent these changes, which would be subsumed in a single, straight vertical pole up an evolutionary bush.)

For Darwin's finches, the oscillations imply shifts from interspecies competition via finely adjusted beak sizes during periods of seed scarcity (La Niña), to rapid population growth, and hybridization during the rainy, and hence seed-rich (El Niño) periods (Grant 1999).

For the Peruvian anchoveta, this implies very large biomasses during the productive La Niña periods, small body sizes and a large *phytoplankton-feeding population at the northern limit of the system, near the Peru–Equator border. Conversely, after El Niño events, there will be an emphasis on *zooplankton feeding by small populations of large anchoveta off southern Peru and Northern Chile (see contributions in Pauly and Tsukayama 1987, and Pauly et al. 1989).

Another example is provided by Trinidadian *guppies, in which the males oscillate between the coloured forms that *sexual selection allows in the absence of predators, and the drab forms that promote survival when predators lurk (Houde 1997).

Oscillatory shifts of this sort induce an evolutionary mode which differs conceptually from a 'Red Queen' situation (Van Valen 1973), wherein a species evolves to keep up with its prey and/or predators, and which tend to result in long-term directed changes, detectable in the fossil record. Rather, oscillatory shifts imply rapid adaptations (of the order of 10^3 to 10^4 darwins), whose direction frequently changes, and which are compatible with both *natural selection (adaptive changes) and punctuated equilibrium (apparent stasis). The integrative view proposed here, while incompatible with the *constant and low* rates of change that CD assumed, shows that punctuated equilibrium is compatible with natural selection working at *constant and high* rates. All we have to do is trade some of Darwin for a few darwins.

Quote Short extract from a published work or speech, reproduced as originally presented, and meant to document a point made by the original author.

Because quotes are *always* taken out of the general context provided by the work from which they were extracted, quotes will always be somewhat biased (Aoki 2000). Keeping this bias minimal is the job of the quoting author, and 'quoting out of context' is what authors do who use quotes to make points not supported by the cited work, usually by deletions of inconvenient parts. Throughout my reading of CD's sources, I never found him quoting anyone 'out of context' as defined here.

In this book, quotes *by* CD are distinguished as follows from quotes *of* CD:

start of CD's own words 'text that CD quotes' end of CD's own words.

I found numerous instances of CD tacitly changing one or two words (especially at the beginning) of his quotes, such as to make them easier to fit into the flow of his own prose. There were also a few cases of 'quotes' assembled from parts not identified as such (see, for example, **Wrasses**). This seems to me rather innocent, though some scholars may disagree.

Far worse, I fear, is the use I made of CD's words, which I adjusted to fit the flow of definitions, or stories, thus generating numerous *non-sequiturs*, and many instances of what I called 'allogenous self-*plagiarism'.

Race A lineage of organisms, usually below the level of the *subspecies, and thus without taxonomic status. Used frequently by CD to describe *breeds of domesticated animals including fish (*see* **Carp; Goldfish**), or populations of a wild species (*see* **Cichlidae; Jaguar**), but in a meaning different from that of *varieties.

Rain *See* **Showers of fish**.

Rainbow runner Common name of *Seriola bipinnulata*, now *Elagatis bipinnulata* (Quoy & Gaimard, 1825), an extremely active *jack (Family Carangidae).

A specimen of this species was collected by CD in the *Cocos Islands; it had a Band on the side azure blue; above a duller greenish blue; beneath two greenish metallic stripes: lower half of the body snow white, but Jenyns noted no trace of the longitudinal stripes on the dried skin into which this specimen was turned (*Fish*, pp. 72–3).

Jenyns also inferred, from the similarity of this specimen with one described from Papua New Guinea by Quoy and Gaimard (1825), that this species "probably [. . .] has a considerable range over the Indian Ocean."

He never knew how right he was: this species has a global distribution, from Massachusetts, USA, to northeastern Brazil in the Western Atlantic, from Côte d'Ivoire to Angola in the Eastern Atlantic, around Italy in the Mediterranean, throughout the western Indian Ocean (though rare or absent in the Persian Gulf), from Fiji and Tuvalu to Australia and New Zealand in the Western Pacific, and from the Gulf of California to Ecuador; including the Galápagos Islands, in the Eastern Pacific (sources in *FishBase). A considerable range indeed!

Range A term partly congruent with the term 'geographic *distribution,' but emphasizing that a wide expanse of land, or sea, is involved.

Many entries in this book attest to such usage by CD, also exemplified in the following quote: It would be very important to show wide range of fish & shell in tropical sea. it would demonstrate: not distance, makes species, but barrier (*Notebook C*, p. 244).

And, as well: if all the organisms, which are now common to Europe & America, could flourish under the present climate between the Arctic circle & 70° (& a great majority do now live there) I can see no insuperable difficulty to their having in the course of ages circulated round the polar regions by this course. No doubt the distance is very great, viz in the parallel of 70°, between 6000 & 7000 miles; but we know that most of the *productions on this long line are now the same; & many species of fish & marine shells have even a wider range in the Indo-Pacific ocean.[1] (*Big Species Book*, p. 540; n. 1 states: For Fish see Report to Brit. Assoc. for 1845 on the Ichthyology of the Seas of China by Sir Richardson, p. 190, 191; Richardson 1846; note that many of these allegedly wide ranges are artefacts due to *lumping, i.e. caused by valid *species being described as *subspecies (Gill 1999)).

Richardson's (1846) discussion of the ranges of fishes is not only correctly interpreted by CD, but also well worth reading in the original. Thus, after correctly noting that tropical fish species, in the Atlantic, range as far north as 32°, to the Bermudas (see Smith-Vaniz *et al.* 1999), he infers from the presence of tropical fishes in Southern Japan (32°N) that "there is probably a current [similar to the Gulf Stream] setting to the northward on the coasts of China" (p. 190). This current exists: it is the 'Kuroshio', or 'Japan Current', and the contributions in Marr (1970) describe its various aspects and impacts on oceanography and fisheries in Northeast Asia.

Rattails Members of the Family Macrouridae, also known as 'Grenadiers' (not to be mistaken for Crust. Macrouri i.e. crabs; *see* **Petrels**). The rattails, whose nearly 300 species are mostly benthic, range from the Sub-Arctic to Antarctica, at depths usually ranging from 200 to 2000 m, where they may contribute the bulk of the fish biomass (Cohen *et al.* 1990).

The common name of rattails is due to the shape of their body (of up to 150 cm), which is posteriorly attenuated, with the anal and second dorsal fins confluent with an elongated tail, often broken in sampled specimens. CD mentions this group, as *Macrourus*, in the context of *submergence: Notacanthus & Macrourus [are] two very remarkable Greenland genera, which inhabit deep water, [and] have recently been discovered on the coast of New Zealand & S. Australia. (*Big Species Book*, p. 555; Richardson 1846, pp. 189, 191). According to the distribution maps in Cohen *et al.* (1990) and text in Gomon *et al.* (1994), at least two species of rattail, the Abyssal grenadier *Coryphaenoides armatus* (Hector, 1875) and the Globehead whiptail *Cetonurus globiceps* (Vaillant, 1884), occur in the North Atlantic, off New Zealand and off southern Australia as well. This, along with the fact that '*Macrourus*' previously *lumped species of rattails now included in different genera, fully confirms CD's statement on the submergence of macrourids.

Rays Dorsoventrally flattened, *cartilaginous fishes, usually feeding on benthic organisms.

CD's *Beagle* collection included two rays, but they were not well preserved, and could not be identified (*Fish in Spirits*, nos. 396, from *Bahia Blanca; 751 from *Tierra del Fuego). On the other hand, the *electric organs of rays were later to provide key evidence for *natural selection.

Rays – or rather skates – also provide evidence for *sexual selection, as documented in the quote below: The males of *Plagiostomous fishes (*sharks, rays) and of *Chimaeroid fishes are provided with claspers which serve to retain the female, like the various structures possessed by many of the lower animals. Besides the claspers, the males of many rays have clusters of strong sharp spines on their heads, and several rows along '*the upper outer surface of their pectoral fins.*' These are present in the males of some species, which have other parts of their bodies smooth. They are only temporarily developed during the breeding-season; and Dr Günther suspects that they are brought into action as prehensile organs by the doubling inwards and downwards of the two sides of the body. It is a remarkable fact that the females and not the males of some species, as of *Raia clavata*, have their backs studded with large hook-formed spines.[1] (*Descent II*, pp. 330–1; n. 1 reads: Yarrell's *Hist. of British Fishes*, vol. ii, 1836, pp. 417, 425, 436. Dr. Günther informs me that the spines in R. clavata are peculiar to the female.).

In the thornback (*Raia clavata*) the adult male has sharp, pointed teeth, directed backwards, whilst those of the female are broad and flat, and form a pavement; so that these teeth differ in the two sexes of the same species more than is usual in distinct *genera of the same *family. The teeth of the male become sharp only when he is adult: whilst young they are broad and flat like those of the female. As so frequently occurs with secondary sexual characters, both sexes of some species of rays (for instance *R. batis*), when adult, possess sharp pointed teeth; and here a character, proper to and primarily gained by the male, appears to have been transmitted to the offspring of both sexes. The teeth are likewise pointed in both sexes of *R. maculata*, but only when quite adult; the males acquiring them at an earlier age than the females. We shall hereafter meet with analogous cases in certain birds, in which the male acquires the plumage common to both sexes when adult, at a somewhat earlier age than does the female. With other species of rays the males even when old never possess sharp teeth, and consequently the adults of both sexes are provided with broad, flat teeth like those of the young, and like those of the mature females of the above-mentioned species.[10] As the rays are

Reproduction

bold, strong and voracious fish, we may suspect that the males require their sharp teeth for fighting with their rivals; but as they possess many parts modified and adapted for the prehension of the female, it is possible that their teeth may be used for this purpose. (*Descent II*, pp. 334–5; n. 10 reads: See Yarrell's account of the rays in his *Hist. of British Fishes*, vol. ii, 1836, p. 416, with an excellent figure, and pp. 422, 432.).

Remora Common name of *Remora remora* (Linnaeus, 1758), one of the eight extant species of sharksuckers, Family Echeneidae, whose dorsal sucking disc enables them to attach themselves to larger animals such as *sharks, other fishes, turtles, marine *mammals or scuba divers (pers. obs.).

CD's only mention of this fish is Sucking fish 'Echeneis Remora' off a shark near *St Paul's [Rock.] (*Fish in Spirits*, no. 114).

Whether their behaviour qualifies *Remora* as phoretic *parasites (jargon for 'hitchhikers') is debatable. It can safely be assumed, on the other hand, that their behaviour and the accompanying morphological adaptations are at the root of the ancient story that remoras are able to slow down and even stop ships by attaching themselves to their hulls (Plinius the Elder, in Cotte 1944). This story is neatly encapsulated in the Spanish word *rémora*, meaning a hindrance, or loss of time. Note, however, that invertebrates can do what remoras cannot: *barnacles do slow down ships, and 'shipworms' (*Teredo* spp.) really stop them ... by sinking them.

Reproduction The process through which organisms transfer their genes from one generation to the next – or, as Dawkins (1989) would put it: the process though which genes transfer themselves from one body to the next.

Except, possibly, in the *hermaphroditic *Rivulus marmoratus* Poey, 1880, reproduction, in fishes, implies both female and male participation. Fortunately, The manner in which Frogs copulate & fish shows how simply instinctive the feeling of other sex being present is – it also shows that semen. must actually reach the ovum (*Notebook D*, p. 388).

Therefore, in most fishes, as in most other organisms, reproduction involves *sexual selection and the related development of *secondary sexual characters. The text below, excerpted from *Descent* (I, pp. 342–4) summarizes various aspects of fish reproduction as seen by CD, with emphasis on sexual selection and survival: To return to our more immediate subject. The case stands thus: female fishes, as far as I can learn, never willingly spawn except in the presence of the males; and the males never fertilize the ova except in the presence of the females. The males fight for the possession of the females. In many species, the males whilst young resemble the females in *colour; but when adult become much more brilliant, and retain their colours throughout life.

In other species the males become brighter than the females and otherwise more highly ornamented, only during the season of love. The males sedulously court the females, and in one case, as we have seen, take pains in displaying their *beauty before them. Can it be believed that they would thus act to no purpose during their courtship? And this would be the case, unless the females exert some choice and select those males which please or excite them most. If the female exerts such choices, all the above facts on the ornamentation of the males become at once intelligible by the aid of Sexual selection.

We have next to enquire whether this view of the bright colours of certain male fishes having been acquired through Sexual selection can, through the law of the equal transmission of characters to both sexes, be extended to those groups in which the males and females are brilliant in the same, or nearly the same degree and manner. [....] With some fishes, as with many of the lowest animals, splendid colours may be the direct result of the nature of their tissues and of the surrounding conditions, without the aid of selection of any kind. [...].

We can see that one sex will not be modified through Natural selection for the sake of protection more than the other, supposing both to vary, unless one sex is exposed for a longer period to danger, or has less power of escaping from such danger than the other; and it does not appear that with fishes the sexes differ in these respects. As far as there is any difference, the males, from being generally smaller and from wandering more about, are exposed to greater danger than the females; and yet, when the sexes differ, the males are almost always the more conspicuously coloured.

The ova are fertilized immediately after being deposited; and when this process lasts for several days, as in the case of the *salmon,[34] the female, during the whole time, is attended by the male. After the ova are fertilized they are, in most cases, left unprotected by both parents, so that the males and females, as far as oviposition is concerned, are equally exposed to danger, and both are equally important for the production of fertile ova; consequently the more or less brightly-coloured individuals of either sex would be equally liable to be destroyed or preserved, and both would have an equal influence on the colours of their offspring. (n. 34 refers to Yarrell, *British Fishes*, vol. ii, p. 11.)

Reproductive drain hypothesis The widespread notion that in organisms, somatic *growth and *reproduction work as antagonists, and hence that organisms, once they reach maturity, cease to grow 'because of the energy drain of reproduction'.

CD presents this as follows: The antagonism which has long been observed,[55] with certain exceptions, between growth and the power of sexual *reproduction [...] is partly explained by the *gemmules not existing in sufficient numbers for these processes to be carried on simultaneously. (*Variations II*, p. 379; n. 55 reads: Mr Herbert *Spencer (*Principles of Biology*, vol. ii, p. 430) has fully discussed this antagonism). Spencer's characteristic verbiage on this appears in a chapter titled "Antagonism between growth and sexual genesis," which starts as follows "[i]n so far as it is a process of separation, sexual genesis is like asexual genesis; and is therefore, equally with asexual genesis, opposed to that aggregation which results in growth". Further below, Spencer claims to present "leading illustration of this truth", but all he does is restate his premise, even when he stumbles into a comparison between *Cod and *Stickleback, with the former exceeding the latter in terms of both fecundity *and* growth.

Moreover, in spite of this "truth", it is in fishes usually the females that make the largest reproductive investment (the *seahorses and *pipefishes represent two of the few exceptions), while also growing faster, toward larger sizes than the males (Pauly 1984, 1994a).

Interestingly, CD, who studied *sexual dimorphism in great detail, had access to sufficient material to see that something was wrong with Spencer's truth. Unfortunately, he didn't. (*See also* **Growth; Oxygen**.)

Retronym A word modified from an existing one because of the emergence of a new concept or item. A good example is the 'watch,' which became an 'analogue watch' when digital watches became popular.

Similarly, 'Atlantic *salmon' is a retronym of 'salmon', required because of the appearance of 'Pacific salmon' on the English cultural horizon. The alert reader will find many retronyms in this book.

Rhombus The genus *Rhombus* Walbaum, 1792, pertaining to Brill and *Turbot, includes *flatfishes of the Family Scophthalmidae, which have their eyes on the left side.

Jenyns assigned a specimen (Above pale purplish brown, with rounder darker markings.) sampled by CD in *Bahia Blanca to that genus,

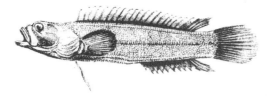

Fig. 28. Roundhead *Acanthoclinus fuscus*, Family Plesiopidae. Based on a specimen collected by CD in the Bay of Islands, New Zealand (*Fish*, Plate XVIII).

although it had its eyes on the right side (*Fish*, p. 139; *Fish in Spirits*, no. 395). The species in question appears to have been the Remo flounder *Oncopterus darwinii* Steindachner, 1874 (Menni *et al.* 1984), an *eponym.

Rio de Janeiro City in Southern Brazil, known for its carnival and beach visuals, all missed by CD, who came in the wrong century. All he did there was to sample specimens later described in *Fish*: *Prionotus punctatus* (p. 28); *Gerres gula* (p. 58); *Pimelodus gracilis* (p. 110); *Pimelodus exsudans* (p. 111); *Poecilia unimaculata* (p. 114); *Tetragonopterus scabripinnis* (p. 125); *Tetragonopterus taeniatus* (p. 126); and *Muraena ocellata* (p. 145).

Roach Common name of *Rutilus rutilus* (Linnaeus, 1758), a member of the Family *Cyprinidae feeding on *zoobenthos, and tolerant of a wide range of conditions, including turbid waters. Abundant in European waters where it is both despised (Germany) and highly appreciated (France). CD mentions this fish only once, casually, in an enumeration of fish that change colours during the breeding season, and which included the *Tench.

Roundheads Members of the Plesiopidae, small coastal fishes represented here by the Olive rockfish, described in *Fish* (pp. 92–3) as *Acanthoclinus fuscus* Jenyns, 1842, a New Zealand *endemic. The original description was based on a specimen of "3 inc. 8 lin." sampled by CD in December 1835 in the Bay of Islands (*Fish in Spirits*, no. 1344; Fig. 28). As a 'line' is one twelfth of an inch, and an inch is 2.54 cm, the *holotype has a *length of 9.3 cm, i.e. about half the maximum observed in this species.

According to Paulin and Roberts (1992), Olive rockfish build their nests beneath boulders where dense, gelatinous clusters of about 10 000 *eggs are deposited. The male guards the nest after sealing it from the inside, using small stones and mud.

Ruff The Australian ruff *Arripis georgianus* (Valenciennes, 1831) belongs to the ill-named Australian 'salmons' (Arripidae), related to the *perches.

CD collected a specimen in *King George's Sound, Australia, in March 1836, at "the same place in which it was discovered by MM. Quoy and Gaimard." (*Fish*, pp. 13–15; Quoy and Gaimard 1824).

These comments, along with others scattered throughout *Fish*, nicely illustrate the key role played by the *Histoire naturelle des poissons* in documenting fish biodiversity, and providing a sound baseline for the work of European, including English, ichthyologists and other *naturalists. Thus our score: France 1 : England 0 (but see next page for an extra-ichthyologic score).

Saint Helena Island Perhaps the most remote tropical island in the world, 1870 km to the coast of Angola, Southwest Africa, St Helena was visited by the *Beagle* near the end of its voyage, 25 years after the death of its most famous resident. CD described his impressions of the place, especially its vegetation, based on excursions from lodgings within a stone's throw from Napoleon's tomb (*Journal*, May 9, 1836). England 1: France 0.

CD apparently did not use this opportunity to sample any marine life, thus missing an ichthyofauna with an estimated percentage of *endemic fishes (12.5% for bottom-dwelling neritic species; Edwards 1990, p. 43), similar to estimates for the *Galápagos [which, however, included wide-ranging pelagics; the percentage of endemic 'bottom-dwelling neritic' fishes is higher] and for Ascension Island, located 1290 km to the northeast of St Helena.

This did not prevent CD, however, from commenting later, if briefly, on the affinities of St Helena fishes (*see* **Ascension**).

Saint Paul's Rock A small, isolated group of barren islets on the Mid-Atlantic Ridge, now belonging to Brazil, where they are known as 'Penedos de São Pedro e São Paulo' (Edwards and Lubbock 1980).

Saint Paul's Rock was visited by the *Beagle on February 16, 1832, and CD reports for that date a description from which the following was excerpted: This cluster of rocks is situated in 0° 58′ north latitude, and 29° 15′ west longitude [and] rises abruptly out of the depths of the ocean (*Journal*, p. 7).

We only observed two kinds of birds – the booby and the noddy. The former is a species of gannet, and the latter a tern. [...] The booby lays her eggs on the bare rock; but the tern makes a very simple nest with sea-weed. By the side of many of these nests a small *flying-fish was placed; which, I suppose, had been brought by the male bird for its partner. (*Journal*, p. 10; also in *Birds*, p. 145; there are *two* species of noddy on Saint Paul's Rock, the Common or Brown noddy *Anous stolidus*, and the Black or White-capped noddy *A. minutus*, Edwards *et al.*, 1981).

Confirming part of CD's observations, however, Edwards *et al.* (1981) report that the Black noddy "made conspicuous bracket-like nests of *Caulerpa clavifera* cemented with guano." Also, *Anous stolidus* is known to occasionally consume flying fish (Dorward and Ashmole 1963).

Thus, we may go on with CD's observations: It was amusing to watch how quickly a large and active crab (*Graspus*), which inhabits the crevices of the rock, stole the fish from the side of the nest, as soon as we had disturbed the birds. Not a single plant, not even a lichen, grows on this island; yet it is inhabited by several insects and spiders. The following list completes, I believe, the terrestrial *fauna: a species of Feronia and an acarus, which must have come here as *parasites on the birds; a small brown moth, belonging to a genus that feeds on feathers; a staphylinus (*Quedius*) and a woodlouse from beneath the dung; and lastly, numerous spiders, which I suppose prey on these small attendants on, and scavengers of the waterfowl. (*Journal*, p.10; 'Graspus' refers to the rock crab *Grapsus grapsus*, Edwards and Lubbock 1980).

The often-repeated description of the first colonists of the coral islets in the South Sea, is not, probably, quite correct: I fear it destroys the poetry of the story to find, that these little vile insects should thus take possession before the cocoa-nut tree and other noble plants have appeared. (*Journal*, Jan. 16, 1832).

What is interesting about this account is that it nicely documents how small oceanic islands, in contrast to larger islands and continents, *import* biomass and nutrients from the sea (Polis and Hurd 1996). Indeed, it is the extent of these imports that determines the level of production in these islands' terrestrial ecosystems.

Salarias spp. *See* **Blennies**.
Salmo eriox *See* **Trout**.

Salmo lycaodon See **Salmon (III: Pacific)**.

Salmon (I) A name originally applied to members of the North Atlantic species *Salmo salar* Linnaeus, 1758, but now also used for nine species of the genus *Oncorhynchus*, the (North) *Pacific salmon (the two other species in that genus are Rainbow and Cutthroat *trout). Both Atlantic and Pacific salmon now also occur in the Southern Hemisphere (e.g. in Chile and New Zealand) owing to *introductions.

To the naturalist, the most interesting feature of salmon is their migratory habits, which link *eggs and early stages in rivers with the marine habitat of the adults, and the physiological and morphological changes associated with sexual maturation and their associated upriver migrations, both covered in the entries below.

Salmon (II: Atlantic) Common name of *Salmo salar* Linnaeus, 1758, previously known simply as 'salmon'. Atlantic salmon is a typical *retronym, which became necessary to avoid confusion with Pacific salmon (see below).

The colours of Atlantic salmon can be quite lively during the spawning season: The male salmon at this season is 'marked on the cheeks with orange-coloured stripes, which give it the appearance of a *Labrus, and the body partakes of a golden orange tinge. The females are dark in colour, and are commonly called black-fish.[22] (*Descent II*, p. 340; n. 22 cites Yarrell (1836), Vol. 2, pp. 10, 12 and 35).

Salmon (III: Pacific) A group of seven species of the genus *Oncorhynchus*, Family Salmonidae: Sockeye *O. nerka*; Pink *O. gorbuscha*; Chum *O. keta*; Chinook *O. tshawytscha*; Coho *O. kisutch*, Masu *O. masou* and Amago *O. rhodurus* (Groot and Margolis, 1991), characterized by an *anadromous life cycle, wherein the parents ascend rivers to spawn and then die ('terminal' spawning). The up-river migration is accompanied by strong physiological and morphological changes, particularly in the males, whose jaws develop into a 'hook' even more pronounced than in the Atlantic salmon.

CD commented on this as follows: In our salmon this change of structure lasts only during the breeding-season; but in the *Salmo lycaodon* of N.-W. America the change, as Mr. J. K. Lord[8] believes, is permanent, and best marked in the older males which have previously ascended the rivers. In these old males the jaw becomes developed into an immense hook-like projection, and the teeth grow into regular fangs, often more than half an inch in length. With the European salmon, according to Mr. Lloyd,[9] the temporary hook-like structure serves to strengthen and protect the jaws, when one male charges another with wonderful violence; but the greatly developed teeth of the male American salmon may be compared with the tusks of many male mammals, and they indicate an offensive rather than a protective purpose. (*Descent II*, pp. 332–4; n. 8 cites Lord (1866), p. 54; n. 9 Lloyd (1854), pp. 100 and 104, but the cited bit is on pp. 97–8 in the second, 1854 edition).

It is true that Sockeye *Oncorhynchus nerka* (Walbaum, 1792), i.e. the *Salmo lycaodon* Pallas, 1814 mentioned above, develops an immense hook-like projection. However, this structure starts developing when the fish return to estuaries and river mouths (Burgner 1991), i.e. when they get ready for a migration which invariably ends in death. Thus, it is not appropriate to speak of their hook-shaped jaws as 'permanent' features, especially as they preclude feeding. They are, rather, 'terminal features' (Fleming and Gross 1994), whose role can be understood only in the context of *sexual selection.

Salmon (III: Not pacific) Male salmon fight a lot during the spawning season, and CD was clearly fascinated by this and related morphological correlates of reproduction.

Here are some quotes: Male salmon have been observed fighting all day long [. . .]. The males of carnivorous animals already well armed; though to them and to others, special means of defence may be given through means of *Sexual Selection, as the mane of the lion, and the hooked jaw of the male salmon; for the shield may be as important for victory, as the sword or spear. (*Origin*, VI, p. 69). Mr Shaw saw a violent contest between two male salmon which lasted the whole day; and Mr R. Buist, Superintendent of Fisheries, informs me that he has often watched from the bridge at Perth the males driving away their rivals, whilst the females were spawning. The males 'are constantly fighting and tearing each other on the spawning-beds, and many so injure each other as to cause the death of numbers, many being seen swimming near the banks of the river in a state of exhaustion, and apparently in a dying state'.[6] (*Descent II*, p. 332; n. 6 reads: [Buist] The Field, 29 June, 1867. For Mr. Shaw's statements, see Edinburgh Review, 1843. Another experienced observer (Scrope's Days of Salmon Fishing, p. 60) remarks that like the stag, the male would, if he could, keep all other males away. Anon. 1843; Scrope 1843).

Mr Buist informs me, that in June, 1868, the keeper of the Stormontfield breeding-ponds visited the northern Tyne and found about 300 dead salmon, all of which with one exception were males; and he was convinced that they had lost their lives by fighting. The most curious point about the male salmon is that during the breeding-season, besides a slight change in colour, 'the lower jaw elongates, and a cartilaginous projection turns upwards from the point, which, when the jaws are closed, occupies a deep cavity between the intermaxillary bones of the upper jaw'.[7] (*Descent II*, p. 332; n. 7 cites Yarrell (1836), Vol. 2, p. 10; see also Fig. 29).

Lots of fighting indeed.

Salmon (IV: Migrations) CD often referred to salmon migrations in his writings, usually implying Atlantic salmon. Thus we have from *Notebook C* (p. 306): instincts in young animals, well developed, just like, habits easily gained in child hood. – Young salmons. first a species

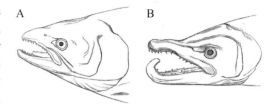

Fig. 29. Heads of Salmon *Salmo salar* during the breeding season. **A** Female: **B** Male. Woodcuts by Mr G. Ford, originally published in *Descent II*, p. 335.

which lived in estuaries its taste. taught it to go to <sea> salter water (& its necessities teach it taste, but that is much more general argument) & therefore down the stream followed ebb tide, therefore got into habit of going down stream which would last were the stream 1000 miles long. – (*See also* **Homing**.)

Concerning the possible origin of this behaviour, CD offers analogies: <Horses> Colts cantering in S. America capital instance of heredetary habit: – there must, however, be a *mental impulse* (though unconscious of it) to move its legs so, as much as in the young salmon to go towards the sea. or down the stream; which it does unconsciously of any end. – N B. There is a wide difference, between the means by which an animal performs an instinct, & its impulse to do it. – [the means must be present on any hypothesis whatever] an animal may so far be said to *will* to perform an instinct that it is uncomfortable if it does not do it. (*Notebook N*, p. 582).

Also: Every instinct must, by my theory, have been acquired by slight changes <illegible> of former instinct, each change being useful to its then species. [. . .] Take migratory instincts, faculty distinct from instinct, animals have notion of time [. . .] geological changes – fishes in river. . . . (*Foundations*, p. 19).

McDowall (1997), who reviewed the evolution of *diadromy in fishes, could not identify any consensus regarding the origin of anadromy in salmonids, though he discusses several current hypotheses, all compatible with CD's hope that it occurred by slight changes from non-diadromous life histories.

Salmonids (-idae) The fish Family of which *Salmo* is the type genus, and to which the *Salmon, *Trout, *chars and *Pollan belong, all discussed by CD, along with other groups now considered to belong to other families. Hence CD's reference to South American 'salmon' (*see* Ichthyology), i.e. *characins, or to the *capelin, previously one of Salmonidae.

The broader definition of '*Salmo*' in earlier times also explain the quote Besides small fresh water fish (526), I have good reason to believe the genus Salmo exists (*Zoology Notes*, p. 134), which refers to Tierra del Fuego (see *Fish in Spirits*, no. 526), where the genus *Salmo* surely did not 'exist' – though this may have changed since, given the *introduction (for aquaculture purposes) of Atlantic salmon into Chilean waters.

Sand-eels Small planktivorous fishes of the Family Ammodytidae, better known as sandlance, and including 18 species, notably *Ammodytes tobianus* Linnaeus, 1758. Sand-eels are named after their habit of burrowing themselves in the sand, which does not prevent them from being a major prey to a vast number of predators, notably *Cod (which watch sand-eels bury themselves, then go after them, or wait until they leave the sand to feed, as they must).

CD thought sand-eels lacked *complexity, noting that: naturalists have observed that in most of the great classes a series exists from very complicated to very simple beings; thus in Fish, what a range there is between the sand-eel and *shark. (*Foundations*, p. 227). What range?

Sandperches Fishes of the Pinguipedidae and allied Families.

CD collected specimens of *Pinguipes brasilianus* Cuvier, 1829, and *Pinguipes chilensis* Valenciennes, 1833 in southeastern South America (by hook and line at a depth of 14 fathoms in Northern *Patagonia, lat. 38°20′S, and in *Mal-

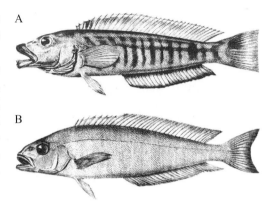

Fig. 30. Sandperches, Family Pinguipedidae. **A** *Pinguipes brasilianus* (*Pinguipes fasciatus* in *Fish*, Plate V), from the coast of Northern Patagonia, Argentina. **B** *Caulolatilus princeps* (*Latilus princeps* in *Fish*, Plate XI), from Chatham (San Christobal) Island, Galápagos.

donado), and in *Valparaiso, respectively (*Fish in Spirits*, nos. 354, 1012). *Jenyns described the latter species as *Pinguipes fasciatus*, a junior *synonym. CD described the former species as: Above pale, 'chestnut brown' so arranged as to form transverse bands on sides, head, fins, with a black tinge; beneath irregularly white; under lip, pink; eyes with pupils black, with yellow iris (*Fish in Spirits*, no. 354); also: When cooked, was good eating (*see Fish*, pp. 22–4; Fig. 30A).

The other two sandperch species, assigned by taxonomists either to the Pinguipedidae or to the Malacanthidae, are *Prolatilus jugularis* (Valenciennes, 1833) and *Caulolatilus princeps* (Jenyns, 1840), both earlier part of the genus *Latilus*. The specimens were collected by CD in *Valparaiso and *Galápagos, respectively (*Fish in Spirits*, nos. 1017, 1268). The former was Beneath brilliant white; head and back clouded with purplish and carmine red; longitudinal and transverse irregular bands of the same, the latter Above, and the fins, obscure greenish; sides obscure coppery, passing on the belly into salmon-colour. Pectorals edged with dull blue. Iris yellowish brown: pupil black-blue. (*Fish*, pp. 51–4; *Fish in Spirits*, no. 1268; Fig. 30B).

Scale A word that CD uses in three different meanings:

(1) The bony structures covering all, or at least parts of the body of most fishes and earlier used to define their affinities (*see* **Scales**);
(2) The range of *colour plates in Syme (1821), which he used to describe fish and other organisms; and
(3) All extant and extinct organisms, arranged in a manner reminiscent of the *Great Chain of Being.

The present volume, which generally cites CD only in conjunction with fish, includes several instances of CD using the last meaning of 'scale' – proof of its pull, notwithstanding CD's understanding of life as a branching bush, not a linear array of increasingly complex forms.

Scales Fish scales, the usually thin and flexible replacement, in modern fishes, for the thick, body-covering armour of ancient fishes, come in different shapes and sizes. Some are so small and deeply embedded in the skin that their owners (e.g. *eels) look naked; others are large and heavy, making their owners resemble pine cones. The basic types of scale commonly recognized are 'placoid' (as in *sharks, which are covered by the remains of ancient armour), 'ganoid' or rhombic (heavy, as occur in sturgeons and other *ganoid fishes), '*cycloid' and light, as occur in *Carp and many other bony fishes, and '*ctenoid' (i.e. with a comb-like trailing edge), as occurs mostly in advanced teleosts such as perch (Lagler *et al.* 1977, pp. 108–9).

Following a practice implied by *Agassiz' early classification of fishes, which grouped them according to type of scale, CD, in his annotations to Agassiz (1850), Pictet (1853–4) and Sedgwick (1850), uses the words 'ctenoids' and 'cycloids' to identify broad groupings of fishes now seen as paraphyletic.

Scarus spp. *See* **Parrotfishes**.

Scissor-beak Common name of *Rhyncops nigra*, a South American bird whose appearance and behaviour CD thought very extraordinary. Only excerpts from his description are reproduced here: [it] has short legs, web feet, extremely long-pointed wings, and is of about the size of a tern. The beak is flattened laterally, that is, in a plane at right angles to that of a spoonbill, or duck. It is as flat and elastic as an ivory paper-cutter, and the lower mandible, differently from every other bird, is an inch and a half longer than the upper. [...].

The specimen now at the Zoological Society was shot at a lake near Maldonado, from which the water had been nearly drained, and which, in consequence, swarmed with small *fry. [...] In their flight they frequently twist about with extreme rapidity, and so dexterously manage, that with their projecting lower mandible they plough up small fish, which are secured by the upper half of their scissor-like bills. [...] Being at anchor, as I have said, in one of the deep creeks between the islands of the Parana, as the evening drew to a close, one of these scissor-beaks suddenly appeared. The water was quite still, and many little fish were rising.

The bird continued for a long time to skim the surface, flying in its wild and irregular manner up and down the narrow canal, now dark with the growing night and the shadows of the overhanging trees. At Monte Video I observed that some large flocks during the day remained on the mud banks at the head of the harbour, in the same manner as on the grassy plains near the Parana; and every evening they took flight direct to seaward. From these facts, I suspect that the Rhyncops generally fishes by night, at which time many of the lower animals come most abundantly to the surface. (*Journal*, October 15, 1833; a similar account may be found on p. 144 of *Birds*; both are based on CD's *Ornithological Notes*, pp. 221–2).

Scorpaena histrio *See* **Scorpionfishes**.

Scorpionfishes Members of the Family Scorpaenidae, in which the dorsal, anal, and pelvic spines are often elongated, and/or bear venom glands. The scorpionfish family contains some of the world's most venomous fishes; moreover, the skin also contains various alkaloids, released during cooking. This is the soul of bouillabaisse, of which the 'rascasse', as scorpaenids are called in French, is the key ingredient (Liebling 1962). Actually, the best bouillabaisse includes 'capons', i.e. *Scorpaena elongata* Cadenat, 1943 and/or *S. scrofa* Linnaeus, 1758, but we shall let dear Liebling believe that any old scorpionfish would do.

Correct identification of scorpionfishes can indeed be a problem. Thus, we have in *Notebook C* (p. 247, referring to Lesson (1830–1)) p. 71 *Chimera – Antarctica <<Tæniatole austal>> <<caught>> Chile, Van Diemen's Land & Cape of Good Hope. The insertion of Tæniatole austal pertains to a phrase by Lesson, on p. 73, which, translated, reads as follow: "At the end of the three great capes of the southern hemispheres lives the 'taeniatote austral, bristling with fins ['*tout herissé de nageoires*'] and which reaches a large size."

Lacepède (1802, vol. 4, p. 303) describes "Les taeniatotes" as a group of species of which many were and still are assigned to the Scorpaenidae, e.g. the Leaf scorpionfish *Taeniatotus triacanthus* Lacepède 1802. Lesson probably followed Lacepède in this, and thus CD's Tæniatole austral would be a scorpionfish. The species cannot be identified, however, in spite of Lesson's brief description, as no scorpionfish species appears to occur in Tasmania (=Van Diemen's Land), in Southern Chile *and* at the Cape of Good Hope. Moreover, while many scorpionfish are covered by long, spiny fins (especially in *Pterois volitans* Linnaeus 1758, known even to James Bond in *The Spy Who Loved Me*), most of them remain small.

The other scorpionfishes that CD dealt with are easier to identify: the first of these

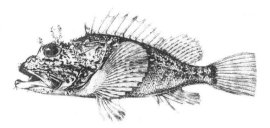

Fig. 31. Bandfin scorpionfish *Scorpaena histrio*, Family Scorpaenidae (*Fish*, Plate VIII). Lithograph by B. Waterhouse Hawkins, based on a specimen from Chatham (San Christobal) Island, Galápagos.

is the Bandfin scorpionfish *Scorpaena histrio* Jenyns, 1840, sampled by CD at Chatham (San Christobal) Island, Galápagos. Its appearance: Whole body scarlet red, fins rather paler; small irregularly-shaped light black spots; very extraordinary; are they distinct? (*Fish*, pp. 35–7; *Fish in Spirits*, nos. 1282, 1283; Fig. 31). And yes, they were distinct.

Then we have the Patagonian redfish *Sebastes oculatus* Valenciennes, 1833, obtained by CD in *Valparaiso, Chile, from Claude *Gay, and described, partly based on Syme (1821), as follows: [u]nder surface, sides, branchial covering, and part of the fins, 'tile and carmine red'; dorsal scales pale yellowish dirty brown. Jenyns' description (*Fish*, pp. 37–8) is less colourful, being based on a "uniform brown", dried skin that nobody would want to put into her or his bouillabaisse.

Sea scorpion Earlier common name of the Shorthorn sculpin *Myoxocephalus scorpius* (Linnaeus, 1758), a member of the Family Cottidae or sculpins, distributed in coastal waters throughout much of the northeast Atlantic and reaching a reported maximum length of 90 cm (*see* FishBase). CD described their *sexual dimorphism: The male of the *Cottus scorpius*, or sea-scorpion, is slenderer and smaller than the female. There is also a great difference in colour between them. It is difficult, as Mr. Lloyd[15] remarks, 'for any one, who has not seen this fish during the spawning-season, when its hues are brightest, to conceive the admixture of brilliant colours with which it, in other respects so ill-favoured, is at that time adorned'. (*Descent II*, p. 337; n. 15 cites Game birds of Sweden, i.e. Lloyd (1867), p. 466).

Seahorses (I) A group of aberrantly shaped fishes belonging to the genus *Hippocampus* and the Family Syngnathidae ('fused-jaws'), which also includes *Syngnathus* and other *pipefishes.

A recent revision identifies 32 mainly tropical species of seahorse, with temperate forms ranging from the British Isles, New England and Japan in the North to South Africa and New Zealand in the South (Lourie *et al*. 1999; see also Kuiter 2000).

Seahorses not only have strange shapes, but also extremely interesting reproductive adaptations, as noted by CD: Mr Lockwood believes (as quoted in Quart. Journal of Science, April, 1868, p. 269), from what he has observed of the development of Hippocampus, that the walls of the abdominal pouch of the male in some way afford nourishment. (*Descent I*, p. 163, n. 30).

In most of the *Lophobranchii (pipefish, Hippocampi, etc.) the males have either marsupial sacks or hemispherical depressions on the abdomen, in which the ova laid by the female are hatched. The males also show great attachment to their young.[39] The sexes do not commonly differ much in colour; but Dr Günther believes that the male Hippocampi are rather brighter than the females (*Descent II*, p. 346; n. 39 cites Yarrell, *Hist. of British Fishes*, vol. ii, 1836, p. 329, 338; the statement that the males show great attachment to their young refers to *pipefishes).

Recent research has confirmed that indeed, male seahorses become pregnant; however, (post-release) parental care does not occur in *Hippocampus* (Lourie *et al*., 1999, p. 20). In a few species at least, the males are more brightly coloured than the females (Lourie *et al*. 1999, pp. 67–131), which implies that the complete reversal of sexual roles that might be inferred from male pregnancy and the reasoning behind CD's *sexual selection does not occur in seahorses, although it does in their relatives, the ghost *pipefishes. Indeed, as stated by the title of Vincent (1994): "Seahorses exhibit conventional sex roles in mating competition, despite male pregnancy".

Presently, the most important feature of seahorses, in terms of their evolution, and even of the survival of their few species, is that there is in China an immense, booming market in dried seahorses, taken – you guessed correctly – as a male aphrodisiac (though they are also taken, for example, as a cure for asthma and heart diseases). This has generated intense, export-oriented fisheries throughout the world, several of which have collapsed, following the extirpation of the underlying populations from their shallow habitats (Vincent 1996).

Project Seahorse, documented in Lourie *et al*. (1999), addresses these, and related issues of trade and conservation of seahorses. *Project Seahorse* should run a campaign mocking men who believe that ingesting dried-up, male, pregnant, and strictly monogamous little fishes is going to do anything for them.

Seahorses (II) Strangely enough, seahorses serve in *Origin* as a device to explain the development of the organs after which the *mammals are named. Here is CD's case: The mammary glands are common to the whole class of mammals, and are indispensable for their existence; they must therefore, have been developed at an extremely remote period, and we can know nothing positively about their manner of development. Mr Mivart asks: 'Is is conceivable that the young of any animal was ever saved from destruction by accidentally sucking a drop of scarcely nutritious fluid from an accidentally hypertrophied cutaneous gland of its mother? And even if one was so, what chance was there of the perpetuation of such a variation?' But the case is not here put fairly. It is admitted by many evolutionists that mammals are descended from a marsupial form; and if so, the mammary glands will have been at first developed within the marsupial sack. In the case of the fish (Hippocampus) the eggs are hatched, and the young are reared within a sack of this nature; and an American naturalist, Mr Lockwood, believes from what he has seen of the development of the young, that they are nourished by a secretion from the cutaneous glands from the sack. Now with the early progenitors of mammals, almost before they deserved to be so thus designated, is it not at last possible that the young might have been similarly nourished? (Sixth edition, p. 189; Mivart 1871; Lockwood 1868; CD defines 'cutaneous' as of or belonging to the skin *Origin VI*, p. 432).

Recent research has confirmed that young seahorses require for their development and survival the chemical environment of the brood pouch, notably for the supply of calcium and enzymes secreted by the males into the 'marsupial fluid,' though it does not seem to supply nourishment proper, i.e. food energy (Huot 1902; Linton and Soloff 1964; Boisseau and Le Menn 1967). Thus, CD's case holds. Yet specious arguments such as put forward by Mivart continue to this day. Being based not on evidence, but on lack of scientific creativity, and unwillingness to even look at the animals being pontificated on, these arguments will forever be there for ignoramuses (or even 'ignorami') to give themselves a critical touch (*see also* **Non-Darwinian evolution**).

Seasquirts A group of chordate animals, also known as ascidians, used by CD to infer *vertebrate origins, and which he defined as invertebrate, hermaphrodite, marine creatures permanently attached to a support.

They hardly appear like animals, and consist of a simple, tough, leathery sack, with two small projecting orifices. They belong to the Molluscoida of *Huxley – a lower division of the great kingdom of the Mollusca; but they have recently been placed by some naturalists among the Vermes or worms. Their larvae somewhat resemble *tadpoles in shape,[23] and have the power of swimming freely about.

M. Kovalevsky[24] has lately observed that the larvae of Ascidians are related to the *Vertebrata, in their manner of development, in the relative position of the nervous system, and in possessing a structure closely like the *chorda dorsalis* of vertebrate animals; and in this he has been since confirmed by Professor Kupffer. M. Movalevsky writes to me from Naples, that he has now carried these observations yet further, and should his results be well established, the whole will form a discovery of the very greatest value. Thus, if we may rely on *embryology, ever the safest guide in *classification, it seems that we have at least gained a clue to the source when the Vertebrata were derived.[25]

We should then be justified in believing that at an extremely remote period a group of animals existed, resembling in many respects the larvae of our present Ascidians, which diverged into two great branches – the one retrograding in development and producing the present class of Ascidians, the other rising to the crown and summit of the animal kingdom by giving birth to the Vertebrata. (*Descent I*, pp. 159–60; n. 23 reads: At the Falkland Islands I had the satisfaction of *seeing, in April, 1833, and therefore some years before any other naturalist, the locomotive larvae of a compound Ascidian, closely allied to the Synoicum, but apparently generically distinct from it. The tail was about five times as long as the oblong head, and terminated in a very fine filament. It was, as sketched by me under a simple microscope, plainly divided by transverse opaque partitions, which I presume represent the great cells figured by Kovalevsky. At an early stage of development the tail was closely coiled round the head of the larva; n. 24 refers to *Mémoires de l' Acad. des Sciences de St Petersbourg, vol. x,* *no. 15, 1866*; Professor Kupffer is probably Kupffer (1869); n. 25 refers to Giard (1872), p. 281; *Synoicum* is a genus of seasquirts (*see also* **Vertebrate origins**).

Sebastes spp. *See* **Scorpionfishes**.

Secondary sexual characters Features, other than the gonads themselves, by which the females and males of a given species differ, either permanently, upon reaching maturity, or during the mating season.

Having read Hunter (1837, p. 47), CD jotted the comment that Fishes have no secondary characters (*Notebook D*, p. 369). Hunter's report ("in fishes, there are no great difference . . .") was very wrong. Indeed CD gave later prominence to fishes when, in *Descent*, he assembled numerous descriptions of secondary sexual characters, both morphological and behavioural, in different groups of animals.

In fact, the section of that review dealing with fishes is the longest text written by CD pertaining exclusively to this group. Readers are invited to read *Descent* (II, pp. 330–47) for a complete version of this account, here broken up, because of its length, into entries dealing with a species of fish, or a group of species (e.g. *sticklebacks, *salmon, *Trout), originally serving as examples for some generalization.

Because CD emphasized sexual differences in coloration, we end this entry with his summary of the effect of such colour differences: On the whole we may conclude, that with most fishes, in which the sexes differ in colour or in other ornamental characters, the males originally varied, with their variations transmitted to the same sex, and accumulated through Sexual selection by attracting or exciting the females. In many cases, however, such characters have been transferred, either partially or completely, to the females. In other cases, again, both sexes have been coloured alike for the sake of protection; but in no instance does it appear that the female alone has had her colours or other characters specially modified for this latter purpose. (*Descent II*, p. 347).

Seeing The act of perceiving through vision, involving the interpretation, by the *brain, of the raw images it receives. This interpretation unavoidably overemphasizes the small part of what we 'see' that is relevant to some pre-existing theory. Or, as put by CD, in flagrant contradiction to the cant in his *Autobiography*, [h]ow odd it is that anyone should not see that all observation must be for or against some view if it is to be of any service. (Darwin and Seward 1903, Vol. II, p. 195). Therefore, the experienced practitioners of each scientific discipline, operating within a dominant *paradigm, must teach their students to 'see' what it is that is currently relevant to that discipline and its various theories when looking at a specimen, or through some apparatus (telescope, microscope, etc.). This is nicely illustrated by the *quote below, although it has a dubious character (Louis *Agassiz) being cited by another one (Ezra Pound 1934; as cited in Scholes 1984):

No man is equipped for modern thinking until he has understood the anecdote of Agassiz and the fish:

A post-graduate student equipped with honours and diplomas went to Agassiz to receive the final and finishing touches. The great man offered him a small fish and told him to describe it.

Post-Graduate Student: 'That is only a *sunfish.'

Agassiz: 'I know that. Write a description of it.'

After a few minutes the student returned with a description of the Ichtus Heliodiplodokus, or whatever term is used to conceal the common sunfish from vulgar knowledge, family of Heliichthinkerus, etc., as found in textbooks of the subject.

Agassiz again told the student to describe the fish.

The student produced a four-page essay. Agassiz then told him to look at the fish. At the end of three weeks the fish was in an advanced state of decomposition, but the student knew something about it.

It is hard to have to agree with Pound, given his identity as Ezra the Thrice Committed [first to a particularly obscurantist form of Poetry, then – in Italy, no less – to Mussolini's original brand of Fascism, and finally, back in the USA, to another lunatic asylum, so he could escape the consequence of his second commitment]. Yet what Pound so light-heartedly describes is the very way *Grant taught CD to look at his *Lumpfish, and which led to CD's attempts to relate what he saw to previously established patterns. This eventually led to CD being able to see new, larger patterns, one of these being *natural selection.

Scholes (1984, p. 654), a postmodernist by the standards of Gross and Levitt (1994), sees this as reprehensible (as would Aoki 2000). He complains that "Agassiz' student [. . .] seems to be reporting about a real and solid world in a perfectly transparent language, but actually, he is learning how to produce a specific kind of discourse, controlled by a particular scientific paradigm, which requires him to be constituted as the subject of that discourse in a particular way and to speak through that discourse of a world made visible by that same controlling paradigm. The teacher's power over the student is plain in Pound's example, and the student's ritual suffering as he endures the smell of the decomposing fish and the embarrassment of the teacher's rebuffs is part of an initiation process he must undergo to enter a scientific community."

Actually, what Scholes doesn't know is that the "ritual suffering" over the stinky fish is what senior ichthyologists use to identify smart students (of both sexes, not only 'he'): they keep the smart students, i.e. those that put their fish in alcohol long before they start decomposing.

Selection A process experienced by all living organisms, and which determines whether they survive (and reproduce) or not.

As first noted by CD, selection occurs in two basic forms, as *natural selection, wherein natural circumstances determine survival and reproduction (the latter also affected by *sexual selection), and as *artificial selection, wherein *humans* choose whether an organism should survive and reproduce (or not). Whereas variation, due to reshuffling of genes, and to mutations, is a truly random process, selection is definitely non-random and reflects either a conscious will (that of the breeder in case of artificial selection), or physical, ecological or other constraints (in the case of natural selection).

Hence CD's theory of natural selection does not represent, as often asserted, a view of *evolution dominated by 'randomness'. Indeed, the history of the biological sciences is, to a large extent, a chronicle of more and deeper constraints to selection being discovered, and thereby allowing the explanation of more phenomena (*see also* **Hox**; **Oxygen**; **Parsimony**).

Seriola bipinnulata *See* **Rainbow runner.**

Serranus *See* **Groupers (II).**

Sex ratio The relative numbers of females and males in (samples taken from) a population, from which various inferences on their distribution, behaviour (including their catchability by sampling gear; Arreguin-Sánchez, 1996), reproduction, growth and mortality can be drawn. Of course, when sex ratios consistently differ from the 1 : 1 value that is to be expected from first principles (see, for example, Dawkins 1995), an explanation is required which makes sense in terms of natural selection. CD saw the problem, and did propose explanations: there are certain animals (for instance, fishes and *cirripedes) in which two or more males appear to be necessary for the fertilization of the female; and the males accordingly largely preponderate, but it is by no means obvious how this male-producing tendency could have been acquired. (*Descent II*, p. 259).

With fish the proportional numbers of the sexes can be ascertained only by catching them in the adult or nearly adult state; and there are many difficulties in arriving at any just conclusion.[69] Infertile females might readily be mistaken for males, as Dr *Günther has remarked to me in regard to *trout. With some species the males are believed to die soon after fertilizing the ova. With many species the males are of much smaller size than the females, so that a large number of males would escape from the same net by which the females were caught. (*Descent II*, p. 249; n. 69 reads: Leuckart quotes Bloch (Wagner, *Handwörterbuch der Phys.*, Vol. iv, 1853, p. 775), that with fish there are twice as many males as females; Leuckart 1853; Bloch 1782–4, p. 117, states, without comparing their body sizes, that at least twice as many males as females are required, under natural conditions, to fully fertilize the freshly extruded eggs of 'fish,' i.e. *cyprinids. Hence his recommendations for fish farmers to use four males per female for fertilizing fish eggs under controlled conditions [p. 12], or 3–4 for *Bream [p. 77], and 2–3 for common *Carp [p. 102]).

M. Carbonnier,[70] who has especially attended to the natural history of the *pike (*Esox lucius*), states that many males, owing to their small size, are devoured by the larger females; and he believes that the males of almost all fish are exposed from this same cause to greater danger than the females. (*Descent, II*, p. 249; n. 70 refers to the Farmer, 18 March, 1869, p. 369, i.e. Carbonnier (1869a)).

Nevertheless, in the few cases in which the proportional numbers have been actually observed, the males appear to be largely in excess. Thus Mr R. Buist, the superintendent of the Stormontfield experiments, says that in 1865, out of 70 *salmon first landed for the purpose of obtaining the ova, upwards of 60 were males. In 1867 he again 'calls attention to the vast disproportion of the males to the females. We had at the outset at least ten males to one female.' Afterwards females sufficient for obtaining ova were procured. He adds, 'from the great proportion of the males, they are constantly fighting and tearing each other on the spawning-beds'.[71] (*Descent II*, p. 249; n. 71 refers to Buist (1866), p. 23, to Buist (1867), and to a letter sent to CD on March 5, 1868; *Calendar*, no. 5984).

This disproportion, no doubt, can be accounted for in part, but whether wholly is doubtful, by the males ascending the rivers before the females. Mr F. Buckland remarks in regard to trout, that 'it is a curious fact that the males preponderate very largely in number over the females. It *invariably* happens that when the first rush of fish is made to the net, there will be at least seven or eight males to one female found captive. I cannot quite account for this; either the males are more numerous than the females, or the latter seek safety by concealment rather than flight.' He then adds, that by carefully searching the banks, sufficient females for obtaining ova can be found.[72] Mr H. Lee informs me that out of 212 trout, taken for this purpose in Lord Portsmouth's park, 150 were males and 62 females. (*Descent II*, p. 249; n. 72 refers to Buckland (1868), with additional information from a letter sent by Buckland on Feb. 27, 1868, in response to a query by CD; see *Calendar*, nos. 5866 and 5946; Lee's information stems from a letter sent Jan. 21, 1868; *Calendar*, no. 5793; for details on the naturalist Henry Lee, see Tort 1996, p. 2607).

The males of the *Cyprinidae likewise seem to be in excess; but several members of this Family, viz., the *carp, *tench, *bream and *minnow, appear regularly to follow the practise, rare in the animal kingdom, of polyandry; for the female whilst spawning is always attended by two males, one on each side, and in the case of the bream by three or four males. This fact is so well known, that it is always recommended to stock a pond with two male tenches to one female, or at least with three males to two females. With the *minnow, an excellent observer states, that on the spawning-beds the males are ten times as numerous as the females; when a female comes among the males, 'she is immediately pressed closely by a male on each side; and when they have been in that situation for a time, are superseded by other two males'.[73] (*Descent II*, pp. 249–50; n. 73 reads: Yarrell, *Hist. British Fishes*, vol. i, 1826 [should be 1836], p. 307; on the *Cyprinus carpio*, p. 331; on the *Tinca vulgaris*, p. 331; on the *Abramis brama*, p. 336. See, for the minnow (*Leuciscus phoxinus*), London's Mag. of Nat. Hist., vol. v, 1832, p. 682.).

The problem posed by sex ratios different from 1:1 has now largely been solved, even for *barnacles, *hermaphroditic fishes such as *wrasses, or *groupers, and for social insects. The solution relies largely on the differential costs of producing and raising daughters and sons, relative to the probability that either will themselves produce offspring (Maynard Smith 1978; Dawkins 1995; Barlow 2000, pp. 52–53). Here, we have gone way beyond CD, while remaining firmly rooted in *natural selection.

Sexual dimorphism Dimorphism is the condition of the appearance of the same species under two dissimilar forms (*Origin VI*, p. 433). Thus, sexual dimorphism pertains to sex-related differences in body shape or proportions, maximum size, and size at age of organism, usually the result of *sexual selection.

Thus, as noted by CD [i]n many kinds of fish the males are much smaller than the females, and they are believed often to be devoured by the latter, or by other fishes. (*Descent II*, p. 216). In regard to size, M. Carbonnier[11] maintains that the female of almost all fishes is larger than the male; and Dr *Günther does not know of a single instance in which the male is actually larger than the female.

With some Cyprinodonts the male is not even half as large. As in many kinds of fishes the males habitually fight together, it is surprising that they have not generally become larger and stronger than the females through the effects of *Sexual selection. The males suffer from their small size, for according to M. Carbonnier, they are liable to be devoured by the females of their own species when carnivorous, and no doubt by other species. Increased size must be in some manner of more importance to the females, than strength and size are to the males for fighting with other males; and this perhaps is to allow of the production of a vast number of ova. (*Descent II*, p. 335; n. 11 cites The Farmer, 1868, p. 369, presumably referring to Carbonnier (1869a)).

In fishes, the females are larger than the males in about three quarters of all species (Pauly 1994a, p. 172; *FishBase can be used for within-group estimations of this ratio). The Nile tilapia *Oreochromis niloticus*, of the family *Cichlidae, is one representative of the rest, and hence the importance in aquaculture, of methods for sexing fingerlings, and/or methods for reversing the sex of the females (see, for example, contributions in Pullin *et al.* 1996).

Interestingly, CD did not notice that male fishes being generally smaller than their female conspecifics contradicts the notion that growth and reproduction operate in direct antagonism to each other, as stated in *Variations* (*see* **Growth; Reproductive drain hypothesis; Oxygen**).

Sexual selection A process distinguished from *natural selection, resulting from competition for mates, and which explains most instances of *sexual dimorphism, as well as many differences of *colour patterns and of mating behaviour between the sexes of animal species.

Here is CD's definition: This form of selection depends, not on a struggle for existence in relation to other organic beings or to external conditions, but on a struggle between the individuals of one sex, generally the males, for the possession of the other sex. The result is not death to the unsuccessful competitor, but few or no offspring. Sexual Selection is therefore less rigorous than Natural Selection.

(*Origin*, II, p. 69). Though earlier authors had discussed aspects of what CD called sexual selection (Thornhill 1986), it is he who put it on a solid footing – although this did not prevent it from sinking into near oblivion, following two decades of debates after the publication of *Descent*, where most of CD's case is made.

Perhaps the notion of female choice, an essential component of sexual selection, was just too much for the very male zoologists of Victorian England. Be that as it may, it was only in the mid twentieth century that sexual selection was revived. It is now a very active area of research (Andersson 1994).

Interestingly, fish play a prominent role in this research, just as they did in *Descent*, whose Chapter XII is one of only two in all of CD's opus where the word 'fish' appears in the title (the other is Chapter VIII in *Variation*, and it emphasizes *goldfish).

Given the format of this book, Chapter XII in *Descent* was chopped up into species-sized chunks, which are then commented upon separately. The astute reader, however, will want to read that chapter as a whole, if only to quantify the extent of allogenous self-*plagiarism thus induced.

Shad Common name of *Alosa sapidissima* (Wilson, 1811), a member of the Family Clupeidae, naturally occurring along the Eastern coast of the USA and Canada (Florida to Newfoundland), and introduced to the West Coast, where it now ranges from Baja California to Alaska, and beyond to Kamchatka.

Alosa sapidissima and its congeners are famous for their *anadromy, and their spawning in rivers, where they supported major fisheries (Leim 1924), most now gone.

CD, based on Eaton (1831), himself based on a personal communication from 'Mr. Adams', mentions shad in the context of their presumed competitive interactions with a related species, the Alewife *Alosa pseudoharengus* (Wilson, 1811).

The quote in question reads: Fish with allied habits must chiefly affect fish; & thus the shad (Clupea sapidissima) has increased in the Hudson, in parts full twenty-fold, owing to the erection of a dam, & the consequent decrease chiefly of another species of Clupea.[6] (*Big Species Book*, p. 200; n. 6 reads: Mr. Adams in N. American Journal of Science Vol. 20, p. 150.).

Sharks CD often refers to 'sharks' in the *Diary*, in the *Journal*, in his *Notebooks, and in subsequent writing, and the following gives, in rough chronological order, his references to sharks neither further identified, nor used elsewhere in this book: A large shark followed the ship, & was first struck by a harpoon; after this he was hooked by a bait & again being struck broke the hook & escaped. Such an adventure creates great interest all over the whole ship (*Diary*, March 19, 1832, near Bahia); The greatest event of the day has been catching a fine young shark with my own hook. It certainly does not require much skill to catch them, yet this in no way diminishes the interest. In this case the hook was bigger than the palm of my hand & the bait only a bit of salted pork just sufficient to cover the point. Sharks when they seize their prey turn on their backs; no sooner was the hook astern, than we saw the silvery belly of the fish & in a few moments we hauled him on deck. (*Diary*, March 21, 1832; between coast of Brazil and *Abrolhos). We have seen great quantities of shipping; & what is quite as interesting, Porpoises, Sharks & Turtles (April 3, 1832, harbour of *Rio de Janeiro).With respect to Sharks distributing *fossil remains: Sharks followed Capt. Henry's vessel from the Friendly Isles. to Sydney; know by having been seen & from the contents of its maw, amongst which were things pitched over board early in the passage!! (*Notebook R*, p. 23).

Also, we have: In Phillips. p. 90. it seems the most organized fishes lived far back, fish approaching to reptiles at *Silurian age. (*Notebook B*, p. 213; Phillips 1839, pp. 90–1), and My idea of propagation almost infers, what we call improvement, – All Mammalia from one stock, & now that one stock cannot be supposed to be <<most>> perfect (according to our ideas of perfection); but intermediate in character, the same reasoning will allow of decrease in character. (which perhaps is) Case with fish – as some of the most perfect kinds of shark. lived in remotest epochs. (*Notebook B*, p. 222).

Finally, we have from *Notebook C* (p. 247) Sharks very generally distributed:[8] Mem of great geological age (n. 8 cites Lesson (1830–1), p. 69). (*See also* **Basking shark; Galápagos shark; Great white shark; Hippopotamus; Megatooth shark**.)

Shoal A word used in CD's writings to refer to (1) areas of shallow waters, and (2) social groups of fishes. Here is an example of the former use: In shoaler water, at the distance of a few miles from the coast, very many kinds of crustacea and some other animals were numerous (*Journal*, December 6, 1833; see also **Abrolhos**).

The latter use is illustrated by the following quote: large shoals of two species of *Scarus, one inhabiting the surf outside the reef and the other the *lagoon, subsist entirely [. . .] by browsing on the living polypifers (*Coral Reefs*, p. 20).

This gives us the opportunity to define, with Pitcher (1983), the difference between shoals and schools of fish. "Shoal is [. .] a general term for groups of fish, [. . .], the only criterion being that the shoal is a social, rather than a purely exogenously-determined assembly." The fish within a school, however, exhibit synchronized swimming. Thus, all schools are shoals, but the converse does not apply. There is no evidence of CD having dealt with this or related issues of shoaling/schooling, though his wonderment is evident in his *Marginalia* to Agassiz (1850), wherein this staunch supporter of *Creationism writes on shoals created as shoals.

Showers of fish Fish falling from the sky, as an inevitable consequence of having earlier been lifted by hurricanes or other freak meteorological events, fascinated CD, who saw in such falls a means of dispersal, affecting the *distribution of freshwater fishes.

Thus, *Notebook E* (p. 453) mentions a Long attested account of fall of fish in India.[2] (n. 2 refers to Prinsep (1833), p. 650). Also: Considering our ignorance of the many strange chances of diffusion by birds (which occasionally wander to immense distances) and quadrupeds swallowing seeds and ova (as in the case of the flying *water-beetle which disgorged the *eggs of a fish), and of whirlwinds carrying seeds and animals into strong upper currents (as in the case of volcanic ashes and showers of hay, grain and fish), [. . .] we ought to be very slow in admitting the probability of double creations. (*Foundations*, p. 169; *see* **Morris, Margaretta Hare**, for neat contradiction with CD's later perception of fish egg dispersal by water-beetles).

Showers of fishes are important to explain their distribution. Some, however, doubt the existence of such showers, even though many reports attest to them. [Note, however, that this is also true of Loch Ness monsters (Sheldon and Kerr 1972).] Thus, the Reuters Press Agency reported on such an event in early June 2000: "drought-stricken farmers tending their field in southern Ethiopia got a nasty shock when the heaven opened and they were pelted by fish, a local newspaper reported yesterday. 'The unusual rain of fish which dropped from the air – some dead, some still struggling – created panic among the mostly religious farmers,' the weekly Amharic newspaper said. Saloto Sodoro, a fish expert in the region, attributed the phenomenon to heavy storms in the Indian Ocean which swept up the fish before shedding them on the unsuspecting farmers."

A letter of Hensleigh Wedgwood to CD mentions another case: "I don't know whether you admit those showers of fishes as an undoubted fact, but I ascertained that this was certainly the case with the one mentioned in the Times last week. The fishes fell in a violent shower of rain in the yard of our colliery over a space of about 100 yards long by 14 or 15 broad. A large proportion fell on the roof of our engineers shed about 25 high in the ridge. Others in a puddle in the yard which is dry in dry weather & these were living & were gathered by several of the people. The overlooker had a lot of them in a goldfish jar which were *minnows and *stickleback" (*Correspondence*, March 13–19, 1859; the event reported in *The Times* (p. 7, March 10, 1859) had occurred at the village of Mountain Ash, Glamorganshire), and CD annotated the letter with Shower of fishes.

This theme was picked up again in *Origin*; see **Freshwater fishes (III).**

Silurian Geological period of the central Palaeozoic era, named after the Silures, an ancient tribe living in southeastern Wales, and which pertains to times between 443 and 418 million years ago, right before the *Devonian. The most advanced fishes in the Silurian belong to the now extinct Acanthodii, the earliest known fishes with true jaws (Nelson 1994, p. 66). Should we also note that the first 'Siluri', i.e. *catfishes (from Latin '*silurus*,' referring to a freshwater fish) did not evolve in the Silurian, but much later, in the *Cretaceous?

Siluri(-oid) *See* **Catfishes.**

Silurus CD's term for *catfishes. The genus *Silurus* in the narrow sense is represented here by *Silurus glanis* (*see* **Wels catfish**).

Silversides Members of the Family Atherinidae, small fishes represented here by *Basilichthys microlepidotus* (Jenyns, 1841), *Odontesthes argentinensis* (Valenciennes, 1835) and *O. incisa* (Jenyns, 1841), all previously attributed to the genus *Atherina* (*Fish*, pp. 77–80). *Odontesthes argentinensis* was sampled by CD at *Maldonado (Uruguay), and described as silvery, with a silver lateral band, above bluish grey; very common, (also in brackish waters) (*Fish in Spirits*, no. 693); *B. microlepidotus* was

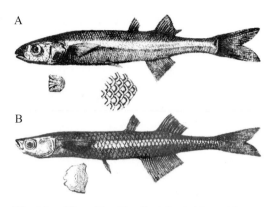

Fig. 32. Silversides, Family Atherinidae, with inserts showing magnified scales. **A** *Basilichthys microlepidotus* (*Atherina microlepidota* in *Fish*, Plate XVI), based on a specimen collected by CD in freshwater in Valparaiso, Chile. **B** *Odontesthes incisa* (*Atherina incisa*, Plate XVI).

sampled in *Valparaiso (Chile; *Fish in Spirits*, no. 1056; Fig. 32A).

The specimens of *Odontesthes incisa*, of about 5 cm, were semitransparent, colourless; with a bright silver band on each side; also marked with silvery about the head, and were sampled in September 1832, at 39°S and 61°W (i.e. off *Bahia Blanca), some miles from the land (*Fish in Spirits* no. 367; Fig. 32B). Jenyns notes that this would "indicate that the specimens were not so very young, as the *fry of most fish keep close in shore" (*Fish*, pp. 79–80). This inference is appropriate, given that this species reaches a maximum (total) *length of 15 cm (Cervigón and Fischer 1974; Froese and Binohlan 2000), and that Jenyns' measurement can be assumed to have referred to standard length.

Sirenians A groups of herbivorous *marine mammals, whose odd shape and the possession of mammary glands inspired P. T. Barnum to exhibit some as *mermaids, part of the *Great Chain of Being.

Sirenians nowadays include only the manatee, whose three species feed on seagrass in the (sub)tropical Atlantic, and on macrophytes in rivers of South America and West Africa, and the similarly adapted Indo-Pacific dugong (*see* **Hippopotamus**).

The Steller sea cow (*Hydrodamalis gigas*), which fed on kelp in the North Pacific, was hunted to extinction in 1768 (Steineger 1886; Norse 1993), and the remaining sirenian species are threatened in most of their now much reduced *ranges.

Skate Common name of some members of the Family Rajidae, notably *Raja batis* (Linnaeus, 1758) of the Eastern North Atlantic. Skates differ from *rays in having shorter, stouter tails which lack a venomous barb.

CD sampled skates in Patagonia: Skate, two specimens both sexes Good Success Bay. Colour 'broccoli brown,' and marked (like binding of book) with rings and lines of 'chocolate red.' Iris silvery grey upper part depending fringed, sometimes almost concealing the pupil (*Fish in Spirits*, no. 538; see annotation to Syme (1821) for the bizarre colour names).

As well, he sampled a skate, above muddy cream colour in *King George's Sound (*Fish in Spirits*, no. 1389). Like those from Patagonia, this skate was not described in **Fish*, probably because it belonged to that part of his *collections that rotted away.

The presence of an electric organ in the skate was discovered by Robert M'Donnell (1861), who, given the existence of such organ in *Torpedo, was inspired to search for it in the skate, a relative, after reading CD's point in *Origin* about "unity of type resulting from community of descent" (M'Donnell 1861, p. 58). CD, who had been informed prior to publication by M'Donnell himself, describes this with an enthusiasm matched only by his wild abandon when *spelling the man's name.

First to *Huxley: I had a letter today from R.McDonnell of Dubline (I wonder whether he is the bearded man one sees at B. Assoc.^{n} if so I fear he is rash & wild) & he says owing to passage in my Book on Electric fishes he has been dissecting Rays, & believes he finds in same fish the *homologues of *both* the anterior & posterior proper electric organs of fish: which, if true, seems to me an interesting fact (*Correspondence*, Nov. 16, 1860); again I am so much obliged to you for telling me about Mac donnell […]: Since writing to you I have had a second, such a capital, letter from him. He will in time come round to our view on Species as I believe (*Correspondence*, Nov. 22, 1860).

Then to C. *Lyell: Some friend who is much opposed to me seems to have crowed over McDonnell, who reports that he said to himself that if Darwin is right there must be homologous organs both near the Head & Tail in other non-electric fish. He set to work & by Jove he has found them. So that some of *difficulty is removed, & is it not satisfactory that my hypothetical notions sh^d have led to pretty discovery. (*Correspondence*, Nov. 24, 1860; need we recall that Jove [= Zeus] was the ancient electrician in charge of lightning?). Next to Jeffries Wyman: … Speaking of transitions Mr MacDonnell of Dublin writes to me that he has made some curious discoveries on the electrical organs of the Rays, being led thereto by trying to make the case of the Electrical organs, already so very difficult to me, still more unpleasant; but as I understand him his new facts help my views considerably (*Correspondence*, Dec. 3, 1860).

The finale went to Huxley: MacDonnels, of course, pleases me greatly (*Correspondence*, Jan. 3, 1861).

Sleepers Fishes similar to *gobies, but belonging to the related Family Eleotridae.

The only species of the family considered here is the Giant bully *Gobiomorphus gobioides* (Valenciennes, 1837), whose common name suggests it must be quite a handful in terms of its behaviour, given its small size (up to 4 cm).

Jenyns' description of this fish – as *Eleotris gobioides* – is based on a specimen taken by CD in fresh water, near the *Bay of Islands, New Zealand (*Fish in Spirits*, nos. 1339, 1340; *Fish*, p. 98).

Smelt Common name of *Osmerus eperlanus* Linnaeus, 1758, a member of the Osmeridae, a group of small pelagic fishes also including *capelin. CD, based on Harmer (1767), discussed the fecundity of this and other fishes (*see* **Eggs of fishes**).

Snappers Fishes of the Family Lutjanidae, widespread through the tropics and subtropical seas, and occasionally entering fresh water.

The species collected by CD in the *Cocos Islands and described as *Diacope marginata* in *Fish* (p. 13; [u]pper part pale lead colour: pectorals yellow; ventrals and anal orange: sides very pale yellow) is now identified as Blacktail snapper *Lutjanus fulvus* (Forster, 1801).

This species, frequently referred to as *Lutjanus vaigiensis* or *L. marginatus* by recent authors (Allen 1985), has a wide range throughout the Indo-Pacific, from East Africa to the Marquesas and Line Islands, north to southern Japan, and south to Australia. Interestingly, it is one of the few species of marine fish successfully introduced from one country to another (Randall 1987a), i.e. from the Marquesas into the Hawaiian Islands.

Social Darwinism An ideological construct, initiated in part by Herbert *Spencer, which misused scientific concepts to assign to certain humans more worth than to others, and which quickly evolved into aggressive 'eugenism' and various forms of racism (Gould 1996a).

Social Darwinism was discredited by its most ardent proponents, the Nazis. Indeed, after the war they instigated (i.e. the Second World War – just in case), the notion that 'Darwinism', i.e. evolution by natural selection, provides a useful framework for interpreting data on human behaviour and culture was largely eclipsed by the opposite notion, that the human *brain is, at birth, a sort of empty vessel into which a given culture can and will pour any arbitrary content. This too turned out to not fit the facts, such as the overwhelming similarities among humans in the way in which natural languages are acquired (Pinker 1994), or in our perception of *beauty. A renewal of interest in applying CD's work to human culture and society was thus inevitable. Major steps in this renaissance, now largely devoid of its original stench, are Wilson's *Sociobiology* (1980), and the *Evolutionary Psychology* of Barkow *et al*. (1992).

As a result, many new, 'humane' interpretations of our behaviour have recently emerged (Singer 2000), though controversy continues (Brown 1999). Interestingly, some of these interpretations connect neatly with the enlightened views on social and racial issues which CD held, in contrast to most of his friends and colleagues. Thus, the following quote, in a language that social Darwinists, now and then, would never use: It is often attempted to palliate slavery by comparing the state of slaves with our poorer countrymen: if the misery of our poor be caused not by the laws of nature, but by our institutions, great is our sin; but how this bears on slavery, I cannot see; as well might the use of the thumb-screw be defended in one land, by showing that men in another land suffered from some dreadful disease. Those who look tenderly at the slave owner, and with a cold heart at the slave, never seem to put themselves into the position of the latter; what a cheerless prospect, with not even a hope of change! picture to yourself the chance, ever hanging over you, of your wife and your little children – those objects which nature urges even the slave to call his own – being torn from you and sold like beasts to the first bidder! And these deeds are done and palliated by men, who profess to love their neighbours as themselves, who believe in God, and pray that his Will be done on earth! It makes one's blood boil, yet heart tremble, to think that we Englishmen and our American descendants, with their boastful cry of liberty, have been and are so guilty: but it is a consolation to reflect, that we at least have made a greater sacrifice, than ever made by any nation, to expiate our sin. (*Journal*, August 19, 1836, second edition, 1845; the last part of the last sentence refers to the abolition of the British slave trade in 1807, of slavery in the English colonies in 1833, and the concomitant suppression of the transatlantic slave trade by the Royal Navy).

Sole Flatfishes of the families Soleidae (Old World) or Achiridae (New World); also common name of *Solea solea* (Linnaeus, 1758).

Fish (p. 140) describes only one 'sole', collected by CD at San Blas, *Patagonia, and the identification is only to the genus level (*Plagusia*; see Fish in Spirits, no. 480). However, Jenyns notes a similarity to *P. brasiliensis*, now *Symphurus tessellatus* (Quoy & Gaimard, 1824). Note that on p. xiv, in the 'Systematic Table of Species' of *Fish*, Jenyns lists "*Plagusia lineatus?,*" while in its Index, p. 171, he refers to "*Plagusia fasciatus?,*" a confusion hard to sort out, though updated identities could be proposed for these names as well (see Eschmeyer 1998; *see also* **Names, updated**).

In the nineteenth century, the flatfishes were considered to belong all to a single, broad family, the Pleuronectidae (*see*, for example, *Fish*, pp. 138–40; Günther 1880, pp. 553–9). Hence CD's statement that the Pleuronectidae are admirably adapted by their flattened and asymmetrical structure for their habits of life, is manifest from several species, such as soles, flounders, etc., being extremely common [...]. The different members, however, of the family present, as Schiödte remarks, 'a long series of forms exhibiting a gradual transition from *Hippoglossus pinguis*, which does not in any considerable degree alter the shape in which it leaves the ovum, to the soles, which are entirely thrown to one side'. (*Origin VI*, p. 186; Schiödte 1868).

But it cannot be supposed that the peculiar speckled appearance of the upper side of the sole, so like the sandy bed of the sea [...is] due to the action of the light. Here Natural selection has probably come into play, as well as in adapting the general shape of the body of [this fish], and many other peculiarities, to their habits of life. (*Origin VI*, p. 188; *see also* **Flatfish controversy (I), (II)**).

Sole, Lined Common name of *Achirus lineatus* (Linnaeus, 1758), a *sole of the Western Atlantic, ranging from Florida to Northern Argentina, and occurring mainly in waters of reduced salinities, e.g. in estuaries.

Carvalho-Filho (1994, p. 238) suggests that the Lined sole reaches 12 cm, but a specimen collected by CD, at a market in Buenos Aires, was 9 inches long, i.e., about 23 cm. This and other fish bought from the market were all edible (*Fish in Spirits*, nos. 738, 739, 745), as one would expect.

Solenostoma *See* **Pipefishes, ghost**.

Sounds Many fishes emit sounds that are easily heard underwater, and some species produce sounds loud enough to be heard above water (review in Fish 1954). However, these sounds are not caused by "fish of about six pounds weight beating [their] tail against the vessel to relieve [themselves] from the pain caused by multitudes of parasitic worms" (Richardson 1836, p. 204).

CD used more reliable sources, notably Dufossé (1858a, b, 1862): fishes are known to make various noises, some of which are described as being musical. Dr Dufossé, who has especially attended to this subject, says that the sounds are voluntarily produced in several ways by different fishes: by the friction of the pharyngeal bones – by the vibration of certain muscles attached to the *swim-bladder, which serves as a resounding board – and by the vibration of the intrinsic muscles of the swim-bladder. By this latter means the *Trigla produces pure and long-drawn sounds which range over nearly an octave. But the most interesting case for us is that of two species of *Ophidium, in which the males alone are provided with a sound-producing apparatus, consisting of small movable bones, with proper muscles, in connection with the swim-bladder.⁴¹

The drumming of the *Umbrinas in the European seas is said to be audible from a depth of twenty fathoms; and the fishermen of Rochelle assert 'that the males alone make the noise during the spawning-time; and that it is possible by imitating it, to take them without bait'.⁴² From this statement, and more especially from the case of *Ophidium, it is almost certain that in this, the lowest class of the Vertebrata, as with so many insects and spiders, sound-producing instruments have, at least in some cases, been developed through Sexual selection, as a means for bringing the sexes together. (*Descent* II, p. 347; n. 41 reads: *Comptes Rendus*, vol. xlvi, [Dufossé] 1858, p. 353; vol. xlvii, [Dufossé] 1858, p. 916; vol. liv, 1862, p. 393. The noise made by the Umbrinas (*Sciaena aquila*), is said by some authors to be more like that of a flute or organ, than drumming: Dr. Zouteveen, in the Dutch translation of this work (vol. ii, p. 36), gives some further particulars on the sounds made by fishes; n. 42 reads: The Rev. C. Kingsley, in *Nature*, May, 1870, p. 40; see annotation to Darwin (1872) for more on Zouteveen and his translation).

Actually, the quote marked '42' is on p. 46 of Kingsley (1870), this being one of CD's rare lapses in this regard. The text in question starts with: "the males alone make it in spawning time; and that . . . . . ." Thus, CD replaced 'it' in the original text by 'the noise,' to make it more readable, but did not indicate the change. He made a lot of such small alterations in his writings. It is OK, as it does not alter the meaning of the *quotes.

As for getting an idea of how fish sounds are, you can *see* selected spectograms in Fish and Mowbray (1970), and in Bright (1972), or *hear* the sounds themselves, via a new online feature of *FishBase.

Sparus *See* **Porgies**.

Speciation The process by which *species are formed, i.e. separated from ancestral and/or sibling species. CD did not write much about this (Pauly 2002a); here may be his best shot: The passage from one stage of difference to another may, in many cases, be the simple result of the nature of the organism and of the different physical conditions to which it has long been exposed; but with respect to the more important and adaptive characters, the passage from one stage of difference to another, may be safely attributed to the cumulative action of natural selection, hereafter to be explained, and to the effects of the increased use or disuse of parts. A well-marked variety may therefore be called an incipient species; but whether this belief is justifiable must be judged by the weight of the various facts and considerations to be given throughout this work. (*Origin*, VI, p. 42).

It seems now accepted that, at least in fishes, most – if not all – speciation involves the physical separation of a group (which may be very small, and consist of a single female with fertilized eggs) of founding organisms from the distribution area of an ancestral species (Barlow 2000, pp. 244–5; *see also* **Punctuated equilibrium**).

Local adaptation, via *natural selection, to the circumstances of the new habitat, jointly with the 'founder effect' (Mayr 1982, pp. 600–6), can then quickly lead to a population different from the ancestral species. The duration of a separation is important: if the barrier that led to an 'incipient species' disappears too early, then the population in question will merge back into the ancestral species. (*See also* **Panama**.)

On the other hand, if the separation is long enough for a new 'mate recognition system' (Paterson 1985) to emerge, then the new species may maintain itself even if sympatry with the ancestral (or sibling) species is re-established, subject of course to the vagaries of competition.

Speciation may be very rapid in some species, and numbers as low as a few thousand years have been suggested for the emergence of new mate recognition systems, especially for *Cichlidae in the great lakes of Africa, where even less than a thousand years (Stiassny and Meyer 1999; Barlow 2000) may suffice for species to become separate.

Species A group of animals or plants having common characteristics and able to breed together to produce fertile offspring, so that they maintain their 'separateness' from other groups. (Note that this definition implies the infertility of *hybrids.) Also, the basic rank of biological nomenclature as implemented in the *Code. Several 'species concepts' exist in addition to that incorporated in the Code, and reviews emphasizing fishes are given by Nelson (1999) and Kullander (1999).

In contrast to what may be expected from its title, CD's most important book, "On the *Origin of Species . . ." does not really cover *speciation, the process by which two daughter species separate from their (single) parent species (Pauly 2002a). CD's reluctance to engage in debates about speciation and species definition may stem from his view, from the 1840s onward, that species were too impermanent, and the limits between them too fluid, for evolutionists to gain from spending much time on attempts to derive a generally applicable species definition. Thus, he writes: I look at the term species as one arbitrarily given, for the sake of convenience, to a set of individuals closely resembling each other, [. . .] it does not essentially differ from the term variety, which is given to less distinct and more fluctuating forms. The term variety, again, in comparison with mere individual differences, is also applied arbitrarily, for convenience' sake. (Origin, VI, p. 42; see also **Subspecies**).

On the other hand CD's work on *barnacles provided him with a sound background in *taxonomy, and convinced him of the need for taxonomists to produce rigorous identification and description of species, which for all practical purposes, can be treated as unchanging entities, with names serving as 'hooks' to which information (on biology, distribution etc.) can be attached, and straightforwardly retrieved (as done in *FishBase). Hence CD's comment on a sloppy practice that makes such information retrieval more difficult than need be: Unfortunately and stupidly, Gærtner does not append authors' names to the species (Correspondence to J. D. Hooker, July 14, 1857; the unfortunate author is Karl Friedrich von Gärtner, whom CD frequently cited; Tort 1996, pp. 1794–5).

Spelling CD's orthography was adventurous at times – e.g. Barrow cooter for *barracuda. Indeed, as noted by Sulloway (1983), his shifts to alternative spellings of the same *words have been used to date some of the entries in his manuscripts.

In this volume, CD's spelling was retained, so far as it had been in the sources that were consulted – as is the case for his *Correspondence and *Notebooks (see also section on conventions used in the text). Also, I abstained from identifying CD's errors, as could have been done, e.g., by adding 'sic' after such errors. I took these decisions in spite of the risk that readers think the misspellings are mine – which of course they will occasionally be: think of the epistemological difficulties involved in checking that something is 'correctly wrong', not to speak of the need to outwit the spellchecker of one's word processor. (See also **Albicore**.)

Spencer, Herbert English engineer (1820–1903), famous in his time for a philosophical system and 'laws' that pigeonholed much of the contemporary knowledge (e.g. biology, in Spencer 1864–7) into an irresistible march toward *progress, culminating in the peculiar arrangements of Victorian society (Tort 1996, pp. 4080–113; Singer 2000). CD disliked him (*Autobiography*, p. 108–109).

Spencer would be largely forgotten today, were it not for his reinterpretation of nascent Darwinism as a *social* doctrine (*social Darwinism). This was used to justify, in his time, the crushing hegemony of the landed and entrepreneurial classes in England, and their imposition, throughout a global system of colonies and vassal states, a form of 'globalization' (then called 'free trade') from which they alone benefited. Spencer also urged his notion of *'survival of the fittest' upon a reluctant CD, who used it only in *Variations*, and in the fifth and sixth editions of *Origin*, as a synonym of *natural selection (*Variations*, I, p. 6; *Origin VI*, pp. 72, 169).

However, contrary to CD's natural selection, Spencer's concept of the survival of the fittest was not much more than a restatement of the *Great Chain of Being, with British men (yes, men!) near the top of the chain, just under the Angels.

Only one of Spencer's many and largely vacuous 'laws' is briefly discussed here, i.e. that now referred to as the *Reproductive drain hypothesis.

Spontaneous generation The notion that 'lower' organisms emerge from non-living matter.

Spontaneous generation is not as ridiculous as it sounds. We do, after all, assume that the ancestors of the most primitive bacteria evolved from non-living matter, if only once, over 3.5 billion years ago. [Unless we believe, with Arrhenius (1908), that spores floated onto Earth from outer space, which only sweeps the problem under a big dark interstellar carpet, or unless we adhere to *creationism, in which case we give up on reason altogether.] CD, though he wrote that My theory leaves quite untouched the question of spontaneous generation (*Notebook E*, p. 446), proposed, in a phrase eerily suggestive of Miller's (1953) famous experiments, that we could conceive in some warm little pond, with all sorts of ammonia and phosphoric salts, light, heat, electricity, &c., present that a proteine compound was chemically formed ready to undergo still more complex change, at the present day such matter would be instantly devoured or absorbed, which would not have been the case before living creatures were formed. (Darwin 1887, Vol. III, p. 18, referring to a note written in 1871).

In the early nineteenth century, before CD proposed his descent with modifications (1859), a theory proposed by *Lamarck held sway, which stated that organisms, through their striving to improve themselves, effectively moved up some evolutionary ladder (the *Great Chain of Being in another guise). This steady upward movement would leave what may be called 'empty niches' in the habitat of the most primitive organisms, and spontaneous generation was the mechanism required to 'fill in' these empty niches.

Hence, refuting Lamarck (and the political consequences of a theory that appeared to imply the possibility of improvement for those at the bottom of the social heap), required that spontaneous generation be refuted. This was done by Louis Pasteur in 1864, in a series of 'scientific evenings' at the Sorbonne before audiences composed of the French social elite (Dubos 1988, pp. 47–50). This expression of interest by the members of a social group then as frivolous as they are now suggests that it was the political, rather than the scientific, implications of the demonstration that attracted them.

Squalus Type genus of the family Squalidae, or *dogfish. CD assumed that a small shark he sampled in *Patagonia belonged to this genus: Squalus. (very small specimen). However, Jenyns found it too badly preserved to describe (*see Fish in Spirits*, no. 415; also no. 505). As it happens, this may have been a specimen of the Piked dogfish *Squalus acanthias* Linnaeus, 1758, which occurs in Argentina.

CD gives a poignant description of another, female *Squalus* he caught by hook in Patagonia: Body bluish grey; above with rather blacker tinge, beneath much whiter; Its eyes the most beautiful thing I ever saw, pupil pale, 'verdegris green,' but with lustre of jewel, appearing like a Sapphire or Beryl. Iris pearly edge dark. Sclerotica pearly; in stomach was remains of large fish. In the uterus, the young ones for a long time after the viscera were opened continued to move; good specimen for dissecting (*Fish in Spirits*, no. 359).

Squirrelfishes A group of usually nocturnal reef fishes, also known as 'soldierfishes,' belonging to the Family Holocentridae.

CD mentions squirrelfishes only once: p. 309. says [. . .] that the phalanges have separate movements in the Holocentrus ruber (a fish) (*Notebook J*, p. 634), referring to Macculloch (1837, Vol. 1, p. 309) which states that "in the *Holocentrus ruber*, the rows of phalanges become as independent as fingers."

The current name of the Redcoat squirrelfish is *Sargocentron rubrum* (Forsskål, 1775), a tropical species originally ranging from East Africa and the Red Sea to Southern Japan. However, the redcoat has recently clawed its way through the Suez canal (a 'Lessepsian *migration' *sensu* Por 1978), and now inhabits the coast of the southeastern Mediterranean (Golani *et al.* 1983) from Libya to Turkey and the Dodecanese waters of Greece.

Stegastes imbricatus *See* **Damselfishes**.

Sticklebacks Members of the Family Gasterosteidae, which includes the well-studied Three-spined stickleback *Gasterosteus aculeatus* Linnaeus, 1758, actually a complex of (sub)species that includes 'G. *leiurus*' and 'G. *trachurus*' (Nelson 1994, p. 297). *G. aculeatus* has been shown to be capable of quickly evolving from a generalized form into separate limnic and benthic forms (Schluter 1996), at rates of thousands of *darwins (*see* Fig. 27 and **Punctuated equilibrium**).

CD summarized as follows the knowledge of his time on stickleback reproduction: Too little is known of the habits of reptiles and fishes to enable us to speak of their marriage arrangements. The stickle-back (Gasterosteus), however, is said to be a polygamist;[17] and the male during the breeding season differs conspicuously from the female. (*Descent II*, p. 220; n. 17 cites "Noel Humphreys, *River Gardens*, 1857").

Also: the male stickleback (Gasterosteus leiurus) has been described as 'mad with delight,' when the female comes out of her hiding-place and surveys the nest which he has made for her. 'He darts round her in every direction, then to his accumulated materials for the nest, then back again in an instant; and as she does not advance he endeavours to push her with his snout, and then tries to pull her by the tail and side-spine to the nest.'[3] The males are said to be polygamists;[4] they are extraordinarily bold and pugnacious, whilst 'the females are quite pacific.' Their battles are at times desperate; 'for these puny combatants fasten tight on each other for several seconds, tumbling over and over again, until their strength appears completely exhausted.' With the rough-tailed stickleback (G. trachurus) the males whilst fighting swim around and round each other, biting and endeavouring to pierce each other with their raised lateral spines.

The same writer adds,[5] 'the bite of these little furies is very severe. They also use their lateral spines with such fatal effect, that I have seen one during a battle absolutely rip his opponent quite open, so that he sank to the bottom and died.' When a fish is conquered, 'his gallant bearing forsakes him; his gay colours fade away; and he hides his disgrace among his peaceable companions, but is for some time the constant object of his conqueror's persecution'. (*Descent II*, pp. 331–2; n. 3 cites Warington (1852, 1855);

n. 4 Humphreys (1857); n. 5 Humphreys (1830), p. 331, with some details omitted).

Concerning brood care, CD adds: the males of certain fishes [. . .] take exclusive charge of the young. This is the case [. . .] with the sticklebacks (Gasterosteus), in which the males become brilliantly coloured during the spawning season. The male of the smooth-tailed stickleback (*G. leiurus*) performs the duties of a nurse with exemplary care and vigilance during a long time, and is continually employed in gently leading back the young to the nest, when they stray too far. He courageously drives away all enemies, including the females of his own species. It would indeed be no small relief to the male, if the female, after depositing her *eggs, were immediately devoured by some enemy, for he is forced incessantly to drive her from the nest.[37] (*Descent II*, p. 345; n. 37 refers to Warington (1855)).

Another striking instance out of many is afforded by the male stickleback (*Gasterosteus leiurus*), which is described by Mr Warington,[25] as being then '*beautiful beyond description.*' The back and eyes of the female are simply brown, and the belly white. The eyes of the male, on the other hand, are 'of the most splendid green, having a metallic lustre like the green feathers of some humming-birds. The throat and belly are of a bright crimson, the back of an ashy-green, and the whole fish appears as though it were somewhat translucent and glowed with an internal incandescence.' After the breeding-season these colours all change, the throat and belly become of a paler red, the back more green, and the glowing tints subside. (*Descent II*, pp. 340–1; n. 25 cites Warington (1852)).

Stromateus maculatus *See* **Butterfishes**.

Submergence The phenomenon wherein marine organisms of high *latitude regions tend to occur in deeper areas as latitude decreases, with the deepest records occurring within the intertropical belt. This may lead to disjunct distributions in some groups (see Ekman 1967).

CD alludes to submergence as follows: Turning now to marine *productions, we hear from Sir J. Richardson,[1] that Arctic forms of fishes disappear in the seas of Japan & of northern China, are replaced by other assemblages in the warmer *latitudes & reappear on the coast of Tasmania, southern New Zealand & the antarctic islands.

He further states that the southern cod-fish are 'much like those of the north, & *Notacanthus & *Macrourus, two very remarkable Greenland genera, which inhabit deep water, have recently been discovered on the coasts of New Zealand & S. Australia'. (*Big Species Book* p. 555; n. 1 cites Richardson (1846), p. 189, 191; *see also* **Madeira**).

In a reprise, we also have, from *Origin* VI (p. 338): Sir J. Richardson, also, speaks of the reappearance on the shores of New Zealand, of northern forms of fish.

Subspecies According to the *Code, "a taxonomically and geographically distinct subgroup within one *species; a somewhat distinct morphological and reproductively isolated subgroup of a species".

Another definition is given by Mayr and Ashlock (1991): "A subspecies is an aggregate of phenotypically similar populations of a species inhabiting a geographic subdivision of the range of that species and differing taxonomically from other populations of that species." Thus, a subspecies is not an evolutionary unit, and is different from what is implied in the *Code*, which only uses 'subspecies' for anything that has a trinomen. (*See also* **Speciation**.)

Sunfishes Members of the Family Centrarchidae, *endemic to North America, but now widely distributed throughout the world, owing to introduction of species such as Largemouth bass *Micropterus salmoides* (Lacepède, 1802) and Smallmouth bass *Micropterus dolomieui* Lacepède, 1802. (*See* FishBase for updated accounts of sunfish introductions to various countries).

CD never encountered sunfishes, as these were first imported into the British Isles in 1879 (Lever 1996). Sunfishes are mentioned in CD's writings only in a quote taken from a letter by *Agassiz (July 22, 1868; *Calendar* no. 6286) who, then based in the USA, compares "our Pomotis" to members of the Family *Cichlidae.

We also have in Notebook M, p. 639, p. 268 [of Macculloch 1837, vol. I] <<grinding>> in <stomach of> sun-fish, in mouth of swine & in stomach of lobster.

Here we must disagree: fishes in general, and sunfishes in particular do not grind food items in their *stomach*, but rather in the back of their throat, using pharyngeal teeth whose shape and size depend, at least in part, on the food type to which the juveniles are exposed. Thus, for example, Wootton (1991, p. 39, based on Keast 1978) reports that, in Lake Opinicon (Ontario, Canada), the Pumpkinseed sunfish *Lepomis gibbosus* (Linnaeus, 1758) feeds on molluscs and isopods, and thus has "stout, flattened pharyngeal teeth [which] act as a grinding mechanism."

Surgeonfishes Members of the Family Acanthuridae, whose common name derives from a sharp spine on their *caudal peduncle. The family is represented here by two species, the Convict surgeonfish and the Orangespot surgeonfish.

The Convict surgeonfish *Acanthurus triostegus* (Linnaeus, 1758), originally assigned to 'Chaetodon' and named for its bars, has a very wide distribution (which real convicts are not supposed to have), ranging from South Africa in the east to the Pacific coast of Colombia in the west, and from Southern Japan in the north to the Great Barrier Reef, Australia, in the south. Convict surgeonfish are dubious characters: themselves not territorial, they form "roving gangs [..] whose goal is to overcome the defences of [..] territory holders" (Barlow 2000, p. 96).

Noting that "the reader might wonder how such a wide-ranging species as *A. triostegus* failed to reach the Hawaiian Islands" Randall (1998) pointed out that the '*Teuthis sandvicensis*' of previous authors should be seen as a *subspecies of the Convict surgeonfish, *A. triostegus sandvicensis*, previously described as a *variety (Streets 1877).

A specimen of Orangespot surgeonfish was collected by CD in early April 1836 from a coral reef of the *Cocos Islands. This was listed, but not described in *Fish* (p. 75), as *Jenyns considered the species sufficiently well documented by previous authors, notably in the *Histoire Naturelle des Poissons*, his key reference. The Orangespot surgeonfish *Acanthurus olivaceus* Bloch and Schneider, 1801, on the other hand, is described (as *A. humeralis*), based on a specimen splendid verditer blue and green sampled by CD in *Tahiti (*Fish in Spirits*, no. 1317; *Fish*, pp. 75–7).

Randall *et al.* (2001) point out that such vivid and distinct coloration in surgeonfishes is one reason why it is relatively straightforward to identify *hybrids among them.

Survival of the fittest While concerned with the effects of various adaptations on the survival of various organisms, CD did not coin the phrase 'survival of the fittest,' which originated with Herbert *Spencer (Beddall 1988b), the first *social Darwinist.

Moreover, contrary to Spencer, CD used 'survival of the fittest' as a strict synonym of *natural selection. Could we please not use this empty phrase when dealing with CD?

Swimbladder An organ – also called 'air bladder' – whose origin CD had problems explaining. Indeed, this is where he committed his biggest blunder concerning fishes.

First, the stage is set: It may be objected such perfect organs as eye and ear, could never be formed, in latter less difficulty as gradations more perfect; at first appears monstrous and to <the> end appears difficulty. But think of gradation, even now manifest [. . .]. [S]wimming bladder by gradation of structure is admitted to belong to the ear system. (*Foundations*, pp. 15–16).

Also: it should be here borne in mind, that a part having originally a wholly different function, may on the theory of gradual selection be slowly worked into quite another use; the gradations of forms, from which naturalists believe in the hypothetical metamorphosis of part of the ear into the swimming bladder of fishes, and in insects of legs into jaws, shows the manner in which this is possible. (*Foundations*, p. 129).

Then comes the blunder: [t]he proper function of the swim-bladder in fish is explained by its name, but in some fish it becomes divided by vascular partitions & has an air passage or ductus pneumaticus into the oesophagus, & certainly aids respiration but these fish have, also, *branchiae. There can be no doubt[5] that the lungs of the higher vertebrata are *homologous or "ideally similar" with the swim-bladder of fish; & according to our theory the progenitor of all the vertebrate animals having lungs, had a swim-bladder, <& that the *transition was effected by> the swim bladder having been perfected for respiration through natural selection; – the ductus pneumaticus having become the wind-pipe or trachea, – whilst the branchiae have been atrophied. As the Branchiae became useless for respiration, they might have been slowly converted for some other purpose. (*Big Species Book*, p. 356; n. 5 reads: See the most interesting account of the use & homologies of the swim-bladder in Prof. *Owen's Hunterian Lectures on Fish, p. 278–28; Owen 1846).

The blunder – which CD restated all the way to the sixth edition of *Origin* – is, of course, that in fishes, as pointed out by Liem (1988), it is the swimbladder that evolved from lungs, not lungs from the swimbladder.

But this persistence can be excused: swimbladders as ancestors of lungs was just too good a story. As CD emphasized If a few fish were extinct, who on earth would have ventured even to conjuncture that lungs had originated in a swim-bladder? (cited in Browne 2002, p. 330, with reference to Darwin and Seward 1903; "3: 135"). Moreover, many trustworthy biologists believed this at the time. Even *Haeckel (*Calendar*, no. 6040), as late as 1868, confirmed the erroneous sequence to CD.

Syngnathus spp. *See* **Pipefishes.**

Synonym According to the *Code, an invalid scientific name of an organism proposed later than the accepted *name. Many of the binomens used by CD were synonyms of the presently accepted names. Taxonomists (e.g. Eschmeyer 1990, 1998) distinguish different types of synonym, but this was not done here; interested readers will find this information in *FishBase.

Tadpoles The larvae of *amphibians. (*See also* **Seasquirts**.)

Tahiti Port of call of the *Beagle* from November 15 to 26, 1835. This stopover allowed CD a trip inland, on November 18, that he fondly remembered (*see* **Food-fish; Otter**).

CD sampled in Tahiti a number of marine fishes, which Jenyns described as *Dules leuciscus* (*Fish*, p. 17); *Upeneus trifasciatus* (p. 26); *Caranx torvus* (p. 69); *Acanthurus humeralis* (p. 76); *Scarus globiceps* (p. 106); *Scarus lepidus* (p. 108); *Muraena* sp. (p. 146); *Syngnathus conspicillatus* (p. 147); *Balistes aculeatus* (pp. 155–6); and *Ostracion punctatus* (p. 158).

However, when later reading Lesson (1826), CD did not mention any of these, nor his visit to *Mauritius, although he noted that many fish of Taiti found at Isle de France (*Notebook C*, p. 243; *see also* CD's annotation to Schlegel 1843).

Taphonomy The discipline whose adepts study the manner in which animals and plants are fossilized and become potentially available to the *fossil record. Eldrege (1995, pp. 95–6) regards CD as the founding father, while the paleontologist Anna Behrensmeyer is celebrated as the mother of taphonomy (Schiebinger 1999, p. 47, citing an interview published in *Science*, March 12, 1992, in an amazing case of asynchronous parenthood).

Parts of a letter by CD are reproduced here which indicate the kind of questions a taphonomist would ask, concerning skeletons being found somewhere:

1. Were any bones of any of the skeletons found in proper relative <p>osition, as if animal had been washed in whole.-
2. Are bones rounded or broken.
3. Are there any stones, angular or rounded, of limestone or of foreign rock in mud. – Are there horizontal line of deposition in mud.
4. Distance of Caldy Is^d from main & depth of channel.-
5. *Particularly describe whether beaks of birds were embedded actually with Elephant bones, or chiefly in upper part of fissure.*
6. Describe same for fishes bones. (*Correspondence* to Gilbert Nicholas Smith, Aug. 15, 1840).

Taxonomy The study of organisms with regard to their kinds, nature, diversity and evolutionary relationships. A part of systematics, taxonomy is the science of *classification of animals, plants and other living organisms.

CD's contributions to taxonomy in the narrow sense were his work on *barnacles and his participation in the committee set up by Strickland to codify the rules used by taxonomists (Strickland *et al.* 1843), which eventually led to the *International *Code of Zoological Nomenclature* (ICZN 1999).

CD's contribution to taxonomy in the wider sense is, of course, that he provided the evolutionary framework which allows for making sense of the taxonomic features of organisms.

Teleosts/Teleostean fishes Fishes of the kind familiar to us in the present day, having the skeleton usually completely ossified and the scales horny (Glossary of *Origin VI*, p. 440; *see also* **Bony fishes**).

Tench Common name of *Tinca tinca* Linnaeus, 1758, of the Family *Cyprinidae.

The reproductive habits of this species, once important in European aquaculture, are mentioned twice in CD's writings. Thus: I hear from Professor *Agassiz and Dr. *Günther, that the males of those fishes, which differ permanently in colour from the females, often become more brilliant during the breeding-season. This is likewise the case with a multitude of fishes, the sexes of which are identical in colour at all other seasons of the year. The tench, *roach, and *perch may be given as instances (*Descent II*, p. 340; note that 'golden' tench exist, which are kept as ornamental fish).

The other explicit mention of Tench by CD relates to their *sex ratio, and may be found under that entry (*See also* **Experiments (IV)**, and **Fossils**).

We may note, as well, that tench have been found to devote up to 40% of their (low) routine metabolism to breathing itself (Schumann and Piiper 1966), thereby providing strong support for the notion that *oxygen supply to fish, being limited by gill area, generally limits their *growth and the size they ultimately reach.

Territoriality A form of behaviour, frequent among fishes, wherein an area or 'territory' is defended, mainly against conspecifics, and often in the context of *sexual selection.

Examples of territorial behaviours in fishes discussed in this volume are provided by the *blennies, the *damselfishes and the *mudskippers (see also Gibson 1993; Zeller 1998).

Tetragonopterus **spp.** *See* **Characins**.

Tetraodon (Tetrodon) **spp.** *See* **Puffers**.

Thornfishes Members of the Family Bovichthyidae, a group of eleven species of small fish with exclusively marine habitats in southern South America, southern Australia, and New Zealand, and both marine and freshwater habitats in southeastern Australia and Tasmania.

The thornfishes are represented here by two species, initially assigned to the genus *Aphritis*: *Pseudaphritis porosus* (Jenyns, 1842), and *P. undulatus* (Jenyns, 1842). Their description is based on specimens left among the mud banks at Port Desire, Central *Patagonia (*Fish in Spirits*, no. 776; Fig. 33), and from Lowe's Harbour, Chonos Archipelago (Chile), respectively (*Fish*, pp. 160–2; *Fish in Spirits*, no. 1138).

Threefold parallelism A concept formally proposed by *Agassiz (1857a), but implicit earlier (e.g. in Agassiz 1844), wherein he considered it proven "that the embryo of a fish during its development, the class of living fish in its numerous families, and the fish type in its planetary history, in every respect go through analogous phases, throughout which one can

Fig. 33. Thornfish *Pseudaphritis undulatus*, Family Bovichthyidae (*Aphritis undulatus* in *Fish*, Plate XXIX), based on a specimen from the Chonos Archipelago, Chile.

always trace the same creative idea" (Agassiz 1846, p. 37).

The concept means that *evolution had been staring Agassiz in the face all along. Need we add, as well, that threefold parallelism is but a form of the 'law' stating that '*ontogeny recapitulates phylogeny' (*see also* **Haeckel**), with *creationism thrown in?

Tierra del Fuego Large island at the tip of the South American mainland, uneasily shared between Chile and Argentina, and so named because of the many fires ('fuegos') lit by the now extinct local people, the Fuegians (Campbell 1997, p. 154).

In Tierra del Fuego, CD sampled fish specimens later described by Jenyns: *Mesites alpinus* (*Fish*, p. 121); *Aplochiton taeniatus* (p. 132); *Clupea fuegensis* (p. 133); *Conger punctus* (p. 143); and *Myxine australis* (p. 159).

Lloris and Rucabado (1991) reviewed the *ecology and ichthyology of the Beagle Channel, the narrow passage between Tierra del Fuego and the smaller Navarino Island, where CD apparently sampled all his Fuegian fishes.

Tigerperches Fishes of the Family T(h)eraponidae, occurring both in fresh and sea water, and called 'grunters' in Australia owing to their ability to generate *sounds from a *swimbladder connected to the skull by specialized muscles (Gomon *et al.* 1994, p. 500).

CD collected a specimen of *Pelates octolineatus* (Jenyns, 1840) with a *length of "unc. 9.

lin. 9" in King George's Sound, Australia. This became the *holotype of a new species, the Striped perch, described in *Fish* (p. 18) as *Helotes octolineatus*.

Toadfishes Members of the Family Batrachoididae, consisting of about 70 species of generally small, ugly-looking fish, occurring on sandy and muddy bottoms.

The toadfishes are represented here by *Porichthys porosissimus* (Cuvier, 1829), *"cast upon the beach"* in *Bahia Blanca, *Patagonia, described as *Batrachus porosissimus* in *Fish* (p. 99–100), and with colours Above purple – coppery; sides pearly; beneath yellowish with silver dots in regular figures; iris coppery; not uncommon (*Fish in Spirits*, no. 402).

Toads See **Amphibians**.

Torpedo A genus of *rays (Family Torpedinidae), whose members, e.g. the Common torpedo *Torpedo torpedo* (Linnaeus, 1758), are characterized by *electric organs, used to stun prey and deter would-be predators.

Trachipterus arcticus See **Deal fish**.

Transition When discussing *evolution, the problem is that if we look to an organ in a very isolated being, as the duck-like bill of the ornithorhynchus; or to an organ common to the greater part of a great class, as to the swimbladder in fishes, the web-secreting organs in Spiders & a thousand such cases, we are very seldom able to indicate intermediate states, & therefore are not able even to conjecture how such structures could have been produced through natural selection. (*Big Species Book*, p. 359).

However, [w]e have seen some cases, as that of the *eye, most difficult from its transcendant perfection; some from no transitional stages being known, and some from our not seeing as with Electric fishes, how any transition is possible; but I think facts enough have been given to show how extremely cautious we ought to be in ever admitting that a transition is not possible. (*Big Species Book*, p. 374. This transition is not only possible, but occurred numerous times in teleosts; see Alves-Gomes 2001).

Indeed, the argument that something "cannot be" because one doesn't know of it, or cannot conceive of it, has been used much too long by people (such as Behe 1996) who, strangely, believe that their ignorance is an argument in *their* favour.

Triassic A period of the Mesozoic era, lasting from 220 to 180 million years ago, right after the Permian, and followed by the *Jurassic (*see also* **Bony fishes**).

Triggerfishes Members of the Family Balistidae, consisting of highly derived fishes, represented here by two species.

One was the Queen triggerfish *Balistes vetula* Linnaeus, 1758, taken on March 21, 1832 about 65 miles from land, 14°20′S (*Fish*, p. 155; *Fish in Spirits*, nos. 147, 149), i.e. north of Bahia, Brazil.

The other triggerfish sampled by CD was *Balistes aculeatus*, now *Rhinecanthus aculeatus* (Linnaeus, 1758), from *Tahiti (*Fish in Spirits*, no. 1329; *Fish*, pp. 155–6).

***Trigla* spp.** See **Gurnards**.

Tripterygion capito See **Blennies**.

Trout Smaller salmonids of the genera *Salmo* and *Oncorhynchus*. When referring to 'trout', CD means the Brown trout *Salmo trutta* Linnaeus, 1758, whose various populations tend to differ in their morphology and other features, leading to local 'forms' (see Izaak *Walton's *Compleat Angler*), many with local common *names (Trewavas 1953).

The scientific names corresponding to these *varieties now tend to be viewed as *synonyms and this includes the two major forms of the British Isles, 'resident trout' (*S. trutta fario*) and 'sea trout' (*S. trutta trutta*). Trewavas (1953) suggested that even these two groups interbreed sufficiently for this distinction to be meaningless; other authors disagree. This debate is summarized in Elliott (1994, p. 9), who believes that there are "sympatric, reproductively isolated populations that qualify at least as *races or perhaps *subspecies".

CD's numerous jottings on Brown trout also deal with this issue: M^r. Bunbury says has heard the Trout from different lakes of N. Wales can be distinguished – & Jackson here (Capel-Curig) says that he certainly tell Trout from Ogwen, Capel-Curig & some other lakes, (different waters) He cannot, however, tell them from L. Groznerat, <<on road to Bethgellert>> wh flows by Tremadoc. but can tell them from lake S. of Moel Siabod. wh. flows into Conway by Bettws & there joins streams from Capel-Curig (*Notebook S*, p. 473; pers. comm. from C. J. F. Bunbury, whom CD met on June 18, 1842, during a 'geologising' trip to Capel Curig).

A double blind organoleptic test would perhaps have been required to verify the claims of Messrs. Bunbury and Jackson. Boasts such as theirs are easy to make, but hard to substantiate.

As for the colours of trout, CD added, after referring to prespawning colour changes in Salmon, that [a]n analogous and even greater change takes place with the *Salmo eriox* or bull trout (*Descent* II, p. 340; *S. eriox* is a junior synonym of *S. trutta*).

Trout-perch A member of the family Percopsidae, or trout-perches, freshwater fishes of North America (Alaska to Quebec), sporting both *cycloid and *ctenoid scales (Nelson 1994, p. 220). The family (though not the genus *Percopsis* itself) is well represented in fossils from the Eocene and Miocene (Müller 1985, p. 343).

Thus CD, upon reading p. 285 of Agassiz (1850), wrote: Excellent case of Percopsis of Chalk, which combined characters, which soon diverged, intermediate between *Ctenoids and *Cycloids; I wonder whether this agrees with Müllers classification, as seen in *Owen Lectures; if Fish properly classed whether so related to geologi. Formation (*Marginalia* 11–12; Müller 1844; Owen 1846).

Turbot Common name of *Scophthalmus maximus* (Linnaeus, 1758), a *flatfish of the Family Scophthalmidae, closely related to the Pleuronectidae.

CD mentions this fish only once: But it cannot be supposed that the peculiar speckled appearance of the upper side of the sole, so like the sandy bed of the sea [. . .] or the presence of bony tubercles on the upper side of the turbot, are due to the action of the light. (*Origin* VI, p. 188).

The name 'turbot' is also used, in Canada, for the *Greenland halibut, a species which CD used to illustrate evolutionary trends in flatfishes, and a cache of which Canada used, in March 1995, as *post hoc* justification for its martial stand against a Spanish fishing vessel.

Type locality According to the *Code*, "the geographical place of capture or *collection of the name-bearing type of a nominal *species or *subspecies. If the name-bearing type was captured or collected after being transported by boat, vehicle, aircraft, or other human or mechanical means, the type locality is the place from which it, or its wild progenitor, began its unnatural journey".

Type species According to the *Code*, "the species of a genus with which the generic name is permanently associated; the description of a genus is based primarily on its type species, being modified and expanded by the features of other included species". An example is *Salmo salar* Linnaeus, 1758, the type species of the genus *Salmo*.

Umbrina spp. *See* **Croakers**.
Upeneus spp. *See* **Goatfishes**.

Valparaiso: City in Chile where, in April 1834, CD sampled a number of specimens later described in Jenyns' *Fish*: **Pinguipes chilensis* (p. 22); **Sebastes oculata* (p. 37); **Latilus jugularis* (p. 51); **Heliases crusma* (p. 54); **Atherina microlepidota* (p. 78); **Hippoglossus kingii* (p. 130); and **Syngnathus acicularis* (p. 147).

Variation(s) Inherited difference(s) in features of living organisms. CD's theory of *natural selection requires random variations to occur in the anatomy, behaviour or performance of animals and plants, and his writings document many of these, or point to literature that does. (*See also* **Variety**.)

Variations Short title of The Variations of Animals and Plants under Domestication (Darwin 1875), whose two volumes contributed to many entries in the present book, notably on *goldfish.

Variety(-ies) Refers to different forms of the same species. Thus, CD noted that: Fish of the same species are well known to present distinguishable differences in different lakes: Sir H. Davy² states that red-fleshed dark-banded *trout were taken from one Scotch lake & put into another, where the trout were white-fleshed; the young here produced had their flesh less red, & in 20 years the variety was lost. Laying on one side the probability of crosses having taken place, we see here that the red flesh was in some degree inherited; & some would assert that that if the red trout in their own lake had transmitted their character for some additional hundred-thousand generations, the character would have kept truer.

From these & similar considerations I have thought it advisable to use only the term "variety," & where it is known or almost known to be strictly inherited "race:" and I use the term variety loosely, simply in accordance with common acceptation, as I do the term *species. <for the same reason in both cases> If the distinction could be drawn between hereditary & temporary *variation in a state of nature it would be of great importance for our object; for variations in a state of nature which are not inherited are of little signification, & deserve notice, (perhaps) only as showing the possibility of change in structure. (*Big Species Book*, p. 100; n. 2 refers to Davy (1828)).

We have seen that in the best known countries there is much uncertainty in deciding what to call species & what varieties [. . .] look at the King of beasts, as popularly called, whether or not the Maneless lion of Persia is a distinct species [. . .]. So with Fish, it is certain that the *salmon of many different rivers can be distinguished by fishermen; & the *Herring which has been so closely studied, is found to present a vast range of variation.[6] (*Big Species Book*, pp. 116–17; n. 6 reads: Wilson's voyage round Scotland [1842,] vol 2, p. 206. The *Herring fishery was one of the points especially attended to in this voyage).

Velvet belly A small shark of the Northeast Atlantic, reaching up to 60 cm, originally described by *Linnaeus as *Squalus spinax*, and now assigned to the genus *Etmopterus*.

CD noted that it serves as host to *barnacles, as does the *Basking shark.

Vertebrate origins A problem CD felt needed to be looked at seriously, and hence his dim view of Macdonald (1839), who linked fish to insects, and his surprise that any one may believe in such rigmarole about analogies and numbers (*Notebook D*, p. 354; the numbers refer to the now rejected 'Quinarian' and 'Quartenary' systems of classification, wherein the taxa of given levels were presumed to contain a fixed number of taxa of the next lower level).

Indeed, CD, who had earlier surmised that we have not the slightest right to say there never was common progenitor to Mammalia & fish, when there now exist such strange form as ornithorhyncus (*Notebook B*, p. 194), later felt he had solved the problem of vertebrate origins. Thus his comment: [i]t amused me to see Sir R. Murchison quoted as judge of affinities of animals; it gave me a 'cold shudder' to hear of anyone speculating about a true Crustacean giving birth to a true Fish! (*Correspondence* to A. Gray, Aug. 11, 1860; Parson (1860) reports that Roderick Impey Murchison, a geologist, had proposed the Devonian ur-fishes *Cephalaspis* and *Pterichthys* as intermediate between ancient crustaceans and modern fishes; also note that *Pterichthys* is not a genus of "fossil flying fishes", as stated on p. 318, n. 12 of Burkhardt *et al.* (1993)).

Consequently, CD was confident his colleagues would see the light: I have had letter from Carpenter this morning [. . .]. He is convert, but does not quite go as far as I – but quite far enough; for he admits that all Birds from one progenitor; & probably all fishes & reptiles from another parent. But the last mouthful chokes him – he can hardly admit all Vertebrates from one parent – He will surely come to this from *Homology & *Embryology. (*Correspondence* to C. Lyell, Dec. 3, 1859).

And indeed, CD's views of vertebrate origins, as articulated, for example, in *Descent* (I, pp. 164–5) largely anticipate present concepts (Erwin *et al.* 1997): The most ancient progenitors in the kingdom of the *Vertebrata, at which we are able to obtain an obscure glance, apparently consisted of a group of marine animals,[32] resembling the larvae of existing *Ascidians. These animals probably gave rise to a group of *fishes, as lowly organized as the *lancelet; and from these the *Ganoids, and other fishes like the *Lepidosiren, must have been developed. From such fish a very small advance would carry us on to the *Amphibians. We have seen that *birds and reptiles were once intimately connected together; and the Monotremata now connect *mammals with reptiles in a slight degree. But no one can at present say by what line of descent the three higher and related classes, namely, mammals, birds, and reptiles, were derived from the two lower vertebrate classes, namely, amphibians and fishes.

Note 32 starts as follows The inhabitants of the sea-shore must be greatly affected by the tides; animals living either about the *mean* high-water mark, or about the *mean* low water mark, pass through a complete cycle of tidal changes in a fortnight. Consequently, their food supply will undergo marked changes week by week. The vital functions of such animals, living under these conditions for many generations, can hardly fail to run their course in regular weekly periods. Now it is a mysterious fact that in the higher and now terrestrial Vertebrata, as well as in other classes, many normal and *abnormal processes have one or more weeks as their periods; this would be rendered intelligible if the Vertebrata are descended from an animal allied to the existing tidal Ascidians.

Actually, although it is correct that tidal (i.e. lunar) rhythmicity occurs throughout the vertebrate class, from the lunar spawning periodicity of fish (Johannes 1981, Chapter 3) to menstruation in *Homo sapiens*, this doesn't by itself prove we descend from tidal Ascidians: this rhythmicity could be the result of the same external forcing that affects palolo worms (Johannes 1981).

On the other hand, it is largely true that: [o]ur ancestor was an animal which breathed water, had a *swim-bladder, a great swimming tail, an imperfect skull & undoubtedly was an *hermaphrodite! Here is a pleasant genealogy of mankind! (*Correspondence* to Lyell, Jan. 10, 1860. O.K., but we had no swim-bladders).

Wallace, Alfred English naturalist (1823–1913). His key paper (Wallace 1858) was seen by many earlier commentators as having made him co-discoverer of *natural selection, on equal terms with CD, while a few authors even suggested *plagiarism by CD.

Detailed analysis of the sequence of events that led to the formal presentation, on July 1, 1858, of Wallace's paper, and of a summary of CD's current work (Darwin 1858) refutes these claims (Beddall 1988a), as do CD's *Notebooks and *Foundations. Indeed, the best refutation of these claims is provided by Wallace himself, who used the term "Darwinism" as title for his own opus on the topic (Wallace 1889; see also Wallace 1905), and who in 1890 dedicated to CD his famous book on *The Malay Archipelago*.

On the other hand, Camerini (1994) has shown that CD, in his correspondence with Wallace, pointed out to him the role of water depth in separating the fauna and flora of the 'Malay Archipelago' (present day Malaysia, Philippines and Western Indonesia) from those of Southwestern Oceania (Eastern Indonesia, Papua New Guinea and Australia). CD may thus have helped him define what later became known as 'Wallace's line'. Quite a reversal!

Walton, Izaak English author (1593–1683), known for his *Compleat Angler* (1653), a now classic guide on why and how one should become a 'Brother of the Angle', an expression picked up by CD (see **Angling**).

The *Compleat Angler* presents, in the form of charmingly unrealistic exchanges among an all-knowing '*Piscatorius*' (the '*compleat*', or perfect, angler), a naïve '*Venator*' (hunter), and a haughty '*Auceps*' (falconer), various aspects of the biology of freshwater fish common in England (*Salmon, *Trout, *Minnow, *Pike, etc.), as well as folklore and cooking recipes about these same *species. (See **Darwin's Bass** for another, contemporary mix of angling lore and reflections on the human condition).

The *Compleat Angler* also repeats the old myth that the rotten planks of ships turn into *barnacles. Wisely, CD does not cite this in his own work on barnacles.

Water-beetles Members of the Coleoptera, the most speciose order of insects, and of the Family Dytiscidae.

CD, in various writings refers to both *Dytiscus*, the type genus, and to *Colymbetes*, both of which prey on the eggs of fish. Thus: In those animals which produce an astonishing number of *eggs, the destruction probably chiefly falls on the eggs, as is known to be the case with Fish, from other fish, water-beetles &c. But when the old can protect their young few are generally produced . . . (*Big Species Book*, p. 186; see also **Morris, Margaretta Hare; Showers of fish**).

Waterhouse, Benjamin Hawkins See **Hawkins**.

Waterhouse, George English naturalist (1810–88). Author of *Mammalia* (Waterhouse 1839), one of the volumes of *Zoology, and, along with CD and others, a member of the Strickland Commission (see **Taxonomy**).

This Waterhouse, who also published papers on CD's insect *collection (Porter 1983), is mentioned here explicitly to prevent him from being mistaken for B. Waterhouse *Hawkins, who drew many of CD's fishes.

Weber, Ernst Heinrich German anatomist and physiologist (1795–1878). CD mentions him once, but without reference: I may just allude to Webers curious discovery of the* swim-bladder in certain fish being brought in connection with the organ of hearing by a chain of little bones & cavities, & so aiding this function. Indeed, in some fish (*Owen, Hunterian Lectures. Fish p. 210) as the *Cobitis barbatula, the swim-bladder apparently subserves no other function. (*Big Species Book*, p. 356, n. 4; Owen 1846).

The missing reference is Weber (1820), and the 'Weberian ossicles' – so the *eponym – indeed make up a chain of [four] little bones, vibrating within gas-filled diverticula and linking the *swim-bladder to the *brain case. In fishes sporting them (e.g. in the *carps and *catfishes), the Weberian ossicles have been unequivocally demonstrated to function as an organ of hearing (Grassé 1958), similar to the middle ear of humans.

Wels catfish One of the common names of *Silurus glanis* (Linnaeus, 1758), Family Siluridae. A predominantly Eastern European freshwater catfish, originally not occurring in the British Isles (Jenyns 1835, p. 421), but introduced in 1864 (Welcomme 1988; see also *FishBase), and whose common name stems from German ('Wels'). The most important feature of Wels is the enormous size it can reach, i.e. "longer than a large man" (Hartmann 1827, p. 84), and up to 330 kg (Nelson 1994, p. 157).

CD asked Samuel Pickworth Woodward, in a letter dated June 5, 1861, for information on this fish: I want to beg a favour, if Dr *Günther is in London, to get him to give me a note, or reference to any paper, on the Silurus escaping from the Danube. How was it ascertained?

Woodward complied, and Günther, who had thoroughly studied the ichthyofauna of Southern Germany (Günther 1853, 1855) replied to Woodward: "It appears that the 'Wels' has been accidentally transferred by inundations into the Lake of Constance from the minor lakes north of it. This, certainly, happened more than once, but after some time the individuals which had not been introduced in sufficient number, died out. Thus you might find, that already old Gesner mentions the Wels as a fish of the Lake of Constance" (*Correspondence*, June 14, 1861; Gesner 1558).

Günther's letter contains further details on the introduction of this nest-building fish into Lake Constance, and cites Hartmann (1827) and Rapp (1854). The letter reached CD, who annotated it with Dr Günther On Fish repeatedly introduced by inundation. I believe a Silurus. However, CD does not appear to have followed up on this.

Whales A group of *marine mammals, also known as cetaceans, including the large baleen whales and the toothed whales, the smallest of which are also known as dolphins (Jefferson *et al.* 1993). Their relatively late arrival on the evolutionary scene has made it possible for them to muscle their way onto the top of most marine *food webs (*see also* **Complexity**; and Pauly *et al.* 1998).

Understanding the evolution of whales requires understanding their true affinities, which perhaps explains one of CD's more opaque jotting: Would not relationship express, a real affinity & affinity – whales & fishes (*Notebook B*, p. 211).

In any case, the evolution of whales from ancestral carnivorous ungulates related to the ancestors of the two extant species of *Hippopotamus*, is now well documented (Zimmer 1998, p. 212), with previously large gaps in the *fossil record now replaced by a series of superb specimens. Indeed, the scenario which CD proposed in the first edition of *Origin* for the evolution of whales, but deleted from subsequent editions because of (unwarranted) criticism, has been largely vindicated (*see* **Jaguar**).

Words CD wrote a lot, another setback for those who believe that great scientists limit themselves to a few major contributions while minor characters attempt to offset absent quality by sheer quantity (Pauly 1986).

I present below an attempt to quantify CD's total word output, by category, well aware that some words (e.g. in the first edition of *Origin*) may count far more than some of his others (e.g. in the letter documenting his payment for the purchase of *cod liver oil (*Correspondence*, Jan. 26, 1862)).

Item(s)	Words
The Works of Charles Darwin. Vol. 1–29[a]	3 100 000
The Correspondence of Charles Darwin. Vol. 1–12[b]	2 000 000
The Collected Papers of Charles Darwin. Vol. 1–2[c]	260 000
Charles Darwin's *Notebooks, 1836–1844[d]	153 000
Marginalia to CD's reprint collection[e]	250 000
Marginalia to CD's books[f]	84 000
Miscellaneous other writings[g]	375 000
Total	**6 222 000**

[a] Approximated by counting the number of words in Barrett and Freeman (1989) for at least five randomly selected lines in each volume, and raising by the number of lines (40) and the number of 'full page equivalents' (i.e. excluding front and back matters provided by editors, figures, etc.).

[b] As in (a), excluding letters *to* CD, and using the *Calendar* of Burkhardt *et al.* (1985a) to account for letters still unpublished.

[c] As in (a), based on Barrett (1977).

[d] As in (a), based on Barrett *et al.* (1987).

[e] As given in Vorzimmer (1963).

[f] As in (a), based on Di Gregorio and Gill (1990).

[g] *Big Species Book*, *Coral Islands* and *Zoology Notes*.

The present volume contains *quotes by CD which sum up to about 45 000 words, i.e. about 0.7% of the above total. This may provide an idea of the degree to which CD cared about fishes.

Works of Charles Darwin A complete edition of CD's books, comprising 29 volumes, edited by Barrett and Freeman (1989), and used extensively in the preparation of *Darwin's Fishes*.

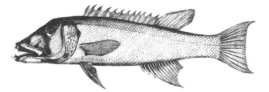

Fig. 34. The eponymous wrasse *Pimelometopon darwini*, Family Labridae (*Cossyphus darwini* in *Fish*, Plate XX), from Chatham Island, Galápagos.

Worms Short title of the last book published by CD, The Formation of Vegetable Mould through the Action of Worms with Observations on their Habits (Darwin 1881).

Worms provided material for only two entries in this book, one dealing with Mr *Fish, the other being the last appearance, in CD's work, of Professor Möbius' stupid *pike.

Given his admiration for worms, the book dedicated to their work would have been an excellent place for CD to mention, as he did in his *Autobiography* (p. 27), that while *angling, he came to hate spitting live worms on hooks. He missed here a great opportunity to present himself as the sensitive male he was.

Wrasses Members of the widespread Family Labridae, whose individuals, in most species, change colour as they grow, and also turn from females to males, thus inducing large deviations from the 'normal' *sex ratio of 1 : 1 (*see also* **Hermaphroditism**).

Only two species of wrasse are mentioned in *Fish*. One is the Pacific red sheephead *Pimelometopon darwini* (Jenyns, 1842), described as *Cossyphus darwini* by Jenyns (*Fish*, pp. 100–2; *Fish in Spirits*, no. 1269; Fig. 34). This was listed by Freeman (1978) as the sole Darwin *eponym among fishes, though at least six existed at the time Freeman assembled his list – this perhaps adding an argument for the creation of global, easy-to-access databases such as *FishBase.

The *holotype of *S. darwini*, sampled by CD in Chatham (= San Christobal) Island,

*Galápagos, is a dry skin, now kept at the Natural History Museum – formerly *British Museum (Natural History) – under catalogue number '1918, 1.31.11' (a superfluous detail?). *S. darwini* occurs in Ecuador, *Peru and Northern Chile (Merlen 1988; De Buen 1966). Grove and Lavenberg (1997, pp. 495–6) mention that the adults reach up to 67 cm, live in deeper waters, and feed on benthic invertebrates, notably gastropods.

The other wrasse in *Fish* (pp. 102–4) is *Cheilio ramosus*, and its description is based on a specimen said to come from the Japan Sea, and given to CD in *Valparaiso, Chile, by the surgeon of a whaling ship (*Fish in Spirits*, no. 1002). For all their trouble, the species in question, described as new by Jenyns, turned out to be synonymous with *Cheilio inermis* (Forsskål, 1775).

CD commented on the behaviour of European wrasses: Both sexes of the Labrus mixtus, although very different in colour, are beautiful; the male being orange with bright blue stripes, and the female bright red with some black spots on the back. (*Descent II*, p. 337). Also, [w]e have [. . .] to enquire whether [. . .] the bright colours of certain male fishes having been acquired through *Sexual selection can, through the law of the equal transmission of characters to both sexes, be extended to those groups in which the males and females are brilliant in the same, or nearly the same degree and manner. In such a genus as Labrus, which includes some of the most splendid fishes in the world – for instance, the Peacock Labrus (*L. pavo*), described,[28] with pardonable exaggeration, as formed of polished scales of gold, encrusting lapis-lazuli, rubies, sapphires, emeralds, and amethysts – we may, with much probability, accept this belief; for we have seen that the sexes in at least one *species of the genus differ greatly in colour. (*Descent II*, p. 342; n. 28 cites Bory de Saint-Vincent (1822–31)).

Certain fishes [. . .] make nests [. . .]. Both sexes of the bright coloured *Crenilabrus massa* and *melops* work together in building their nests with sea-weed, shells, etc.[35] (*Descent II*, pp. 344–5; n. 35 reads: According to the observations of M. Gerbe; see Günther's *Record of Zoolog. Literature*, 1865, p. 194; based on Gerbe 1864). Mr W. S. Kent says that the male of the *Labrus mixtus*, which, as we have seen, differs in colour from the female, makes 'a deep hollow in the sand of the tank, and then endeavours in the most persuasive manner to induce a female of the same species to share it with him, swimming backwards and forwards between her and the completed nest, and plainly exhibiting the greatest anxiety for her to follow'. (*Descent II*, p. 341; the 'quote' is assembled from bits and pieces in Saville-Kent 1873a; the Cuckoo wrasse *L. mixtus* Linnaeus 1758 has kept its name, but the Ornate wrasse *L. pavo* is now *Thalassoma pavo* (Linnaeus 1758), the Grey wrasse *Crenilabrus massa* is now *Symphodus cinereus* (Bonnaterre 1788), and the Corkwing wrasse *C. melops* is now *S. melops* (Linnaeus, 1758).

One cannot but wonder whether the reason why these male wrasses work together so nicely with the females on their domestic arrangements is because, as protogynous hermaphrodites, they can draw on insights from their previous life as females . . .

Xiphophorus helleri *See* **Livebearers.**

Yarrell, William: English zoologist (1784–1856), author of highly successful works on British birds and fishes.

CD's many citations of his ichthyological work (Yarrell 1836, 1839), partly listed in Tort (1996, pp. 4715–16), are all included in this volume.

Zoarces viviparus *See* **Eelpouts.**

Zoobenthos The community of marine animals living just above, on, or within the sea bottom. The invertebrate zoobenthos is comprised of numerous groups which CD mentions in passing, and consists of crabs and other motile groups, as well as sessile groups, e.g. sea anemones (*Zoophites) and bryozoans (Flustra).

Numerous fish species collected by CD may be assigned to the zoobenthos, e.g. the *blennies, *gobies, and *pipefishes (*see* **Dredging; Möbius**).

Zoology Short title of the set of four volumes documenting the bulk of the animals collected by CD during the voyage of the *Beagle (1832–6).

The British government funded the publication of descriptions of this rich material, and hence the quote: Pray tell Leonard, that my government work is going on smoothly & I hope will be prosperous. – He will see in the prospectus his name attached to the fish. I set my shoulder to the work with a good heart (*Correspondence* to J. S. Henslow, Nov. 4, 1837).

The books containing these descriptions consist of *Mammalia* (Waterhouse, 1839); *Fossil Mammalia* (Owen 1840); *Birds (Gould et al. 1841), *Fish (Jenyns 1842), and *Reptiles (Bell 1843), all edited and including notes or sections by CD.

However, the invertebrates collected by CD were not documented, except in scattered publications (Porter 1985; *see also* Collections).

Zoology Notes Short title of *Charles Darwin's Zoology Notes & Specimen Lists from H.M.S. Beagle*, edited by Richard Darwin Keynes (2000), based on four bound volumes kept in the Darwin Archive at Cambridge University (DAR 30.1, 30.2, 31.1, and 31.2) and reproducing, in chronological order, CD's field observations, the results of dissections, and his zoological catalogues. Parts of this material have been used in a number of CD publications, and by the authors of *Zoology* (Burkhardt *et al.* 1985b, p. 546). Thus, CD's notes on fishes were extensively used by *Jenyns when he wrote *Fish, though only after they had been extracted by Syms *Covington to yield "*Fish in Spirits of Wine*."

The *Zoology Notes*, whose longer sections largely overlap with text in CD's other contributions (notably the *Diary and *Journal) were used here to complement entries based on these contributions, and to verify the original transcription of "*Fish in Spirits of Wine*" presented in Appendix I.

Zoophite CD's *spelling for 'Zoophyte,' i.e. an aquatic invertebrate whose outward appearance resembles that of a plant. Examples are the sea anemones, often the home of anemonefishes, a group of *damselfishes.

Zooplankton Aquatic animals too small and/or too weak to swim against water currents, and hence drifting with water masses; the animal component of plankton, composed mainly of small herbivorous crustaceans (copepods and other *Entomostraca, *Pteropoda, etc.) but also including comb-jellies such as *Beroe.

CD lacked a concept for *plankton, and was unaware of the existence and ecological role of *phytoplankton as the basis of most marine *food webs, and of herbivorous zooplankton as primary consumers. The *ad hoc* hypothesis he proposed to compensate for this was that zooplankton photosynthesizes, i.e. that it perhaps possesses the power of decomposing carbonic acid gas, like the members of the vegetable kingdom. (*Journal*, December 6, 1833; *see also* **Food webs**).

Zooplankton 1 : CD 0.

ZZZ The letters used, according to my copy of the *Random House Dictionary*, to represent the sound of a person snoring.

CD probably snored, like a *croaker, i.e. '*roncador*' (snorer in Spanish; *see also* **Sleepers**). However, 'ZZZ' was inserted here only to congratulate readers who should have reached this far, and to thank them for their patience.

Appendix I Fish in Spirits of Wine

JACQUELINE McGLADE
European Environment Agency, Copenhagen

The document below is a transcription of the part devoted to fishes of the *Zoological Notes* compiled by Charles Darwin (CD) during the voyage of the *Beagle*. This was originally copied by Syms *Covington in 1838 or thereabouts. CD contributed corrections and additional text to the manuscript of 20 pages, kept in the Darwin Archive at the Library of Cambridge University as DAR 29.1.

I have in the main followed for this transcription the practice established by the editors of CD's *Correspondence (see also **Conventions used in the text**). Thus, this transcription was as literal as possible, though in a few cases, Jenyns' *Fish* (which often cites 'Fish in Spirits of Wine' verbatim) and Keynes' (2000) *Zoology Notes*, were used to select from a range of options. Also, I have attempted to reproduce the *form* of the original document: as already noted by Porter (1985), the right side of *Fish in Spirits of Wine* reproduces CD's field notes, while the left side mainly contains the species names resulting from Jenyns' identifications (including preliminary identifications, and names he subsequently rejected).

The inside page of the front cover of *Fish in Spirits of Wine* bears in CD's writing:
+ signifies the fish the names of which I am anxious to know.

Fish in Spirits of Wine

1832 Fish in Spirits of Wine "Rio de Janerio"

"Exd 17„ Porto Praya caught by hook "C. de Verd" [Isl] "Serranus goriensis Val."

"Exd 18„, Hab. (ditto) D° vermilion, with streaks of iridiscent blue "Upeneus prayensis"

"Exd 19„, Fish. Quail Island; they bite very severely; having driven teeth through Mr Sullivans finger. "Salarius atlanticus."

"Exd 20„, Do "Salarius Atlant."   D° Porto Praya C.Verd Isl"

"Exd 21„, Do   D° "Salarius vomerinus Z Cuv & Vals"

"Exd 43„, Fish, dark greenish, black above, beneath lighter, sides marked with light emerald green tips of anal, caudal, and hinder parts of dorsal tipped saffron yellow. Tip of pectorals"

"+" orpiment orange Jany 25th Quail Is

+ serranus aspersus Jan.

"Exd 44, 45, 46. Fish

114 Sucking fish "Echeneis Remora" off a shark near St Pauls.

126 Fishes. St Jago. Feb. and March.

"+„ 132 Diodon "antennatus U P: 22 Bahia D°

137 Fish.   D°   D° "(Thrown away: bad)„

138 Fish "(Syngnathus crinitus, Jenn.)„ D° D°

147, 149 Fish caught on the 21° of March

  Lat. 14° 2′0 S. Long 3°. ′S West. About 65 miles from land; became of a pinker colour from spirits of wine" (Balistes vetula. B1. Young)„,

154 Fish. (very small) Lat 17. 12. S. Long 36.33W.

Pages 156.-157 Fish. In the above latitude and longitude, Caught two specimens of a fish; belly silvery white. Mottled with brownish black, side blueish with dusky greenish markings. Iris yellow with dark blue pupil

Stegastes imbricatus   ···Muraena

"180	Pimelodus gracilis, D'Orb" (Val. In D'Orb. Voy. Dans l'Amer. Merid. Al. Lehth. [?] Pl.2.fig.5 and Cuv. Et Val. Hist.des Poiss. Tom. xv.p.134.) [Fishes] (D'Orbigny)	"Exd 180	Fish. Running block. Socêgo? Not common;p pectoral fin causes Painful pricks. April
"181 & 182	These were probably the Pimelodus exsudans, Jen. - a h.d. of two specimens were found in the collections without labels Tetragonopterus taeniatus	181 182	Another specimen from same site. Ditto Ditto Ditto
"Nos 195, 210 & 228 - Are the same Poecilia unimaculata, Val"		195 Ex^d	Fresh water fish, in great numbers in a small ditch. Rio de Janeiro. April
		210 Ex^d	Fish out of a salt lagoon in great num= bers, precisely the same as (195) those taken in fresh water. D° D° May
		228 Ex^d	Fish, same as (195) Fresh water D° D°
"269	Prionotus punctatus."	269	Fish, swimming on surface. Rio bay. above and sides olive brown with red spots and marks, beneath silvery white; edges of pectoral fin Prussian blue, emitted a sound like a croak D° D° June
"288	Tetragonopterus scabripinnis, Jen."	288	Fish fresh water, same as (195) D° D°
"301	Gymnothorax ocellatus Spix. & Agas. " * (Spix et A Pisces Brazil. p. 91 tab. 50b)	301	Fish. Rio Harbour. D° D°
		309	Fish D° D° D°
		334	Fish, little pools near the river Monte Video, August.
"347.	Percophis brasilianus	347	Fish. Coast of Patagonia. Lat 38° 20'. August 26th 14 fathom, caught by hook and line Above pale, regularly and symmetrically marked with "brownish red" (by the tip of each scale being so coloured). Beneath silvery white; side with faint coppery tinge; ventral fins yellowish. Pupil of eye intense black. When cooked was good eating.
"348	Plectropoma patachonica, Jen"	348	Fish. Hab same as last. Many specimens exceeded a foot in length Above aureous – coppery; with wave like lines of dark brown, these often collect into 4 or 5 transverse bands, fins leaden

1832 Fish in Spirits of Wine

"354	Pinguipes fasciatus, Jen."	Colour, beneath obscure; pupil dark blue, when caught vomited up small fish and a Pilumneonus. Mr Earle states these fish are plentiful at Tristan da A cünha," where it is called the Devils fish; from the bands being supposed the marks of the Devils fingers. Was tough for eating, but good. This sort was taken in very great numbers. Coast of Patagonia. August.
354		Fish. Hab. same as last. Above pole "chestnut brown" so arranged as to form transverse bands on sides; sides, head, fins, with a black tinge; beneath irregularly white; under lip, pink; eyes with pupil black, with yellow ris. From D°
"358	Plectropoma patachonica, Jen.	Fish. Hab. same as (347) Colour above salmon coloured.
359	Great shark (bad?)	Squalus. August 28th. Lat 3° S. 2'5 South. Soundings 14 fathoms. Caught by a hook a specimen of genus Squalus; Body "blueish grey," above with rather blacker tinge; beneath much whiter; Its eye was the most beautiful thing I ever saw, pupil pale "verdegris green" but with lustre of a jewel, appearing like a Sapphire or Beryl. Iris pearly edge dark. Scelrotica pearly; In stomach was remains of large fish. In the uterus, the young ones for a long time after the viscera were opened continued to move; good specimen for dissecting D° D°
367	Atherina incisa, Jen N.I	Fish. Lat. 39 Long 61 W. Body semitransparent, colourless; with a bright silver band on each side; also d° marked about the head; taken some miles from the land. Sept D°

		1832 Fish in Spirits of Wine - Bahia Blanco. Sep^t
"371	Clupea arcuata, Jen. NS."	371 Fish. Body silvery, excepting back "greenish blue.
"390	(underneath another sp. Clupea?) Clupea (Alosa) pecttinata, Jen. N.V	390, 391, 392, 394 Fish
		390 Fish. Caught on a sand bank in the net: body silvery, dorsal scales iridescent with green and copper; head "greenish;" tail "yellow."
		391 Fish. Body pale. Darker above: band on sides: common - broad. Silver." This is probably the old fish of the small ones (367) taken at sea.
"392	Umbrina arenata, C.&V."	392 Fish. Body mottled, with "silver and green;" dorsal and caudal fins "Read colour;" common.
"393	Mugil Liza? C.&V. died, & in bad condit. (thrown away.)"	393 Fish. Back coloured like Labrador feldspar; iris coppery; plentiful.
"394	Platessa P. Orbignyana, Val. "	394 Fish. Above dirty "reddish brown;" beneath faint "blue." iris "yellow;" plentiful.
"395	Rhombus - ?" [in panel underneath Platessa (?) - ? London]	395 Fish. Above pale "purplish brown," with rounded darker Markings.
396	Young ray - (bad)	"396 Fish flat. D°"
"402	Batrachus porosissimus, Val.?"	402 Fish. Cast up on the beach. Above purple – coppery; sides pearly; beneath yellowish with silver dots in regular figures; iris coppery; not uncommon.
"415	Young Shark (bad)"	415 Squalus. (very small specimen)
"416	Clupea arcuata, Jen. N.S."	416 Fish; back blue, belly silvery.
"431	Great Conger? Eel at Society (bad)"	431 Fish; above reddish lead colour - October B. Blanca

"449	Paropsis signata, Jen. / N.Gen/"	449	Fish. Uniform bright silvery, ridge of back blueish; black patch on gill-cover, and another under pectoral fin.
450	Same as 390 - Cuv. Va. -	450	Fish; scales silvery indescent, back especially greenish; caudal fins yellow; remarkable from circular dark green patch behind gill cover.

Fish in Spirits of Wine
M. Video. November

458.	Corvina adusta, Agaz. Dry - 459 - Mugil Liza? Dry & bad.	458. 459.	Fish. Monte. Video. Octob.r 29th.
		465..466.	Fish. "M.Video
		467.	Fish D°
470	Lebias multidentata. Jen. N.S.	470.	Fish. Fresh Water D°
479	Unkn - Dry and in bad order	479.	Fish. San Blas.. From the "Schooners, coast of Patagonia.
480.	Achirus (Plagusia)? Dry & in bad state -	480	Fish
		494	Fish. In coral. "Tierra del Fuego" 35 fathoms, about 30 miles off. Northern Tierra del Fuego. Dec.r 16th Lat. 53.S. Head coloured purple (colour a "bru" ptly truncate posteriorly) with a white line over the nose, belly purplish; rest of body dirty yellow.
		504.	Fish. Very common in the Kelp. Coppery orange, with dark transverse markings; pectorals and verticals reddish orange.
		505.	Squalus. Gore 1833 _ Sound: same Hab. as last. Above, with white and dark spots and transverse marks; [erased something] breast, and pectoral and ventral fins clouded with "scarlet red." Jan. 7.
		511.	Fish, with irregular bands of pale "reddish brown," the pale parts with a most beautiful metallic violet coloured glitter along the sides. Grows to be one foot long. Jany

515	513.	Fish. dusky orange red, above dosave. Kelp.
	514.	Fish. coppery orange above dosure, common in the Kelp.
	"+ names of genus?"	Fish. 504. 514. 520. And others form chief subsistence to the Fuegians. D°
Myxine australis, Jen.	515	Fish. (Cyclostomes?) Caught by hook amongst the Kelp. Goree Sound and other parts of Tierra del Fuego.
"Myxinnus"		Above coloured like an earth worm but more leaden; beneath yellowish and head purplish; very vivacious
	1833	Fish in Spirits of Wine
		- Goree Sound Jan^y "Tierra del Fuego" And retained its life for a long. Had great powers Of twisting itself and could swim tail first; when irritated struck at any object with its teeth; and by protruding them ,,in its manner,,, much resembled an adder striking with its fangs. Head most curiously ornamented with tentacles; vomited up a sipunculus. When caught. this fish is abundant amongst the rocky islets, having found one on the beach nearly dead. I observed a milky fluid transcending through the row of internal pores or orifices. It would appear to be Myxinidus.
	516.	Fish. D°
	517.	Fish. Above curiously marked with reddish purple, grey, and black.
	519.	Fish. Uniform yellow.
	520.	Fish. Anal and ventral fins black. Pectoral orange; three orange stripes on the sides. All these fish caught by hook in the Kelp.
	523.	Fish. Colour "crimson red." Kelp.
	524. 525.	Fish. Caught in Kelp.

526.	Aplochiton taeniatus. Jen. N.S.	
	526.	Fish. (3 specimens) these were caught in the mouth of a fresh water stream; when the water was quite fresh upon being placed in salt water: they immediately died.
	531.	Fish. Beagle Channel - Tierra del Fuego
	535.	Fish. Abdomen with a fine red.
536.	Mesites alpinus, Jen. N.S.	
	536.	Fish. Alpine fresh water fish in lake; Hardy Peninsula.
	1833	Fish in Spirits of Wine Tierra del Fuego. N.B. In the same cash, as the Skate there is a Gadus, caught in Good Success Bay and common about Cape Fairweather; it leaves the coast in March.
	538.	Skate, two specimens both sexes Good Success Bay. Colour "broccoli brown" and marked (like binding of book) with rings and lines of "chocolate red." Iris silvery grey upper part depending fringed, sometimes almost concealing the pupil.
	542.	Fish. In pools left by the tide.
	543.	Fish, fresh water brook. Hardy Peninsula.
543.	Mesites maculatus, Jen. N.S	
	546.	Merlus; caught in Good Success Bay, by hook and line; colours reddish brown and white variously marked. Eye with singular fleshy appendage; to the fish are sewn another pair from another specimen. March
	549. 550.	Fish. On the rocks on coast Good Success Bay. March.

553, 4, 5.	Aplochiton zebra. Jen. N.S (three spes.)	
	553. 554. 555	Fish from fresh water lake (Silurus?); dull leaden colour; good eating grow about as half as large again; common; Falkland Islands. This lake is not far from the sea, and connected by a brook
	+	" " — Falkland Island. March.
	558.	Fish; under stones; sea coast.
	562.	Fish (two species). Rocks. Sea coast.
	578.579.580 .581	Small rocky pools; at low water.
	592.	Fish, fresh water, embourchure of brook.
598	Phucocoetes latitans, N.S	
	598. 599.	Fish. Caught amongst Kelp.
	636.	Fish. Hab. same as (634) Darwin Fresh water "brook" Maldonado. May
		"Rio Plata"
	1833	Fish in Spirits of Wine
	660.	Fish, fresh water lake. Maldonado "Mardonado" May. Lake left dry by breaking of bank; Lake sometimes a little brackish; above greenish black sides paler, Slightly iridescent.
661. No. 669. *	Hydrocyon Hepsetus - Poecilia decem-maculata. Jen. N.S. - (2). Lebias lineata, Jen. N.S. – (3) Tetragonopterus interruptus, Jen. N.S	
	661.	Fish. Fresh water lake; blueish silvery.
	669.	Five species of fish from a lake which was suddenly drained. The fish with beard I have seen 8 or 9 inches long","", The „smallest fish, with black spots on side I think is full grown - I have taken them so repeatedly in brooks &c of the same size.
	692.	Fish. May and June Maladonado

692.	Percophis brasilianus.	Exd	692.	Fish. Mottled with red, beneath beautiful white; common, good eating.
693	Atherina argentinensis? Cuv & Val. .	Exd	693.	Fish, silvery, with silver lateral band, above blueish grey; very common, (also in brackish water.)
694.	Otolithus Guatucupa, Cuv & Val. -.	Exd	694.	Fish, silvery white, above iridescent with violet purple and blue.
695.	Corvina adusta, Agaz. - .	Exd	695.	Fish. Above more coppery, with irregular transverse bars of brown; beautifully iridescent with violet. June.
696.	Dules auriga, Cuv. .	Exd	696.	Fish. Sides with numerous waving longitudinal lines of brownish red; intermediate species greenish - silvery, so figured as to look mottled; head marked with lines of dull red & green; ventral, and "anal ", fins coloured dark greenish blue.
				The above five fish Maldonado Bay, June.
714	Umbrina arenata, Cuv. & Val. .	Exd	(+) 710 714	Fish. Colour blueish silvery, fins, darker Maldonado. June
723.	Diodon rivulatus, Cuv.	End	"723"	Diodon. Picked up on shore of R. Plata. Maldonado"
			1833	Fish in Spirits of Wine
739.	Achirus lineatus, D'Orbig.		738. 745. 741. +	Fish bought in market of Buenes Aires and all edible Is a fish excessively abundant high up the Rio Parana. In like manner is (745) the Armado. This fish is peculiar by the very loud harsh grating noise which it can make, heard even before hauled out of water. Is able to seize very firm hold of any object with the serated pectoral bones and dorsal fins.
			746. (July)	Fish. High up the Parana, "near to Razjano" and [not] very abundant. Upper part of body with its fins with tint of yellow, but stronger on the head, with dorsal clouds of black, tip of tail black. Beneath silvery white, pupil black, iris white, usual size sometimes larger.

Tetragonopterus abramis. Jen. NS	747.	Fish same locality. Caught in October in the Rio Parana as high as Rozario. The "four first," fish "747: 48: 49: 50,", are the commonest of the river fry. Back blueish silvery, with silver band on side, blueish black spot behind the Branchiae. Fins pale orange, tail with central band black.. D°
Tetragonopterus rutilus, Jen. NS	748.	Fish. Back iridescent greenish brown,,, silver band on side. Fins dirty orange, tail with central black band, above and below bright red and orange.
	749	Fish. Silvery; eyes fine black, "Hab,, peculiar form of belly; grows to twice the size of this specimen. "Hab" D°
	750.	Fish. Called salmon grows to one or two feet long. Above blueish gradually shading down, on sides; fins tipped with fine red, especially the tail, which latter organ has central black band. "Hab" D°
P St Julian & Port Desire Central Patagonia	1833	Fish in Spirits of Wine
Port Famine Tierra del Fuego	751.	Fish. Jar with the latter and B. Ayres fish also a cod and one sex of a Ray from Good Success Bay Tierra del Fuego.
	775.	Fish, in rocky pools of salt water - Beneath dirty white; back with dive brown darker in the middle. Port Desire, Jan 1834
Aphritis porosus, Jen. NS.	776.	Fish, left among the mud banks. D°
	785.	Fish, rocky pools: Port St Julian. D°
	788.	Fish, whole body silvery, upper part of back iridescent blue, lower greenish, spotted with coppery - head, circular patches; common size.
Stromateus maculatus, Cuv. & Val.? In bad condition & thrown away	789.	Fish. Back blackish; centre of each scale greenish white; reaches to one and two feet long. P. St Julian. At Port Famine one was 2^{ft} 4^{inc} in length. A Pescado del Rey was ther"re,,, likewise 20 inches long and wonderfully numerous.

	791.	Small fish. P. St Julian D°
	818.	Fish. Above coppery yellow with 5 or 6 transverse brown bands; caught by hook and line. Port Famine.
	819.	Dog fish yellowish brown clouded with "cochineal red." P. Famine.
	829.	Fish pale yellowish brown, with figure of S (or muscles) on sides pale coppery; about mouth bronchial covering tips of pectorals and ventrals, reddish orange; caught by hook, uncommon. D°
	835.	Fish. Hab. as "S of C. Penas east coast of T. del Fuego beautifully silvery with raised lateral line; upper part of back, pale most beautiful Auricular purple Straits of Megellan.
	1834	Fish in Spirits of Wine Tierra del Fuego. Feb.
Clupea Fuegensis, Jen.	838.	Fish. Caught at night. 2 miles from shore. Cape Ines. 19th "13 fathoms"
	840.	Dog-fish. Colour pale "Lavender purple," with cupreous gloss, sides silvery D°, above with irregular quadruple chain of circular and oblong snow white spots tip of dorsal and caudal blackish under part of caudal reddish, iris pearly white. Length of old specimen tip to tip 2 feet 3 inches. breadth from tip of pectoral to tip of other 8 inches. young specimen out of belly; with it is posterior spine of old specimen.
	847.	Fish, above greenish black, beneath yellowish white, sides iridescent where the dark back shades away. N.B. bought of and cleaned by the Fuegians - Kelp fish – east entrance of Beagle Channel.

838.

	848.	Fish. Pectorals. Ventrals red orange; anal caudal dorsal blackish; back and sides mottled, reddish and greenish, & blackish &:on kelp. East entrance of Beagle Channel.
	849.	Fish. Pecteral. Dorsal and Caudal. "Tile and vermilion red" side of head. 4 or 5 very irregular rows D° colour. Anal ventral and branchial covering dark blue black. Hab. as above.
	(March)	
866.	866.	Fish; very active; roots of Fucus. Hab: D°
870 Conger punctu^slatus, Jen. <u>NS</u>. -	870.	Fish. Hab: D° (Young of 866)
	876.	Kelp Fish. Back mottled with dirty red and green; fins with orange eyes coppery. Beagle Channel.
		"Tierro del Fuego"
	1834.	Fish in Spirits of Wine. March
	877.	Gadus. Back "yellowish and chestnut brown" dorsal fins "Silver brown." Tierra del Fuego.
	882.	Dog-fish; upper part coppery "Brownish purple and Cochineal red" with small white: "Hab do., spots and large blackish ones: Ponsonby Sound.
	883.	Kelp. Fish. Hab: D° Mottled with orange: pectoral and part of caudal D° colour; anal ventral, Dorsal blackish green.
		"
		E. Falkland Island. March.
	896, 897, 898	Fish roots of Kelp. Berk^eley Sound
	905.	Fish. General colour "gallstone and honey yellow" browner on the back.

	906.		Fish. More brown on back; same general colour as above; with small irregular patches on sides of body, head and branchial covering, of pale silvery blue.
	907.		Fish D° Pecteral. Ventral and Caudal fins mottled with orange; body with brown blade; much more tenacious of life than latter two; all caught in Kelp.
			" "
			Santa Cruz. May
947.	947.	Ex^d	Fish found dead high up river of S. Cruz; "Patagonia" pale yellowish brown, with black motlings
952. Perca laevis, Jen. NS.	952.		Fish, numerous in streamlets and creeks high up river; pale greenish brown with small irregular transverse bars of black, belly snow white.
952. Mesites maculatus, Jen. N.Gen.			
957.	957.		Fish found dead on beach. Cape Virgins. Saw many skeletons in estuary of S. Cruz.
1002. One of the Gadidae; but in very bad condition, & thrown away –	1002.	Ex^d	Fish. Said to come from the Japan Sea. A whaler. Chiloe. July.
1002. Cheilio ramosus, Jen. NS.?	1007.		Fish. Above dirty "gallstone and honey yellow" posterior half of the body becoming reddish, while spots on the side and smaller ones above head. Valparaiso. July.
	1834.		Fish in Spirits of Wine
	1008.		"Valparariso, Valparoriso. July 25th "(„Fish bought in Market.")„ Above blackish grey, indistinct bands of d° on sides; beneath white. Are found 3 or 4 times as large.
	1009.		A uniform pale greenish tinge, most thickly mottled with "greenish black."
	1010		Uniform pale flesh colour (especially beneath) mottled with deep "reddish brown" and transverse dorsal bands of D°; Branchial covering yellowish. Inferior edge of Pectoral pink.

1011.	Heliases crusma, Cur. & Val.	Above leaden colour, beneath paler; grows considerably larger.
1012	Pinguipes chilensis, Val.	"Ex^d 1011, fins dark
		"Ex^d 1012,, D° D° D° slightly iridescent D° grow to 2 or 3 times this
1013.		size.
1014	Sebastes oculata, Val? (Dried & in bad order)	Under surface. Sides; Branchial covering part of fins, "Tile and carmine red," dorsal scales pale yellowish dirty brown.
1015		Uniform tinge pale dirty yellow with numerous angular spots of black. Above clouded with pale brown. Ventral and tips of pectoral and anal "reddish orange." Common size.
1016.		Sides "Cochineal red mixed with grey," an indescribable tint, belly strongly tipped with yellow, fins pale "blackish green," posterior half of body with numerous small scarlet dots.
1017	Latilus jugularis, Cuv. & Val. (Gony.)	"Ex^d 1017,, Beneath brilliant white; head and back clouded with "purplish and carmine red" longitudinal and transverse irregular bands of D°
1018.		Whole body silvery, back and fins with few clouds of leaden colour. Grows to 3 or 4 feet long. All the above bought in Market.
1834		Fish in Spirits of Wine Valparaiso. August.
1026.		Fish. Above "wood & yellow brown" with white and dark brown spots, grows more than a foot long, July.

1056.	Atherina microlepidota, Jen. N.S	Ex^d	Fresh water fish.
1075.	Syngnathus acicularis, Jen. N.S		D° D° (Most extraordinary) Tadpole!!
			Fish. The two last Sep^t & Octob^r
			Archiüago of Chileo. Nov^r & Dec^r
1080	Ilucoetes fimbriatu, Jen. N.S		Fish. Island of Lemuy Octob^r
1081	Gobiesox marmoratus, Jen. N.S		Blennius under stones
			Sucking fish D°
			Fish.
			Fish tidal rocks Island of Incho North part of Cape
			Ires Montco. Dec^r
1123	Agriopus hispidus, Jen. N.S	Ex^d	Fish Hab. same as (1122) caught in middle of cove. Coloured pale "reddish orange" with black spots on the fins, and a dusky shade on back; strange appeare with its bony face. D°
			Chonus Archipelago Jan 1135 "(South of Chiloe)
1138	Aphritis undulatus, Jen. N.S	Ex^d	Two sorts of marine fish. Lowes Harbour
1139	Gobius ophicephalus, Jen. N.S	Ex^d	Fish tidal; pale lead-colour coarsely reticulated with brown
			D°
1141			Fish; silvery, bright, back blue. Lowes Harbour.
			Chiloe Jan.
1145			Cabora del Cavallo. Fish same as at Chonos.
↑			"Cz,,
1146	Stromateus maculatus, Cuv. & Val. ?		Fish. Silvery, blue above with regular circular leaden spots.
1147			Fish. All silvery
1148.			D° Above with fine tint of purple.
1149			D° Mottled reddish above, beneath white.
1150			D° D°
1151			D° Silvery, irregular leaden coloured marks

1153	Aspidophorus chiloensis	Ex^d	D° Above dusky
		1835	Fish in Spirits of wine Chiloe Jan.
1154	Syngnathus - ? - In very bad condition, & thrown away. -	1154	Fish
		1161	D° under stones on sea beach. Chanques Is^d
		1188	Fish. Disc of body yellowish brown with minute spots. 4 transverse bands in front part and superior convex edge, most beautiful Cobalt blue; gody generally dark and yellowish brown - "Coast of Chili"
	1202 Blennechis fasciatus, Jen. NS.	Ex^d 1201	Fish & 1202 Marine Concepcion "Chile
1204	Clinus crinitus, Jen. NS	Ex^d 1204	Fish Coquimbo D°
1211	Blennechis ornatus, Jen. NS	Ex^d 1211	Fish etc etc D°
		1215	Sucking fish. R. Maule "Chili,,
1218, & 1220	Umbrina ophicephala, Jen. NS (Bad)	1216 to 1222	Fish. Coquimbo
1229	Engraulis ringens, Jen. NS	1229	Fish Iquique Peru
1232	Exocoetus exsiliens, Bl.	1232	Flying fish Lat 18° S. July
			"
			- Lima San Lorenzo July -
1238	Unkn. Dry & in bad order -	1238.	Fish dull lead colour. with pale transverse bands. Pectoral ventral annual caudal fins pale vermilion.
		1239	Fish dull coloured blue with numerous red spots like a trout. Pectoral and caudal orange.
		1240	Fish pale greyish blue. with blacks specks and clouds of D°. tips of fins pinkish.
		1241. 1242	Small silvery fish.

1244	One of Siluridae - very bad & thrown away -	1244	Fish ^back indescdent green, belly white. all Fins and Barbillons reddish purple.
		1246	Fish, above nearly black.
1251	pen ↓ ↓pencil↓ Cuv. & Val ~ Clupea sagax;? Jen., N.S	1251. 1252. 1253 1254. 1255	Fish. D° D° D°
1260	Otolithus angusticeps, Jen. N.S	1258 1260 1261	Fish Coquimbo D° Callao D° D°
		1835	Fish in Spirits of Wine Sepr Galapagos. Chatham Island
1265	Tetrodon aerostaticus, Jen. N.S	1265	Fish. Above dull green, base of Pectorals and Dorsal black, a white patch beneath the Pectorals, inflatable.
1266 Ex^d	Pristipoma miles cantharhinum, Jen. N.S	1266	Fish. Blueish silvery.
1267 Ex^d	Prionotus ruber? Jen. N.S ↑	1267	Fish. Beneath silvery white, above mottled brilliant "Tile red"
1268	Latilus princeps, Jen. N.S	1268	Fish. Above and fin greenish; sides coppery. passing in belly into Salmon colour. Pectoral fins edge with dull blue. Iris yellowish brown, pupil black blue.
1269 Ex^d	Cossyphus Darwini, Jen. N.S	1269	Fish. Centre of each scale pale "vermilion red" lower jaw quite white, large irregular patch above the pectorals bright yellow. Iris red, pupil blue black.
1273 Ex^d	Serranus labriformis, Jen. N.S -	1273	Fish. Mottled brown-yellow, black and white upper and lower edge of tail, edges of ventral and dorsal ("Art. And purplish red")

1274	Chrysophrys taurina, Jen. N.S	Ex^d	1274.	Fish. White with four dark brown much interrupted bands giving mottled appearance, d° coloured above head, top of d°, ridge of back, edges of dorsal, tail, and ventral fins, tinted with fine Azure blue.
1275	Serranus Galapegensis, ^olfax, Jen. N.S	Ex^d	1275.	Fish common large mottled brown fish
1282, 3	Scorpaena histrio, Jen. N.S	Ex^d	1282, 1283	D° Whole body "scarlet red," fins rather paler; small irregularly shaped light black spots; very extraordinary; are they distinct?
1284	Prionodes fasciatus, Jen. N.S	Ex^d	1284	Fish, pale yellowish brown, with numerous transverse bars of which upper part reddish black; lower "vermilion red," gill covers head, and fins tinted with d°. "~~Fish~~, – CD
1286	Muraena lentiginosa (Same as 1299.)		1286	Eel, tidal pools, colour dark reddish purple brown, with pale, or whiteish brown spots, eyes blueish.
			1835	Fish in Spirits of Wine Sep^t Galapagos. Chatham Island
1287	Gobius lineatus, Jen. N.S. 1288. Gobiesox poecilophthalmos, Jen.		Ex^d 1287, 1288	Fish tidal pools
1293	Tetrodon annulatus; Jen – N.S		1293.	Diodon, Beneath snow white, above dark brownish black; this colour is placed in broad rings one within the other on the back, so that on the side they form oblique ones which point both ways: whole upper surface spotted with darker black circular spots. Pectoral and dorsal fins yellowish brown; Iris inner edge clouded with orange. Pupil dark green – blue – made a loud grating noise. Cha^les Island
1299	Muraena lentiginosa, Jen. N.S		1299	Fish, fine dark purplish brown with yellow circular spots.

1302	Pristipoma cantharhinum, Jen. (Young)	Exd. 1302	Fish silvery above, shaded with brown and indescent with blue; fins and iris sometimes edged with blackish brown. Flap of gill cover edged with black.
1304	Serranus albomaculatus, Jen. N.S	Exd 1304	Fish varys[ies] much in colour. Above pale blackish green, belly white; Fins, gills covers and part of sides dirty reddish orange; on side of back 6 or 7 good-sized snow white spots, not very regular outline. In some specimens, the blackish green above becomes "dark" & is separated by a straight line from the paler under parts; Again other ,, specimen coloured dirty "reddish orange and gallstone yellow"CD upper parts only rather darker, but in all white spots clear. (5 or 6 in one row and one placed above). Sometimes fins banded with orange and the CD black green, length ways. Otahiti Novr
1317	Acanthurus humeralis Cuv. & Val. Exd	Exd. 1317	Fish. Splendid verditer blue and green.
1318	Scarus lepidus Jen. N.S	1319 Scarus globiceps, C.&V Exd 1320 Upeneus trifasciatus	Exd. 1318, 1319 1320	Fish
		1835	Fish in Spirits of Wine Otahiti Novr
1321	Caranx torvus, Jen. N.S Exd	1321 1324.	Fish. Extraordinary Fish
1325, 1326	Ostracion meleagris, Shaw	1325 1326	Fish
1327	Muraena 1328 Syngnathus conspicillatus, Jen	1327 1328	D°
1329	Balistes aculeatus, Bl. -	1329	D°

1330, 1331	Dules leuciscus, Jen. N.S	Ex^d	1330 1331	Fresh water Fish CD
				Bay of Islands N. Zealand Decemb.
1337	Anguilla australis, Rich. bad spec.		1335.	Fresh water Fish
1337			1337	Fresh water Eel
1339	Eleotris gobioides, Val (1340 / Mesites attenuatus, Jen. N.S	Ex^d	1339 1340	D° F.water" Fish
1341	Trigla kumu, Less et Garn.	Ex^d	1341	Fish whole body bright red
1343	Tripterygion capito, Jen. N.S	Ex^d	1343	Fish. Tidal rocks
1344	Acanthoclinus fuscus, Jen. New Gen.	Ex^d	1344	D° D°
			1345	D° D°
	Caranx Caranx			King Georges Sound March 1836 "New Holland"
	declivis georgianus			
Ex^d	1371 / 1372 / 1368 / 1374 / 1365, 1366 /		1365 to 1386	Various fish caught by net in Princess Royal Harbour.
			1389	Skate, above muddy cream colour.
			1390	Fish above varied dull green, with pale D°, beneath snow white.
			1391	Fish, pale copperish brown, with water marks of a fine darker brown.
			1392	Fish, very pale brown, fins pale orange.
			1393	Fish, mottled with pale blackish green, leaning white spots
			1394	Fish. Sides fine dark green, and pale silvery green, fins tipped with red. Iris fine green, handsome, fish.

Appendix II Fish of the *Beagle* in the BMNH

Extracts from the Accession Catalogue of the Ichthyology Section, Museum of Natural History, London, for the period from July 14–25, 1917, pertaining to specimens originally listed under: " "Beagle" collection. Recd. from the University Museum of Zoology, Cambr[idge]."

7.14	1	*Engraulus ringens*	(Type)	In spirit	Coast of Peru.
	2	*Haplochiton zebra*	(")	"	Fresh Water Lake, Falkland Is.
	3	" "	(Type)	"	Falkland Is.
	4–5	" *taeniatus*	(")	"	Goree Sound, Tierra del Fuego.
	6	*Galaxias maculatus*	(")	"	Hardy Peninsula
	7–9	" "	(Co-types)	"	Santa Cruz, Patagonia.
	10	" *alpinus*	(Type)	"	Hardy Peninsula, Tierra del Fuego.
	11	" *attenuatus*	(")	"	Bay of Islands, New Zealand.
	12	*Congromuraena punctus*	(")	"	Tierra del Fuego.
	13	*Muraena lentiginosa*	(")	"	Charles Isd., Galapagos Archipelago
	14	*Tetragonopterus rutilus*	(Type)	"	Rio Panana, S. America.
	15	" *scabripinnis*	(Type)	"	Rio de Janeiro, S. America.
	16–17	*Chirodon interruptus*	(")	"	Maldonado, S. America.
	18–19.19a	*Callichthys paleatus*	(")	"	S. America. [Note to the right of 'S. America:' "Specimen i.a. 19a received about 1912, but not registered then"]
	20–23	*Jenynsia lineata*	(")	"	Maldonado, S. America.
	24	" "	(type of *Fitzroyia multidentata*)	"	Monte Video, "
	25–26	*Cnestrodon decemmaculatus*	(type)	In spirit	Maldonado.

27	*Syngnathus conspicillatus*	(")	"	Tahiti.
28	" *acicularis*	(")	"	Valparaiso.
29	" *crinitus*	(")	"	Bahia Blanca, N. Patagonia.
30	*Caranx declivis*	(")	"	New Holland.
31	" *torvus*	(type)	"	Otaheiti
32	*Paropsis signata*	(")	"	North Patagonia.
33	*Percichthys laevis*	(")	"	River Santa Cruz.
34–35	*Acanthistius patachonicus*	(types)	"	E. coast of Patagonia.
36	*Epinephelus ascensionis*	(type of *Serranus aspersus*)	"	Cape Verde Islands.
37	*Acanthoclinus littoreus*	(type of *A. fuscus*)	"	Bay of Islands.
38–40	" "	(" " ")	"	"
41	*Helotes octolineatus*	(type)	In spirit	King George's Sound.
42	*Pristipoma cantharinum*	(Co-type)	"	Charles Is. Galapagos Archipelago
43	" "	(")	"	Chatham Is. " "
44	*Otolithus analis*	(type)	"	Callao, W. coast of S. America.
45	*Glyphidodon luridus*	(type of *Stegastes imbricata*)	"	Quail Is. Porto Praya.
46	*Cheilio inermis*	(type of *C. ramosus*)	"	
47	*Pseudoscarus lepidus*	(type)	"	Tahiti
48–49	*Eleginops maclovinius*		"	Chonos Archipelago.
50	" "	(type of *Aphrites porosus*)	"	Coast of Patagonia.
51	*Pinguipes fasciatus*	(type)	In spirit	Coast of Patagonia
52	*Atherinichthys argentinensis*	(")	"	S. America
53	*Gobius lineatus*	(")	"	Galapagos Archipelago
54	*Gobiosoma ophicephalum*	(")	"	Chonos Archipelago
55	*Clinus crinitus*	(")	"	Coquimbo, Chile
56–57	*Tripterygium nigripinne*	types of *T. capito*)	"	Bay of Islands
58–59	*Petroscirtes fasciatus*	(types)	"	Concepcion, Chile
60–64	" "	(types)	"	Coquimbo, Chile

Appendix II

	65–66	*Salarias quadricornis*	(types)	"	Keeling Is.
	67–68	*Phucocoetes latitans*	(types)	"	Falkland Is.
	69	*Iluocoetes fimbriatus*	(types)	"	Archipelago of Chiloe, S. America
	70	*Gobiesox poecilophthalmos*	(type)	"	Chatham Is., Galapagos Archipelago
	71–72	" *marmoratus*	(type)	"	Archipelago of Chiloe, S. America
	73	*Scorpaena histrio*	(co-type)	"	Chatham Is.
	74	" "	(")	"	" "
	75	*Prionotus miles*	(")	"	" "
	76	*Platycephalus inops*	(")	"	King George's Sound, New Holland
	77–78	*Agonus chiloensis*	(type)	"	Is. of Chiloe, W. coast of S. America
	79	*Tetrodon hispidus*	(type of *Implutus*)	"	Keeling Is.
	80	" *annulatus*	(type)	"	Chatham Is.
	81	*Porichthys porosissimus*	(type)	"	Bahia Blanca, N. Patagonia
7.25	1	*Brevoortia pectinata*	(type)	dry	Bahia Blanca, N. Patagonia
	2	*Sardina sagax*	(")	"	Lima, San Lorenzo
	3	*Solea jenynsii*	(")	"	Buenos Ayres
	4	*Acanthistus patachonicus*	(type)	(dry)	South America
	5	*Serranus labriformis*	(type)	(dry)	Chatham Is.
	6	" *albomaculatus*	(")	(")	Charles Is.
	7	*Serranus olfax*	(")	(")	Chatham Is.
	8	*Latilus princeps*	(")	(")	"

Appendix III
Checklist of fish specimens, identified as collected by Charles Darwin on the *Beagle* voyage, that ought to be present in the collections of the University Museum of Zoology, Cambridge

ADRIAN FRIDAY
Curator of Vertebrates, University Museum of Zoology, Cambridge

This list is taken from a full manuscript list of *Beagle* fish that was transcribed from the relevant catalogues. **The list below includes only those specimens that should *currently* be in the Museum collections.** Many specimens were transferred to the British Museum (Natural History) in 1917 [see Appendix II].

The entries are mostly as found in the catalogues. Vagaries of spelling, style and punctuation have generally been retained, except where this would introduce confusion or ambiguity. Generic names have been expanded from their single-letter abbreviations where this is all that is given in the catalogue. Generic and specific names have been converted to italics for ease of reading.

Roman numerals refer to the volume numbers of the British Museum fish catalogues that were used as the basis of the Cambridge catalogues. The numbers that precede each entry are based on this volume number. Numbers refer to pages, not to individual specimens: several specimens may therefore be listed separately under the same number.

Round and square brackets are as used in the catalogue; curly brackets denote a comment included during the preparation of the list.

Most specimens were noted as present in the 1939 stock-take. Four ('dry') specimens not noted as present at the 1939 check remain missing. A few specimens not noted as present in 1939 are noted as present in subsequent years (and that information is included). Several specimens have no confirming marks from stock-taking.

The 'EXHIBITED SERIES' does not order the entries according to the British Museum catalogue. The three entries are given in the order in which they occur in the Cambridge catalogue.

Numbers in bold at the end of the entries are the field numbers assigned by Charles Darwin as reported in Jenyns' manuscript *Notes on the Fishes Collected by Chas. Darwin Esq. In the Voyage of H.M.S. Beagle Between the years 1826 and 1836; by the Rev. Leonard Jenyns F.L.S. etc. 1842*. This volume is in the archives of the University Museum of Zoology, Cambridge. Other comments on or from the Jenyns MS are also given in bold. Most specimens could be unambiguously united with their field numbers.

Appendix III

Catalogue of 'NON-EXHIBITED SERIES'

I

I.133 *Serranus goreensis*, Cuv. & Val. C. Verde Islands. {1939} **17**

I.181 *Genyroge marginata*. Keeling Is. {1939} **1433 As *Diacope marginata*.**

I.253 *Arripis (Centropristes) georgianus*. King George's Sound. New Holland. {1939} **1371**

I.266 *Dules auriga*, C. & V. Maldonado Bay. S. America. {1939} **696**

I.270 *Dules malo* = "*D. leuciscus*", Jenyns (2) River Matavai, Tahiti. "The correct name for this species is *Kuhlia marginata*, Cuv. & Val. C. Tate Regan", Mar. 1. 1913. {1939} **1331**

I.403 *Mulloides flavolineatus*, Lacep. = "*Upeneus flavolineatus*, Jenyns". Keeling Is. Indian Ocean. {1939} **1441 As *Upeneus flavolineatus*.**

I.407 *Upeneus trifasciatus*, Lacep. Tahiti, Pacific Ocean. {1939} **1320**

I.409 *Upeneus prayensis*, C. & V. Porto Praya. C. Verde Islands. {1939} **18**

I.434 *Haplodactylus punctatus*. ? Valparaiso, Coast of Chile. {1939} **No field number or locality given. As *Aplodactylus*.**

II

II.6 *Chaetodon setifer* (2). Keeling Islands. Indian Ocean. 1836. {1939} **1431**

II.105 *Sebastes oculatus*, C. & V. Dry. Valparaiso. {Not ticked, 1939; not found 1999} **1014**

II.130 (genus only) *Apistus* sp. Dry. [skeleton] King George's Sound. New Holland. {Not ticked, 1939; but noted as Mus. drawers 23/1. 1980.} **1391 & 1386 both given.**

II.139 *Agriopus hispidus*, Jenyns. (2). TYPE. Archipelago of Chiloe. W. coast of S. America. {1939} **1123**

II.193 *Prionotus punctatus*, C. & V. Bay of Rio de Janeiro, E. coast of S. America. {1939} **269**

II.248 *Percophis brasilianus*, Quoy & Gaim. Coast of Brazil. {1939} 347 or 692 As P. brazilianus. **347 is given as 'Coast of Patagonia. In Mus. Zool. Soc. Lond.' 692 is given as 'Maldonado.'**

II.252 *Pinguipes chilensis*, Molina. Valparaiso, Chile. {1939} **1012**

II.253 *Latilus jugularis*, Cuv. & Val. Valparaiso, Chile. {1939} **1017**

II.253 *Latilus jugularis*, Cuv. & Val. Valparaiso, Chile. {1939} **? also 1017**

II.276 *Umbrina arenata*, C. & V. Bahia Blanca, E. coast of S. America. {1939} **392**

II.276 *Umbrina arenata*, C. & V. Maldonado, Rio Plataq. {?} S. America. {1939} **714**

II.289 *Sciaena adusta*, Agass. & Spix. = "*Corvina adusta*, Jenyns." Dry. Montevideo. {1964} **As *Corvina adusta*. No specimen with this locality listed.** {This must, however, be field number **458** from Darwin's entry.}

II.289 *Sciaena adusta*, Agass. & Spix. Maldonado Bay, Rio Plata, South America. F.2601 {1939} **695 As *Corvina adusta*.**

II.309 *Otolithus guatucupa*, C. & V. Maldonado Bay, Rio Plata, South America. {1939} **694**

II.398 *Stromateus maculatus*, C. & V. Id of Chiloe, W. Coast of S. America. {1939} **1146**

II.440 *Caranx georgianus*, C. & V. (2). King Georges Sound. New Holland. {1939} **1365 & 1366**

II.468 *Seriolichthys* ("*Seriola*") *bipinnulatus*, Quoy & Gaim ["Freycinet, Voy. Zool. pi. 61, fig. 3"] Dry. Keeling Id. Indian Ocean. {1980} **1423 As *Seriola bipinnulatus*.**

II.495 *Psenes leucurus*, ? {the query probably refers to the species identification} C. & V. S Atlantic Ocean. {1939} **156 & 157 Not identified to species.**

III

III.114 *Eleotris gobioides*, Cuv. & Val. New Zealand. {1939} **1339 Bay of Islands, N. Zealand. Fresh-water.**

III.218 *Blennius sanguinolentus* = "*B. palmicornis*, Cuv. & Val." Quail Id. {1939} **44 *parvicornis* or *palmicornis*** {i.e. Jenyns offers a choice of spelling.}

III.242 *Salarias atlanticus* (2) [Described by Jenyns] [Jenyns MS pp. 71 & 72. C.Verde Is.]{1939} **19 & 20**

III.254 *Salarias vomerinus*, C. & V. Porto Praya, C. Verde Is. {1939} **21**

III.327 *Acanthurus triostegus*, L. Keeling Is. Indian Ocean. {1939} **1413**

III.336 *Acanthurus olivaceus* = "*A. humeralis*, C. & V." [Specimen described by Jenyns] Otaheiti. {1939} **1317 As *humeralis*.**

III.403 *Atherinichthys microlepidota*, Jenyns. TYPE. Valparaiso. {1939} **1056 As *Atherina microlepidota*.**

III.405 *Atherinichthys incisa*, Jenyns. TYPE. N. Patagonia. {1939} **367**

III.409 genus only. *Mugil* sp. Dry. Keeling Is. {1980} **1440**

III.423 *Mugil liza*, C. &V. Dry. Montevideo. {1980} **393**

III.465 *Agonostoma forsteri*. Dry. = "*Dajaus diemensis*, Richards." King George's Sound, New Holland. {1980} **1378 As *Dajaus diemensis* Richardson.**

IV

IV.61 *Heliastes crusma*, C. &V. Chili. {1939} **1011 As *Heliasus*.**

IV.208 genus only. *Scarus*? "Not named in Zool. of 'Beagle'." Jenyns MS p. 94. Labelled "Voyage of Beagle, p. 109. Keeling Is. Indian Ocean. 1836". {1939} **1430**

IV.224 *Pseudoscarus globiceps*, C. & V. [Tahiti] TYPE of Species. {1939} **1319 As *Scarus globiceps*.** {Jenyns does not give this as a TYPE, but gives 'C. & V.' The later addition of the words 'TYPE of Species' to the Cambridge catalogue must be an error.}

IV.256 *Gerres gula*, C. & V. Rio de Janeiro, S. America. {1939} **179**

IV.290 "*Chromis facetus*, Jenyns." TYPE. = *Heros facetus*. Maldonado, mouth of Rio Plata. {1939} **660 As *Chromis facetus*.**

IV.407 genus only. *Rhombus* sp. Dry. Bahia Blanca. {Not ticked 1939; not found 1999} **395**

IV.438 *Pleuronectes* (*Platessa*) sp. Dry. King George's Sound, New Holland. {Not ticked 1939; not found 1999} **1381 As *Platessa*.**

IV.491 *Plagusia* sp. Dry. Coast of Patagonia. {Not ticked 1939; not found 1999} **480 As *Achirus* {*Plagusia*}.**

V

V.121 *Pimelodus gracilis*, Val. Rio de Janeiro, S. America. {1939} **180 'Running Brooks' added to locality.**

V.132 *Pimelodus exsudans*, Jenyns. TYPE. Rio de Janeiro, S. America. {1939} **No field number given.**

V.321 *Tetragonopterus abramis*, Jenyns. TYPE. Rio Parana, S. America. {1939} **747**

V.329 *Tetragonopterus taeniatus*, Jenyns. TYPE. Rio de Janeiro, S. America. {1939} **181 & 182**

Appendix III

V.356 *Xiphorhamphus jenynsii*, Gunth. = "*Hydrocyon hepsetus*, Jenyns. TYPE". Fresh-water lake, Maldonado, Mouth of the Plata River, S. Am. {1939} **661 As *Hydrocyon hepsetus*.**

VI

VI.346 *Poecilia unimaculata*, Val. (several) Rio de Janeiro, S. America. {1939} **? 210 or 228**

VI.346 *Poecilia unimaculata*, Val. (2) Rio de Janeiro, S. America. {1939} **? 210 and/or 228**

VII

VII.413 [doubtful species No. 24] *Clupea fuegensis*, Jenyns. TYPE. Tierra del Fuego. {1939} **838**

VII.442 *Clupea arcuata*, Jenyns. (2) TYPES. Bahia Blanca, N. Patagonia. {1939} **416**

VIII

VIII.36 *Anguilla australis*, Richards. B. of Islands, New Zealand. {1939} **1337 Authority given as 'Richardson'.**

VIII.93 genus only. *Muraena* sp. [Jenyns MS. p. 133] Quail Id., B. of Porto Praya, C. Verde Is. [Labelled "Voyage of Beagle, p. 145"] {1939} **46**

VIII.93 genus only. *Muraena* sp. [Jenyns MS. p. 134] Tahiti. ["Voyage of Beagle, p. **146**"] {1939} **1327**

VIII.102 *Muraena ocellata* = *Gymnothorax ocellatus*, Spix & Agass. Rio de Janeiro, S. America. {1939} **301**

VIII.215 *Balistes vetula*. S. Atlantic Ocean. [Labelled "Wrongly named, but important as type of Jenyns' description".] {1939} **147 & 149**

VIII.223 *Balistes aculeatus*. Tahiti. {1939} **1329**

VIII.229 *Aleuterius maculosus*, Richards. King George's Sound, New Holland. {1939} **1393 As *Aleuteres maculosus*. {Jenyns does give 'Richards.'}**

VIII.244 *Monacanthus auraudi* (2) TYPE of "*Aluteres velutinus*, Jenyns". King George's Sound, New Holland. {1939} **1392 & 1380 As *Aluteres velutinus*.**

VIII.261 *Ostracion punctatus*, Schn. (2) Tahiti. {1939} **1325 & 1326**

VIII.287 *Tetrodon angusticeps*, Jen. TYPE. Chatham Id. Galapagos Archip'o {1939} **1265**

VIII.294 *Tetrodon stellatus*, juv. = "*T. aerostacius*, Jen." TYPE. {1939} **No field number or locality given.**

VIII.310 *Chilomycterus geometricus*, Bl. {?} Schn. = "*Diodon rivulatus*, Cuv." Maldonado, S. America. {1939} **723 As *Diodon rivulatus*.**

VIII.311 *Chilomycterus* ("*Diodon*") *antennatus*, Cuv. Bahia, Brazil. {1939} **132**

VIII.315 *Atopomycterus nychthemerus* = "*Diodon nichthemerus*, Jenyns". {1939} **No field number or locality given. As *Diodon nichthemerus*.**

VIII.511 *Myxine australis*, Jenyns. TYPE. Goree Sound, Tierra del Fuego. AG29.21/1 {1939} **515**

Catalogue of 'EXHIBITED SERIES'

IV.261 *Gerres oyena* (F.2349) Spirit. Keeling Island, Indian Ocean. [Spirit Stores] **1432**

II.289 *Sciaena adusta* (F.2601) {that no. crossed out} Spirit. Maldonado Bay, Rio Plata, S. America. Re-exhibited as F.2601 {See duplicated entry above} **695**

II.204 *Trigia kumu* (F.3093) Dried. New Zealand. {Drawers 23/15, 1980} **1341**

Bibliography

The list of references below includes contributions cited by CD (marked +), contributions annotated by CD, but not cited within the CD *quotes included here (marked #), and the (unmarked) references that I added. The last group includes all contributions in which the original descriptions are given of the species listed here as Darwin's Fishes. This was done to provide a cross section of ichthyology in CD's time, and thus contribute to defining the intellectual climate surrounding him. Though I have seen only about 85% of the references cited here, I have wherever I could expanded their authors' initials into given names. Also, I have expanded the titles and other particulars of most references, these often providing interesting details about the intentions associated with, or the circumstances leading to, their publication.

The references are followed by annotations, providing (a) occasional comment on the reference, and/or an English translation of its title; (b) CD's marginalia or other comments, if available (*in this font*); and (c) the alphabetic entry or entries (**in bold**) where the reference is cited (if it is). Asterisks (*) provide links to other entries, and thus another level of indexing.

Agassiz, Jean-Louis Rodolphe. 1829. *See* Spix and Agassiz (1829, 1831).

+Agassiz, Jean-Louis Rodolphe. 1833–44. *Recherches sur les poissons fossiles.* 5 volumes, with Atlas of 369 plates and supplement. Neuchâtel, 1420 pp.
  ["*Research on fossil fishes*"; **Agassiz; Ganoid(s); Megatooth shark; Scales**]

Agassiz, Jean-Louis Rodolphe. 1835. [Affinities and distribution of the fishes of the family Cyprinidae.] *Proceedings of the Zoological Society of London.*
  [As summarized by Richard *Owen; **Beards**]

Agassiz, Jean-Louis Rodolphe. 1844. On the classification of fishes. *The Edinburgh New Philosophical Journal* **37**: 132–43.
  [**Agassiz**]

Agassiz, Jean-Louis Rodolphe. 1846. On the Ichthyological Fossil Fauna of the Old Red Sandstone. *The Edinburgh New Philosophical Journal* **41**: 17–49.
  [Original published in 1844/45 as "Monographie des poissons fossiles du vieux grès rouge, ou système dévonien des Iles Britanniques."; **Threefold parallelism**]

Agassiz, Jean-Louis Rodolphe. 1849. On the differences between progressive, embryonic, and prophetic types in the succession of organized beings through the whole range of geological time.

Bibliography

Proceedings of the American Association for the Advancement of Science. First Meeting, held in Philadelphia, September 1848: 432–8.

[**Threefold parallelism**]

#Agassiz, Jean-Louis Rodolphe. 1850. *Lake Superior: its character, vegetation, and animals, compared with those of other similar regions.* Gould, Kendall & Lincoln, Boston.

[p. 33 *Gar-pike *Ganoid of F.W. in N. America; p. 34 Another ancient Fish in F.W.; 36 On <u>lowness</u>, because like Embryo; 252 On embryonic forms not deserving a separate class; 255 Ganoid &c in F. W.; 260 Reptilian characters of Ganoid, 'embodying prospective view of another class'; 285 Excellent case of Percopsis of Chalk, which combined characters, which soon diverged, intermediate between *Ctenoids and *Cycloids; I wonder whether this agrees with Müllers [1844] classification, as seen in Owen [1846] Lectures; if Fish properly classed whether so related to geologi. Formation; 289 Hardly one Family in which some species are not both Marine and & F.W.; 292 case of variability in a *Perch, good as for Agassiz; 317 *Esox boreus is made distinct by Agassiz; 327 Account of uniformity of *Salmonidae by uniformity of conditions; 352 Range of Cyprinoids p363; 374 Are F.W. Fish of N. America distinct; p.375 On F.W. Fish being <u>analogous</u> with those of Europe & Asia; 377 On *shoals created as shoals, *Marginalia* 11–12; **Ctenoid; Cycloid; Trout-perch**]

+Agassiz, Jean-Louis Rodolphe. 1851. Observations on the Blind Fish of the Mammoth cave. *American Journal of Science and Arts*, 2nd Series **11**: 127–8.

[**Cavefishes**]

+Agassiz, Jean-Louis Rodolphe. 1857a. On some young gar pikes from Lake Ontario. *American Journal of Science and Arts*, 2nd Series **23**: 284.

[**Gar-pike**]

#Agassiz, Jean-Louis Rodolphe. 1857b. *Contribution to the natural history of the United States of America.* Vol. 1. Part 1: Essay on Classification. Little, Brown and Co., Boston, Treubner and Co., London.

[p. 15 Amblyopsis very remote affinities; 30 isolated Fam. of Fishes . . . What? Abnormal? Amblyopsis is so; 37 admits that conditions do not explain distribution; 82 On classification of Fishes *Marginalia* 9, 10]

Agassiz, Jean-Louis Rodolphe. 1865. Lettre de M. Agassiz relative à la faune ichthyologique de l'Amazonie datée d'Ega, du 22 Septembre, 1865. *Annales des Sciences Naturelles. Zoologie et Biologie animale*, Sér. 5 **4**: 382–3.

[**Mouth-brooding**]

Agassiz, Jean-Louis Rodolphe. 1883. The Tortugas and the Florida Reefs. *Memoirs of the American Academy of Arts and Sciences* **11**: 107–34.

[Where Agassiz finds that CD's theory of coral reef formation does not apply to Florida reefs. He later extended this rejection to all coral reefs, thus ending up, here as well, on the losing side of a major scientific debate; **Agassiz; Cocos Islands**]

+Agassiz, Jean-Louis Rodolphe and Elizabeth Cary Agassiz. 1868. *A Journey in Brazil*, 4th edition. Ticknor and Fields, Boston, 540 pp.

[For details on the expedition of which this book is a popular account, see http://research.amnh.org/ichthyology/neoich/expeditions/Thayer; **Agassiz; Cichlidae (II)**]

#Agassiz, Jean-Louis Rodolphe and Augustus Addison Gould. 1848. *Principles of Zoology*. Part 1: Comparative Physiology. Gould, Kendall & Lincoln, Boston.

[p. 31 Blind cavern fish & Crabs; 165 Arctic Region not one bright bird or Fish with varied hue proof of action of external conditions; 179 Rivers of U. States some fish in common, some distinct. *Marginalia* 14]

+Alison, William Pulteney. 1847. Instinct. *In*: Robert B. Todd (ed.), *The Cyclopaedia of Anatomy and Physiology*. Vol. 3. London.

[**Altruism**]

Allen, Gerald R. 1985. *Snappers of the world. An annotated and illustrated catalogue of lutjanid species known to date*. FAO Species Catalogues. Rome, Vol. 6, vi + 208 pp.
[**Snappers**]

Allen, Gerald R. and William F. Smith-Vaniz. 1994. Fishes of the Cocos (Keeling) Islands. *Atoll Research Bulletin* (Smithsonian Institution, Washington, D.C.). No. 412, 21 pp.
[**Cocos Islands**]

Alvarez, Walter. 1998. *T. rex and the crater of doom*. Random House, New York. xii + 185 pp.
[A fascinating account of the research that provided the clinching evidence that the impact of an extraterrestrial object, 65 million years ago in the Yucatan Peninsula, did the dinosaurs in, along with many other organisms; **Extinction; Lyell**]

Alves-Gomes, J. A. 2001. The evolution of electroreception and bioelectrogenesis in teleost fish: a phylogenetic perspective. *Journal of Fish Biology* **58**(6): 1489–1511.
[**Transition**]

Andersson, Malte. 1994. *Sexual selection*. Princeton University Press, Princeton, 599 pp.
[**Sexual selection**]

Andrew, John C. and Patrick Gentien. 1982. Upwelling as a Source of Nutrients for the Great Barrier Reef Ecosystem: A Solution to Darwin's Question? *Marine Ecology Progress Series* **8**: 257–69.
[**Food webs**]

+Anon. 1843. [Review of Scrope 1843]. *Edinburgh Review* **79** (Art. iv): 87–113.
[Also reviews the work of Shaw 1840; **Salmon (III: Not pacific)**]

Anon. 1854. *Descriptive Catalogue of the Fossil Organic Remains of Reptilia and Pisces Contained in the Museum of the Royal College of Surgeons of England*. London.
[This museum, which received some of the *Beagle* samples, is also known as the 'Hunterian'; **Megatooth shark**]

+Anon. 1871. Spawning of the capelin. *American Naturalist* **5**: 119–20.
[**Capelin**]

Anon. 1986. *Sharks: silent hunters of the deep*. Reader's Digest, Sydney, 208 pp.
[**Hippopotamus**]

Aoki, Douglas Sadao. 2000. The thing never speaks for itself: Lacan and the pedagogical politics of clarity. *Harvard Educational Review* **70**(3): 347–69.
[A strange work, which the author insists is *not* a postmodernist critique of clarity, and an argument in favor of opaque writing, though the manner the case is made, and its articulation around the utterly inaccessible work of Jacques Lacan (who actually *bragged* about his prose being incomprehensible), strongly suggest the opposite. On the other hand, the argument that quotes are *always* taken out of context is convincing; **Quote; Seeing**]

Arancibia, Hugo, R. Alarcón, L. Caballero and R. Concha. 2000. Nuevas pesquerías para Chile Central. Anguila babosa (*Eptatretus polytrema*). Proyecto FONDEF D971–1058, Desarollo de nuevas pesquerías en recursos marinos bentónicos, pelágicos y demersales de Chile Central. Documento Técnico No 1, UNITEP, Departamento de Oceanografía, Universitad de Concepción, 16 pp.
["New fisheries for Central Chile: Fourteen-gill hagfish"; **Hagfishes**]

Aristotle. 1962. *Historia Animalium*. Translated by D'Arcy Wentworth Thompson. Vol. 4. *In*: J. A. Smith and W. D. Ross (eds.), *The Work of Aristotle*. The Clarendon Press, Oxford, 633 pp. + index.
[Quote is on p. 570; **Eel**]

Armstrong, Patrick H. 1985. *Charles Darwin in Western Australia: a young scientist's perception of an environment*. University of Western Australia Press, Nedlands, 80 pp.
[**King George's Sound**]

Bibliography

Armstrong, Patrick H. 1990. Darwin in Mauritius: natural history, landscape and society. *Indian Ocean Review* **3**(4): 11–13.
[**Mauritius**]

Armstrong, Patrick H. 1991a. Three weeks at the Cape of Good Hope 1836: Charles Darwin's African Interlude. *Indian Ocean Review* **4**(2): 8–13, 19.
[**Cape of Good Hope**]

Armstrong, Patrick H. 1991b. *Under the Blue Vault of Heaven: Charles Darwin's sojourn at the Cocos (Keeling) Islands*. Indian Ocean Centre for Peace Studies, Nedlands, ix + 120 pp.
[**Cocos Islands**]

Armstrong, Patrick H. 1992a. *Darwin's Desolate Islands: a Naturalist in the Falklands, 1833 and 1834*. Picton Publishing, Chippenham, xii + 147 pp.
[**Falkland Islands**]

Armstrong, Patrick H. 1992b. Charles Darwin's visit to the Bay of Islands, December 1835. *Auckland-Waikato Historical Journal* **60**: 10–24.
[**Bay of Islands**]

Armstrong, Patrick H. 1992c. Charles Darwin's last Island: Terceira, Azores, 1836. *Geowest* (University of Western Australia) no. 27, 64 pp.
[**Azores**]

Armstrong, Patrick H. 1992d. Human impacts on Australia's Indian Ocean Tropical Island ecosystems: a review. *The Environmentalist* **12**(3): 191–206.
[**Cocos Islands**]

Armstrong, Patrick H. 1993a. An ethologist aboard *HMS Beagle*: the young Darwin's observations on animal behavior. *Journal of the History of the Behavioral Sciences* **29**: 339–44.
[**Cormorants**]

Armstrong, Patrick H. 1993b. Darwin's Perception of the Bay of Islands, New Zealand, 1835. *New Zealand Geographer* **49**(1): 26–9.
[**Bay of Islands**]

Armstrong, Patrick H. 1994. Human impact on the Falkland Islands environment. *The Environmentalist* **14**(3): 215–31.
[**Falkland Islands**]

Armstrong, Patrick H. 2000. *The English parson-naturalist*. Gracewing, Leominster, Herefordshire, 208 pp.
[Presents numerous characters mentioned in this book, and shows how several nineteenth-century thinkers argued that the Biblical creation myth, properly translated, is compatible with the evolutionary sequence; **Creationism; Jenyns; FishBase**]

Arreguin-Sánchez, Francisco. 1996. Catchability: a key parameter for fish stock assessment. *Reviews in Fish Biology and Fisheries* **6**(2): 221–42.
[**Sex ratio**]

Arrhenius, Svante. 1908. *Worlds in the Making: the Evolution of the Universe*. Harpers & Brothers Publishers, New York and London, 230 pp.
[Translated by H. Born; **Spontaneous generation**]

Artedi, Petrus (Arctaedius). 1738. *Ichthyologia sive opera omnia piscibus scilicet: Bibliotheca ichthyologica. Philosophia ichthyologica. Genera piscium. Synonymia specierum. Descriptiones specierum. Omnia in hoc genere perfectiora, quam antea ulla*. Posthuma vindicavit, recognovit, coaptavit & edidit Carolus Linnaeus. Leiden, Conrad Wishoff.

[Reprint 1962 by J. Cramer, Weinheim, 554 pp. "*Ichthyology, or Complete Work on Fish, namely: Biography. Philosophy of Ichthyology. Genera of Fishes. Synonymy of the Species. Descriptions of the Species. All more complete than previous work in the field.* The final work posthumously taken over, reviewed, edited and published by Carl Linnaeus"; **Linnaeus**]

Arthur, Wallace. 2002. The emerging conceptual framework of evolutionary developmental biology. *Nature* **414** (6873): 757–64.

[**Ontogeny**]

Ashworth, J. H. 1935. Charles Darwin as a student in Edinburgh, 1825–1827. *Proceedings of the Royal Society of Edinburgh* **55**(II): 97–113.

[**Edinburgh; Grant**]

Auerbach, Paul S. 1991. *A medical guide to hazardous marine life*, 2nd edition. Mosby Year Book, St Louis, 61 pp.

[**Barracuda; Ciguatera; Parrotfishes**]

Bacon, Francis. 1620. *Novum Organum*. London.

[The version consulted here was translated and edited – with other parts of *The Great Instauration* – by Peter Urbach and John Gibson, 1994. Chicago, Open Court, 334 pp. Aphorism 95 in 'Book I' shows that Bacon did not suggest, as many believe, that science, the 'new tool' of his title, should consist only of accumulating facts (i.e. 'things'), and ignore theories ('webs'): "[t]hose who have handled the science have been either Empiricists or Rationalists. Empiricists, like ants collect things and use them. The Rationalists, like spider webs spin out webs by themselves. The middle way is that of the bee, which gathers its material from the flowers in the garden and field, but then transforms and digests it by a power of its own." **Autobiography; Lizards; Popper**]

#Baer, Karl Ernst von. 1828–37. *Untersuchungen über die Entwickelungsgeschichte der Fische; nebst einem Anhang über die Schwimmblase*. Leipzig.

[The contents of this contribution, '*Studies on the ontogeny of fishes, with an appendix on the swim-bladder*', were summarized by T. Huxley in a letter to CD, which he annotated (*Correspondence*, Sept. 17, 1858); however, CD does not cite von Baer, nor C. Vogt, also mentioned in the letter, when discussing this organ in **Origin**]

+Baer, Karl Ernst von. 1828 & 1837. *Über Entwickelungsgeschichte der Thiere: Beobachtung und Reflexion*. Königsberg. Erster Theil, 271 pp. Zweiter Theil, 315 pp.

["*On the ontogeny of animals: observations and reflections.* In two parts." CD probably read a partial translation, cited as follows on the von Baer web page: "*Fragments relating to Philosophical Zoology. Selected from the Works of K. E. von Baer. – Scientific memoirs, natural history*. Ed. by A. Henfrey and Thomas H. Huxley. London, 1853, Vol. 7, pp. 176–238"; **Progress**]

Bagnis, Raymond, E. Loussan and S. Thevenin. 1974. Les intoxications par poissons perroquet aux Iles Gambier. *Médecine Tropicale* **34**(4): 523–7.

["Poisonings through parrotfishes in the Gambier Islands"; **Ciguatera; Parrotfishes**]

Bakun, Andrew. 1996. *Patterns in the oceans: ocean processes and marine population dynamics*. California Sea Grant, La Jolla, and Centro de Investigaciones Biologicas del Noroeste, La Paz, xv + 323 pp.

[**Punctuated equilibrium**]

Balon, Eugene K. 1995a. Origin and domestication of the wild carp, *Cyprinus carpio*: from Roman gourmets to the swimming flowers. *Aquaculture* **129**: 3–48.

[**Carp; Koi**]

Balon, Eugene K. 1995b. The wild carp, *Cyprinus carpio*: its wild origin, domestication, and selection as colored nishikigoi. *Guelph Ichthyology Reviews* no. 3: 1–55.
[**Carp**; **Koi**]

Balon, Eugene K., Michael N. Bruton and Hans Fricke. 1988. A fiftieth anniversary reflection on the living coelacanth, *Latimeria chalumnae*: some new interpretations of its natural history and conservation status. *Environmental Biology of Fishes* **23**(4): 241–80.
[**Bats**; **Extinction**]

Banack, Sandra Anne. 1998. Diet selection and resource use by flying foxes (genus *Pteropus*). *Ecology* **76**(6): 1949–67.
[**Bats**]

Bancroft 1834: *see* Griffith and Smith (1834).

Banse, Karl. 1990. Mermaids – their biology, culture, and demise. *Limnology and Oceanography* **35**(1): 148–53.
[Note that real mermaids are fish, not *Sirenians; **Mermaids**]

Barkow, Jerome K., Leda Cosmides and John Tooby (eds.) 1992. *The adapted mind: evolutionary psychology and the generation of culture*. Oxford University Press, Oxford, xii + 666 pp.
[**Social Darwinism**]

Barlow, George W. 2000. *The Cichlid Fishes: Nature's grand experiment in evolution*. Perseus Publishing, Cambridge. xvi + 335 pp.
[**Beauty**; **Cichlidae (I, II)**; **Cuckoo-fish**; *Geophagus*; **Hermaphroditism**; **Mouth-brooding**; **Obstinate nature**; **Sex ratio**; **Speciation**; **Surgeonfishes**]

Barlow, Nora (ed.) 1933. *Charles Darwin's Diary of the Voyage of* H.M.S. Beagle, *edited from the MS*. Cambridge University Press, Cambridge, xxx + 451 pp.
[*Diary*]

Barlow, Nora (ed.) 1958. *The Autobiography of Charles Darwin 1809–1882, with original omissions restored*. Edited with Appendix and Notes by his grand-daughter. Harcourt, Brace and Company, New York, 253 pp.
[*Autobiography*]

Barlow, Nora (ed.) 1963. Darwin's Ornithological Notes. *Bulletin of the British Museum (Natural History)*, Historical Series 2: 203–78.
[**Birds**; *Ornithological Notes*]

Barlow, Nora (ed.) 1967. *Darwin and Henslow: the growth of an idea. Letters 1831–1860*, edited by Nora Barlow for the Bentham-Moxon Trust. University of California Press, Berkeley and Los Angeles, xii + 251 pp.
[*Correspondence*; **Henslow**]

Barrett, Paul H. (ed.) 1977. *The collected papers of Charles Darwin*. University of Chicago Press, Chicago, Vol. 1, xviii + 277 pp.; Vol. 2, viii + 326 pp.
[*Collected papers*; **Words**]

Barrett, Paul H. and Howard E. Gruber (eds.) 1980. *Metaphysics, Materialism, & the Evolution of Mind: early writings of Charles Darwin*. Transcribed and annotated by P. Barrett, with a commentary by H. E. Gruber. The University of Chicago Press, Chicago, xxiv + 228 pp.
[*Notebooks*]

Barrett, Paul H., and Richard Broke Freeman (eds.) 1989. *The Works of Charles Darwin*. Pickering & Chatto, London, 29 volumes.

[*Fish*; **Jenyns (I)**; **Words**; *Works of Charles Darwin*]

Barrett, Paul H., Peter J. Gautrey, Sandra Herbert, David Kohn and Sydney Smith (eds.) 1987. *Charles Darwin's Notebooks, 1836–1844: geology, transmutation of species, metaphysical enquiries*. British Museum (Natural History), London, and Cornell University Press, Ithaca and New York, viii + 747 pp.

[**Notebooks**]

Barrett, Paul H., Donald Jerome Weinshank and Timothy T. Gottleber (eds.) 1981. *A concordance to Darwin's* Origin of Species, *first edition*. Cornell University Press, Ithaca, xv + 834 pp.

[**Concordance**]

Barrett, Paul H., Donald J. Weinshank, Paul Ruhlen, Stephan J. Ozminski and Barbara Newell-Berghage. 1986. *A concordance to Darwin's* The expression of the emotions in man and animals. Cornell University Press, Ithaca, ix + 515 pp.

[Refers to the first edition (1872); **Concordance**]

Barrett, Paul H., Donald J. Weinshank, Paul Ruhlen and Stephan J. Ozminski. 1987. *A concordance to Darwin's* The Descent of man, and selection in relation to sex. Ithaca, Cornell University Press, x + 1137 pp.

[**Concordance**]

Barrie, Annemarie. 1992. *The Professional's Book of Koi*. Tropical Fish Hobbyist Publications, Neptune City, NJ, 160 pp.

[**Koi**]

Basalla, George. 1963. The voyage of the Beagle without Darwin. *Mariner's Mirror* **49**: 42–8.

[**Beagle**]

Bauchot, Marie-Louise, Jacques Daget, and Roland Bauchot. 1990. L'ichtyologie en France au début du XIXe siècle: l'*Histoire naturelle des poissons* de Cuvier et Valenciennes. *Bulletin du Muséum National d'Histoire Naturelle*. Sect. A: Zool. 4e Série, Vol. 12 (1) Supplément, 142 pp.

[English version "Ichthyology in France at the beginning of the 19th century: The *Histoire naturelle des poissons* of Cuvier (1769–1832) and Valenciennes (1794–1865)" is included as chapter 2 in Pietsch and Anderson (1997); **Gay**; **Hamilton-Smith**; *Histoire naturelle des poissons*; **Naturalists**]

Becker, Robert O. and Gary Selden. 1985. *The body electric: electromagnetism and the foundation of life*. William Morrow, New York, 364 pp.

[**Polydactylism**]

Beddall, Barbara G. 1988a. Darwin and Divergence: the Wallace connection. *Journal of the History of Biology* **21**(1): 1–68.

[**Wallace**]

Beddall, Barbara G. 1988b. Wallace's annotated copy of Darwin's *Origin of Species*. *Journal of the History of Biology* 21(2): 265–89.

[**Freshwater fishes (II)**; **Survival of the fittest**]

Behe, Michael, J. 1996. *Darwin's Black Box: the biochemical challenge to evolution*. Simon & Schuster, New York, 307 pp.

[A recent representative of this large class of pamphlets and books whose authors, after pointing out *their* inability to solve simple evolutionary problems (how to get from feature X to feature Y) loudly proclaim that it can't be done. In this case, the author, while magnanimously granting that evolution did occur and that natural selection works at the level of organism and above, denies that it applies to cellular and biochemical processes. This is a particularly sad case of NIMBY (Not In My Backyard) in view of the thousands of people (particularly children) killed every year by resurgent malaria epidemics, brought about by mosquitoes whose body cells, thanks to a well understood instance of natural selection, now favour newly evolved biochemical pathways, capable of neutralizing the insecticides that previously killed them; **Difficulties; Non-Darwinian evolution; Transition**]

Bell, Kim N. I. 1999. An overview of goby-fry fisheries. *Naga, The ICLARM Quarterly* **22**(4): 30–6.
[**Gobies**]

Bell, Thomas. 1843. Reptiles. *In*: *The Zoology of the voyage of H. M. S. Beagle*, *under the command of Captain FitzRoy, R. N., during the years 1832 to 1836*. Edited and Superintendented by Charles Darwin. Part V. Smith, Elder, and Co., London.
[**Zoology**]

Bennett, Edward Turner. 1828. On some fishes from the Sandwich Islands. *Zoological Journal* **4**: 31–43.
[**Blennies**]

Bennett, Edward Turner. 1832. Characters of some new species of fishes, collected by Mr. Cuming. *Proceedings of the Zoological Society, London* (Part 2): 4–5.
[**Pigfishes**]

Bennett, John Hughes. 1841. *Treatise on the oleum jecori aselli, or cod liver oil as a therapeutic agent in certain forms of gout, rheumatism and scrofula; with cases*. London.
[**Cod liver oil**]

Berg, Leo Simonevich. 1958. *System der rezenten und fossilen Fischartigen und Fische*. VEB Verlag der Wissenschaften, Berlin, 310 pp.
["Classification of recent Fish-likes and Fishes"; German translation of the second Russian edition of a contribution whose first edition (1944) was translated into English and widely used and cited; **Cichlidae**]

Berglund, Anders and Gunilla Rosenqvist. 1993. Selective males and ardent females in pipefishes. *Behavioral Ecology and Sociobiology* **32**: 331–6.
[**Pipefishes (I)**]

Berlin, Brent. 1992. *Ethnobiological classifications: principles of categorization of plants and animals in traditional societies*. Princeton University Press, Princeton, 335 pp.
[**Names, common**]

Beverton, Raymond J. H. and Sidney J. Holt. 1957. *On the Dynamics of Exploited Fish Populations*. Ministry of Agriculture, Fisheries and Food, Fishery Investigations, Series II, Volume XIX, 533 pp.
[Reprint by Chapman & Hall, London, 1993, with a foreword by Daniel Pauly; **Gear selectivity**]

#Bianconi, Giovanni Giuseppe. 1875. La Théorie darwinienne et la création dite indépendante: lettre à M. Ch. Darwin. Nicola Zanichelli, Bologna.
["Darwinian theory and so-called independent creation. Letter to Mr CD"; translated from Italian by G. A. Bianconi; nothing of importance; p. 186 why three bones & not in fin of fish or water Beetle; *Marginalia* 58]

Bigelow, H. B. and W. C. Schroeder. 1948. Fishes of the Western North Atlantic. Sharks. *Memoir Sears Foundation on Marine Research* no. 1, Part One: Lancelets, Cyclostomes, Sharks, pp. 59–546. New Haven.
[**Great white shark**]

Bird, Eric C. F. 1994. Physical setting and geomorphology of coastal lagoons. *In*: Björn Kjerfve (ed.) *Coastal Lagoon Processes*, pp. 9–39. Elsevier Science Publishers, Amsterdam.
[**Lagoon**]

Blanc, M., J. L. Gaudet, P. Banarescu and Jean-Claude Hureau. 1971. *European inland water fish: a multilingual catalogue*. Fishing News (Books) Ltd, London.
[**Fossils**]

Bleeker, Pieter. 1855. Over eenige visschen van Van Diemensland. *Versl. Akad. Amsterdam* **2**: 1–30 + 31.

["On a few fishes from Van Diemensland" (= Tasmania), Australia; **Flatfishes**]

+Bloch, Marcus Eliezer. 1782–84. *Oekonomische Naturgeschichte der Fische Deutschlands, vorzüglich derer in den Preußischen Staaten*. Berlin. In three parts, and an Atlas.

["*Economic Natural History of the Fishes of Germany, especially those in the Prussian States*." All three parts reprinted in facsimile as volume 1 of '*Bloch's Atlas*,' 1999, Mergus Verlag, Melle, Germany, whose volumes 2 and 3 include Bloch (1785–95). These three volumes thus correspond to what Bloch called his "*Allgemeine Naturgeschichte der Fische*", i.e. a general natural history of fishes; **Carp, Prussian; Sex ratio**]

Bloch, Marcus Eliezer. 1788. Ueber zwey merkwürdige Fischarten. *Abhandlungen der Böhmischen Gesellschaft der Wissenschaften* **3**: 278–82.

["On two noteworthy fish species"; **Deep-sea spiny eels**]

Bloch, Marcus Eliezer. 1785–95. Naturgeschichte der ausländischen Fische. Berlin.

["Natural History of Foreign Fishes"; **Burrfishes (I); Climbing fish;** *Gastrobranchus*; **Gurnards; Jacks; Pipefishes (II)**]

Bloch, Marcus Eliezer and Joseph Gottlob Schneider. 1801. *M. E. Blochii, Systema Ichthyologiae iconibus CX illustratum*. Post obitum auctoris opus inchoatum absolvit, correxit, interpolavit Jo. Gottlob Schneider, Saxo. Berolini Sumptibus Auctoris Impressum et Bibliopolio Sanderiano Commissum. p. i–lx + 1–584.

["*M. E. Bloch's System of Ichthyology, illustrated by 110 figures*. An unfinished work, completed, corrected and improved after the author's death by Johann Gottlob Schneider, of Saxony. Printed in Berlin at the author's expense and bound by the Sander Book Company." This volume includes fish descriptions now credited to Johann Reinhold Forster (1729–98), a German naturalist who accompanied Capt. James Cook on his second voyage around the world. **Bacalao; Blennies; Boxfishes; Jacks; Livebearers; Puffers; Surgeonfishes**]

Blunt, Wilfrid. 1971. *The Compleat Naturalist: a Life of Linnaeus*. The Viking Press, New York, 256 pp.

[**Beards; Linnaeus**]

+Blyth, Edward. 1845. Notices of various Mammalia, with Descriptions of Many New Species. *Annals and Magazine of Natural History* **15**: 449–75.

[**Bats**]

Blythe, Neil X. 1996. Fish oil as an alternative fuel for internal combustion engines. *ICE/ASME Spring Technical Conference* **26**(3): 85–92.

[**Capelin**]

Bocek, Alex. 1982. Rice terraces and fish: integrated farming in the Philippines. *ICLARM Newsletter* **5**(3): 24.

[**Loaches**]

Boisseau, Jean-Paul and Françoise Le Menn. 1967. Origine et modification des protéines du liquide marsupial d'*Hippocampus guttulatus* Cuv.: étude électrophorétique. *Comptes rendus de l'Académie des Sciences de Paris* **265** (Sér. D): 2036–9.

[No translation of title required, as it nicely illustrates how scientific French and English converge; **Seahorses (II)**]

Bonnaterre, Joseph Pierre. 1788. *Tableau encyclopédique et méthodique des trois règnes de la nature . . . Ichthyologie*. Paris. i–lvi + 1–215 pp.

["*Encyclopedic and systematic review of the three kingdoms of nature . . . Ichthyology*"; **Albicore; Gurnards; Wrasses**]

+Bory de Saint Vincent, Jean Baptiste Marie Georges (ed.) 1822–31. *Dictionnaire Classique d'Histoire Naturelle* Rey & Gravier, Paris, 17 volumes.

[A set of this up-to-date encyclopedia travelled with CD on the *Beagle*; Vol. 5, p. 274 on Cyprinus; p. 277 on varieties of Gold Fish; *Marginalia* 196; Vol. 9, pp. 150–3 on Labrus; **Goldfish (I); Wrasses**]

Boulenger, George Albert. 1908. Diagnoses of new fishes discovered by Capt. E. L. Rhoades in Lake Nyassa. *Annals and Magazine of Natural History* (Ser. 8). **2** (9): 238–3.

[**Parasites**]

Boustany, Andre M., Scott F. Davis, Peter Pyle, Scot D. Anderson, Burney J. LeBoeuf and Barbara A. Block. 2002. Expanded niche for white shark. *Science* **415** (6867): 35–6.

[**Hippopotamus**]

Briggs, John C. 1955. A monograph of the clingfishes (Order Xenopterygii). *Stanford Ichthyological Bulletin* **6**: 1–244.

[**Clingfishes**]

+Briggs, John Joseph. 1862. Proved facts in the natural history of the salmon. *The Field, The Country Gentleman's Newspaper* **20**: 412.

[On regrowth of salmon fins; **Polydactylism**]

Bright, T. J. 1972. Bio-acoustic studies on reef organisms. *In*: Bruce B. Collette and Sylvia A. Earle (eds.) Results of the Tektite Program: ecology of the coral reef fishes. *Science Bulletin* No. 14, pp. 45–70. Natural History Museum of Los Angeles County, Los Angeles.

[Distributed with one 45 r.p.m. vinyl record, reproduced in *FishBase; **Sounds**]

+Bronn, Heinrich Georg. 1843. *Handbuch der Geschichte der Natur*. Zweiter Band. III Theil: Organisches Leben. Ergebnisse hauptsächlich aus der lebenden Welt über Entwicklung, Verbreitung und Untergang der früheren Bevölkerungen der Erde. Stuttgart.

["Handbook of natural history", Vol. 2. Part III. Organic life. Results mainly from the living world on development, distribution and extinction of the earlier populations of the Earth"; 1841 edition: p. 56 on mixing of Salt & FW Fish in Baltic; Cyprinus in F. & salt water; 58 Caspian Fauna genera of salt & fresh fish & Crust & Shell; 107 most important case of variation in fish; the intermediate form between these two supposed species [*Cyprinus carassius* and *C. gibelio*; see **Carp, Prussian**], found in a ditch where one species has been turned in. Yarrell. Vol. I alludes to these two fishes and gives summary of their differences; 117 It is important to consider whether the males . . . can propagate the sportive tendency, because if so it will show, that the varying tendency in the generative system, under domestication, is the effect of impregnation & not the womb influence. In fact if fish & silkworms vary much, it cannot be foetal influence; 132 case of carp (which breeds true) with 4 times larger scales in lines, with some places bare – call Looking-Glass Carp; 173 Fish; 235/6 Fish & Crab Rain; 237 Fish Rain; 241 Fish eggs perhaps stick to birds; 278 no fish eggs keep more than 2 months dry. account of a disconnected pool annually dry & annually repeopled with Fish; 284 number of seeds – kind of animals which have most – number of eggs in Crab – Fishes; 286 One is always astonished at geometric increases, *Marginalia* 78, 81, 82, 84, 87, 89, 90; **Carp, Prussian**]

Brown, Andrew. 1999. *The Darwin Wars: the scientific battle for the soul of man*. London, Simon & Schuster, 241 pp.

[**Social Darwinism**]

Browne, Janet. 1995. *Charles Darwin: Voyaging*. Volume I of a Biography. Alfred A. Knopf, New York, 605 pp.

[**Angling; Darwin, Charles Robert; Experiments (IV); Jenyns (I); Loaches**]

Browne, Janet. 2002. *Charles Darwin: The Power of Place*. Volume II of a Biography. Alfred A. Knopf, New York, xi + 591 pp.

[**Beards; Darwin, Charles Robert; Correspondence; Experiments (IV); Flatfish controversy (I); Food-fishes; Günther; Swimbladder**]

Brünnich, Martin-Thomas. 1788. Om de islandske fisk, bogmeren, *Gymnogaster arcticus*. *Det Kongelige Danske Videnskabernes Selskab – Naturvidenskabelige og mathematiske Afhandlinger* **3**: 408–13.

["On an Icelandic fish, the Ribbonfish *Gymnogaster arcticus*"; **Deal fish**]

+Brullé, Auguste. 1844. Recherches sur les transformations des appendices chez les articulés. *Annales des Sciences Naturelles. Série Zoologie* **2**: 271–374.

["Research on the limbs of the articulated animals"; **Brains; Ontogeny**]

+Buckland, William. 1836. *Bridgewater Treatise No. 6: Geology and mineralogy considered with reference to natural theology.* Pickering, London, 2 volumes.

[One of a series of books (see also Kirby 1835) by leading scientists and moralists, funded from the estate of the late Earl of Bridgewater, administered by the Royal Society, and devoted to illustrating "the power, wisdom, and goodness of God as manifested in the Creation"; CD's abstract in DAR **71**: 125–7; **Extinctions**]

+Buckland, Francis Trevelyan. 1868. [Columns on marine and freshwater fishes and fisheries.] *Land and Water*, London.

[The son of William, 'Frank' Buckland contributed a weekly column to *Land and Water*, from its first issue of Jan. 27, 1867; this column is cited by CD: p. 41 on Trout culture; p. 62 on Cod fertility; p. 377 on *Chimaera monstrosa*; **Chimaera; Cod; Sex ratio**]

Budker, Paul. 1971. *The life of sharks*. Columbia University Press, New York, xvii + 222 pp.

[**Hippopotamus**]

+Buist, Robert. 1866. *The Stormontfield Piscicultural Experiment. 1853–1866.* Edmonston and Douglas, Edinburgh, 32 pp.

[**Sex ratio**]

+Buist, Robert. 1867. The Stormontfield Experiments. *The Field, The Country Gentleman's Newspaper*, June 29: 491–2.

[**Salmon (IV); Sex ratio**]

Burgess, G. H. O. 1967. *The curious world of Frank Buckland*. John Baker, London, xii + 242 pp.

[This book discusses, among other things, the strong opposition of Frank Buckland – and his father William – to CD's idea, owing to their belief in a recent, created world; **Chimaera**]

Burgner, Robert, L. 1991. Life history of Sockeye salmon (*Oncorhynchus nerka*). *In*: Cornelis Groot and Leo Margolis (eds.), *Pacific Salmon Life Histories*, pp. 3–117. University of British Columbia Press, Vancouver, 564 pp.

[**Salmon (III)**]

Burkenroad, Martin D. 1943. A possible function for bioluminescence. *Sears Foundation Journal of Marine Science* **5**(2): 161–4.

[**Bioluminescence**]

Burkhardt, Frederick, Sydney Smith, David Kohn and William Montgomery (eds.) 1985a. *A Calendar of the Correspondence of Charles Darwin, 1821–1882.* Garland Publishing, New York, 690 pp.

[A second edition, of 1994, updates this; ***Calendar; Correspondence; Words***]

Burkhardt, Frederick, Sydney Smith, David Kohn and William Montgomery (eds.) 1985b. *The Correspondence of Charles Darwin*. Vol. 1, 1821–1836. Cambridge University Press, Cambridge, xxix + 702 pp.

[***Correspondence; Zoology Notes***]

Bibliography

Burkhardt, Frederick, Sydney Smith, Janet Browne, David Kohn and William Montgomery (eds.). 1986. The *Correspondence of Charles Darwin*. Vol. 2, 1837–1843. Cambridge University Press, Cambridge, xxxiii + 603 pp.
[*Correspondence*]

Burkhardt, Frederick, Sydney Smith, Janet Browne, David Kohn and William Montgomery (eds.) 1987. *The Correspondence of Charles Darwin*. Vol. 3, 1844–1846. Cambridge University Press, Cambridge, xxix + 523 pp.
[*Correspondence*]

Burkhardt, Frederick and Sydney Smith (eds.) 1988. *The Correspondence of Charles Darwin*. Vol. 4, 1847–1850. Cambridge University Press, Cambridge, xxxiii + 711 pp.
[*Correspondence*]

Burkhardt, Frederick and Sydney Smith (eds.) 1989. *The Correspondence of Charles Darwin*. Vol. 5, 1851–1855. Cambridge University Press, Cambridge, 705 pp.
[*Correspondence*]

Burkhardt, Frederick and Sydney Smith (eds.) 1990. *The Correspondence of Charles Darwin*. Vol. 6, 1856–1857. Cambridge University Press, Cambridge, xxix + 673 pp.
[*Correspondence*]

Burkhardt, Frederick, Sydney Smith, Janet Browne and Marsha Richmond (eds.) 1991. *The Correspondence of Charles Darwin*. Vol. 7, 1858–1859 + Supplement to the Correspondence 1821–1857. Cambridge University Press, Cambridge, xxv + 671 pp.
[*Correspondence*; **Lungfish**]

Burkhardt, Frederick, Duncan M. Porter, Janet Browne and Marsha Richmond (eds.) 1993. *The Correspondence of Charles Darwin*. Vol. 8, 1860. Cambridge University Press, Cambridge, xxxvii + 766 pp.
[*Correspondence*; **Lancelet**]

Burkhardt, Frederick, Duncan M. Porter, Joy Harvey and Marsha Richmond (eds.) 1994. *The Correspondence of Charles Darwin*. Vol. 9, 1861. Cambridge University Press, Cambridge, xxxiii + 609 pp.
[*Correspondence*]

Burkhardt, Frederick, Duncan M. Porter, Joy Harvey and Jonathan R. Topham (eds.) 1997. *The Correspondence of Charles Darwin*. Vol. 10, 1862. Cambridge University Press, Cambridge, xxxviii + 909 pp.
[*Correspondence*]

Burkhardt, Frederick, Duncan M. Porter, Sheila Ann Dean, Jonathan R. Topham and Sarah Wilmot (eds.) 1999. *The Correspondence of Charles Darwin*. Vol. 11, 1863. Cambridge University Press, Cambridge, xxxiv + 1038 pp.
[*Correspondence*; **Polydactylism**]

Burkhardt, Frederick, Duncan M. Porter, Sheila Ann Dean, Paul S. White, Sarah Wilmot, Samantha Evans and Alison M. Pearn (eds.) 2001. *The Correspondence of Charles Darwin*. Vol. 12, 1864. Cambridge University Press, Cambridge, xxxvii + 694 pp.
[*Correspondence*]

#Burmeister, Herman. 1834. *Beiträge zur Naturgeschichte der Rankenfüsser*. G. Rainer, Berlin.
["*Contributions to the natural history of the cirripedes*;" p. 29 fish-bones, *Marginalia* 103]

Burstyn, Harold L. 1975. If Darwin wasn't the *Beagle*'s naturalist, why was he on board? *The British Journal for the History of Science* **8**(28): 62–9.

[*Beagle*; **Naturalists**]

Cadenat, Jean. 1943. Les Scorpaenidae de l'Atlantique et de la Méditerranée. Première note. Le genre *Scorpaena*. *Revue des Travaux de l'Institut des Pêches Maritimes* **13**(1–4): 525–63.

[Actual publication date may have been 1945; **Scorpionfishes**]

Cain, Arthur J. 1957. Logic and memory in Linnaeus's system of taxonomy. *Proceedings of the Linnaean Society of London* **169**: 144–63.

[**Linnaeus**]

Caldecott, Randolph. 1878. *The House that Jack built*. With 33 Illustrations. Routledge Shilling Toy Books, John Routledge and Sons, London.

[**Dace**]

Camerini, Jane R. 1994. Evolution, Biogeography and Maps: an early history of Wallace's Line. *In:* Rod MacLeod and Philip F. Rehbock (eds.), *Darwin's Laboratory: evolutionary theory and natural history in the Pacific*, pp. 70–109. University of Hawai'i Press, Honolulu.

[**Wallace**]

Campbell, John. 1997. *In Darwin's Wake: Revisiting Beagle's South American Anchorages*. Sheridan House, Bobbs Ferry, New York, 271 pp.

[*Beagle*; **Tierra del Fuego**]

+Candolle, Alphonse de. 1855. *Géographie botanique raisonnée*. 2 volumes. J. Kessman, Paris.

["Critical review of botanical geography"; Vol. 1, p. 599, on range of Marsh plants [. . . :] speculations, which I shall introduce on Fish, bears on this; change of River courses: most lakes connected with streams; Vol. 2: Isolation most important, as preventing migration & so altering conditions, & making gaps in economy of nature, & quite secondarily causing organisms to vary. . . . Possibly isolation not long enough for many cases, as in Alps & F.W. Fish. – Few individuals for isolation, & this gives bad chance for new forms, but time wd make up for that; 447 Struggle between Fish & Water Plants; 635 If herons eat fish with seed, such means wd have been more energetic formerly, when country wild; 1165 I suspect lower Fams. more broken?? good if I could show as it could be due to increase of number in higher Fams. – No. Higher Reptiles. Higher Mollusc. Higher or more Reptilian Fish most broken: if contest within each Family it would be so; 1078–1079 Edin. New Phil. Journ. 61/70 Changed habits: fish salt water. Zoologist p. 20 d[itt]o; *Marginalia* 120–2, 126, 133, 134; **Experiments (V)**]

Canestrini, Giovanni. 1866. Prospetto critico dei pesci d'acqua dolce d'Italia. *Archivio per la zoologia, l'anatomia e la fisiologia* **4**: 47–187.

["Critical outline of the freshwater fishes of Italy"; **Classification**]

Cantor, Theodore Edward. 1842. General features of Chusan, with remarks on the flora and fauna of that island. *Annals and Magazine of Natural History* (N.S.) **9**: 481–93.

[**Loach**]

Cantor, Theodore Edward. 1849. Catalogue of Malayan fishes. *Journal of the Asiatic Society of Bengal* **18**(2), pp. i–xii + 981–1443.

[**Gouramy**]

+Carbonnier, Pierre. 1867. Observations sur le brochet. *Bulletin de la Societé Impériale Zoologique d'Acclimatation*. 2ième Série **4**: 574–7.

["Observations on the Pike"; as quoted in *The Farmer*, March 18, 1869, p. 369; **Sex ratio; Sexual dimorphism**]

Bibliography

+Carbonnier, Pierre. 1869a. *See* Carbonnier (1867).

+Carbonnier, Pierre. 1869b. Rapport et observations sur l'accouplement d'une espèce de poisson de Chine. *Bulletin de la Societé Impériale Zoologique d' Acclimatation*. 2ième Série. **6**: 408–14

["Report and observations on the coupling of a species of fish from China"; **Gouramy**]

+Carbonnier, Pierre. 1870. Nouvelle note sur un poisson de Chine appartenant au genre Macropode. *Bulletin de la Societé Impériale Zoologique d' Acclimatation*. 2ième Série **7**: 26–32.

["New note on a Chinese fish of the genus *Macropodus*"; **Gouramy**]

+Carpenter, William B. 1854. *Principles of Comparative Physiology*, 4th revised edition. John Churchill, London.

[p. 79 High Fish. N.B. I think on this subject there is much difference whether we look to Fish alone or to other classes??; 101 Electric organs; 257 On 3 kinds of lungs in Fishes; 258 The Foundation for another kind of Lung; 261 Fish? Lepidosiren; 272, 277 Branchial vessels in loops in young chicks like those of Fish or tadpoles; 320 Gradation in Respiratory organs: . . . Branchiae: Re-use of swimming bladder; 325 . . . So ranks as Reptile & not Fish; 465 The 'proper' electric current of frog has curious analogy with electric discharge of Fish; 467 Electrical Fishes; *Marginalia* 155–7; **Electric fishes**]

Carvalho-Filho, Alfredo. 1994. *Peixes: Costa Brasiliera*. Editoria Marca d'Agua, São Paulo, 304 pp.

["*Fishes of the Brazilian coast*"; **Duckbills; Groupers; Sole, lined**]

Castelnau, François Louis Nompar de Coumont de Laporte de. 1873. *Contribution to the ichthyology of Australia*. Nos. III–IX [with subtitles, indexed as one work]. *Proceedings of the Zoological Acclimatisation Society of Victoria* **2**: 37–158.

[Description of *Tetrodon darwinii* is on pp. 94–5 of No. V (Notes on Fishes from Northern Australia), issued separately in Melbourne, 1842; **Eponyms**]

Casti, John L. 1989. *Paradigm Lost: Images of Man in the Mirror of Science*. William Morrow & Company, New York, 565 pp.

[**Natural selection**]

Castillo, Sady and Jaime Mendo. 1987. Estimation of unregistered Peruvian anchoveta (*Engraulis ringens*) in official catch statistics, 1951–1982. *In*: D. Pauly and Isabel Tsukayama (eds.), *The Peruvian anchoveta and its upwelling ecosystem: three decades of change*. ICLARM Studies and Reviews **15**, pp. 109–16.

[**Peruvian anchoveta**]

Cavolini, Filippo. 1787. *Memoria sulla generazione dei Pesci e dei Granchi*. Napoli.

["Memoir on the (re)production of fishes and crabs"; **Groupers (II)**]

Cervigón, Fernando and Walter Fischer. 1974. INFOPESCA. Catálogo de especies marinas de interes economico actual o potencial para América Latina. Parte 1: Atlántico centro y suroccidental. FAO/UNDP, SIC/79/1. FAO, Rome, 372 pp.

["Catalog of species of commercial or potential interest for Latin America. Part 1: Central and Southwestern Atlantic"; **Silversides**]

Chaisson, Eric J. 2001. *Cosmic evolution: the rise of complexity in nature*. Harvard University Press, Cambridge, xii + 274 pp.

[**Complexity**]

#[Chambers, Robert]. 1994. *Vestiges of the Natural History of Creation and Other Evolutionary Writings*. Chicago University Press, Chicago, 253 pp.

[Reprint of anonymous 1844 edition, of which CD's copy includes the annotations: The idea of a Fish passing into a Reptile (his idea monstrous). – . . . Never use the words higher and lower – use more complicated, as the fish type (& not a mere repetition of parts) where cartilaginous forms are higher for being nearer reptiles & consequently mammalia; p. 235 Yarrells Birds <u>Gull</u> getting thickened stomach Vol. 3 p. 571 Quotes Pennant on Trout in Galway getting thickened stomach. Was it Trout? Yarrell Fishes vol. 2 p. 57 thinks Gillaso only a var. inner cuticle only undirected; 367 Remarks on islds not having mammals & less perfect life but really I need not allude to such rubbish *Marginalia* 164]

Chancellor, Gordon, Angelo DiMauro, Ray Ingle and Gillian King. 1988. Charles Darwin's Beagle Collections in the Oxford University Museum. *Archives of Natural History* **15**(2): 197–231.

[**Oxford University Museum; Parasites**]

Chervova, L. S. 1997. Pain sensitivity and behaviour of fishes. *Journal of Ichthyology* **37**(1): 98–102.

[Demonstrates "analgesic efficiency" in Cod and Steelhead trout, and shows that "the pain sensitivity in fishes significantly decreases under the action of stressors"; **Flying fishes**]

Christensen, Villy. 1996. Managing fisheries involving predator and prey species. *Reviews in Fish Biology and Fisheries* **6**(4): 417–42.

[**Eggs of fish**]

Christensen, Villy and Daniel Pauly (eds.) 1993. *Trophic models of aquatic ecosystems*. ICLARM Conference Proceedings **26**. x + 390 pp.

[Includes quantitative description of food web models constructed with the Ecopath software (www.ecopath.org) and further documented in www.seaaroundus.org; **Food fishes; Food webs**]

Chungue, E., Raymond Bagnis, N. Fusetani and Y. Hasimoto. 1977. Isolations of two toxins from a parrotfish *Scarus gibbus*. *Toxicon* **15**: 89–93.

[**Ciguatera**]

#Clark, Henry James. 1865. *Mind in nature*. D. Appleton & Co, New York.

[p. 263 Lepidosiren, affinities, *Marginalia* 166]

Clément, G. and A. Berthoz. 1994. Influence of gravity in the processing of information by the central nervous system. *In*: *Proceedings of 5th European Symposium on Life Science Research in Space*, Arcachon, France, 1993. European Space Agency, ESA SP-366, pp. 333–8.

[**Motion sickness**]

Clover, Charles. 1991. Denmark: pulping sand eel for power. *Telegraph Magazine*, *London*, October 26, 1991, p. 8.

[**Capelin**]

Cohen, Daniel M., Tadashi Inada, Tomio Iwamoto and Nadia Scialabba. 1990. *Gadiform fishes of the world (Order Gadiformes)*. An Annotated and Illustrated Catalogue of Cod, Hakes, Grenadiers and other Gadiform Fishes Known to Date. FAO Species Catalogue. Vol. 10. FAO, Rome, 442 pp.

[**Burbot; Rattails**]

Coleman, Neville. 1980. *Australian Sea Fishes South of 30° S*. Doubleday, Lane Cove, New South Wales, 302 pp.

[**Burrfishes (I); Pipefishes (I)**]

#Colin, Gabriel Constant. 1854–6. *Traité de physiologie comparée des animaux domestiques*. J. B. Baillière, Paris.

["Treatise on the comparative physiology of domestic animals"; p. 617 my notions not half as off as life of parasite; bred in fish & matured in cormorant; Desfossé on hermaphrodite Fish. Serranus *Marginalia* 168; Desfossé is Dufossé 1854; *see* **Groupers (II)**]

Collett, Robert. 1889. Diagnoses des poissons nouveaux provenant des campagnes de "l'Hirondelle." IV. Description d'une espèce nouvelle du genre *Hoplostethus*. *Bulletin de la Société Zoologique de France* **14**: 306.

["Diagnosis of new fishes from the cruises of the 'Hirondelle:' Part IV. Description of a new species of the genus *Hoplostethus*"; original description of Orange roughy, from a specimen taken off the Azores; **Darwin's roughy**]

#Collingwood, Cuthbert. 1868. *Rambles of a naturalist on the shores and waters of the China Sea*. John Murray, London.

[p. 374 Flying fish habits *Marginalia* 169]

Collins, Timothy. 1996. Molecular comparisons of transisthmian species pairs: rates and patterns of evolution. *In*: Jeremy B. C. Jackson, Ann F. Budd and Anthony G. Coates (eds.), *Evolution and environment in tropical America*, pp. 303–34. The University of Chicago Press, Chicago.

[**Panama**]

Conner, H. M. 1813. *The old woman and her pig*. J. Harris & Sons, Corner of St. Paul's Churchyard, London, 29 pp.

[This edition, which cost "one shilling plain, eighteen pence (hand) coloured" might well have been the very one young CD had as a child. Herein, the cat decided to eat the mouse that wouldn't gnaw the rope that wouldn't hang the butcher that wouldn't kill the ox, and so on all the way to the stick that wouldn't beat the dog that wouldn't bite the pig that wouldn't go over the stile, thus starting off a domino effect that finally made it possible for the old woman and her pig to go home; **Dace**]

Cotte, M. J. 1944. *Poissons et animaux aquatiques au temps de Pline*. Paul Lechevalier, Paris, 265 pp.

["Fishes and aquatic animals in Plinius' time"; **Remora**]

+Couch, Jonathan. 1825. Some particulars of the natural history of fishes found in Cornwall. *Transactions of the Linnean Society of London* **14**: 69–92.

[**Cod; Ling**]

#Crawfurd, John. 1852. *A grammar and dictionary of the Malay language*. Vol. 1. Smith, Elder & Co., London.

[p. ccixii In Marianne group natives use Fish bones for arrow ∴ not deer, *Marginalia* 172]

Crisp, D. J. 1983. Extending Darwin's investigations on the barnacle life history. *Biological Journal of the Linnean Society* **20**: 73–83.

[**Barnacles**]

Cruz, G. A. 1987. Reproductive biology and feeding habits of Cuyamel, *Joturus pichardi* and Tepemechín, *Agonostomus monticola* (Pisces; Mugilidae) from Rio Plátano, Mosquitia, Honduras. *Bulletin of Marine Science* **40**(1): 63–72.

[**Experiments (IV)**]

Cury, Phillippe. 1994. Obstinate nature: an ecology of individuals. Thoughts on reproductive behavior and biodiversity. *Canadian Journal of Fisheries and Aquatic Sciences* **51**(7): 1664–73.

[**Homing; Obstinate nature**]

Cushing, David H. 1975. *Marine ecology and fisheries*. Cambridge University Press, Cambridge, xiv + 278 pp.

[**Herring**]

Cushing, David H. 1988. *The Provident Sea*. Cambridge University Press, Cambridge, ix + 329 pp.

[**Herring**]

Cuvier, Georges. 1816 and 1829. *Le Règne Animal distribué d'après son organisation pour servir de base à l'histoire naturelle des animaux et d'introduction à l'anatomie comparée*. Les reptiles, les poissons, les mollusques et les annélides. Vol. 2. Chez D'Eterville et chez Crochard, Libraires, Paris, xviii + 532 pp.

["*The animal kingdom, arranged in conformity to its organization, to serve as basis for natural history and introduction to comparative anatomy*"; 1st edition 1816, 2nd edition 1829; see Griffith and Smith (1834) for English edition; **Burrfishes (I); Butterfishes; Creationism; Sandperches; Toadfishes**]

Cuvier, Georges. 1828. [*Historical portrait of the progress of ichthyology, from its origin to our own time.*] Edited by Theodore W. Pietsch and translated by Abby J. Simpson. 1995. The Johns Hopkins University Press, Baltimore and London, xxiv + 366 pp.

[First part of Vol. 1 of the *Natural history of fishes*; **Histoire Naturelle des Poissons; Ichthyology**]

Cuvier, Georges and Achille Valenciennes. 1829. *Histoire naturelle des poissons*. Tome quatrième. Livre quatrième. Des acanthoptérygiens à joue cuirassée. Vol. 4, F. G. Levrault, Paris, xxvi + 518 pp.

["On Acanthopterygians with armoured cheeks"; *Apistus*; **Flatheads**]

Cuvier, Georges and Achille Valenciennes. 1830. *Histoire naturelle des poissons*. Tome sixième. Livre sixième. Partie I. Des Sparoïdes; Partie II. Des Ménides. Vol. 6, F. G. Levrault, Paris, xxiv + 559 pp.

[**Porgies**]

Cuvier, Georges and Achille Valenciennes. 1831. *Histoire naturelle des poissons*. Tome septième. Livre septième. Des Squamipennes. Livre huitième. Des poissons à pharyngiens labyrinthiformes. Vol. 7, F. G. Levrault, Paris, xxix + 531 pp.

[**Ruff**]

Cuvier, Georges and Achille Valenciennes. 1832. *Histoire naturelle des poissons*. Tome huitième. Livre neuvième. Des Scombéroïdes. Vol. 8, F. G. Levrault, Paris, xix + 509 pp.

[**Marblefishes**]

Cuvier, Georges and Achille Valenciennes. 1833. *Histoire naturelle des poissons*. Tome neuvième. Suite du livre neuvième. Des Scombéroïdes. Vol. 9, F. G. Levrault, Paris, xxix + 512 pp.

[**Sandperches; Scorpionfishes**]

Cuvier, Georges and Achille Valenciennes. 1835. *Histoire naturelle des poissons*. Tome dixième. Suite du livre neuvième. Scombéroïdes. Livre dixième. De la famille des Teuthes. Livre onzième. De la famille des Taenioïdes. Livre douzième. Des Athérines. Vol. 10, F. G. Levrault, Paris, xxiv + 482 pp.

[**Goatfishes; Silversides**]

Cuvier, Georges and Achille Valenciennes. 1836. *Histoire naturelle des poissons*. Tome onzième. Livre treizième. De la famille des Mugiloïdes. Livre quatorzième. De la famille des Gobioïdes. Vol. 11, F. G. Levrault, Paris, xx + 506 pp.

[**Blennies**]

Cuvier, Georges and Achille Valenciennes. 1837. *Histoire naturelle des poissons*. Tome douzième. Suite du livre quatorzième. Gobioïdes. Livre quinzième. Acanthoptérygiens à pectorales pédiculées. Vol. 12, F. G. Levrault, Paris, xxiv + 507 pp.

[**Sleepers**]

Cuvier, Georges and Achille Valenciennes. 1840a. *Histoire naturelle des poissons*. Tome quatorzième. Suite du livre seizième. Labroïdes. Livre dix-septième. Des Malacoptérygiens. Vol. 14, F. G. Levrault, Paris, xxii + 464 pp.

[**Wrasses**]

Cuvier, Georges and Achille Valenciennes. 1840b. *Histoire naturelle des poissons*. Tome quinzième. Suite du livre dix-septième. Siluroïdes. Vol. 15, F. G. Levrault, Paris, xxxi + 540 pp.

[**Catfishes**]

Bibliography

Dalzell, Paul. 1993. Management of ciguatera fish poisoning in the South Pacific. *Memoirs of the Queensland Museum* **34**(3): 471–80.

[**Ciguatera**]

+Dampier, William. 1703. *A new voyage round the world*. Vol. 3: A Voyage to New Holland, &c. in the Year 1699. London.

[Dampier was a learned buccaneer, and the pilot of Capt. Woodes Roger, the pirate who rescued Alexander Selkirk, later used by Daniel Defoe as model for his 'Robinson Crusoe'; **Abrolhos; Hippopotamus**]

Daniel, Frank. 1922. *The Elasmobranch fishes*. University of California Press, Berkeley, xi + 334 pp.

[**Bering pike; Lumpfish**]

Darwin, Charles. 1839. *Journal of Researches into the Geology and Natural History of the Various Countries visited by H.M.S. Beagle, under the Command of Captain FitzRoy, R.N. from 1832 to 1836*. Henry Colburn, London, 637 pp.

[*Journal*]

Darwin, Charles. 1840. On the formation of mould. *Transactions of the Geological Society of London*. ser. 2, pt. 3, **5**: 505–9.

[Reprinted in Vol. 1 of Barrett (1977), pp. 49–53; **Cretaceous**]

Darwin, Charles. 1842. *The Structure and distribution of Coral Reefs*. Being the First Part of the Geology of the Voyage of the Beagle, under the Command of Capt. FitzRoy, R.N., during the years 1832 to 1836. Smith, Elder & Co, London, 214 pp. + 2 pp. with maps.

[*Coral Reefs*]

Darwin, Charles. 1846. The geology of the Falkland Islands. *Quarterly Journal of the Geological Society of London*. Proceedings of the Geological Society, Part 1, **2**: 267–74.

[Reprinted in Vol. 1 of Barrett (1977), pp. 203–21; **Distribution**]

Darwin, Charles. 1849. Geology. *In:* J. F. W. Herschel (ed.), *A manual of Scientific Enquiry; Prepared for the Use of Her Majesty's Navy: and Adapted for Travellers in General*, Section VI, pp. 156–95. John Murray, London, xi + 448 pp.

[*Catalogue*]

Darwin, Charles. 1851. *A Monograph on the Sub-Class Cirripedia, with Figures of all the species*. The Ray Society, London. Vol. I: The Lepadidae; or pedunculated Cirripedes, 400 pp. Vol. II, Part I & II: The Balanidae; or sessile Cirripedes, 684 pp.

[The attention devoted by CD to barnacles, both living and fossil – over eight years of meticulous work – which established his reputation as a serious taxonomist, has long intrigued Darwin scholars, as it seemed to imply a deliberate departure from the trajectory CD appeared to be on, following his completion of *Foundation. Some have suggested that CD's long attachment to barnacles was due to his fear of the public outcry that publishing his new discovery (evolution driven by natural selection) would entail. CD's *Autobiography* (pp. 116–17) and his other writings suggest a more prosaic reason: *Cirripedia* probably resulted from the Law of Unintended Consequences, i.e. one thing just followed on another, as for the writing of this book; **Barnacles; Growth**]

Darwin, Charles. [Read July 1, 1858]. On the tendency of species to form varieties; and on the perpetuation of varieties and species by natural means of selection. *Journal of the Proceedings of the Linnean Society of London* (Zoology) **3** (1859): 45–62.

[The paper assembled from existing material of CD's, never intended for publication, and therefore not written with care (*Collected papers II*, p. 18), to prevent his being anticipated by Wallace 1858; ***Big Species Book***; **Wallace**]

258

Darwin, Charles. 1859. *The origin of species*. John Murray, London, ix + 502 pp.

[**Darwin, Charles Robert;** ***Origin***]

Darwin, C. 1872. *De afstamming van den mensch en de seksueele teeltkeus*. Uit het Engelsch vertaald en van aanteekeningen voorzien door Dr. H. Hartogh Heys van Zouteveen. Tweede deel. Joh. Ijkema, Delft.

["*Descent of man and selection in relation to sex*. Translated from English and annotated by Dr. H. Hartogh Heys van Zouteveen"; in 2 volumes. van Zouteveen was a lawyer, doctor of mathematics and physics, blunt freethinker and editor-in-chief of a provincial newspaper. His 'enriched translation' includes lots of material not in the first English edition, notably details on fishes that produce sounds, as noted by CD in the second English edition; **Groupers (II); Sounds**]

Darwin, Charles. 1875. *The variations of animals and plants under domestication*. John Murray, London, 2nd edition, Vol. 1, xiv + 473 pp; Vol. 2, x + 495 pp.

[***Variations***]

Darwin, Charles. 1877a. *The various Contrivances by which Orchids are Fertilized by Insects*. John Murray, London, xvi + 300 pp.

[Revised 2nd edition; **Butterflyfishes**]

Darwin, Charles. 1877b. *Descent of man, and selection in relation to sex*. Revised and augmented 2nd edition, including an essay by T. H. Huxley. John Murray, London, Vol. 1, pp. x + 1–206; Vol. 2, pp. xiv + 207–693.

[First edition in 1871, ***Descent***]

Darwin, Charles. 1881. *The formation of vegetable moulds through the action of worms, with observations on their habits*. London, John Murray. vii + 326 pp.

[***Worms***]

Darwin, Charles. 1890. *The expression of the emotions in men and animals*. 2nd edition, edited by Francis Darwin. John Murray, London, viii + 394 pp.

[First edition (1872) contained 106 912 words, 7515 different; ***Expression*; Kingfisher**]

Darwin, Charles. 1909. *The foundation of the origin of species: two essays written in 1842 and 1844, edited by his son Francis Darwin*. Cambridge University Press, Cambridge, xxx + 263 pp.

[***Foundations***]

#Darwin, Erasmus. 1794–6. *Zoonomia*. J. Johnson, London. 2 Vols.

[p. 97 close interbreeding cause of mutants? . . . Fish & Dogs are; 504 *Lamarck concisely forestalled by my Grandfather *Marginalia* 186, 187]

#Darwin, Erasmus. 1800. *Phytologia*. J. Johnson, London.

[p. 559 Fish? *Marginalia* 185]

Darwin, Francis (ed.) 1887. *The Life and Letters of Charles Darwin, including an autobiographic chapter*. London, John Murray. In 3 Volumes.

[***Autobiography*; Spontaneous generation**]

Darwin, Francis and A. C. Seward (eds.) 1903. *More Letters of Charles Darwin: a record of his work in a series of hitherto unpublished letters*. London. Vol. 1, xxiv + 494 pp.; Vol. 2, viii + 508 pp.

[***Correspondence*; Electric fishes; Mammals; Swimbladder**]

David, Lore. 1936. Deux espèces de *Synodontis* du Moyen-Congo. *Revue de Zoologie et de Botanique Africaines* **28**(3): 416–18.

[**Dohrn**]

+Davy, Humphry. 1828. *Salmonia, or Days of Fly-Fishing; in a Series of Conversations: with some Account of the Habits of Fishes belonging to the Genus Salmo*. London, 273 pp.

[CD listed this work as read in his *Reading Notebooks*; *Correspondence*, Vol. 4, p. 444 (Burkhardt and Smith 1988); **Variety(-ies)**]

#Davy, John. 1855. *The angler and his friend; or piscatory colloquies and fishing excursions*. London.

[CD recorded in his *Reading Notebooks* on March 25, 1855 that he had read The Angler Piscator Dr. Davy; *Correspondence*, Vol. 4, p. 492; the same day, he mentioned this to the author: I arrived home late last night, (bringing with me the Angler & his Friend). *Correspondence*, March 25, 1855]

+Davy, John. 1856. Some observations on the ova of salmon, in relation to the distribution of species; in a letter addressed to Charles Darwin. *Philosophical Transactions of the Royal Society of London* **146**: 21–9.

[This paper, read in March 1855, responded publicly to questions which CD had asked privately. He was pleased with this, writing: I assure you I feel much pleased & flattered by the form you have done me the honour of putting the communication under; *Correspondence*, March 25, 1855; **Egg development**]

Dawkins, Richard. 1989. *The Selfish Gene*, 2nd edition. Oxford University Press, Oxford, xi + 352 pp.

[**Altruism; Reproduction**]

Dawkins, Richard. 1995. *River out of Eden: a Darwinian View of Life*. Basic Books, New York, xiii + 172 pp.

[**Sex ratio**]

Dawkins, Richard. 1996. *Climbing Mount Improbable*. Viking, Penguin Books, London, xi + 308 pp.

[**Difficulties; Extinction; Eyes; Non-Darwinian evolution**]

Dawkins, Richard. 1998. *Unweaving the Rainbow: science, delusion and the appetite for wonder*. Houghton Mifflin, Boston, xiv + 336 pp.

[**Extinction; Non-Darwinian evolution**]

+Day, Francis. 1878. Remarks on Mr. Whitmee's paper on the manifestation of fear and anger in fishes. *Proceedings of the Zoological Society* **1878**: 214–21.

[**Expression**]

De Buen, Fernando. 1966. Resultados de una campaña ictiólogica in las Provincias del Norte. *Estudios Oceanograficos en Chile* **2**: 5–9.

["Results of an ichthyological survey of the Northern provinces"; **Pipefishes (I); Wrasses**]

DeKay, James Ellsworth. 1842. Zoology of New-York, or the New-York fauna; comprising detailed descriptions of all the animals hitherto observed within the state of New-York, with brief notices of those occasionally found near its borders, and accompanied by appropriate illustrations.... *In*: *Natural History of New York*. Zoology, Part 4. Fishes, pp. 1–415. Printed by Carroll and Cook, printers to the Assembly, Albany, NY.

[**Cavefishes**]

De Queiroz, Kevin. 1988. Systematics and the Darwinian revolution. *Philosophy of Science* **55**: 238–59.

[**Classification**]

De Robertis, E. M. and Yoshiki Sasay. 1996. A common plan for dorsoventral patterning in bilateria. *Nature* **380**: 37–40.

[**Dohrn**]

Descartes, René. [1637] 1998. *Discours de la méthode pour bien conduire sa raison, et chercher la vérité dans les sciences*. Translated by Donald A. Cress. Hackett Pub. Co, Indianapolis, xv + 44 pp.

[The "*Discourse on the method for conducting one's reason well, and for seeking truth in the sciences*" is where this great French philosopher (1596–1660) showed that doubt, at some point, must also doubt itself ('*cogito ergo sum*'). In other works, Descartes established, among other things, the deep connection between algebra and geometry, thus making modern quantitative science possible, to the great ire of pseudo-scientists and would-be gurus, who use the adjective 'Cartesian' not as the compliment it should be, but as a term of opprobrium. Non-francophones who want to move beyond these people will find the references of numerous translations of Descartes' published work in the above-cited contribution; **Popper**]

Deshaye, Gérard Paul. 1836. [Tertiary fossil shells from the Crimea.] *Bulletin de la Société Géologique de France* **8**: 215–18.

[**Caspian Sea**]

Desmond, Adrian. 1984. Robert E. Grant: the social predicament of a pre-Darwinian transmutationist. *Journal of the History of Biology* **17**: 189–223.

[**Grant**]

Desmond, Adrian. 1994. *Huxley: from Devil's disciple to evolution's high priest*. Helix Books/Perseus Books, Reading, MA, 820 pp.

[**Huxley**]

Desmond, Adrian and James R. Moore. 1992. *Darwin*. Michael Joseph, London. xxi + 807 pp.

[**Darwin, Charles Robert**]

De Sylva, Donald P. 1963. Systematics and life history of the Great Barracuda *Sphyraena barracuda* (Walbaum). *Studies in Tropical Oceanography*. Institute of Marine and Atmospheric Sciences, University of Miami. No. 1, 179 pp.

[**Barracuda; Ciguatera**]

De Vries, T. J. and W. G. Pearcy. 1982. Fish debris in sediments of the upwelling zone off Central Peru: a late quaternary record. *Deep-Sea Research* **28**(IA): 87–109.

[**Punctuated equilibrium**]

Di Gregorio, M. A. and N. W. Gill. 1990. *Charles Darwin's* Marginalia, *1809–1882*. Garland Publishing, New York.

[Volume 1, referring to CD's books; the marginalia for his large reprint collection are still pending; **Marginalia; Words**]

Dohrn, Anton. 1875. *Der Ursprung der Wirbelthiere und das Prinzip des Funktionswechsel*. Wilhelm Engelmann, Leipzig, xv + 87 pp.

["*The origin of vertebrates and the principle of functional change*"; **Dohrn**]

Dorward, D. F. and N. P. Ashmole. 1963. Notes on the biology of the brown noddy *Anous stolidus* on Ascension Island. *Ibis* 103b: 447–57.

[**Saint Paul's Rock**]

Douglas, Michael Edward and John C. Avise. 1982. Speciation rates and morphological divergence in fishes: tests of gradual versus rectangular modes of evolutionary change. *Evolution* **36**(2): 224–32.

[**Punctuated equilibrium**]

Dubos, René. 1988. *Pasteur and modern science*. Springer Verlag, New York, 168 pp.

[**Spontaneous generation**]

Dufossé, Léon. 1854. De l'hermaphroditisme chez certain vertebrés. *Annales des Sciences Naturelles* (Zoologie) 4ième sér. **5**: 295–332.

["On hermaphroditism in certain vertebrates"; also published in 1856 by V. Masson, Paris, 40 pp; **Groupers (II)**]

+Dufossé, Léon. 1858a. Des différents phénomènes physiologiques nommés voix des poissons. *Comptes Rendus Hebdomadaires des Séances de l'Académie des Sciences* **46**: 352–6.

["On the different physiological phenomena called voice of fishes"; **Sounds**]

+Dufossé, Léon. 1858b. De l'ichtyopsophie, ou des différents phénomènes physiologiques nommés voix des poissons (deuxième partie). *Comptes Rendus Hebdomadaires des Séances de l'Académie des Sciences* **46**: 916.

["On Ichthyopsophy, or the different physiological phenomena called voice of fishes (Second part)." (The learned term is derived from Greek *psophos*, for strong sound, noise); **Sounds**]

+Dufossé, Léon. 1862. Sur les différents phénomènes physiologiques nommés voix des poissons, ou sur l'Ichthyopsophose (troisième partie). *Comptes Rendus Hebdomadaires des Séances de l'Académie des Sciences* **54**: 393–5.

["...(Third part)"; **Gurnards; Sounds**]

Duméril, Auguste H. A. 1865. *Histoire naturelle des poissons ou ichthyologie générale*. Tome Premier. I. Elasmobranches, Plagiostomes et Holocéphales ou Chimères. Roret, Paris, 720 pp.

[**Chimaera**]

Dunbar, M. J. 1980. The blunting of Occam's Razor, or to hell with parsimony. *Canadian Journal of Zoology* **58**(2): 123–8.

[**Occam's Razor**]

Dutra, Guilharme F. 1999. Coral Reefs of the Abrolhos Bank, Brazil. *In*: Ian Dight, Richard Kenchington and John Baldwin (eds.), *Proceedings of the International Tropical Marine Ecosystem Symposium*, Townsville, Australia, November 1998, pp. 363–5. Great Barrier Reef Authority, Townsville.

[**Abrolhos**]

Eaton, Amos. 1831. Fishes of Hudson River. *American Journal of Science and Arts* **20**: 150–2.

[**Shad**]

Eckhardt, George H. 1950. A woman in a garden. *Frontiers – Magazine of Natural History* **15**(2): 57–8.

[**Morris (II)**]

+Eckström, C. U. 1840. Beobachtungen über die Formveränderung bey der Karausche (*Cyprinus carassius* L.). *Oken's Isis*: 143–53.

["Observations on the change of shape in Crucian carp (*Cyprinus carassius* L.)"; as cited by Bronn 1843; see p. 601 in Stauffer 1975; **Carp, Prussian**]

Edcoff, Nancy. 1999. *Survival of the prettiest: the science of beauty*. Abacus Books, London, 344 pp.

[**Beauty**]

Edwards, Alasdair J. 1990. *Fish and Fisheries of Saint Helena Island*. Education Department, Government of Saint Helena and Centre for Tropical Coastal Management Studies, University of Newcastle upon Tyne, 152 pp.

[**Ascension Island; Saint Helena**]

Edwards, Alasdair J. and Roger Lubbock. 1980. Voyage to St. Paul's Rock. *Geographical Magazine* **52**(8): 561–7.

[**St. Paul's Rock**]

Edwards, Alasdair J. and Roger Lubbock. 1982. The shark population of Saint Paul's Rock. *Copeia* **1982**(1): 223–5.

[**Galápagos shark(s)**]

Edwards, Alasdair J., K. Wilson and Roger Lubbock. 1981. The sea bird populations of St Paul's Rock. *Ibis* **123**: 233–8.
[**St. Paul's Rock**]

Edwards, C. 1966. *Velella velella* (L.): the distribution of its dimorphic forms in the Atlantic ocean and the Mediterranean, with comments on its nature and affinities. *In*: Harold Barnes (ed.), *Some Contemporary Studies in Marine Sciences*, pp. 283–96. George Allen and Unwin, London.
[An organism, also known as 'by the wind sailor', which sails either 'left' or 'right' under the wind; **Asymmetry; Food webs**]

#Edwards, William Frédéric. 1832. *On the influence of physical agents on life*. London.
[Many facts given of high temperature at which fish &c can live; *Notebook E*, p. 435]

+Eichwald, Carl Eduard von. 1834–8. *Reise auf dem Caspischen Meer und in den Caucasus. Unternommen in 1825–1826*. Stuttgart, Tübingen.
["Travel on the Caspian Sea and the Caucasus, undertaken in 1825–26"; **Caspian Sea**]

+Eichwald, Carl Eduard von. 1838. Sur la faune de la Mer Caspienne. *L' Institut – Journal Général des Sociétés et Travaux Scientifiques de la France et de l'Etranger* **6**(259), 13 December, p. 412.
["On the fauna of the Caspian Sea"; **Caspian Sea**]

+Eichwald, Carl Eduard von. 1839. Remarks on the Caspian Sea. *Annals of Natural History* **2**: 135–6.
[**Caspian Sea**]

Eiseley, Loren. 1959. *Darwin and the Mysterious Mr. X; new light on the evolutionists*. Gollancz, London, 372 pp.
[**Blyth, Edward; Plagiarism**]

Ekman, Sven. 1967. *Zoogeography of the Sea*. Sidgwick and Jackson, London, xiv + 417 pp.
[**Distribution; Latitude; Submergence**]

Eknath, Ambekar E. and Roger W. Doyle. 1990. Effective population size and rate of inbreeding in aquaculture of Indian major carp. *Aquaculture* **85**: 293–305.
[**Carp, Indian**]

Eldredge, Niles. 1995. *Reinventing Darwin: The Great Debate at the High Table of Evolutionary Theory*. John Wiley and Sons, New York, 244 pp.
[**Non-Darwinian evolution; Punctuated equilibrium**]

Eldredge, Niles. 2000. *The triumph of evolution . . . and the failure of creationism*. W. H. Freeman and Co, New York, 223 pp.
[**Creationism**]

Eldredge, Niles and Stephen J. Gould. 1972. Punctuated equilibria: an alternative to phyletic gradualism. *In*: T. J. M. Schopf (ed.), *Models in Paleobiology*, pp. 82–115. Freeman, Cooper & Co, San Francisco.
[**Non-Darwinian evolution; Punctuated equilibrium**]

Ellegård, Alvar. 1958. *Darwin and the general reader: the reception of Darwin's theory of evolution in the British periodical press, 1859–1872*. Gothenburg Studies in English 8, Göteborg, 394 pp.
[Concludes, based on analysis of nearly 100 reviews of CD's *Origin* in popular outlets, that half were actively hostile, one quarter neutral, and only one quarter positive, or at least polite, and that periodicals were far more negative than dailies; **Caviller**]

Elliott, J. M. 1994. *Quantitative ecology and the brown trout*. Oxford Series in Ecology and Evolution. Oxford University Press, Oxford, xi + 286 pp.
[**Trout**]

+Ercolani, Giovan Battista. 1871. Del perfetto ermafroditismo delle Anguille. *Memorie dell' Accademia delle Scienze dell'Istituto di Bologna* Ser. III, vol. I, p. 529–52.
  ["On the perfect hermaphroditism of the eel"; **Eels**]

Ercolani, Giovan Battista. 1872. Sullo sviluppo degli organi genitali delle Anguille. *Resoconti della Sessione dell' Accademia delle Science dell' Istituto di Bologna*, Anno Accademico 1871/1872: 43–5.
  ["On the development of the reproductive organs of the eel"; **Eels**]

Erdman, C. 1999. *Schmerzempfinden und Leidensfähigkeit bei Fischen: eine Literaturübersicht*. Fachgebiet Fischkrankeiten und Fischhaltung, Tierärztliche Hochschule, Hannover, 154 pp.
  ["*The ability of fishes to feel pain and to suffer: a review of the literature*"; **Flying fishes**]

Erdmann, Mark V., Roy L. Caldwell and M. Karim Moosa. 1998. Indonesian 'king of the sea' discovered. *Nature* **395**: 335.
  [**Extinctions**]

Erwin, Douglas, James Valentine and David Jablonski. 1997. The Origin of Animal Body Plans. *American Scientist* **85** (March–April): 126–37.
  [**Vertebrate origins**]

Eschmeyer, William N. 1969. A systematic review of the scorpionfishes of the Atlantic Ocean (Pisces: Scorpaenidae). *Occasional Papers of the California Academy of Science* no. 79: i–iv + 1–143.
  [**Azores**]

Eschmeyer, William N. 1990. *Genera of Recent Fishes*. California Academy of Science, San Francisco, 697 pp.
  [**FishBase; Galaxiidae;** *Macropus*; **Pigfishes; Synonym**]

Eschmeyer, William N. 1998. *Catalog of Fishes*. California Academy of Science, San Francisco, 3 volumes, 2903 pp.
  [**Preface and Acknowledgments; Flatfishes; FishBase; Greenland halibut; Sole; Synonym**]

+Eudes-Deslongchamps, Jacques-Amand. 1835. Notes sur une anguille retirée d'un puits, au mois de juillet 1831. *Mémoires de la Société Linnéenne de Normandie* **5**: 47–51.
  ["Notes on an eel found in a well, in the month of July 1831"; **Eel**]

+Eudes-Deslongchamps, Jacques-Amand. 1842. Paragraph *in*: Resumé des travaux. *Mémoires de la Société Linnéenne de Normandie* **7**: xxix.
  [**Eel**]

Evans, Ivor H. 1981. *Brewer's dictionary of phrase and fable*. Revised Edition. Cassell Ltd, London, 1213 pp.
  [**Mother Carey's chickens**]

Eydoux, Joseph-Fortuné-Théodore and François Louis Paul Gervais. 1837. Voyage de "La Favorite." Poissons. *Revue et Magazine de Zoologie*. 7ième année: 1–4.
  [**Pipefishes (I)**]

Fabricius, Othon. 1824. Om en ny, og 2de lidet bekjendte, Flynderarten, nemlig: 1) Stenn-Sueren (*Pleuronectes quadridens*). 2) Den lille Hellefisk (*Pleuronectes pinguis*) og 3) Den Grönlandske Flynder (*Pleuronectes platessoides*). *Det Kongelige Danske Videnskabenes Selskab-Naturvidenskabelige og Mathematiske Afhandlinger* **1**: 39–55.
  [Translated and updated, this reads: "On one new and two little-known flatfish species, namely (1) Lemon sole *Microstomus kitt* (Walbaum 1792); (2) Greenland halibut (*Reinhardtius hippoglossoides*); and (3) American plaice (*Hippoglossoides platessoides*)"; **Greenland halibut**]

Falconer, J. D. 1937. Darwin in Uruguay. *Nature* **140**: 138–9.
  [**Maldonado**]

Falkner, Thomas. 1774. *A description of Patagonia and adjoining parts of South America with a grammar and a short vocabulary, and some particulars relating to Falkland's Islands*. Hereford.
  [The author (1707–1784) was a surgeon, and *naturalist on a slavers' ship, and a Jesuit important enough to be listed in the 'Catholic Encyclopedia'. He lived 30 years in the areas he describes. The papers containing his observations were initially published in a compilation by William Combe, then republished in 1935. A 1976 reprint of the 1935 edition was published by AMS Press, New York; **Jaguar**]

Fenton, G. E., S. A. Short and D. A. Ritz. 1991. Age determination of orange roughy *Hoplostethus atlanticus* (Pisces: Trachichthyidae) using 210 Pb/226Ra disequilibria. *Marine Biology* **109**: 197–202.
  [**Darwin's roughy**]

Fernholm, Bo. 1990. Myxinidae. *In*: O. Gon and Phillip C. Heemstra (eds.), *Fishes of the Southern Ocean*, pp. 77–8. J. L. B. Smith Institute of Ichthyology, Grahamstown, South Africa.
  [**Hagfishes**]

Fernholm, Bo. 1998. Hagfish systematics. *In*: Jørgen Mørup Jørgensen, Jens Peter Lomholt, Roy E. Weber and Hans Malte (eds.), *The Biology of Hagfishes*, pp. 33–44. Chapman & Hall, London.
  [*Gastrobranchus*; **Hagfishes**]

Ferreira, C. E. L. and J. E. A. Gonçalves. 1999. The unique Abrolhos Formation (Brazil): need for specific management strategies. *Coral Reefs* **18**: 352.
  [**Abrolhos**]

Fischer, Walter. 1963. Die Fische des Brackwassergebietes Lenga bei Concepción, Chile. *Internationale Revue der gesamten Hydrobiologie* **48**(3): 419–511.
  ["The fishes of the Lenga brackish water area near Concepcion, Chile"; **Galaxiidae**]

Fish, Marie Poland. 1954. The character and significance of sound production among fishes of the Western North Atlantic. *Bulletin of the Bingham Oceanographic Collection* **14**(3): 1–109.
  [**Sounds**]

Fish, Marie Poland and William H. Mowbray. 1970. *Sounds of Western North Atlantic fishes: a reference file for underwater records*. The John Hopkins Press, Baltimore and London, 207 pp.
  [All incorporated in *FishBase; **Grunts; Gurnards; Sounds**]

Fisher, Ronald Aylmer. 1930. *The genetical theory of natural selection*. The Clarendon Press, Oxford, xiv + 272 pp.
  [**Handicap principle**]

Fisher 1965: *see* Mendel 1866.

Fitch, John E. and Robert J. Lavenberg. 1968. *Deep-water fishes of California*. Californian Natural History Guides: 25. University Press of California, Berkeley and Los Angeles, 155 pp.
  [**Bioluminescence**]

Fitzinger, Leopold Joseph Franz Johann. 1837. Vorläufiger Bericht über eine höchst interessante Entdeckung Dr. Natterers in Brasil. *Oken's Isis* **30**: 379–80.
  ["Preliminary report on a highly interesting discovery by Dr. Natterer in Brazil"; **Lungfishes; Lungfish, South American**]

FitzRoy, Robert. 1837. Extracts from the Diary of an attempt to ascend the River Santa Cruz, in Patagonia, with the boats of His Majesty's sloop *Beagle*. *Journal of the Geographical Society of London* **7**: 114–26.
  [**Creole perch**]

FitzRoy, Robert. 1839. *Narrative of the surveying voyages of His Majesty's Ships* Adventure *and* Beagle *between the years 1826 and 1836, describing their examination of the southern shores of South America*. Vol. II. *Proceedings of the Second Expedition*. Henry Colburn, London.

[This book, although rather dull, nicely complements CD's *Diary* and *Journal*. Notably, it attests, if need be, to the incredible amount of energy that CD extended to his sampling, even during periods where he did not himself report on any particular activities. Thus, for example, p. 343 of the *Narrative* reports that in Patagonia, near the mouth of the Santa Cruz River, "Mr. Darwin tried to catch fish with a casting net, but without success; so strong a stream being much against successful fishing. A very sharp frost this night. The net and other things, which has occupied but little room in the boat, were frozen so hard as to become unmanageable and very difficult to stow." **Covington; FitzRoy; King George's Sound**]

Fleming, Ian and Mart R. Gross. 1994. Breeding competition in a Pacific salmon (Coho: *Oncorhynchus kisutch*): measure of natural and sexual selection. *Evolution* **48**(3): 637–57.

[**Salmon (III)**]

+Fleming, John. 1822. *The Philosophy of Zoology, or A general view of the Structure, Function, and classification of Animals*. 2 Vols. Edinburgh.

[Vol. 2, p. 356 Fecundity of Fish *Marginalia* 233; **Dogfish**]

+Forbes, Edward. 1851. Report on the Investigation of British Marine Zoology by Means of the Dredge. Part I. The Infra-littoral Distribution of Marine Invertebrates on the Southern, Western, and Northern Coast of Great Britain. In: *Report of the Twentieth Meeting of the British Association for the Advancement of Science*, held at Edinburgh in July and August 1850, pp. 193–263. John Murray, London.

[**Burbot**]

Forsskål, Peter. 1775. *Descriptiones animalium avium, amphibiorum, piscium, insectorum, vermium; quae in itinere orientali observavit . . . Post mortem auctoris edidit Carsten Niebuhr*. Hauniae. xxxiv + 164 pp.

["Descriptions of the animals: birds, amphibians, fish, insects, worms observed during a voyage to the Orient . . . Edited after the author's death by Carsten Niebuhr. Copenhagen"; **Butterflyfishes; Mojarras; Wrasses**]

+Forster, Johann Reinhold. 1771. A letter from Mr. John Rheinhold Forster, F.A.S., to the Hon. Daines Barrington . . . . on the management of carp in Polish Prussia. *Philosphical Transactions of the Royal Society of London* **61**(1772): 310–25.

[**Hybrids**]

Foster, Johann Reinhold. 1801. *See* Bloch and Schneider 1801.

+[Franklin, Benjamin]. 1755. Observations concerning the Increase of Mankind, Peopling of Countries, &c. In: William Clarke. *Observations on the late and present conduct of the French, with regards to their encroachment upon the British colonies of North America, together with remarks on the importance of these colonies to Great-Britain*, pp. 42–54. Boston; reprinted for John Clarke, under the Royal Exchange, Cornhill, London.

[Originally an anonymous text; **Eggs of fish**]

Freeman, Richard Broke. 1977. *The works of Charles Darwin, an annotated bibliographical handlist*, 2nd edition. Wm Dawson & Sons Ltd, Folkestone, 235 pp.

[**Darwin; Evolution; FishBase**]

Freeman, Richard Broke. 1978. *Charles Darwin: A Companion*. Wm Dawson & Sons Ltd, Folkestone, England, 309 pp.
[*Beagle*; **Eponym; Kelp; Wrasses**]

Freeman, Richard Broke. 1978/79. Darwin's Negro bird stuffer. *Notes and Records of the Royal Society of London* **33**: 83–6.
[**Edmonston**]

+Fries, Bengt Frederik. 1839a. On the genus *Syngnathus*. *Annals of Natural History* **2**: 96–105.
[**Pipefishes (I)**]

+Fries, Bengt Frederik. 1839b. Metamorphosis observed in *Syngnathus lumbriciformis*. *Annals of Natural History* **2**: 225.
[**Pipefishes (I)**]

+Fries, Bengt Frederik. 1839c. Metamorphosis observed in the small pipe-fish *(Syngnathus lumbriciformis)*. *Annals of Natural History* **2**: 451–5.
[**Pipefishes (I)**]

Froese, Rainer and Crispina Binohlan. 2000. Empirical relationships to estimate asymptotic length, length at first maturity and length at maximum yield per recruit in fishes, with a simple method to evaluate length frequency data. *Journal of Fish Biology* **56**: 758–73.
[**Silversides**]

Froese, Rainer and Daniel Pauly (eds.) 2000. *FishBase 2000: Concepts, Design and Data Sources.* Los Baños, Philippines, xvii + 344 pp.
[Distributed with four CD-ROMs; see www.fishbase.org for online version and updates; **Preface; FishBase; Sunfishes**]

Fryer, Geoffrey and T. D. Iles. 1972. *The cichlid fishes of the Great Lakes of Africa: their biology and evolution*. Oliver & Boyd, Edinburgh, xvi + 641 pp.
[**Cichlidae (II)**]

Garber, Janet. 1994. Darwin's correspondents in the Pacific: through the looking glass to the Antipodes. *In*: Rod MacLeod and Philip F. Rehbock (eds.), *Darwin's Laboratory: evolutionary theory and natural history in the Pacific*, pp. 169–211. University of Hawai'i Press, Honolulu.
[**Covington; Distribution**]

Garfield, Eugene. 1980. Premature discovery or delayed recognition – why? *In*: *Essays of an information scientist*, vol. 4, pp. 488–93. ISI Press, Philadelphia.
[**Mendel**]

Garfield, Eugene. 1982. More on the ethics of scientific publications: abuses of authorship and citation amnesia undermine the reward system of science. *In*: *Essays of an information scientist*, vol. 5, pp. 621–6. ISI Press, Philadelphia.
[**Plagiarism**]

Gaskell, Walter Holbrook. 1908. *The origin of vertebrates*. Longmans, Green & Co, London.
[**Dohrn**]

Gaudichaud-Beaupré, Charles. 1826. Botanique. *In*: Louis Claude Desauslces de Freycinet. 1824–1844. *Voyage autour du Monde, Entrepris par Ordre du Roi, Exécuté sur les Corvettes de S.M. 'L'Uranie' et 'La Physicienne,' pendant les années 1817, 1818, 1819 et 1820*. Paris, Vol. VII, 522 pp. and Atlas, 96 pp.
[**Kelp**]

Geberinch, J. B. and M. Laird. 1968. Bibliography of papers relating to the control of mosquitoes by the use of fish: an annotated bibliography for the years 1901–1966. *FAO Fisheries Technical Paper* no. 75, Rome, 70 pp.

[**Livebearers**]

Gee, Henry. 1994. Return of the amphioxus. *Nature* **370**: 504–5.

[**Dohrn**]

+Gegenbaur, Carl. 1870a. Ueber das Gliedmaassenskelet der Enaliosaurier. *Jenaische Zeitschrift für Naturwissenschaft* **5**: 332–49.

["On the limbic skeleton of Enaliosauria"; **Lungfish, Australian**]

#Gegenbaur, Carl. 1870b. Ueber das Skelet der Gliedmaassen der Wirbelthiere im Allgemeinen und der Hintergliedmaassen de Selachier insbesondere. *Jenaische Zeitschrift für Naturwissenschaft* **5**: 397–447.

["On the limbic skeleton of Vertebrates in general, and on the hind limbs of Selachians in particular"; CD asked Günther for his opinion on this paper; *Calendar*, no. 8316]

Génermont, Jean, Michel Delsol and Janine Flatin. 1996. Cécité des cavernicoles. *In*: Patrick Tort (ed.) 1996. *Dictionnaire du Darwinisme et de l'Evolution* vol. 1, pp. 551–3. Presses Universitaires de France, Paris.

["Troglobiont blindness"; **Cavefishes**]

+Geoffroy-Saint Hilaire, Etienne. 1807. Observations sur l'affection mutuelle de quelques animaux, et particulièrement sur les services rendus au requin par le pilote. *Annales du Museum d'Histoire Naturelle* **9**: 469–76.

["Observations on the mutual affection of some animals, particularly on the services rendered to sharks by pilot-fish"; CD cites this paper twice in his *Big Species Book* (pp. 497, 519), but without reference to pilot-fish helping their sharks]

Geoffroy-Saint Hilaire, Etienne. 1822. Considérations générale sur la vertèbre. *Mémoire du Museum National d'Histoire Naturelle* **9**: 89–119.

["General considerations on the vertebra"; **Dohrn**]

+Geoffroy-Saint Hilaire, Etienne. 1830. *Principes de philosophie zoologique*: discutés en mars 1830 au sein de l'Académie royale des sciences, Paris. Pichon & Didier, Rousseau, 226 pp.

[CD commented in *Notebook B*, pp. 196–7, on two issues raised by the Principes de Zool: Philosop, (Cuvier 1828) which discusses its author's famous public debate with Cuvier: (i) the author proposed, though in different words, that "ontogeny recapitulates phylogeny"; and (ii) that 'flipped over' invertebrates have their major organ systems arranged in a fashion similar to those of vertebrates, i.e. I deduce from extreme difficulty of hypothesis of connecting Mollusca & vertebrata, that there must be great gaps. – yet some analogy. CD, unsurprisingly, also noted that Cuvier on opposite side. 1^s Volume of Fish, referring to Vol. 1 of the *Histoire Naturelle des Poissons*, where Cuvier, indeed, refuses to accept analogies between the organ systems of different organisms; (Cuvier 1828, p. 237); **Dohrn**]

+Geoffroy-Saint Hilaire, Isidore. 1832–7. *Histoire genérale et particulière des anomalies de l'organisation chez l'homme et les animaux, ouvrage comprenant des recherches sur les charactères, la classification, l'influence physiologique et pathologique, les rapports genéraux, les lois et les causes des monstruosités, des variétés, et des vices de conformation, ou traité de tératologie.* 4 vols. Paris.

["*General and specific history of structural anomalies in humans and animals, including research on the character, the classification, the influence of physiology and pathology on, the general relationships, the laws and the causes of monstrosities, their varieties, and defects, or treatise of teratology*"; Vol. 2, p. 110 On Carps with imperfect female organs like neuters; 285 Carp *Marginalia 310*; **Monstrosity**]

+Geoffroy-Saint Hilaire, Isidore. 1841. *Essais de zoologie génerale, ou mémoires et notices sur la zoologie génerale, l'anthropologie, et l'histoire de la science*. De Roret, Paris.

["*Essay on general zoology, anthropology, and on the history of science*"; Vol. 1, p. 90 old Geoffroy states never new organ – in relation to Electric Fishes 94 'Nature always works with same materials' *Marginalia* 301; **Monstrosity**]

Geoffroy-Saint Hilaire, Isidore and Henri-Marie Ducrotay de Blainville. 1834. Partie zoologique. *In*: Rapport sur les résultats scientifiques du voyage de M. Alcide d'Orbigny dans l'Amérique du Sud pendant les années 1826, 1827, . . . et 1833. *Nouvelles Annales du Museum d'Histoire Naturelle de Paris* **3**: 84–115.

[See also Orbigny (1834–47) and Valenciennes (1847); **Distribution; Ichthyology**]

George, Wilma. 1982. *Darwin*. Fontana Paperbacks, London, 160 pp.

[**Creationism**]

+Gerbe, Z. 1864. Observations sur la nidification des Crénilabres. *Revue et Magazine de Zoologie* 2ième Sér. **16**: 255–8; 273–9; 337–40.

[Here is Günther's summary of "On nest-making in the genus *Crenilabrus*": "M. Gerbe has made the interesting observation that species of this genus make nests of seaweeds, shells, &c., in which the ova are deposited; both sexes are engaged in the construction. The species observed are determined as *C. massa* [*griseus*] and *C. melops*." **Wrasses**]

Gesner, Conrad. 1558. *De piscium & aquatilium animantium natura*. Vol. 4 of *Historiae Animalium*. Tiguri.

[The English translation of this is "*On the nature of fishes and aquatic animals*; Zürich"; but it is likely that Günther was referring to its German translation (published in 1670, and reprinted in 1981 by Schlütersche Verlagsanstalt und Druckerei, Hannover), which indeed, on pp. 190–1, describes the "awful Syluri" or Wels, caught in many German lakes and rivers, where they "tyrannize" the other fishes, and reach large sizes ("7 to 8 ells"), and of which a specimen was found with a human head and hand in its stomach. However, there is no mention of Lake Constance, as Günther asserts; **Wels catfish**]

Ghiselin, Michael T. 1969. *The Triumph of the Darwinian Method*. Berkeley: University of California Press, x + 287 pp.

[2nd edition, of 1991, included on CD's CD; ***Autobiography***; **Difficulties; Experiments (I)**]

Ghiselin, Michael T. and Pete Goldie (eds.) 1997. *The Darwin Multimedia CD-ROM*, (2nd edition). Lightbinders, Inc., San Francisco.

[**CD**]

Giard, Alfred Mathieu. 1872. Etude critique des travaux d'embriogénie relatifs à la parenté des Vertebrés et des Tuniciers. *Archives de Zoologie* **1**: 233–88.

["Critical study of embryological studies in relation to the ancestry of the vertebrates and tunicates"; **Seasquirts**]

Gibson, Robin N. 1993. Intertidal teleosts: life in a fluctuating environment. *In*: Tony J. Pitcher (ed.), *Behaviour of teleost fishes*, pp. 513–36. Chapman & Hall, Fish and Fisheries Series 7, London.

[**Blennies; Mudskippers; Territoriality**]

Gilbert, Charles Henry. 1890. A preliminary report on the fishes collected by the steamer *Albatross* on the Pacific coast of North America during the year 1889, with descriptions of twelve new genera and ninety-two new species. *Proceedings of the United States National Museum* **13** (797): 49–126.

[**Flying fish(es)**]

Gilchrist, J. D. F. 1913. Review of the South African Clupeidae (herrings) and allied families of fishes. *Union of South Africa Marine Biological Report* no. 1: 46–66.
[**Cape of Good Hope**]

Gill, Anthony C. 1999. Subspecies, geographic forms and widespread Indo-Pacific coral-reef fish species. *In*: Bernard Séret and Jean-Yves Sire (eds.), *Proceedings of the 5th Indo-Pacific Fish Conference, Nouméa, New Caledonia, 3–8 November 1997*, pp. 79–87. Societé Française d'Ichtyologie, Paris.
[**Range**]

Gill, Theodore Nicholas. 1883. Deep-sea fishing fishes. *Forest and Stream*, November 8, 1883: 284.
[**Parasites**]

Gillois, Michel. 1991. *Mathematical population dynamics*. Marcel Dekker, New York.
[**Non-Darwinian evolution**]

Gillois, Michel. 1996. Neutralisme. *In*: P. Tort (ed.), *Dictionnaire du Darwinisme et de l'Evolution*, pp. 3205–26. Presses Universitaires de France, Paris.
["Neutral selection", In "Dictionary of Darwinism and Evolution"; **Non-Darwinian evolution**]

#Godron, Dominique Alexandre. 1859. *De l'espèce et des races dans les êtres organisés et spécialement de l'unité de l'espèce humaine*. 2 Vols. J. B. Baillière et Fils, Paris.
["On the concept of species and on races among organisms, and especially on the unity of the human species"; Fish same in different kinds of Water, *Marginalia* 331]

Golani, Daniel. 1996. The marine ichthyofauna from the Eastern Levant: history, inventory and characterization. *Israel Journal of Zoology* **42**: 15–55.
[Lists *Scorpena azorica* Eschmeyer 1969, previously thought an Azorian endemic; **Azores**]

Golani, Daniel, A. Ben-Tuvia and B. Galil. 1983. Feeding habits of the Suez Canal migrant squirrelfish, *Sargoncentron rubrum*, in the Mediterranean Sea. *Israel Journal of Zoology* **32**: 194–204.
[**Squirrelfishes**]

Goldschmidt, Tijs. 1996. *Darwin's Dreampond: Drama in Lake Victoria*. The MIT Press, Cambridge, MA, 274 pp.
[**Cichlidae (II)**]

Gomon, Martin F., J. C. M. Glover and R. H. Kuiter. 1994. *The Fishes of Australia's South Coast*. The Flora and Fauna of South Australia Handbooks Committee. State Print, Adelaide, 992 pp.
[*Apistus*; **Darwin's roughy**; **Flatfishes**; **Flatheads**; **Rattails**; **Tigerperches**]

Goodsir, John. 1844. On the anatomy of the *Amphioxus lanceolatus*, lancelet of Yarrell. *Transactions of the Royal Society of Edinburgh* **15**: 247–63.
[**Lancelet**]

Goodsir, John. 1855. On the electrical apparatus in *Torpedo*, *Gymnotus*, *Malapterus*, and *Raja*. *Edinburgh Medical Journal* 1855/56, **1**: 139–42; 277–82; 563–5.
[**Electric fishes**]

Goolish, Edward M. 1991. Aerobic and anaerobic scaling in fish. *Biological Reviews* **66**: 33–56.
[**Length**]

Gosse, Philip Henry. 1851. *A naturalist's sojourn in Jamaica*. Longman, Brown, Green & Longmans, London.
[Wool of sheep – Color of Cows – Sea Horse rabbit *Marginalia* 335]

Gosse, Philip Henry. 1857. *Omphalos: an attempt to untie the geological knot*. London, J. Van Voorst, viii + 376 pp.

[Wherein Gosse attempts to convince us that God created the world about 6000 years ago, but bearing signs of great antiquity, because he had in mind the complete possible "life-history of the globe". Hence, for any period, the creatures thus created should bear the attributes they would have had, had the preceding stage been real (including a belly button – *omphalos* in Greek – for Eve, and even for poor, wombless, Adam). Thus, if God had decided, e.g. to create the world as it was in 1857, then it would have "cities filled with swarms of men, [...], houses half-built, castles fallen into ruins..." (p. 352). Hence fossils, and even fossil coprolites. Gosse was nothing if not consistent, if in a crazy sort of way; **Fossils**]

Gottfried, Michael D., Leonard J. V. Compagno and S. Curtiss-Bowman. 1996. Size and skeletal anatomy of giant "Megatooth" shark *Carcharodon megalodon*. *In*: A. Peter Klimley and David G. Ainley (eds.), *The great white shark: the biology of* Carcharodon carcharias, pp. 55–66. Academic Press, San Diego.

[Authors present evidence, based on teeth, that the largest Megatooth sharks may have reached 16–17 m and 50 tonnes; **Megatooth shark**]

Gould, John, Charles Darwin and Thomas Campbell Eyton. 1841. *Birds: described by John Gould, with a notice on their habits and ranges by Charles Darwin, and with an anatomical appendix by T. C. Eyton*. Smith, Elder and Co., London, 164 pp.

[*Birds*; **Cormorants; Darwin's Finches;** *Zoology*]

Gould, Stephen Jay. 1977. *Ontogeny and Phylogeny*. The Belknap Press of Harvard University Press, Cambridge, MA, 501 pp.

[**Gar-pike; Ontogeny**]

Gould, Stephen Jay. 1989. *Wonderful life: The Burgess shale and the nature of history*. W. W. Norton, New York, 347 pp.

[**Divine intervention**]

Gould, Stephen Jay. 1996a. Triumph of the root-heads. *Natural History*, January: 10–17.

[Reprinted in Gould 1998a; **Complexity**]

Gould, Stephen Jay. 1996b. *Full house: the spread of excellence from Plato to Darwin*. Harmony Books, New York, 244 pp.

[A book which readers without an obsession for baseball may find a bit tedious, and whose English edition, titled "Life's Grandeur", adds a "Baseball Primer" to the tedium; **Complexity**]

Gould, Stephen Jay. 1996c. *The Mismeasure of Man*. Revised and Expanded Edition. W. W. Norton and Company, New York, 444 pp.

[An excellent refutation of... **Social Darwinism**]

Gould, Steven Jay. 1997a. Darwinian fundamentalism. *The New York Review of Books* **44**(10): 34–7.

[**Non-Darwinian evolution**]

Gould, Steven Jay. 1997b. Evolution: the pleasures of pluralism. *The New York Review of Books*. **44**(11): 47–57.

[**Non-Darwinian evolution**]

Gould, Steven Jay. 1998a. *Leonardo's Mountain of Clams and the Diet of Worms*. Harmony Books, New York, 422 pp.

[**Dohrn; Owen**]

Gould, Steven Jay. 1998b. An awful, Terrible Dinosaurian Irony. *Natural History* **107**(1): 24–68.

[**Owen**]

Gould, Steven Jay. 2002. *The structure of evolutionary theory*. The Belknap Press/Harvard University Press, Cambridge, MA, 1433 pp.
[**Divine intervention**]

Grant, Peter R. 1999. *Ecology and evolution of Darwin's Finches*. 2nd edition. Princeton University Press, Princeton, NJ, xx + 492 pp.
[**Punctuated equilibrium**]

Grassé, Pierre-Paul. 1958. L'oreille et ses annexes. *In*: Pierre-Paul Grassé (ed.), *Traité de Zoologie: Anatomie, Systématique, Biologie*. Tome XIII: *Agnathes et Poissons: Anatomie, Ethologie, Systématique*. Fascicule II, pp. 1063–98. Masson, Paris.
["The ear and its annex organs"; **Weber**]

Gray, John Edward. 1860. Descriptions of new species of Salamanders from China and Siam. *Annals and Magazine of Natural History* **5** (3rd ser.): 151–2.
[**Goldfish (I)**]

Greenwood, P. H. 1974. *Cichlid fishes of Lake Victoria, East Africa: the biology and evolution of a species flock*. John Wright and Sons Ltd., Stonebridge Press, Bristol, 134 pp.
[Reprinted from the *Bulletin of the British Museum (Natural History)*, Zoology series, Supplement 6; **Cichlidae (II)**]

Griffith, Edward and Charles Hamilton Smith. 1834. *The class Pisces, with supplementary additions*. G. B. Whittaker, London, 680 pp.
[Vol. 10 of the translated and augmented edition of Cuvier 1829; contains the original description of Mountain mullet, by E. N. Bancroft; **Experiments (III); Hamilton Smith**]

Grigg, Richard W. 1982. Darwin Point: a threshold for atoll formation. *Coral Reefs* **1**: 29–34.
[*Coral Reefs*; **Eponym**; **Latitude**]

Groeben, Christiane (ed.) 1982. *Charles Darwin (1809–1882)/Anton Dohrn (1840–1909): Correspondence*. Gaeteno Macchiaroli Editore, Naples, 118 pp.
[*Autobiography*; *Correspondence*; **Dohrn**]

Groot, Cornelis and Leo Margolis (eds.) 1991. *Pacific Salmon Life Histories*. University of British Columbia Press, Vancouver, 564 pp.
[**Jacks**; **Salmon, Pacific**]

Gross, Mart R., Ronald M. Coleman and Robert M. McDowall. 1988. Aquatic productivity and the evolution of diadromous fish migration. *Science* **239**: 1291–3.
[**Diadromy**]

Gross, Paul R. and Norman Levitt. 1994. *Higher Superstition: the Academic Left and its Quarrels with Science*. The Johns Hopkins University Press, Baltimore, xv + 328 pp.
[**Seeing**]

Grove, Jack Stein. 1985. Influence of the 1982–1983 El Niño event upon the ichthyofauna of the Galápagos archipelago. *In*: G. Robinson and E. M. del Pino (eds.), *El Niño in the Galápagos Islands: the 1982–1983 event*, pp. 191–8. Publication of the Charles Darwin Foundation for the Galápagos Islands, Quito.
[**Punctuated equilibrium**]

Grove, Jack Stein and Robert J. Lavenberg. 1997. *The Fishes of the Galápagos Islands*. Stanford University Press, Stanford, 863 pp.
[**Agassiz**; **Clingfishes**; **Flying fish**; **Galápagos**; **Groupers**; **Morays**; **Scorpionfishes**; **Wrasses**]

Groves, Paul. 1998. Leafy Sea Dragons. *Scientific American* **279**(6): 84–9.

[**Pipefishes (I)**]

Gruber, Howard E. 1981. *Darwin on Man: a psychological study of scientific creativity*. Second Edition. The University of Chicago Press, Chicago, 310 pp.

[**Preface**]

Gruber, Jacob W. 1969. Who was the *Beagle's* naturalist? *British Journal for the History of Science* **4**(15): 266–82.

[**Naturalist**]

Günther, Albert Charles Lewis Gotthilf. 1853. *Die Fische des Neckars, untersucht und beschrieben von A. Günther*. Verlag von Ebner & Seubert, Stuttgart, 136 pp.

[*"Fishes of the Neckar River, studied and described by A. Günther"*; **Wels catfish**]

Günther, Albert Charles Lewis Gotthilf. 1855. Beiträge zur Kenntnis unserer Süsswasserfische. *Archiv für Naturgeschichte* **21**: 197–212.

["Contribution to the knowledge of our freshwater fishes"; **Wels catfish**]

+Günther, Albert Charles Lewis Gotthilf. 1859–70. *Catalogue of Acanthopterygian Fishes in the collection of the British Museum*. 8 volumes. London.

[**Blennies; Catfishes; Dragonets; Freshwater fishes (III); Great white shark; Günther; Guppy; Livebearers**]

+Günther, Albert Charles Lewis Gotthilf. 1865a. *Record of the Zoological Literature*, p. 194.

[**Wrasses**]

+Günther, Albert. 1865b. On the Pipefishes belonging to the Genus *Phyllopteryx*. *Proceedings of the Scientific Meetings of the Zoological Society of London* **1865**: 327–8.

[**Pipefishes (I)**]

+Günther, Albert Charles Lewis Gotthilf. 1868a. Description of freshwater fishes from Surinam and Brazil. *Proceedings of the Zoological Society of London* **1868**: 229–47.

[**Beards; Catfishes; Mouth-brooding**]

+Günther, Albert Charles Lewis Gotthilf. 1868b [Abstract of Malm 1868]. *Record of the Zoological Literature* **5**: 155–7.

[**Deal fish; Flatfish controversy (I)**]

+Günther, Albert Charles Lewis Gotthilf. 1869. An account of the Fishes of the States of Central America. *Transactions of the Zoological Society of London* **6**: 377–494.

[**Panama; Livebearers**]

+Günther, Albert Charles Lewis Gotthilf. 1871. Description of *Ceratodus*. *Philosophical Transactions of the Royal Society of London* **161**: 511–72.

[Wherein Günther demonstrated *Ceratodus* to be a wonderfully interesting creature; *Calendar*, # 7983; **Lungfish, Australian**]

+Günther, Albert Charles Lewis Gotthilf. 1872a. On *Ceratodus*. *Popular Science Review* **11**: 257–66.

[In a letter of Nov. 23, 1872, CD thanks the author for issue with this article, which CD had already read with great interest; *Calendar* # 8645; **Lungfish, Australian**]

Günther, Albert Charles Lewis Gotthilf. 1872b. Report on several collections of fishes recently obtained for the British Museum. *Proceedings of the Zoological Society of London*. 1871 (part 3): 652–75.

[**Eels**]

Bibliography

+Günther, Albert Charles Lewis Gotthilf. 1880. *Introduction to the Study of Fishes*. Adams & Charles Black, Edinburgh, 720 pp.

[**Bonito; Cusk eels; Filefishes; Great white shark; Grouper (II); Günther; Linnaeus; Sole**]

Guichenot, Alphonse. 1848. Fauna Chilena: Peces. *In*: Claudio Gay (ed.), *Historia fisica y politica de Chile*. Vol. 2: *Zoologia*, pp. 137–370. Paris & Santiago.

[**Merlu**]

Guimarães, Ricardo Zaluar Passos. 1999. Chromatic and morphological variations in *Halicampus crinitus* (Jenyns) (Teleostei: Syngnathidae) from Southeastern Brazil with comments on its synonymy. *Revue Française d'Aquariologie* **26**(1/2): 7–10.

[Including errata; **Pipefishes (I)**]

Gunner(-us), Johan Ernst. 1765. Brugden (*Squalus maximus*). *Det Trondhiemske Selskabs Skrifter* **3**: 33–49.

["The basking shark (*Squalus maximus*)"; **Basking shark**]

Guppy, Plantagenet Lechmere. 1922. A naturalist in Trinidad and Tobago. *The West India Committee Circular*, October 26, 1922, **37**: 492–3.

[The author can be assumed to have been a descendent of the Reverend John Lechmere Guppy; **Guppy**]

Haeckel, Ernst. 1866. *Generelle Morphologies der Organismen: Allgemeine Grundzüge der organischen Formen-Wissenschaft, mechanisch begründet durch die von Charles Darwin reformierte Descendenz-Theorie*. Vol. 2, Georg Reimer Verlag, Berlin, clx + 462 pp.

["*General morphology of the organisms. General principles of the science of organic form, mechanically founded on the theory of descent, as revised by CD*"; **Gar-pike; Haeckel; Ontogeny**]

#Haeckel, Ernst. 1868. *Natürliche Schöpfungsgeschichte*. Georg Reimer Verlag, Berlin.

["*A natural history of creation*"; p. 438 embryology of Amphioxus; 439 I shd say creations like larvae of Ascidians gave rise to Vertebrata. Both groups out of the same source; 442 Selachians parent-form of all chief Vertebrata; 443 Selachii only in remnant; 445 Selachians parent form; 446/7 Rivers separate intermediate classes; 448 Lepidosiren, *Marginalia* 359]

Haeckel, Ernst. 1890. *Plankton-Studien: vergleichende Untersuchungen über die Bedeutung und Zusammensetzung der pelagischen Fauna und Flora*. Verlag von Gustav Fischer, Jena, viii + 105 pp.

["*Plankton studies: comparative investigations on the importance and composition of pelagic fauna and flora*." A misguided polemic against Victor Hansen's finding, of which only the proposal of the word 'plankton' is accepted, which now replaces the clumsy expression in Haeckel's title; **Plankton**]

+Halbertsma, Hidde Justuszoon. 1864. Normaal en abnormaal hermaphroditismus bij de visschen. *Versl. Akad. Wet. Amsterdam, Wis. Nat. Afd.* **16**: 165–78.

[A. Günther's summary of this contribution, on "Normal and abnormal hermaphroditism in fishes", is on p. 147 of the *Zoological Record*, Vol. **1** (1864); it does not indicate disagreement with the author's conclusions; **Groupers (II)**]

Haldane, John Burton Sanderson. 1949a. Suggestions as to the quantitative measurement of rates of evolution. *Evolution* **3**: 51–6.

[**darwins**]

Haldane, John Burton Sanderson. 1949b. Disease and evolution. *La Ricerca Scientifica*, Suppl. A**19**: 68–76.

[**Beauty**]

Halstead, Bruce Walter. 1978. *Poisonous and venomous marine animals of the world*. Revised edition. The Darwin Press, Princeton, 1043 pp. + 283 plates.

[**Burrfishes (I); Puffers**]

Hamilton, Francis [Buchanan]. 1822. *An account of the fishes found in the river Ganges and its branches*. Edinburgh & London, vii + 405 pp.
[**Carp, Indian**]

Hamilton, William D. 1964. The genetical theory of social behaviour. *Journal of Theoretical Biology* **7**: 1–52.
[**Altruism**]

Hanken, James. 1999. Beauty beyond belief: the art of Ernst Haeckel transcends his controversial scientific ideas. *Natural History* **107**(12): 56–9.
[**Haeckel**]

Harden-Jones, F. R. 1968. *Fish Migration*. Edward Arnold, London, 325 pp.
[**Migrations**]

#Harlan, Richard. 1842. Notice of two new fossil mammals from Brunswick Canal; with observations on some of the fossil quadrupeds from the United States. *American Journal of Science and Arts* **14**: 142–4.
[CD cites Silliman's Journal as his source, which refers to the name of the founding editor of the journal in question, from which he extracted that Fish one step lower in America (*Notebook S*, p. 473). This must refer to that part of Harlan's text stating: "Indications of the existence of a fossil Sus [i.e. pig] I discovered several years since, in a collection of fossils obtained by Mr. Natall, in Newbern, North Carolina, in the newest tertiary, post-pliocene, – these were the teeth of a Sus, occurring with mastodon, elephant, elk, deer, horse, seal, cetacean, tortoise, shark, skate, snake and fish, – all congregated together as if in the mouth of some antediluvian estuary, and commingled with fossil shells, many of which of existing species."]

+Harmer, Thomas. 1767. Remarks on the very different Accounts that have been given of the fecundity of Fishes with Fresh Observations on that Subject. *Philosophical Transactions of the Royal Society of London* **57**: 280–92.
[**Eggs of fish; Smelt**]

Harris, Michael. 1998. *Lament for an Ocean. The collapse of the Atlantic cod fishery: a true crime story*. McClelland & Stewart, Toronto, 342 pp.
[**Greenland halibut**]

Harrison, Peter. 1987. *A Field Guide to Seabirds of the World*. The Stephen Greene Press, Lexington, MA, 317 pp.
[**Cormorants; Mother Carey's chickens; Petrels**]

Hartmann, Georg Leonard. 1827. *Helvetische Ichthyologie, oder ausführliche Naturgeschichte der in der Schweiz sich vorfindenden Fische*. Drell, Füßli und Compagnie, Zürich, 240 pp.
["Helvetic ichthyology, or comprehensive natural history of the fish occurring in Switzerland"; **Wels catfish**]

Harvey, W. H. and J. D. Hooker. 1845 (1847). Algae. *In*: J. D. Hooker (ed.), *The Cryptogamic Botany of the Antarctic Voyage*, pp. 63–81; 148–96; 237–41. London, L. Reeve.
[Title page is dated 1845, but volume is thought to have appeared in 1847; **Kelp**]

Hastings, John Woodland. 1971. Light to hide by: ventral bioluminescence to camouflage the silhouette. *Science* **173**: 1016–17.
[**Bioluminescence**]

Hastings, John Woodland. 1975. Light for all reasons: versatility in the behavioural repertoire of the flashlight fish. *Science* **190**: 74–6.
[**Bioluminescence**]

Hastings, Philip A. and Christopher W. Peterson. 1986. A novel sexual pattern in serranid fishes: simultaneous hermaphrodites and secondary males in *Serranus fasciatus*. *Environmental Biology of Fishes* **15**(1): 59–68.
[**Groupes**]

Hector, J. 1875. Description of five new species of fishes obtained in the New Zealand seas by H.M.S. "Challenger" Expedition July 1874. *Annals and Magazine of Natural History* (Ser. 4), **15**(85): 78–82.
[**Rattails**]

Helfman, Gene S., Bruce B. Colette and Douglas E. Facey. 1997. *The Diversity of Fishes*. Blackwell Science, Malden, MA, 528 pp.
[**Chimaera; Great white shark; Ichthyology; Megatooth shark**]

Hensen, Victor. 1887. *Ueber die Bestimmung des Planktons oder des im Meer treibende Materials an Pflanzen und Thieren*. Fünfter Bericht der Kommission zur wissenschaftlichen Erforschung der deutschen Meere in Kiel für die Jahre 1882–1886. Berlin: 1–107.
["On the identification of plankton, or the plant and animal matter drifting in the sea"; **Plankton**]

Herbert, Pamela. 1980a. Historical aspects of Salmonellosis. *The Western Journal of Medicine* **133**: 408–17.
[**Cod liver oil**]

Herbert, Sandra (ed.) 1980b. *The Red Notebook of Charles Darwin*. Cornell University Press, London, 164 pp.
[*Notebooks*]

+Heron, Robert. 1842. On the breeding of goldfishes in the author's menagerie. *Annals and Magazine of Natural History* **8**: 533.
[CD's reference to this paper as Proc. Zool. Soc. May 25 appears to have been in error, as the journal in question, in 1842, contains no paper by Heron; **Goldfish (I)**]

Herring, Peter J. (ed.) 1978. *Bioluminescence in Action*. Academic Press, London, 570 pp.
[**Bioluminescence**]

Hewlett, P. S. 1975. Mendel's Achievement. *The Biologist* **22**(3): 117–19.
[**Mendel**]

Hilborn, Ray and Carl J. Walters. 1992. *Quantitative fisheries stock assessment: choice, dynamics and uncertainty*. Chapman & Hall, New York, 570 pp.
[**Gear selectivity**]

Hill, Richard. 1881. The fishes of Jamaica. *Handbook of Jamaica*, pp. 121–6. Government Printing Office, Kingston.
[**Experiments (III)**]

Hobson, Edmund S. 1974. Feeding relationships of teleostean fishes on coral reefs in Kona, Hawaii. *U.S. Fisheries Bulletin* **72**(4): 915–1031.
[**Butterflyfishes**]

Hoffman, R. B., G. A. Salinas and A. A. Baky. 1977. Behavioral analyses of killifish exposed to weightlessness in the Apollo-Soyuz test project. *Aviation, Space and Environmental Medicine* **48**: 712–17.
[**Motion sickness**]

Hollard, H. L. G. M. 1854. Monographie de la famille des Balistidés. Suite 3. *Annales des Sciences Naturelles – Zoologie et Biologie Animale* (Sér. 4) **2**: 321–66.
["A monograph of the Family Balistidae"; **Filefishes**]

Hopkins, Robert S. 1969. *Darwin's South America*. The John Day Company, New York, 224 pp.
[**Beagle**]

Houde, Anne E. 1997. *Sex, Color and Mate Choice in Guppies*. Princeton University Press, Princeton, xii + 210 pp.
[**Punctuated equilibrium**]

Hubbs, Carl L. 1958. *Ogcocephalus darwini*, a new batfish endemic at the Galápagos Islands. *Copeia* 1958 (no. 3): 161–70; Pls. 1–5.
[This fish also occurs in northern Peru, and hence is not a Galápagos endemic, notwithstanding the title; **Eponym**]

Hull, David L. 1973. *Darwin and his Critics: the Reception of Darwin's Theory of Evolution by the Scientific Community*. Harvard University Press, Cambridge, MA, xii + 473 pp.
[A compilation of reviews of CD's books; **Caviller**].

Humboldt, (Friedrich Wilhelm Heinrich) Alexander von. 1805–39. *Voyage aux régions equinoxiales du Nouveau Continent, fait en 1799, 1800, 1801, 1802, 1803 et 1804 par Alexandre von Humboldt et Aimé Bonpland*. Paris.
[The different parts of this immense work, covering the geography, natural history, and the history and sociology of South America, and which greatly shaped the way CD planned to work on the *Beagle*, were published from 1805 in various editions, a process lasting until 1839; **Darwin and ichthyology**]

Humphreys, John. 1995. The Laws of Lamarck. *Biologist* **42**(3): 121–5.
[**Lamarck**]

+Humphreys, Noel. 1830. Some account of the Stickleback Fish (*Gasterósteus aculeàtus*). *Loudon's Magazine of Natural History* **3** (14): 329–31.
[Author's name listed as 'O.' in original; identified from CD referring to the same author; **Sticklebacks**]

+Humphreys, Noel. 1857. *River Gardens; being an account of the best methods of cultivating fresh water plants in aquaria in such a manner as to afford suitable abodes to ornamental fish, and many interesting kinds of aquatic animals*. Sampson Low, Son and Co., London.
[**Sticklebacks**]

+Hunter, John. 1837. *Observations on certain parts of the animal oeconomy . . . with notes by Richard Owen*. London.
[**Secondary sexual characters**]

Huot, André. 1902. Recherches sur les poissons lophobranches. *Annales des Sciences Naturelles* (Zool.), ser. 8, **14**: 197–288.
["Research on lophobranch fishes" i.e. Syngnathoidea; **Seahorses (II)**]

Hutchings, Jeffrey A. and Ransom A. Myers. 1994. What can be learned from the collapse of a renewable resource? Atlantic cod, *Gadus morhua*, of Newfoundland and Labrador. *Canadian Journal of Fisheries and Aquatic Sciences* **51**(9): 2126–46.
[**Bacalao**]

Hutton, James. 1795. *Theory of the Earth, with Proofs and Illustrations. In Four Parts*. 2 Volumes. Edinburgh.
[Vol. 3 edited from manuscript by Sir Archibald Geikie, London, 1899; **Fossil record; Natural selection**]

Huxley, Julian. 1932. *Problems of relative growth*. The Dial Press, New York, xix + 219 pp.
[Julian was a grandson of T. H. Huxley; **Growth**]

#Huxley, Thomas Henry. 1864. *Lectures on the elements of comparative anatomy*. John Churchill & Sons, London.

[Characters of Fishes; 62, 64 Fishes & Amphibia hardly distinguishable - (good case telling how unlike say a Frog & Salmon.-) *Marginalia* 425]

#Huxley, Thomas Henry. 1869. *An introduction to the classification of animals*. John Churchill & Sons, London.

[p. 65 Digit in Fishes *Marginalia* 424]

+Huxley, Thomas Henry and Robert Etheridge. 1865. *A catalogue of the collection of fossils in the Museum of Practical Geology, with an explanatory introduction*. London.

[**Complexity**]

I(nternational) C(ommission on) Z(oological) N(omenclature). 1999. *International Code of Zoological Nomenclature*, 4th edition. International Trust for Zoological Nomenclature, London, xxix + 306 pp.

[**Code; Names, scientific; Taxonomy**]

Jackson, Jeremy B. C., M. X. Kirby, W. H. Berger, K. A. Bjorndal, L. W. Botsford, B. J. Bourque, R. Cooke, J. A. Estes, T. P. Hughes, S. Kidwell, and 9 others, 2001. Historical overfishing and the recent collapse of coastal ecosystems. *Science* **293**: 629–38.

[**Hippopotamus**]

Jackson, Michael H. 1993. *Galápagos: A Natural History*. University of Calgary Press, Calgary, 315 pp.

[**Galápagos; Lizards; Morays; Penguins**]

#Jaeger, Gustav. 1874. *In Sachen Darwin's, insbesondere contra Wigand*. Schweizerbart, Stuttgart.

["*On the Darwin case, especially against Wigand*"; Sexual S. use of barbs of fishes as exciting organs *Marginalia* 430]

#Jameson, Robert. 1821. *Manual of Mineralogy*. Archibald Constable, Edinburgh.

[p. 304: Common coal is the more common variety ... of older formation than the Browne. – (was formed before Mammalia Aves Amphibia but there existed fishes & shells); 413 Sandstone ... Iron Pyrites, Sulphates of Lime & Iron occurs in this mineral. also sometimes Amber & Browne clay also various seeds of tropical plants. also Crocodiles &c &c & Cocoa Nuts fishes; 458 The teeth of Sharks are found in great plenty in the Limestones of Malta *Marginalia* 438, 439]

Janvier, Philippe and J. L. Welcomme. 1969. Affinités et paléobiologie de l'espèce *Carcharodon mégalodon* Ag. squale géant des Faluns de la Touraine et de l'Anjou. *Revue de la Fédération Française des Sociétés de Sciences Naturelles*, ser. 5, **8**(34): 1–6.

["Affinities and paleobiology of the species *Carcharodon megalodon* Agassiz, a giant shark of the tertiary, shell-rich sedimentary rocks of the Anjou and Touraine regions"; **Megatooth shark**]

Jefferson, Thomas, Stephen Leatherwood and Marc A. Weber. 1993. *Marine Mammals of the World*. FAO, Rome, 320 pp.

[**Mammals; Whales**]

Jenyns, Leonard. 1835. *A Manual of British Vertebrate Animals, or Descriptions of all the Animals belonging to the Classes Mammalia, Aves, Reptilia, Amphibia and Pisces which have been observed in the British Islands: including the domesticated, naturalized and extirpated Species, the whole systematically Arranged*. The Pitt Press, Cambridge, 599 pp.

[**Jenyns (I); Wels catfish**]

Jenyns, Leonard. 1837. Some Remarks on the Study of Zoology, and on the present state of the Science. *Magazine of Zoology and Botany* **1**(1): 1–31.

[**Jenyns (I)**]

+Jenyns, Leonard. 1840–2. Fish. *In: The Zoology of the Voyage of H.M.S. Beagle, under the Command of Captain *FitzRoy, R.N. during the years 1832 to 1836*. Edited and superintended by Charles Darwin.

Smith, Elder and Co., Cornhill, London (in 4 parts): i–xvi + 172 pp. Plates 1–29.

[pp. 1–32: Jan. 1840; pp. 33–64: June 1840; pp. 65–96: April 1841; pp. 97–172: April 1842. *Fish* is still in print, in three editions: (i) volume 6 of the 29-volume edition of The *Works of Charles Darwin* (Barrett and Freeman 1989); (ii) as volume 3 of *The Zoology of the Voyage of the Beagle during the years 1832–1836. Edited by Charles Darwin*. Facsimile Reprint, 1980, Nova Pacifica Publishing, Wellington, New Zealand; and (iii) on the second edition of the *Darwin *CD-ROM* (Ghiselin and Goldie 1997). Edition (i) differs from (ii), and from the original edition of *Fish* in being repaginated, as required by the use of a format corresponding to that of CD's other books, i.e., smaller than the quarto used for all volumes of the *Zoology*. Here are CD's annotations to his copy of *Fish*: Important to find those genera which have no marine species or migratory species. In these distribution must offer great difficulty. – are there many? Feb./56/; Galapagos Fish; p.3 True Perch in S. America; p.18 Dules R Tahiti – other species Java; p.79 Atherina Valparaiso – some in brackish, some salt species; p.114 Poecilia Cyprinidae Lebias S. America; p.120 Mesites Nov. Gen.; Salmonidae; p.123 Tetragonopterus; p.131 Aplochiton F.W. Genus Falklands & T. del Fuego; p.142 Anguilla New Zealand *Marginalia* 182, 183; **cited in many entries**]

Jenyns, Leonard. 1843. *See* White 1843.

+Jenyns, Leonard. 1846. *Observations in Natural History: with an introduction on habits and observing as connected with the study of that science. Also a calendar of periodic phenomena in natural history; with remarks on the importance of such registers*. John van Voorst, London, 440 pp.

[Polecat devouring Eels p. 55 some parallel facts. Rooks p. 150 feeding on Fish; 212 Abnormal Gold Fishes; 278 Flies hatched in gentlemans intestines, *Marginalia* 441; **Goldfish (I)**]

Jenyns, Leonard. 1862. *Memoir of the Rev. John Stevens Henslow*. John Van Voorst, London, 278 pp.

[**Henslow**]

Jenyns, Leonard [writing as L. Blomefield]. 1889. *Chapters of my life: with Appendix Containing Special Notices of Particular Incidents and Persons; also Thought on Certain Subjects*. Reprint with Additions. For Private Circulation, Bath, 133 pp.

[**Jenyns (I)**]

Johannes, Robert E. 1981. *Words of the lagoon: fishing and marine lore in the Palau district of Micronesia*. University of California Press, Berkeley, xiv + 245 pp.

[**Groupers (II); Vertebrate origins**]

John Paul II. 1996. L'origine et l'évolution de la vie. *L'Osservatore Romano* [Edition française] no. 44, 29 October 1996: 4.

["The origin and evolution of life"; note that the *Osservatore Romano* is the Vatican's official mouthpiece; the English version plus commentaries may be found in the *Quarterly Reviews of Biology* **72**: 377–406 (1997); **Creationism; Flatfish controversy (I); Non-Darwinian evolution**]

Johnels, Alf G. and Gustav S. Svensson. 1954. On the Biology of *Protoperus annectens* (Owen). *Arkiv för Zoologi* **7**(7): 131–64.

[Includes detailed comparisons with other lungfishes; **Lungfishes**]

Johnson, James Yate. 1866. Description of *Trachichthys darwinii*, a new species of berycoid fish from Madeira. *Proceedings of the Zoological Society of London* **1866** (2): 311–15.

[**Darwin's roughy; Eponym; Madeira**]

Johnstone, G. W. 1977. Comparative feeding ecology of the Giant petrels *Macronectes giganteus* (Gmelin) and *M. halli* (Mathews). *In:* George L. Llano (ed.), *Proceedings of the Third SCAR Symposium on Antarctic Biology*, pp. 647–68. Smithsonian Institution, Washington, D.C., and Gulf Publishing, Houston.

[**Food webs**]

Jordan, David Starr. 1896. Notes on fishes little known or new to science. *Proceedings of the California Academy of Science* (Ser. 2) **6**: 201–44, Pls. 20–43.

[Reprinted in *Contrib. Biol. Hopkins Seaside Lab*. **5**: 1–48. Plates XX–XLIII; describes '*Sebastodes*' *darwini*; **Eponyms**]

Jordan, David Starr and Charles Henry Gilbert. 1899. The fishes of Bering Sea. *In*: David Starr Jordan, Leonhard Stejneger and Fredric Lucas (eds.), *The Fur Seal and Fur-Seal Islands of the North Pacific Ocean*: Part III. *Special Papers relating to the Fur Seal and the Pribiloff Islands*, pp. 433–510. Government Printing Office, Washington, D.C.

[**Bering pike**]

Junker, Thomas and Masha Richmond. 1996. *Charles Darwin's correspondence with German naturalists*. Basiliken-Press, Marburg an der Lahn, xlii + 266 pp.

[Based on the *Calendar*; demonstrates that Germans were, after the British, CD's most important correspondents by far; **Möbius; Swimbladder**]

+Jurine, Louis. 1825. Histoire abrégée des poissons du lac Léman. *Société de Physique et d'Histoire Naturelle de Genève*. **3**: 149.

["Short history of the fishes of Lake Geneva"; **Burbot**]

Kalm, Peter. 1753. *Travels into North America*. Stockholm.

[Translated and revised edition in Adolph B. Benson's *Peter Kalm's travels to North America* (1937), Wilson-Erickson,

New York; **Cod; Obstinate nature**]

Kass-Simon, Gabriele, Patricia Farnes and Deborah Nash (eds.) 1990. *Women in Science: Righting the Record*. Indiana University Press. Bloomington & Indianapolis, xvi + 398 pp.

[**Morris, Margaretta Hare (II)**]

Kaup, Johann Jakob. 1856. *Catalogue of the apodal fish in the collection of the British Museum*. London, pp. 1–163.

[**Eels**]

Kay, E. Alison. 1994. Darwin's biogeography. *In*: Rod MacLeod and Philip F. Rehbock (eds.), *Darwin's Laboratory: evolutionary theory and natural history in the Pacific*, pp. 49–69. University of Hawai'i Press, Honolulu.

[**Barriers**]

Keast, Allen. 1978. Trophic and spatial interrelationships in the fish species of an Ontario temperate lake. *Environmental Biology of Fishes* **3**: 7–31.

[**Sunfishes**]

Kessler, Karl Fedorovich. 1870. The Volga lamprey, *Petromyzon wagneri* n. sp. *Trudy St. Petersburg Obsh. Estestv*. **1**: 207–14.

[**Caspian Sea**]

Keyes, I. W. 1972. New records of the elasmobranch *C. megalodon* (Agassiz) and a review of the genus *Carcharodon* in the New Zealand fossil record. *New Zealand Journal of Geology and Geophysics* **15**(2): 228–42.

[**Megatooth shark**]

Keynes, Richard Darwin. 1979. *The Beagle Record: Selections from the original pictorial records and written accounts of the voyage of H.M.S. Beagle*. Cambridge University Press, Cambridge, 409 pp.

[**Beagle**]

Keynes, Richard Darwin (ed.) 1988. *Charles Darwin's* Beagle *Diary*. Cambridge University Press, Cambridge, xxix + 464 pp.
[*Diary*]

Keynes, Richard Darwin (ed.) 2000. *Charles Darwin' s zoology notes and specimen lists from H.M.S Beagle*. Cambridge University Press, Cambridge, xxxiv + 430 pp.
[*Catalogue*; *Ornithological Notes*; **Mauritius**; *Zoology Notes*]

Keys, John D. 1963. *Food for the Emperor: Recipes of Imperial China with a Dictionary of Chinese Cuisine*. The Ward Ritchie Press, San Francisco, 121 pp.
[**Bird's nest soup**]

Kimura, Motoo. 1983. *The neutral theory of molecular evolution*. Cambridge University Press, Cambridge, xv + 367 pp.
[**Non-Darwinian evolution**]

King, Philip P., Pringle Stokes and Robert FitzRoy. 1836. Sketch of the surveying voyages of his Majesty's ships *Adventure* and *Beagle*, 1825–1836. *Journal of the Royal Geographical Society of London* **6**: 311–42.
[*Beagle*; **King**]

Kingdon, Jonathan. 1989. *Island Africa: the evolution of Africa' s rare animals and plants*. Princeton University Press, Princeton, 287 pp.
[**Cichlidae (II)**]

+Kingsley, Charles. 1870. Strange noises heard at sea off Grey Town. *Nature* **2**: 40.
[**Sounds**]

Kinsbourne, Marcel. 1978. Evolution of language in relation to lateral action. *In*: M. Kinsbourne (ed.), *Asymmetrical function of the brain*, pp. 553–65. Cambridge University Press, New York.
[Presents several non-fish examples of torsions, and several corollaries of the hypothesis deriving vertebrates from twist-headed ur-bilaterians; **Dohrn**]

+Kirby, William. 1835. *On the power, wisdom and goodness of God as manifested in the creation*. Bridgewater Treatise no. 7: On the history, habits and instincts of animals, vol. II. London.
[**Catfishes**; **Climbing fish**; **Lumpfish**]

Klimley, A. Peter and David G. Ainley (eds.) 1996. *The great white shark: the biology of* Carcharodon carcharias. Academic Press, San Diego, 517 pp.
[**Great white shark**]

Knight, Andrew. 1828. On some circumstances relating to the economy of bees. *Philosophical Transactions of the Royal Society of London*: 319–23.
[**Hybrids**]

Kölmel, Reinhard. 1986. Victor Hensen als Meeresforscher. *Biologie in unserer Zeit* **16**(3): 65–70.
["Victor Hensen as Marine Scientist"; **Plankton**]

Koestler, Arthur. 1971. *The Case of the Midwife Toad*. Hutchinson, London, 187 pp.
[**Non-Darwinian evolution**]

Kogan, Bernard R. (ed.) 1960. *Darwin and his critics: the Darwinian revolution*. Wadsworth Publishing Company, San Francisco, 180 pp.
[Excerpts from both supporting (+) and negative (−) reviews of CD's work, including many of the authors presented here, e.g. Agassiz (−), Owen (−), Lyell (+); Wallace (+); **Caviller**]

Kouril, J., J. Hamácková, Z. Adámek, I. Sukop, I. Stibranyiová and R. Vachta. 1996. The artificial propagation and culture of young weatherfish (*Misgurnus fossilis* L.). *In*: A. Kirchhofer and D. Hefti

(eds.), *Conservation of Endangered Freshwater Fish in Europe*, pp. 305–10. Birkhäuser Verlag, Basel.
[**Loaches**]

+Kovalevsky, Aleksandr Onofrievich. 1866. Entwicklungsgeschichte der einfachen Ascidien. *Mémoires de l'Académie Impériale des Sciences de St. Pétersbourg*, ser. 7, **10**: 1–19.
["Ontogeny of simple ascidians"; **Seasquirts**]

Kraus, Scott D. and G. S. Stone. 1995. Coprophagy by Wilson's storm petrel (*Oceanites oceanicus*) on North Atlantic right whale (*Eubalaena glacialis*) feces. *Canadian Field Naturalist* **109**(4): 443–4.
[**Food webs**]

Krefft, Johann Ludwig Gerhard. 1870. Description of a gigantic amphibian allied to the genus *Lepidosiren*, from the Wide-Bay district, Queensland. *Proceedings of the Zoological Society of London* **1870** (2): 221–4.
[**Lungfish, Australian**]

Krupp, F. and W. Schneider. 1989. The fishes of the Jordan River drainage basin and Azraq Oasis. *Fauna of Saudi Arabia* **10**: 347–416.
[**Fossils**]

Kuhn, Thomas. 1970. *The Structure of Scientific Revolutions*, 2nd edition, revised and enlarged. *International Encyclopedia of Unified Science*, **Vol. II**, No 2. The University of Chicago Press, 210 pp.
[First edition 1962; **Paradigm; Popper**]

Kuiter, Rudie H. 2000. *Seahorses, Pipefishes and their relatives*. TMC Publishing, Chorleywood, UK, 240 pp.
[**Pipefishes (I); Pipefishes, ghost; Seahorses**]

Kullander, Sven O. 1999. Fish species – how and why. *In*: Joseph S. Nelson and Paul J. P. Hart (eds.), *The species concept in fish biology*. Special Issue, Reviews in Fish and Fisheries **9** (4): 325–52.
[**Characins; Species**]

Kullander, Sven O. and Hans Nijssen. 1989. *The Cichlids of Surinam. Teleostei: Labroidei*. E. J. Brill, Leiden, 256 pp.
[*Geophagus*]

+Kupffer, Carl Wilhelm von. 1869. Die Stammverwandtschaft zwischen Ascidien und Wirbelthieren. *Archiv für mikroskopische Anatomie* **5**: 459–63.
["Phylogenic affinities between Ascidians and Vertebrates"; **Seasquirts**]

Kurlansky, Mark. 1997. *Cod: A biography of the fish that changed the world*. Walker and Company, New York, 294 pp.
[**Cod; Cod liver oil**]

Lacepède, Bernard Germain Etienne de. 1798–1803. *Histoire Naturelle des Poissons*. 5 Vols. Chez Plassan, Paris.
[After the French Revolution, Lacepède became 'citoyen La Cépède,' which was more conducive to keeping one's head on, and which some taxononomists insist is the way his name should be written, but I could not force myself to do it in this book; **Darwin's Bass; Goatfishes; Megatooth shark; Pipefishes (I); Scorpionfishes; Sunfishes**]

Lacepède, Bernard Germain Etienne de. 1804. Mémoire sur plusieurs animaux de la Nouvelle Hollande dont la description n'a pas encore été publiée. *Annales du Muséum d'Histoire Naturelle de Paris* **4**: 184–211.
[**Pipefishes(I)**]

Lack, David. 1953. Darwin's Finches. *Scientific American* **188**(4): 66–72.
  [*Birds*; **Darwin's Finches**]
Lagler, Karl F., John E. Bardach, Robert R. Miller and Dora R. May Passino. 1977. *Ichthyology*, 2nd edition. John Wiley & Sons, New York, xv + 506 pp.
  [**Scales**]
Lakatos, Imre. 1970. Falsification and the methodology of research programmes. *In*: I. Lakatos and A. Musgrave (eds.), *Criticism and the Growth of Knowledge*, pp. 91–196. Cambridge University Press, Cambridge.
  [**Evolution; Paradigm**]
Lakoff, George. 1987. *Women, Fire and Dangerous Things: what categories reveal about the mind*. University of Chicago Press, Chicago and London, 614 pp.
  [**Classification**]
+Lamarck, Jean Baptiste Pierre Antoine de Monet, chevalier de. 1830. *Philosophie Zoologique. Nouvelle Edition*. 2 Volumes. Paris, Dentu, 495 pp.
  [Reprinted by J. Cramer, Weinheim, 1960; "*Zoological philosophy. New Edition*". CD comment on 1st edition (1809): Very poor & useless Book; p.148/9 The economy of world would have gone on without Bats or Ostriches. – It can only be following out some great principle. It is clear Birds made preeminently for air. yet if no birds: Mammalia would best take place. There limits to this Adaptation. Fish could hardly have lived out of water. Though Crabs – Spiders in water; p.157 NB Snakes perform the parts of fish, & fish of snakes *Marginalia* 478, 479; **Flying**]
Lankaster, Ian, 1995. Who was . . . Philip Henry Gosse? *The Biologist* **42**(4): 176–7.
  [**Fossils**]
Lay, G. Tradescant. 1829. Observation on a species of *Pteropus* from Bonin. *Zoological Journal* **4** (Art. LVI): 457–8.
  [Where an unfortunate fruit bat indeed swam pertinaciously after a boat; **Bats**]
Lecointre, Guillaume. 2000. What is a fish? *In*: Rainer Froese and Daniel Pauly (eds.), *FishBase 2000: Concept, Design and Data Sources*, Box 3, p. 57. Los Baños, Philippines.
  [Text also available from www.fishbase.org; **Fish(es)**]
LeGrand, Homer Eugene. 1988. *Drifting continents and shifting theories*. Cambridge University Press, Cambridge, 313 pp.
  [**Bony fishes; Latitude**]
Leim, A. H. 1924. The life history of the shad (*Alosa sapidissima* (Wilson)), with special reference to the factors limiting its abundance. *Contributions to Canadian Biology* (n.s.) **2**: 161–284.
  [**Shad**]
+Lereboullet, Dominique Auguste. 1855a. Sur la monstruosité double chez les poissons. *Comptes Rendus de l'Académie des Sciences, Paris* **40**: 854–6.
  ["On double monstrosity among fishes"; **Monsters**]
+Lereboullet, Dominique Auguste. 1855b. Deuxième note sur la monstruosité double chez les poissons. *Comptes Rendus de l'Académie des Sciences, Paris* **40**: 1028–30.
  ["Second note on double monstrosity among fishes"; **Monsters**]
+Lesson, René-Primevère.1826. Du Grand-Océan et des îles océaniennes. *In*: Louis Isidore Dupperrey (ed.), *Voyage autour du monde . . . sur la corvette de Sa Majesté* "la Coquille", *pendant les années 1822, 1823, 1824 et 1825 . . . Zoologie*. Vol. 1, pp. 2–20. P. Pourras Frères, Paris.
  ["On the Pacific and oceanic islands"; **Distribution; Tahiti**]

+Lesson, René-Primevère. 1830–1831. Poissons. *In*: Louis Isidore Dupperrey (ed.), *Voyage autour du monde... sur la corvette de Sa Majesté "la Coquille", pendant les années 1822, 1823, 1824 et 1825... Zoologie.* Vol. 2, part 1, pp. 66–238; Atlas, Pls. 1–38. P. Pourras Frères, Paris.

[Often cited as 'Lesson & Garnot', the latter referring to Prosper Garnot, Chief surgeon on *La Coquille*; **Chimaera; Jenyns (II); Sharks**]

+Leuckart, Karl Georg Friedrich. 1853. Zeugung. *In*: Rudolf Wagner (ed.), *Handwörterbuch der Physiologie mit Rücksicht auf physiologische Pathologie*, pp. 707–1000. Verlag Wieweg und Sohn, Braunschweig. Vol. IV.

["Reproduction"; CD cites Leuckart's interpretation of a statement on p. 117 of Bloch, 1782–1784; **Sex ratio**]

Lever, Chistopher. 1996. *Naturalized Fishes of the World*. Academic Press, San Diego, 408 pp.

[**Introductions; Sunfishes**]

Lewis, Cherry. 2000. *The dating game: one man's search for the age of the earth*. Cambridge University Press, Cambridge, ix + 253 pp.

[The story of Arthur Holmes, who perfected the dating techniques based on the decay of unstable elements, especially uranium, that allowed estimating the age of the Earth at over 4.5 billion years; **Lyell**]

Lewis, Richard J. and R. Endean. 1984. Ciguatoxin from the flesh and viscera of the barracuda, *Sphyraena jello*. *Toxicon* **22**(5): 805–10.

[**Barracuda**]

Lewis, Richard J. and M. J. Holmes. 1993. Origin and transfer of toxins involved in ciguatera. *Comparative Biochemistry and Physiology* **160**C(3): 615–28.

[**Ciguatera**]

Liebling, A. J. 1962. Onward and upward with the arts: the soul of bouillabaisse. *The New Yorker* (**27**): 189–202.

[**Scorpionfishes**]

Liem, Karel F. 1988. Form and function of lungs: the evolution of air-breathing mechanisms. *American Zoologist* **28**: 739–59.

[**Swimbladder**]

Lilienskiold, Hans Hanssen. 1701. *Speculum Boreale*. Manuscript kept at the Royal Library, Copenhagen. Published in 1943–44 as '*Lilienskiolds Speculum Boreale*.' O. Solberg (ed.), Etnografisk Museum, *Nordnorske Samlinger* **IV**(3): 51–337.

[The '*Northern Mirror*' was written while its author was governor of Norway's northernmost district (1684–1701), where he drafted an account of the murderous 'witch trials' conducted there among fishing communities from 1610 to 1692, which had led to 74 death sentences (65 to female witches). The '*Northern Mirror*' also reports on a nefarious, and now still common, practice: the exploitation of small-scale fishers by absentee traders. Not surprisingly, the author was dismissed from his powerful position, in spite of having dedicated his manuscript to the Danish-Norwegian king Frederic IV; **Cod**]

Lindhom, Rolf and J. Garrey Maxwell. 1988. Stock separation of Jack mackerel *Trachurus declivis* (Jenyns 1841) and Yellowtail *T. novaezealandiae* (Richardson 1843) in Southern Australian waters using Principal Component Analysis. CSIRO Marine Laboratories Report no. 189, 7 pp.

[**Jacks**]

Linnaeus, Carolus. 1758. *Systema naturæ per regna tria naturæ, secundum classes, ordines, genera, species, cum characteribus, differentiis, synonymis, locis*. Tomus I. Editio decima, reformata. Impensis Laurentii Salvii. Holmiæ. v. 1, pp. i–ii + 1–824.

["*Natural system of the three kingdoms of nature, by classes, orders, genera and species, with their characteristics, differences, synonyms, and localities.* Published by L. Salvius. Stockholm." The book and edition serving as anchor for classification and nomenclature within the plant and animal kingdoms. The third kingdom of the title pertained to minerals, for which the Linnean system did not work, and was rejected by geologists; this edition describes 16 genera of fishes, and 378 species, and is the first to move the cetaceans from the fishes to the mammals; **Barbel; Barracuda; Bonito; Bream; Burbot; Cape of Good Hope; Carp; Chimaera; Char; Cod; Croaker; Dace; Dogfish; Eel; Flounder; Gar-pike; Goldfish (I); Great white shark; Gouramy; Hagfishes; Lamprey; Ling; Linnaeus; Loaches; Lumpfish; Merlu; Minnows; Names, scientific; Perch; Pike; Plaice; Remora; Roach; Salmon(I, II); Sand-eels; Scorpionfishes; Sea scorpion; Skate; Sole; Sole, lined;** *Squalus*; **Stickleback; Sunfishes; Surgeonfishes; Tench; Torpedo; Triggerfishes; Trout; Turbot; Type species; Velvet belly; Wels catfish; Wrasses**]

+Linnaeus, Carolus. 1762. *Amoenitates Academicæ; seu, Dissertationes variæ Physicae, Medicæ, Botanicæ antehac seorsim editate, nun collectæ et auctæ.* . . . . . Vol. II, 2nd ed. Stockholm, v, 10 bis. X, 103.

["*Academic tracts, or miscellaneous Dissertations in Physics, Medicine, Botany, previously separate, now collected and published.* . . ."; **Altruism**]

Linnaeus, Carolus. 1766. *Systema naturæ per regna tria naturæ, secundum classes, ordines, genera, species.* Impensis Laurentii Salvii. Holmiæ.

[12th Edition; **Electric eel**]

Linnaeus, Carolus. 1771. *Mantissa plantarum altera generum editionis VI et specierum editionis II.* Holmiæ, Impensis Laurentii Salvie, 464 pp.

["*Second supplement of plant genera and species;*" contains a brief description of "*Fucus pyriferus*", noting dichotomous stem, terminal blades, and bladders; **Kelp**]

Linton, J. R. and B. L. Soloff. 1964. The physiology of the brood pouch of the male sea horse *Hippocampus erectus. Bulletin of Marine Science of the Gulf and Caribbean* **14**(1): 45–61.

[Here is a key conclusion: "That calcium is taken up by the developing seahorse embryo suggests that this calcium may be incorporated into the embryonic skeleton." **Seahorses (II)**]

Litchfield, Henrietta (ed.). 1915. *Emma Darwin: a Century of Family Letters. Edited by her Daughter Henrietta Litchfield.* D. Appleton and Company, New York. Vol. I, 289 pp; Vol. II, 326 pp.

[**Beards; Cod liver oil**]

Li Zhen. 1988. *Chinese Goldfish.* Beijing, Foreign Language Press. 100 pp.

[**Goldfish (I)**]

Lloris, Domingo and Jaime Rucabado. 1991. *Ictiofauna del Canal Beagle (Tierra del Fuego), aspectos ecológicos y análisis biogeográfico.* Publicacion Especial del Instituto Español de Oceanografía no. 8: 1–170.

["*Ichthyofauna of the Beagle channel (Tierra del Fuego): ecological aspects and biogeographical analysis*"; **Gastrobranchus; Tierra del Fuego**]

+Lloyd, Llewellyn. 1854. *Scandinavian Adventures, during a Residence of upwards of Twenty Years. Representing Sporting Incidents, and Subjects of Natural History, and Devices for entrapping wild Animals. With some Account of the Northern Fauna.* In two volumes. Richard Bentley, London.

[**Salmon, Pacific**]

+Lloyd, Llewellyn. 1867. *The Game Birds and Wildfowl of Sweden and Norway: together with an account of the Seals and Salt-water Fishes of those Countries.* Day and Son Limited, 2nd edn, London.

[**Sea scorpion**]

Lobban, Richard A. Jr. 1998. Charles Darwin visits Cape Verde. *Cimboa* (Primavera /Spring 1998): 17–21.
[**Cape Verde Islands**]

+Lockwood, Samuel. 1868. The Sea Horse and its Young. *Quarterly Journal of Science* **5**: 269.
[Summary by James Samuelson and William Crocker (eds.) of an article originally published under the same title in *American Naturalist* **1**: 225–34, 1868; **Seahorses**]

Long, John A. 1995. *The Rise of Fishes: 500 million years of evolution*. University of New South Wales Press, Sydney, 223 pp.
[**Fossil fishes**]

Longhurst, Alan R. 1998. *Ecological geography of the sea*. Academic Press, San Diego, 398 pp.
[**Plankton**]

+Lord, John Keast. 1866. *The Naturalist in Vancouver Island and British Columbia*. Richard Bentley, London.
[Vol. I (of II); **Salmon, Pacific**]

Lourie, Sara A., Amanda C. J. Vincent and Heather J. Hall. 1999. *Seahorses: an identification guide to the world's species and their conservation*. Project Seahorse, London, 214 pp.
[**Seahorses**]

Lovejoy, Arthur O. 1936. *The Great Chain of Being: a Study of the History of an Idea*. Harvard University Press, Cambridge, MA, and London, 382 pp.
[**Great Chain of Being**]

+Lovén, Sven. 1844. Ny art af Cirripedia. *Öfversigt af Kongliga Svenska Vetenskaps-Akademiens Förhandlingar* **1**: 192–4.
["A new species of Cirripedia, *Alepas squalicola*"; **Barnacles**]

Lowe, Richard Thomas. 1834. Characters of a new genus *Leirus*, and of several new species of fishes from Madeira. *Proceedings of the Zoological Society of London* **1833** (1): 142–4.
[**Groupers (I)**]

+Lowe, Richard Thomas. 1843. *A history of the fishes of Madeira, with original figures from nature of all the species, by Hon. C.E.C. Norton and M. Young*. London, 196 pp.
[**Madeira**]

Lubbock, Roger and Alasdair J. Edwards. 1981. The fishes of Saint Paul's Rocks. *Journal of Fish Biology* **18**: 135–57.
[**Galápagos shark(s)**]

#Lucas, Prosper. 1847. *Traité philosophique et physiologique de l'hérédité naturelle*. J. B. Baillière, Paris, 2 vols.
["*Treatise on the philosophy and physiology of natural heredity*;" Vol. 2: Turned up snouts in Crocodile, Goldfish & Bull-dogs; 158 Difference between males & females [in rays] Fish, *Marginalia* 516]

Ludwig, Donald, Ray Hilborn and Carl J. Walters. 1993. Uncertainty, resource exploitation and conservation: lessons from history. *Science* **260**: 17–36.
[**Hagfishes**]

+Lund, Carl Frederic 1761. Rön om Fiske-Plantering uti Insjöar. *Öfversigt af Kongliga Svenska Vetenskaps-Akademiens Förhandlingar* **22**: 184–197.
["On stocking fishes in lakes;" according to Stauffer (1975, pp. 591 and 615), CD read a manuscript translation of this article, though he cited Vol. 4 of the English translation of the journal in which it appeared; **Eggs of fish**]

+Lyell, Charles. 1830–1838. *Principles of Geology*. In 3 volumes. John Murray, London.
  [6th edition, 1840, Vol. 3, p. 140: do any fish live on seeds? fish eaten by Herons *Marginalia* 542; 10th edition, 1867/68, Vol. 1, p. 393 Means of Distribution – organisms in borings of Artesian wells – even living fish, *Marginalia* 544; **Lyell**]

+Lyell, Charles. 1838. *Elements of Geology*. John Murray, London.
  [6th edition, 1865: p. 509 On airbreathers in Coal period U. States; 552 On oldest known fossil fish, *Marginalia* 525; **Lyell**]
  [**Morris, Margaretta Hare (II)**]

+Macculloch, John. 1824. Hints on the possibility of changing the residence of certain fishes from salt water to fresh. *Quarterly Journal of Literature, Art and Science, London* **17**: 209–31.
  [**Caspian Sea**]

+Macculloch, John. 1837. *Proofs and illustrations of the attributes of God, from the facts and laws of the physical universe; being the foundation of natural and revealed religion.* 3 vols. James Duncan, London.
  [Some of CD's comments on this were not very charitable: What bosh!! [. . .] What trash (*Notebook*, pp. 634, 637); **Sunfishes; Notebooks; Squirrelfishes**]

+Macdonald, William. 1839. [Verbal communication on the osseous structure of fishes]. *Ann. Mag. Nat. Hist.* **2**: 69–70.
  [**Vertebrate origins**]

+MacGillivray, William. 1837–1852. *A History of British Birds. INDIGENOUS and MIGRATORY: including their organization, Habits and Relations; Remarks on Classification and Nomenclature; An Account of the Principal Organs of Birds, and Observations Relative to Practical Ornithology.* 5 Vols. London.
  [Information on swift is in Vol. 3, 1840; **Bird's nest soup**]

Macleay, William. 1878. The fishes of Port Darwin. *Proceedings of the Linnean Society of New South Wales* **2**(4): 344–67.
  [**Eponym**]

MacLeod, Rod and Philip F. Rehbock. 1994. Introduction. *In*: Rod MacLeod and Philip F. Rehbock (eds.), *Darwin's Laboratory: evolutionary theory and natural history in the Pacific*, pp. 1–18. University of Hawai'i Press, Honolulu.
  [*Beagle*]

Magurran, Anne Elizabeth and Tony J. Pitcher. 1987. Provenance, shoal size and the sociology of predator evasion behaviour in minnows. *Proceedings of the Royal Society* B**229**: 439–65.
  [The second author of this detailed study, which involved Pike as predator, suggests (pers. comm.) that the story Möbius recounted of a Pike painfully learning to ignore the Minnow in its tank is not too credible – unless perhaps the Pike was fed very regularly; **Möbius**]

Maitland, Peter Salisbury and R. Niall Campbell. 1992. *Freshwater fishes of the British Isles*. HarperCollins, London, 369 pp.
  [**Barbel**]

+Malm, August Wilhelm. 1854. De flundre-artade fiskarnes kroppsbyggnad är mere skenbart än verkligt osymmetrisk. *Öfversigt af Kongliga Svenska Vetenskap-Akademiens Förhandlingar*, **51** (7): 173–83.
  ["The structure of the fish in the flounder family is more seemingly than actually asymmetrical"; **Flatfish controversy (II)**]

+Malm, August Wilhelm. 1868. Bidrag till Kännedom af Pleuronektoidernas Utveckling och Byggnad. *Öfversigt af Kongliga Svenska Vetenskaps-Akademiens Handlingar* (N.F.) **7**(4): 3–28.

[CD actually read and cited Günther (1868b), who abstracted this "Contribution to the knowledge of Pleuronectid development and morphology"; **Deal fish; Flatfish controversy (I, II); Flounder**]

+Malthus, Thomas R. 1826. *An Essay on the Principle of Population*, 6th edn., 2 Vols., London.
[**Eggs of fish**]

Margulis, Lynn. 1981. *Symbiosis in cell evolution*. W. H. Freeman, San Francisco, xxii + 419 pp.
[**Non-Darwinian evolution**]

Margulis, Lynn and Dorion Sagan. 1995. *What is life?* Simon and Schuster, New York, 207 pp.
[**Non-Darwinian evolution**]

Marí-Beffa, M., J. A. Santamaŕn, P. Fernández-Llebrez and J. Becerra. 1996. Histochemically defined cell states during tail fin regeneration in teleost fishes. *Differentiation* **60**: 139–49.
[**Polydactylism**]

Marini, T. L. 1933. La merluza Argentina. *Physis (Buenos Aires)* **11**: 321–6.
[**Merlu**]

Marks, Richard Lee. 1991. *Three Men of the Beagle*. Avon Books, New York, 256 pp.
[**Beagle**]

Marquet, Gérard. 1992. L'étude du recrutement et de la physiologie des anguilles de Polynésie française permet-elle de cerner leur aire de ponte? *Bulletin de l'Institut Océanographique de Monaco*. no. Spécial **10**: 129–47.
["Can the study of the recruitment and physiology of eels in French Polynesia help locate their spawning areas?"; **Eels**]

Marquet, Gérard and René Galzin. 1991. The eels of French Polynesia: Taxonomy, distribution and biomass. Societé franco-japonaise d'océanographie. *La mer* **29**: 8–17.
[**Eels**]

Marquet, Gérard and René Galzin. 1992. Systématique, répartition et biomasse des poissons d'eau douce de Polynésie française. *Cybium* **16**(3): 245–59.
["Systematics, distribution and biomass of the freshwater fishes of French Polynesia"; **Eels**]

Marr, John C. (ed.). 1970. *The Kuroshio: A Symposium on the Japan Current*. East-West Center Press, Honolulu, x + 614 pp.
[**Range**]

Marshall, Justin. 1998. Why are reef fishes so colourful? *Scientific American Presents: The Oceans* **9**(3): 54–7.
[**Colours**]

#Martin, William Charles Linnaeus. 1845. *The history of the dog*. Charles Knight, London.
[p. 180 On a dog liking to catch carp & trout *Marginalia* 568]

Martini, Frederic H. 1998. Secrets of the Slime Hag. *Scientific American* **279** (October): 70–5.
[***Gastrobranchus;* Hagfishes**]

Mathis, Alicia, Douglas P. Chivers and R. Jan F. Smith. 1995. Chemical attractants: predator deterrents or predator attractants? *American Naturalist* **145**(6): 994–1005.
[**Minnows**]

Matteucci, Carlo. 1843. Nuovi fatti per stabilire il parallelo fra la funzione dell'organo elettrico della torpedine e la contrazione muscolare. *Atti della quinta Riunione degli Scienziati Italiani*, [Lucca, settembre 1843], pp. 484–6. Tipografia Giusti, Lucca [1844].
["New fact to establish the parallel between the function of the torpedo's electric organ, and muscular contraction"; **Electric organs (I)**]

Matteucci, Carlo. 1847. Electro-physiological researches. Sixth series. Law of the electric discharge of the torpedo and other electric fishes. Theory of the production of electricity by these animals. *Philosophical Transactions of the Royal Society of London* **1847**: 239–41.
[**Electric organs (I)**]

Matthews, Robert A. J. 1997. The science of Murphy's law. *Scientific American* **276**(4): 88–91.
[**Mendel, Gregor**]

Mayers, William Frederick. 1868. Gold fish cultivation. *Notes and Queries on China and Japan* (**8**): 123–4.
[This article starts as follows "The belief mentioned by Darwin (*Variation of Animals and Plants, & vol. I p. 289*) to the effect that gold-fish *'have been kept in confinement from an ancient period in China'* is well founded," and goes on demonstrating this by citing numerous ancient Chinese sources and details on methods and breeds. CD must have liked this conformation, and he used it in the 2nd edition of *Variations*, though he cited the journal as Chinese notes and queries, thus making this already obscure reference even harder to locate; **Goldfish (I, II)**]

Maynard Smith, John. 1978. *The Evolution of Sex*. Cambridge University Press, Cambridge, 222 pp.
[**Hermaphrodite; Sex ratio**]

Mayr, Ernst. 1982. *The Growth of Biological Thought: Diversity, Evolution and Inheritance*. Belknap Press/Harvard University Press, Cambridge, MA, 974 pp.
[**Classification; Speciation**]

Mayr, Ernst and P. D. Ashlock. 1991. *Principles of Systematic Zoology*, 2nd edn. McGraw-Hill, New York, xx + 475 pp.
[**Subspecies**]

+McClelland, John. 1839. Indian Cyprinidae. Part 2. *Journal of Asiatic Researches or Transactions of the Society instituted in Bengal for Enquiring into the History, the Antiquities, the Arts and Science and Literature of Asia*. **19**(2): 217–471.
[Wherein McClelland writes on p. 230, with reference to small cyprinids: "[t]hey are small species of little or no direct utility to man, nor is it possible to account for the particular brilliancy of their colours in any other way than as an instance of that unscrutable design, by which it would seem that in pursuit of aquatic insects on which they subsist along the surface of waters, they become the better marks for Kingfishers, Skimmers, Tern, and other birds which are destined to keep the number of fishes in check, especially in deep waters beyond the reach of the waders;" Good many fish – semi-alpine 4–5000 feet nevertheless no species similar to European – I believe; Good contrast with Fish of Pacific & Indian Oceans – How is this in N . America? pl[ate] 46. Perilampus perseus 229/230 Fish bright to be caught. I must utterly deny this. – If this could be passed – farewell my thesis. Nothing new spec; 266 On domesticated Fishes of India varying so much; 262 On Salmonidae in India – place filled by Cyprinidae; 458 not so much destroyed & therefore not become so prolific *Marginalia* 550; **Altruism; Cyprinidae; Carp, Indian;** *Perilampus perseus*]

McCormick, Harold, Tom Allen and William E. Young. 1963. *Shadows of the Sea: the Sharks, Skates and Rays*. Chilton Books, Philadelphia, xii + 415 pp.
[**Hippopotamus; Megatooth shark**]

McCormick, Sharon, and Gary Polis. 1982. Arthropods that prey on vertebrates. *Biological Reviews* **57**(1): 29–58.
[**Morris, Margaretta Hare (II)**]

McDonald, Roger. 1998. *Mr. Darwin's Shooter*. Atlantic Monthly Press, New York, 363 pp.
[**Covington**]

McDowall, Robert M. 1970. The Galaxiid Fishes of New Zealand. *Bulletin of the Museum of Comparative Zoology* **139**(7): 341–432.
[**Freshwater fishes (III); Galaxiidae**]

McDowall, Robert M. 1988. *Diadromy in fishes: migrations between freshwater and marine environments.* Croom Helm, London, 308 pp.
[**Diadromy; Galaxiidae**]

McDowall, Robert M. 1997. The evolution of diadromy in fishes (revisited) and its place in phylogenetic analysis. *Reviews in Fish Biology and Fisheries* **7**: 443–62.
[**Diadromy; Galaxiidae**]

McDowall, Robert M. 2001. Anadromy and homing: two life-history traits with adaptive strategies in salmonid fishes? *Fish and Fisheries* **2**: 78–85.
[**Diadromy**]

McDowall, Robert M. and Kazuhiro Nakaya. 1987. Identity of the Galaxoid Fishes of the Genus *Haplochiton* Jenyns from Southern Chile. *Japanese Journal of Ichthyology* **34**(3): 377–83.
[**Galaxiidae**]

+M'Donnell, Robert. 1861. On an organ in the skate which appears to be the homologue of the electrical organ on the Torpedo. *Natural History Review* (n.s.) **1**: 57–60.
[CD commented on this issue of *Nat. Hist. Rev.*: M'Donnell on Electric Organs – gradations, good (*Correspondence* **9**, 1861, p. 2)]

McKenzie, R. A. 1935. *Codfish in captivity.* Biological Board of Canada. Progress reports of Atlantic Biological Station, St. Andrews, N.B. Note no. 47/Atlantic Fisheries Experimental Station, Halifax, N.S. no. 16: 7–10.
[**Motion sickness**]

Medawar, Peter. 1982. *Pluto's Republic.* Oxford University Press, Oxford, 351 pp.
[This book, which includes the text of Medawar's earlier works on "*The Art of the Soluble*" and "*Induction & Intuition in Scientific Thought*", and which articulate Popper's philosophy extremely well, also includes two chapters dealing devastating blows to the scientific pretences of Freud's psychoanalysis. Moreover, a chapter is thrown into the bargain which suggests that following his return from the voyage of the *Beagle*, CD suffered from Chagas' disease, owing to his having let himself be bitten by a "bug of the pampas," now known to transmit this disease. This certainly explains the facts of his long illness much better than an alleged Oedipus complex; **Popper**]

Mendel, Gregor. 1866. Versuche über Pflanzenhybride. *Verhandlungen des Naturforschenden Vereins Brünn* **4**: 1–47.
[Available as "*Experiments in plant hybridisation; Mendel's original paper in English Translation, with commentary and assessment by Sir Ronald A. Fisher, together with a reprint of W. Bateson's biographical notice of Mendel*", edited by J. J. Bennett, Edinburgh, Oliver & Boyd, 95 pp.; **Mendel**]

+Mendel, Gregor. 1869. On *Hieracium*-Hybrids obtained by artificial fertilization. *Verhandlungen des Naturforschenden Vereins Brünn* **8**: 26.
[**Mendel**]

Menni, Roberto C., Raul A. Ringuelet and Raul H. Aramburu. 1984. *Peces marinos de la Argentina y Uruguay.* Editorial Hemisferio Sur, Buenos Aires, Argentina, 359 pp.
["Marine fishes of Argentina and Uruguay;" **Groupers (I);** *Rhombus*]

Merlen, Godfrey. 1988. *A field guide to the fishes of Galápagos.* Wilmot Books, London, 60 pp.
[**Puffers; Wrasses**]

Miall, Louis Compton. 1895. *The natural history of aquatic insects.* Macmillan, London, 395 pp.
[**Morris, Margaretta Hare (II)**]

Milinkovich, Michel C., Guillermo Orti and Axel Meyer. 1993. Revised phylogeny of whales suggested by mitochondrial ribosomal DNA sequences. *Nature* **361**: 346–8.
[**Jaguar**]

#Miller, Hugh. 1849. *Footprints of the creator*. Johnton & Hunter, London.
[p. 69 What is embryonic head of a Placoid or ganoid Fish *Marginalia* 587]

Miller, Peter James. 1979. Adaptiveness and implications of small size in teleosts. *In:* P. J. Miller (ed.), Fish Phenology: anabolic adaptiveness in teleosts. *Symposium of the Zoological Society of London* no. 44, pp. 263–306. Academic Press, London.
[**Gobies**]

Miller, Stanley L. 1953. A production of amino acids under possible primitive Earth conditions. *Science* **117**: 528.
[**Spontaneous generation**]

+Milne-Edwards, Henri. 1844. Considérations sur quelques principes relatifs à la classification naturelle des animaux, et plus particulièrement sur la distribution méthodique des mammifères. *Annales des Sciences Naturelles*, ser. 3, **1**: 65–99.
["Considerations on some principles related to the natural classification of animals, and particularly on patterns in the distribution of mammals"; **Ontogeny**]

#Milne-Edwards, Henri. 1849. *Introduction à la zoologie générale*. Victor Masson, Paris.
[*Introduction to general Zoology*; p. 25 Devonian sharks *Marginalia* 583]

Mitchill, Samuel Latham. 1814. *Report, in part, of Samuel L. Mitchill, M. D., on the fishes of New York*. New York, 28 pp.
[A very rare work, in which 38 species are described as new; edited by Theodore Gill and reprinted in 1898; **Perch**]

+Mivart, Saint-George Jackson. 1871. *On the genesis of species*. Macmillan and Co, London/D. Appleton and Company, New York, 314 pp.
[p. 37 Flat fish; 81 Placentae of mammals & sharks; 145 The same fishes in distant continents; 176 Homology between limb & fins; Mr Mivart's book consists of all objections to nat. selection advanced by various authors & myself, expanded & admirably illustrated, with nothing said in favour, except in opening chapter; *Marginalia* 585, 586; **Flatfish controversy (I, II); Seahorses (II)**]

+Möbius, Karl August. 1873. Die Bewegungen der Thiere und ihr psychischer Horizont. Kiel: populärer Vortrag gehalten in Februar 1872 in der Harmonie in Kiel. *Schriften des Naturwissenschaftlichen Vereins für Schleswig-Holstein*. 20 pp.
["The movements of animals and their psychological limitations: a public lecture given in February 1872 at the 'Harmony' concert hall, Kiel"; **Pike**]

Möbius, Karl August. 1877. *Die Auster und die Austernwirschaft*. Wiegand, Hempel und Parey, Berlin.
["The oyster and the oyster industry"; **Möbius**]

Möller, Heino, and Kerstin Anders. 1989. *Diseases and parasites of marine fishes*. Verlag Möller, Kiel, 365 pp.
[**Parasites**]

+Molina, Giovanni Ignacio. 1788. *Compendio de la Historia Geografica, Natural y Civil del Reyno de Chile, escrito en Italiano por el Abate Don Juan Ignacio Molina. Primera Parte, que abraza la Historia Geografica y Natural*. Traducida en Español por Don Domingo Joseph de Arcuellada Mendoza. Antonio de la Sancha, Madrid, xx + 418 pp.

["*Compendium of the geographic, natural and civil history of the kingdom of Chile, written in Italian by Father Don Giovanni Molina. Vol. 1, covering Geographic and Natural History*. Translated into Spanish by Don Domingo Joseph de Arcuellada Mendoza"; fishes are covered on pp. 241–53; **Bacalao**]

Moller, Peter. 1995. *Electric fishes: history and behavior*. Chapman & Hall, London, 584 pp.
[**Electric eel; Electric fishes**]

Monod, Théodore. 1963. Achille Valenciennes et l'Histoire naturelle des poissons. *In: Mélanges ichthyologiques dédiés à la mémoire d'Achille Valenciennes. Mémoires de l'Institut Français d'Afrique Noire*, no. 68, pp. 9–45. Dakar.
[**Histoire Naturelle des Poissons**]

Moore, James R. 1985. Darwin of Down: the evolutionist as squarson-naturalist. *In:* David Kohn (ed.) *The Darwinian heritage*, pp. 435–81. Princeton University Press, Princeton.
[*See* White (1789)]

Moorehead, Alan. 1971. *Darwin and the* Beagle. Penguin Books, London, 224 pp.
[**Beagle**]

Mori, Shigeo, Genyo Mitarai, Akira Takabayashi, Shiro Usui, Manabu Sakakibara, Makoto Nagatomo and Rudolf von Baumgarten. 1996. Evidence of sensory conflict in recovery in carp exposed to prolonged weightlessness. *Aviation, Space and Environmental Medicine* **67**(3): 256–261.
[**Motion sickness**]

Morris, Molly R., Paul F. Nicoletto and Elizabeth Hesselman. 2003. A polymorphism in female preference for a polymorphic male trait in the swordtail fish *Xiphophorus cortezi*. *Animal Behaviour* **65**: 45–52.
[**Asymmetry**]

+Morton, Samuel George. 1821. A communication of a singular fact in natural history. *Philosophical Transactions of the Royal Society of London* **110**: 20–2.
[**Mule**]

#Morton, Samuel George. 1854. *Types of mankind; with contributions by L. Agassiz, W. Usher, and H. S. Patterson*. Ed. by J. C. Nott and G. R. Gliddon. Lippincott & Grambo, Philadelphia, and Treubner & Co, London.
[The kind of book university libraries nowadays hide in their storerooms, lest their students choke on the toxic brew of prejudice and religion masquerading as science that was once taught to justify slavery and colonialism (and which started by assigning the human 'races' to different species, and asserting that only the European 'species' was capable of civilization). This nasty book is listed here only because CD mentioned 'fish' when he annotated it, suggesting in the process that Agassiz should be ashamed of himself: lxxiv Look at same race in United States & S. America oh fish pudor Agassiz!, *Marginalia* 604. This annotation refers to a plate, attached to page lxxvi, consisting of parallel series of eight drawings per human 'species,' each topped by a male (!) head (strongly prognathous, almost simian in all but the European and American 'species') and a skull (similarly distorted), these being followed by six species of mammal somehow representing the habitat of the human 'species' in question. These were drawn 'adhering as closely as possible to the written instructions of Prof. Agassiz.' CD, who did not believe that contemporary humanity consists of several species, also knew that the beak-nosed 'Indian chief' representing all 'Americans' would never do for the United States & S. America; hence that part of his remark. So this all is sorted out, since oh fish is one of these English imprecations, similar to 'ye gods and little fishes']

Motani, Ryosuke. 2000. Rulers of the Jurassic seas. *Scientific American* **283**(6): 52–9.
[**Ichthyosaurs**]

Motani, Ryosuke, Bruce M. Rothschild and William Wahl Jr. 1999. Large eyeballs in diving ichthyosaurs. *Nature* **402**: 742.
  [**Ichthyosaurs; Lumpfish**]
Moy-Thomas, John Allan. 1971. *Paleozoic fishes*, 2nd edn, extensively revised by R. S. Miles. Chapman and Hall, London, 259 pp.
  [**Devonian; Fish; Fossil fishes**]
Müller, Arno Hermann. 1985. *Lehrbuch der Paleontologie*. Band III, Teil 1: *Fische im weiteren Sinne und Amphibien*. 2nd edn. VEB Gustav Fischer Verlag, Jena, 655 pp.
  [*"Textbook of Palaeontology. Vol. III, Part 1: Fish in the broadest sense and Amphibians"*; **Fossils; Megatooth shark; Trout-perch**]
+Müller, Johannes Peter. 1838–1842. *Elements of physiology*. Translated by W. Baly. 2 Vols. & Supplement. Taylor & Walton, London.
  [p. 1569 Membrane of egg agrees with membrane of uterus (Mem Fish coming to have Placenta); 1596 on embryo Torpedo increasing in weight in womb (a sort of Placenta). 1599 great difference in 2 species of Mustelus in placentation; 1610 Relation of Vertebrae in Fish to embryos of higher animals; 1622 Sharks have gills during early part alone of embryonic life, *Marginalia* 617–18; **Embryology**]
+Müller, Johannes Peter. 1844. Über den Bau und die Grenzen der Ganoiden und über das natürliche System der Fische. *Abhandlungen der Akademie der Wissenschaften, Berlin* **1844**: 117–216.
  ["On the body plan and extent of the Ganoids, and on the natural system of fish" is a classic work, mainly because it sorted out the membership of Agassiz' catch-all 'ganoids'; see also annotation to Agassiz (1850); **Ganoid; Trout-perch**]
Müller, Otto Fredrik. 1776. *Zoologiae Danicae prodromus, seu animalium Daniae et Norvegiae indigenarum characteres, nomina, et synonyma imprimis popularium*. Havniae, Hallageri, i-xxxii + 1–282.
  [*"Introduction to the Zoology of Denmark, or on the characters, synonymy and especially on the local names of the animals indigenous to Denmark and Norway*. Copenhagen . . . "; **Capelin**]
Murie, James. 1868. On the supposed arrest of development of the salmon when retained in freshwater. *Proceedings of the Zoological Society of London*, March 26, no. 6: 247–54.
  [**Hybrid vigour**]
+Murie, James. 1870. Additional memoranda as to irregularity in the growth of Salmon. *Proceedings of the Zoological Society of London*, January 13, no. 3: 30–50.
  [Mentions CD as having called his attention to a study of salmon growth by Mr. George Anderson, of Glasgow; also cites *Variations II*; **Hybrid vigour**]
#Murphy, Joseph John. 1869. *Habit and intelligence*. 2 Vols. Macmillan, London.
  [Vol. 1, Carp 258; 304 Surely in fish we have gradation to bone from cartilage? Other tissues?, *Marginalia* 622, 623]
Murphy, Kim E. and Tony J. Pitcher. 1997. Predator attack motivation influences the inspection behaviour of European minnow. *Journal of Fish Biology* **50**: 407–17.
  [**Minnows**]
Muus, Bent J. and Preben Dahlstrøm. 1974. *Collins Guide to the Sea Fishes of Britain and Western Europe*. Collins, London, 244 pp.
  [**Eel; Ling**]
Myers, George Sprague. 1949. Usages of anadromous, catadromous and allied terms for migratory fishes. *Copeia*, June 1949: 89–97.
  [**Diadromy**]

#Nägeli, Carl von. 1866. *Botanische Mitteilungen*. Vol. 2. F. Staub, Munich.

[*"Botanical communications"*, 2nd vol., p.106 Von Baer – believes Bee on its own type higher than fish; 210 He has 2 embryos in his possession that he cannot tell whether they are Mammals or Fish or young Birds – (good to quote); 211 The more different 2 animals, the further back we must go to find similarity; 214 The embryo of higher animals resembles the embryo of lower; 219&220 But embryo of Mammal more like mature fish, than embryo of Fish is like mature Mammal, *Marginalia* 625]

Neill, Patrick. 1808. List of fishes found in the Firth of Forth, and rivers and lakes near Edinburgh, with remarks. *Memoirs of the Wernerian Natural History Society* **1**: 526–55.

[**Angling**]

Neis, Barbara, Richard Haedrich, Larry Felt and David Schneider. 1997. *Old challenges, new methods: the lumpfish roe fishery and fishers' ecological knowledge in the Bonavista Region of Newfoundland*. Eco-Research Program, Occasional Paper. Memorial University, St. Johns', 22 pp.

[**Lumpfish**]

Nelson, Joseph S. 1994. *Fishes of the World*, 3rd Edn. John Wiley and Sons, New York, 600 pp.

[**Cavefishes; Deep-sea spiny eels; Elephantfishes; Groupers (II); Ichthyology; Lumped; Lumpfish; Sticklebacks; Trout-perch; Wels catfish**]

Nelson, Joseph S. 1999. Editorial and introduction: the species concepts in fish biology. *In*: Joseph S. Nelson and Paul J. P. Hart (eds). *The species concept in fish biology*. Special Issue, *Reviews in Fish and Fisheries* 9(4): 277–80.

[**Species**]

Nelson, Joseph S. and Martin J. Paetz. 1992. *The fishes of Alberta*. The University of Alberta Press, Edmonton, xxiv + 437 pp.

[**Fish(es)**]

Neuweiler, Gerhard. 2000. *The biology of bats*. Oxford University Press, New York and Oxford, 310 pp.

[Translated by Ellen Covey. Discusses 'Fisherman bats' and other fishing bats; **Bats**]

Nicholas, Frank W. and Jan M. Nicholas. 1989. *Charles Darwin in Australia, with illustrations and additional commentary from other members of the Beagle's company including Conrad Martens, Augustus Earle, Captain FitzRoy, Philip Gidley King, and Sym Covington*. Cambridge University Press, Cambridge, 175 pp.

[**King George's Sound**]

Nikol'skii, George V. 1961. *Special Ichthyology*. Israel Program of Scientific Translations. Jerusalem, 538 pp.

[**Eel**]

Nilsson, Dan-E. and Susanne Pelger. 1994. A pessimistic estimate of the time required for an eye to evolve. *Proceedings of the Zoological Society of London* B **256**: 53–8.

[**Eyes**]

Nilsson, Gören E. 1996. Brain and body oxygen requirements of *Gnathonemus petersii*, a fish with an exceptionally large brain. *Journal of Experimental Biology* **199**: 603–7.

[**Elephantfishes**]

Norse, Elliott (ed.) 1993. *Global marine biological diversity*. Island Press, Washington, D.C., 383 pp.

[**Sirenians**]

N(ational) R(esearch) C(ouncil). 1999. *Sustaining Marine Fisheries*. National Research Council. National Academy Press, Washington, D.C., 164 pp.

[**Keystone species**]

Nyrop, Richard F. (ed.) 1981. *Peru: a Country Study*. Foreign Area Studies, The American University, Washington, D.C., 302 pp.

[**Iquique**]

Odum, W. S. 1970. Utilization of the direct grazing and plant detritus food chain by the striped mullet *Mugil cephalus*. *In*: John H. Steele (ed.), *Marine Food Chains*, pp. 222–40. Oliver and Boyd, Edinburgh.

[**Mullets**]

Oppenheimer, Jane. 1968. An embryological enigma in the Origin of Species. *In*: Bentley Glass, Oswei Temkin and William L. Straus, Jr (eds.), *Forerunners of Darwin: 1745–1859*, pp. 292–322. The Johns Hopkins University Press, Baltimore.

[**Embryology**]

+Orbigny, Alcide Charles Victor Dessalines d' (ed.) 1834–1847. *Voyage dans l'Amérique méridionale (le Brésil, la République orientale de l'Uruguay, la République Argentine, la Patagonie, la République du Chili, la République de Bolivia, la République du Pérou), exécuté pendant les années 1826, 1827, . . . et 1833*. Paris et Strasbourg, Vols. 1–9.

["*Voyage to South America (Brazil, Uruguay, . . .), performed during the years 1826–1833*." As the text of Valenciennes' final report on d'Orbigny's fishes was published only in 1847, CD, when mentioning these, may have been referring to the plates, published from 1834 to 1842 and authored by Valenciennes, or more likely, to the summary in Geoffroy-Saint Hilaire and de Blainville 1834; Jenyns, on the other hand, mentions d'Orbigny's work being "now in course of publication" (*Fish*, p. x), and a few species, described in Cuvier and Valenciennes' **Histoire Naturelle des Poissons*; **Distribution; Jenyns (II)**]

Orel, Vitězslav. 1996. *Mendel: the first geneticist*. Oxford University Press, Oxford, 363 pp.

[Translated by Stephen Finn; **Mendel**]

Owen, Richard. 1839. Description of the *Lepidosiren annectens*. *Transactions of the Linnean Society of London* **18**: 327–61.

[Contribution read on April 2, 1839, and published in 1841; **Lungfishes; Lungfish, African**]

Owen, Richard. 1840. Fossil Mammalia. *In: The Zoology of the Voyage of H.M.S. Beagle, under the Command of Captain *FitzRoy, R.N. during the years 1832 to 1836*, Part I, Vol. 4. Edited and superintendented by Charles Darwin. Smith, Elder and Co, Cornhill, London, xiii + 101 pp.

[**Lungfishes; Owen**]

+Owen, Richard. 1846. *Lectures on the Comparative Anatomy and Physiology of the Vertebrate Animals, delivered at the Royal College of Surgeons of England in 1844 and 1846*. Part I, *Fishes*. Longman, Brown, Green and Longmans, London.

[CD comments on this: Owen has been doing some grand work in morphology of the vertebrata: your arm & hand are part of your head or rather the processes (ie modified ribs) of the occipital vertebra! He gave me a grand lecture on cod's Head *Correspondence* to Hooker, April 10, 1846, referring to Lecture V, pp. 84–129; **Analogous organs; Electric fishes; Flying; Loaches; Owen; Trout-perch; Weber**]

+Owen, Richard. 1849. *On the Nature of Limbs: a discourse delivered on Friday, February 9, at an evening meeting of the Royal Institution of Great Britain*. John van Voorst, London. 119 pp.

[p.13 Capital comparison of hand of Mole, Bat & Fin; 45 Horse legs & Lepidosiren good contrast if simplicity from abortion & original; in all these cases the tibia & fibula shows that they are simple by abortion & it is rash to argue from. about original simplicity of limb. apparently aboriginal simplicity. The contrast between the 8 almost singly serial bones of Horses leg & appendage of Lepidosiren good instance of rudimentary & primeval or transitory stage; 59 What is relation in Sharks? 82 Lepidosiren realises nearly ideal Archetype; 86 alludes in grandiloquent sentence to some law governing progression, guided by archetypal light – &c *Marginalia* 655, 666; **Lungfishes; Lungfish, Australian**]

+Owen, Richard. 1860. *Palaeontology, or a systematic summary of the extinct animals and their geological relations*. Adams & Charles Black, Edinburgh, xv + 420 pp.

[p. 145 generalised ancient member of Sturgeon Family; 150 The History of Fishes indicates mutation rather than development — good remark, *Marginalia* 646; **Ichthyosaurs; Owen**]

#Owen, Richard. 1866–8. *On the Anatomy of Vertebrates*. 3 Vols. Longman, Brown, Green and Co, London.

[Vol. 1, p. xxxii rudimentary & nascent organs. Case of fins becoming rudimentary in old age; 254 Gradation between homocercle & heterocercle tail – also in embryos; 331 Eyes of Fish in Lancelet & some others as simple as in the lowest crania; 342 Range of gradation great in F[ish] & R[eptiles]; 354 On air bladder in Colitis aiding organ of hearing. Electric organs; 358 similar action with muscle; 378 six modifications of structure of teeth in fishes, 2 sometimes in same fish or each in same tooth; 486 on the persistence of an embryonic structure in the branchiae in certain low fishes; 487 on an accessory breathing organ in the climbing perch; 492 Structure of air bladder in fishes; 497 ditto; 576, 588 Gradation in reproductive organs of fishes; 609 Embryonic characters of fishes permanent in sharks; 611 Metamorphoses in fishes; On transitory tooth in young sharks & lizards for cutting through egg, *Marginalia* 646; read Co<u>b</u>itis for Co<u>l</u>itis]

+Pacini, Filippo. 1853. Sur la structure intime de l'organe électrique de la torpille, du gymnote et d'autres poissons; sur les condition électromotrice de leurs organes électriques, et leur comparaison respectives avec la pile thermo-électrique et la pile voltaïque. *Archives des Sciences Physiques et Naturelles* **24**: 313–36.

[Translated version of an article originally in Italian, "On the detailed structure of the electric organ of the torpedo, the Gymnotus, and other fishes, on the electricity produced by their electric organs, and their comparison with a thermo-electric battery and a voltaic pile"; **Electric organs (II)**]

Paine, Robert T. 1980. Food webs: linkages, interaction strength and community infrastructure. The Third Tansley Lecture. *Journal of Animal Ecology* **49**: 667–85.

[**Keystone species**]

Pallas, Pyotr Simon. 1814. *Zoographia Rosso-Asiatica, sistens omnium animalium in extenso Imperio Rossico et adjacentibus maribus observatorum recensionem, domicilia, mores et descriptiones, anatomen atque icones plurimorum*. Petropoli, Caes. Academiae Scientarum Impress. Vol. 3, i–vii + 428 pp.

["*Russo-Asiatic Zoogeography, being a convenient compilation of all the animals in the Russian Empire and adjacent seas, with descriptions of their habitats, habits, anatomy, with many figures*. St. Petersburg, Imperial Academy of Science Press"; **Blennies; Salmon (III)**]

Palomares, Maria D. Lourdes and Daniel Pauly. 1998. Predicting food consumption of fish populations as functions of mortality, food type, morphometrics, temperature and salinity. *Marine and Fisheries Research* **49**: 447–53.

[This shows, that other things (temperature, salinity, body size, etc.) being equal, fishes with high caudal fin aspect ratio ($A = h^2/s$, where h is the height of the tail, and s its surface area) have higher food consumption than fishes with low caudal fin aspect ratio. This can be used for predictions, for example when constructing models of aquatic food webs; **Caudal (fin)**]

Palomares, Maria D. Lourdes, Cristina V. Garilao and Daniel Pauly. 1999. On the biological information content of common names: a quantitative case study of Philippine fishes. *In*: Bernard

Séret and Jean-Yves Sire (eds.), *Proceedings of the 5th Indo-Pacific Fish Conference, Nouméa, New Caledonia, 3–8 November 1997*, pp.861–6. Societé Française d'Ichthyologie, Paris.

[**Names, common**]

Pancaldi, Guiliano. 1991. *Darwin in Italy: Science across cultural frontiers*. Updated and expanded edition, translated by Ruey Brodine Morelli. Indiana University Press, Bloomington and Indianapolis, 222 pp.

[**Classification; Jenyns (I)**]

Papastavrou, Vassili, Sean C. Smith and Hal Whitehead. 1988. Diving behaviour of the Sperm whale, *Physeter macrocephalus*, off the Galápagos Islands. *Canadian Journal of Zoology* **67**: 839–46.

[**Complexity**]

Parker, T. Jeffery, William A. Haswell and J. J. Marshall. 1964. *A Text-Book of Zoology*, Vol. II (of 2). MacMillan, London, ix + 951 pp.

[7th, reprinted edition; **Lizards**]

Parodiz, Juan Jose. 1981. *Darwin in the New World*. E. J. Brill, Leiden, 137 pp.

[**Maldonado; Montevideo**]

Parson, Theophilus. 1860. On the origin of species. *American Journal of Science and Arts* (ser.2) **30**: 1–13.

[**Vertebrate origins**]

Paterson, Hugh E. H. 1985. The recognition concept of species. *In*: E. S. Vrba (ed.), *Species and Speciation*, pp.21–9. Transvaal Museum no. 4, Pretoria.

[**Speciation**]

Patterson, Colin. 1981. Agassiz, Darwin, Huxley, and the fossil record of teleost fishes. *Bulletin of the British Museum of Natural History (Geology)* **35**(3): 213–24.

[**Ganoid(s); Owen**]

Paulin, Chris and Clive Roberts. 1992. *The rockpool fishes of New Zealand (Te ika aaria o Aotearoa)*. Museum of New Zealand (Te Papa Tongarewa), Wellington, 177 pp.

[**Roundheads**]

Pauly, Daniel. 1974. On some features of the infestation of the mouth-brooding fish *Tilapia melanotheron* Rüppel, 1852 by the parasitic copepod *Paenodes lagunaris* van Banning 1974. *Beaufortia* **22**(287): 9–15.

[**Mouth-brooding**]

Pauly, Daniel. 1975. On the ecology of a small West African lagoon. *Berichte der Deutschen Wissenschaftlichen Kommission für Meeresforschung* **24**(1): 46–62.

[**Lagoon**]

Pauly, Daniel. 1976. The biology, fishery and potential for aquaculture of *Tilapia melanotheron* in a small West African lagoon. *Aquaculture* **7**(1): 33–49.

[**Lagoon; Mouth-brooding**]

Pauly, Daniel. 1977. The Leiognathidae (Teleostei): A hypothesis relating their mean depth occurrence to the intensity of their counter-shading bioluminescence. *Marine Research in Indonesia*, no. 19: 137–46.

[**Bioluminescence**]

Pauly, D. 1978. *A critique of some literature data on the growth, reproduction and mortality of the lamnid shark* Cetorhinus maximus *(Gunnerus)*. International Council for the Exploration of the Sea. Council Meeting 1978/H:17 Pelagic Fish Committee, 10 pp.

[Updated in Pauly 2002b; **Basking shark**]

Pauly, Daniel. 1979. Gill size and temperature as governing factors in fish growth: a generalization of von Bertalanffy's growth formula. *Berichte des Institut für Meereskunde an der Universität Kiel* no. 63, xv + 156pp.
[**Oxygen**]

Pauly, Daniel. 1981. The relationship between gill surface area and growth performance in fishes: a generalization of von Bertalanffy's theory of growth. *Berichte der Deutschen Wissenschaftlichen Kommission für Meeresforschung* **28**(4): 251–82.
[**Branchiae; Oxygen**]

Pauly, Daniel. 1982. Further evidence for a limiting effect of gill size on the growth of fish: the case of the Philippine goby (*Mistichthys luzonensis*). *Kalikasan/Philippine Journal of Biology* **11**(2–3): 379–83.
[**Gobies**]

Pauly, Daniel. 1984. A mechanism for the juvenile-to-adult transition in fishes. *Journal du Conseil International pour l'Exploration de la Mer* **41**: 280–4.
[Reprinted as Chapter 8 in Pauly 1994a; **Oxygen; Reproductive drain hypothesis**]

Pauly, Daniel. 1986. Fisheries Scientists must write. *Naga, The ICLARM Quarterly* **9**(2): 8–9.
[Reprinted as Chapter 18 in Pauly 1994a; **Words**]

Pauly, Daniel. 1991. Growth of the checkered puffer *Sphoeroides testudineus*: postscript to papers by Targett and Pauly & Ingles. *Fishbyte* **9**(1): 19–22.
[**Puffers**]

Pauly, Daniel. 1992. The Peruvian anchoveta, Charles Darwin and us. *Naga, the ICLARM Quarterly* **15**(4): 14–15.
[**Peruvian anchoveta**]

Pauly, Daniel. 1994a. *On the Sex of Fishes and the Gender of Scientists: Essay in Fisheries Science*. Chapman & Hall Fish and Fisheries Series 14, London, 250 pp.
[**Oxygen; Reproductive drain hypothesis; Sexual dimorphism**]

Pauly, Daniel. 1994b. Quantitative analysis of published data on the growth, metabolism, food consumption, and related features of the red-bellied piranha, *Serrasalmus nattereri* (Characidae). *Environmental Biology of Fishes* **41**: 423–37.
[**Characins**]

Pauly, Daniel. 1994c. Resharpening Ockham's razor. *Naga, the ICLARM Quarterly* **17**(2): 7–8.
[**Occam's Razor**]

Pauly, Daniel. 1994d. A framework for latitudinal comparisons of flatfish recruitment. *Netherlands Journal of Sea Research* **32**(2): 107–18.
[**Latitude**]

Pauly, D. 1995a. Dim photography in a Caribbean coral reef. *Annals of Improbable Research* **1**(3): 20.
[Having observed how coral reef fishes often swim upside down, but not made photos, and having the degree that makes this possible, I once doctored two published images of scuba diving scenes, by crudely glueing upside down fishes onto them, thus generating a sort of 'fake forgery' to document a real phenomenon. I then sent the doctored figures and an accompanying text under a vaguely sexual title ("Two new cases of *position obversa* in fishes") to the *Journal of Irreproducible Results* (*JIR*), which promptly split on me: the editorial staff left the editor of *JIR* (taking my manuscript with them), and founded the *Annals of Improbable Research* (*AIR*). A while later, I received a letter from the editor of the *JIR* (remember: he did not have the manuscript) grandly rejecting my submission, reportedly on advice from referees. *AIR*, on the other hand, published one of my figures, but

under a new title, and with a legend added by someone who missed the figures' true nature, and thus the point of the whole thing. There are many lessons in this, some funny, some not; **Dohrn**]

Pauly, D. 1995b. Anecdotes and the shifting baseline syndrome of fisheries. *Trends in Ecology and Evolution* **10**(10): 430.
[**Herring**]

Pauly, Daniel. 1997. Geometric constraints on body size. *Trends in Ecology and Evolution* **12**(11): 442.
[**Branchiae; Length; Oxygen**]

Pauly, Daniel. 1998. Tropical fishes: patterns and propensities. *In*: T. E. Langford, J. Langford and J. E. Thorpe (eds.), *Tropical Fish Biology. Journal of Fish Biology* **53** (suppl. A): 1–17.
[**Branchiae; Oxygen**]

Pauly, D. 2002a. Charles Darwin, ichthyology and the species concept. *Fish and Fisheries* **3**(3): 1–5.
[**Darwin and ichthyology; Species; Speciation**]

Pauly, D. 2002b. Growth and mortality of basking shark *Cetorhinus maximus*, and their implications for whale shark *Rhincodon typus*. In: S. L. Fowler, T. Reid, and F. A. Dipper (eds.), *Elasmobranch biodiversity: conservation and management. Proceedings of an International Seminar and Workshop held in Sabah, Malaysia*, pp.199–208. IUCN, Gland.
[**Basking shark**]

Pauly, Daniel and Rainer Froese. 2001. Fish stocks. *In*: Simon Levin (ed.), *Encyclopedia of Biodiversity*. Vol. 2, pp. 801–14. Academic Press, San Diego.
[**Bony fishes**]

Pauly, Daniel and Roger Stephen Vernon Pullin. 1988. Hatching time in spherical, pelagic marine fish eggs in response to temperature and egg size. *Environmental Biology of Fishes* **21**(2): 261–71.
[**Cuckoo-fish**]

Pauly, Daniel and Isabel Tsukayama (eds.) 1987. *The Peruvian anchoveta and its upwelling ecosystem: three decades of changes*. ICLARM Studies and Reviews no. 15, 351 pp.
[**Iquique; Peruvian anchoveta; Punctuated equilibrium**]

Pauly, Daniel and Alejandro Yáñez-Arancibia. 1994. Fisheries in coastal lagoons. *In*: Björn Kjerfve (ed.), *Coastal Lagoon Processes*, pp. 377–99. Elsevier Science Publishers, Amsterdam.
[**Lagoon**]

Pauly, Daniel, Peter Muck, Jaime Mendo and Isabel Tsukayama (eds.) 1989. *The Peruvian Upwelling Ecosystem: Dynamics and Interactions*. ICLARM Conference Proceedings no. 18, 438pp.
[The book that was to include a CD quote on the Peruvian anchoveta, but which ended instead with an anchoveta cartoon by Leo Cullum, originally published in the *New Yorker*, and reproduced by permission (= US$ 100). Hence, the book that ultimately caused this one to be written; **Peruvian anchoveta; Punctuated equilibrium**]

Pauly, Daniel, Christine Casal and Maria D. Lourdes Palomares. 2000. DNA, cell size and fish swimming. *In*: Rainer Froese and Daniel Pauly (eds.), *FishBase 2000: Concepts, Design, and Data Sources*, Box 34, p. 254. Los Baños, Philippines.
[The DNA content of animal cells is highly variable, but so far, only cell size has been shown to be useful to predict DNA contents. As metabolic rate tends to be inversely proportional to cell size, the aspect ratio of the caudal fin of fish can also be shown to correlate with the DNA contents of their cells, which is rather neat. A short essay on this, and a graph illustrating this interesting correlation, may be found under 'Genetics and Aquaculture' in the *FishBase book, which is available online; **Caudal (fin)**]

Pauly, Daniel, Andrew Trites, Emily Capuli and Villy Christensen. 1998. Diet composition and trophic levels of marine mammals. *ICES Journal of Marine Science* **55**: 467–81.

[With small erratum in *ICES J. Mar. Sci.* **55**: 1153 (1998); **Whales**]

Peckham, Morse. 1959. *The Origin of Species: by Charles Darwin: A Variorum Text*. University of Philadelphia Press, Philadelphia, 816 pp.

[**Electric organs (II)**; *Origin*]

Péronnet, Louise, Rose Mary Babitch, Wladyslaw Cichocki and Patrice Brasseur. 1998. *Atlas linguistique du vocabulaire maritime acadien*. Les Presses de l'Université Laval/Université Saint-Nicholas, Québec, 667 pp.

[*"Linguistic atlas of the acadian maritime vocabulary";* **Tench**]

Peters, Robert H. 1983. *The ecological implications of body size*. Cambridge University Press, Cambridge, 329 pp.

[**Length**]

Peters, Wilhelm. 1852. Diagnosen von neuen Flussfischen aus Mossambique. *Monatsberichte der Akademie der Wissenschaften, Berlin* (1852): 275–6, 681–5.

["Diagnosis of new riverine fishes from Mozambique"; **Freshwater fishes (III)**]

Peters, Wilhelm. 1860. Eine neue vom Herrn Jagor im atlantischen Meere gefangene Art der Gattung *Leptocephalus*, und über einige andere neue Fische des Zoologischen Museums. *Monatsberichte der Akademie der Wissenschaften, Berlin* (1859): 411–13.

["A new species of the genus *Leptocephalus*, caught by Mr. Jagor in the Atlantic, and on some other new fishes of the Zoological Museum"; **Guppy**]

+Phillips, John. 1839. Treatise on Geology. *In*: D. Lardner (ed.), *The Cabinet Cyclopaedia. Natural History*. Longman, Orme, Brown, Green & Longman, London.

[**Sharks**]

#Phillips, John. 1860. *Life on the Earth*. Macmillan, Cambridge and London.

[argue against this; it is not always the perfect types which first appear – Ruminants & Pachyderms. Intermediate Reptiles – Intermediate fish *Marginalia* 665]

#Pictet (de la Rive), François Jules. 1844–45. *Traité élementaire de paléontologie*. 3 Vols. Cherbuliez, Genève.

[*"Elementary treatise on paleontology"*; Vol. 1, p.68 Fish!! 346 Water & Land Birds. How strange not more common in Secondary period – Lobsters Fish in Old Red also – Didelphys. What a gap from Lower Jura to Tertiary. Vol. 3: When we consider the different mineralogical nature of some of the formations, & difference of depth (such as chalk sea probably deep) it is wonderful when an existing genus . . . far more generally appears in Chalk & Tertiary. Fish genera are too short lived for this to appear: but yet I think it holds pretty often; but then the formation for fish are so rare. Connection in Geographical Range: so in space & time. – I did not think of this, till beginning Gasteropods: easy to see it in other orders. In Fish, the law had better be tested by families, *Marginalia* 669, 670]

+Pictet (de la Rive), François Jules. 1853–54. *Traité de paléontologie*. 2nd Edn, 4 Vols + Atlas. J. B. Baillière, Paris.

[*"Treatise of Paleontology"*; Vol. 2, p. 36 So he thinks Teleosteon a recent fish the most perfect; 409 . . . very different in Fish & Mollusca Cephalopods. Vol. 4, p. 231 . . . so many cases of this (leaving out Silurian) that it must be a rule, though exceptions as in Fish Ctenoids coming in; 644 So he brings down to level of Teleostees & before that Fish not very rich *Marginalia* 673, 674; **Bony fishes; Ctenoid; Scales**]

Pictet (de la Rive), François Jules and Alois Humbert. 1866a. *Nouvelles recherches sur les Poissons Fossiles du Mont Liban*. Genève, vii +115 pp.

["*New research on the fossil fishes of Mount Lebanon*"; **Fossils**]

+Pictet (de la Rive), François Jules and Alois Humbert. 1866b. Recent researches on the fossil fishes of Mount Lebanon. *Annals and Magazine of Natural History* (ser. 3) **18**: 237–47.

[Translated by A. O'Shaughnessy from an article in *Archives des Sciences de la Bibliothèque Universelle, Genève*, June 1866; CD refers to this in a letter to *Lyell of Sept. 8/9, 1866 as a capital paper [...] It is capital in relation to modification of species; I would not wish for more confirmatory facts, though there is no direct allusion to the modification of species; Darwin and Seward 1903, p.160; **Fossils**]

Pielou, Evelyn Christine. 1979. *Biogeography*. John Wiley and Sons, New York, 351 pp.

[**Latitude**]

Pietsch, Theodore W. 1976. Dimorphism, parasitism and sex: reproductive strategies among deepsea ceratoid anglerfishes. *Copeia*, 1976 (4): 781–93.

[**Parasites**]

Pietsch, Theodore W. and William D. Anderson, Jr. (eds.) 1997. *Collection building in ichthyology and herpetology*. American Society of Ichthyologists and Herpetologists Special Publication N0. 3. Allen Press, Lawrence, KS, 593 pp.

Pietsch, Theodore W. and D. D. Grobdecker. 1987. *Frogfishes of the world. Systematics, zoogeography, and behavioral ecology*. Stanford University Press, Stanford, xxii + 420 pp.

[**Pipefishes (II)**]

Pinker, Steven. 1994. *The language instinct: how the mind creates language*. W. Morrow & Co, New York, 494 pp.

[Includes CD quotes documenting his prescient views on language acquisition as a human 'instinct,' and on the similarities between the evolution of languages and that of organic form; **Dohrn; Social Darwinism**]

Pitcher, Tony J. 1983. Heuristic definitions of fish shoaling behaviour. *Animal Behaviour* **31**(2): 611–13.

[**Shoal**]

Pitcher, Tony J. 2001. Fisheries managed to rebuild ecosystems? Reconstructing the past to salvage the future. *Ecological Applications* **11**(2): 601–17.

[**Herring**]

+Playfair, Robert Lambert and Albert Günther. 1867. *The fishes of Zanzibar, with a list of the fishes of the whole east coast of Africa*. London.

[Reprinted in 1971, with a new introduction by G. S. Myers. Newton K. Gregg, Publisher, Kentfield, California, i–xix + 153 pp; CD cited an edition of this work dated 1866; **Pipefishes, ghost**]

Poey, Felipe. 1858–61. *Memorias sobra la historia natural de la Isla de Cuba, acompañadas de sumarios Latinos y extractos en Francés*. La Habana, Vol. 2, pp. 97–336.

["Memoirs on the natural history of Cuba, with Latin summaries and extracts in French"; **Experiments (III)**]

Poey, Felipe. 1880. Revisio piscium Cubensium. *Anales de la Sociedad Española de Historia Natural*, Madrid **9**: 243–61.

["Revision of the fishes of Cuba"; **Reproduction**]

Polis, Gary A. and Stephen D. Hurd. 1996. Linking marine and terrestrial food webs: allochthonous input from the oceans supports high secondary production on small islands and coastal land communities. *American Naturalist* **147**: 396–423.

[**Saint Paul's Rock**]

Popper, Karl Raymund. 1934. *See* Popper (1980).

Popper, Karl Raymund. 1979. *Objective knowledge: an evolutionary approach* (Revised edition). Clarendon Press, Oxford, x + 395 pp.
[**Popper**]

Popper, Karl Raymund. 1980. *The logic of scientific discovery*. Unwin Hyman, London, 480 pp.
[Originally: "*Logik der Forschung*", 1934; **Parsimony; Popper**]

Popper, Karl Raymund. 1985. Natural selection and its scientific status. *In*: David Milled (ed.), *Popper: Selections*, pp.239–46. Princeton University Press, Princeton.
[In which Sir Karl presents a recantation of his previous view that natural selection is untestable, and hence part of a 'metaphysical' research programme; **Popper**]

Por, Francis Dov. 1978. *Lessepsian Migrations: the Influx of Red Sea Biota by Way of the Suez Canal*. Springer-Verlag, Berlin, Heidelberg and New York, 228 pp.
[**Squirrelfishes**]

Porep, Rüdiger. 1980. Methodenstreit in der Planktologie – Haeckel contra Hensen. *Medizinisch-Historisches Journal* **7**: 72–83.
["A methodological debate in planktology: Haeckel vs. Hensen"; **Plankton**]

Porter, Duncan M. 1983. More Darwin *Beagle* notes resurface. *Archives of Natural History* **11**(2): 315–16.
[**Waterhouse, George**]

Porter, Duncan M. 1985. The *Beagle* collector and his collections. *In*: David Kohn (ed.), *The Darwinian Heritage*, pp. 973–1010. Princeton University Press, Princeton.
[*Catalogue; Zoology;* **Appendix I**]

+Pouchet, Georges. 1871. Physiologie – système nerveux des poissons. *L'Institut – Journal Universel des Sciences et des Societés Savantes en France et à l' Etranger* **39**: 134–5.
["Physiology – nervous system of fishes" is an abstract of Pouchet (1872); **Flatfish controversy (II); Flounder**]

Pouchet, Georges. 1872. Du rôle des nerfs dans le changement de coloration des poissons. *Journal de l' Anatomie et de la Physiologie* **8**: 71–4.
["On the role of the nerves in colour changes of fishes"; **Flatfish controversy (II)**]

Poulsen, Thomas L. 1963. Cave Adaptations in Amblyopsid Fishes. *American Midland Naturalist* **70**(2): 257–90.
[**Cavefishes**]

Pound, Ezra Loomis. 1934. *ABC of Reading*. Yale University Press, New Haven, 197 pp.
[**Seeing**]

Pouyaud, L., S. Wirjoatmodjo, I. Rachmatika, A. Tjakrawidjaja, R. Hadiaty and W. Hadie. 1999. Une nouvelle espèce de coelacanthe. Preuves génétiques et morphologiques. *Comptes Rendus de l'Academie des Sciences, Paris, Sciences de la Vie/Life Sciences*, no. 322: 261–7.
["A new species of coelacanth (*Latimeria menadoensis*): genetic and morphological evidence"; **Extinction**]

Powell, Jeffrey R. and James P. Gibbs. 1995. A report from Galápagos. *Trends in Ecology and Evolution* **10**(9): 351–4.
[**Galápagos**]

Power, Mary E., David Tilman, James A. Estes, Bruce A. Menge, William J. Bond, L. Scott Mills, Gretchen Daily, Juan Carlos Castilla, Jane Lubchenco and Robert T. Paine. 1996. Challenges in the quest for keystones. *Bioscience* **46**(8): 609–20.
[**Keystone species**]

+Prinsep, J. 1833. Fall of fish from the sky. *Journal of the Asiatic Society of Bengal* **2**: 650–2.
[**Showers of fish**]

Profumo, David and Graham Swift. 1985. *The Magic Wheel: an Anthology of Fishing in Literature*. Picador/Heinemann, London, 460 pp.
[**Fishing**]

Provine, Will. 1997. Darwin condensed: review of The Darwin Multimedia CD-ROM (2nd Edition) edited by Michel T. Ghiselin and Peter Goldie. *Trends in Ecology and Evolution* **12**(8): 329.
[**CD**]

Pullin, Roger Stephen Vernon, Jérome Lazard, Marc Legendre, Jean-Baptiste Amon Kothias and Daniel Pauly (eds.) 1996. *Proceedings of the Third International Conference on Tilapia in Aquaculture, 11–16 November 1991, Abidjan, Côte d' Ivoire*. ICLARM Conference Proceedings no. 41, x + 574 pp.
[Proceedings also available in French; **Sexual dimorphism**]

#Quatrefages De Bréau, Jean-Louis Armand de. 1854. *Souvenirs d' un naturaliste*. 2 Vols. Charpentier, Paris.
["Memories of a naturalist"; p. 121 Sharks & Salmon & Pike The mere fact of being less like Reptiles makes more Fish-like; 122 Lamprey; 137 Fish?, *Marginalia* 693]

+Quatrefages De Bréau, Jean-Louis Armand de. 1856. Physiologie comparée: les métamorphoses, la généagénèse. *Revue des Deux Mondes*. **4**: 55–82.
["Comparative physiology: metamorphoses, ontogeny"; edition of 1855, Estrail de la Reine des Deux Mondes: pp. 138, 140 Serranus Hermaphrodite Fish, *Marginalia* 692; **Groupers (II); Hermaphrodite**]

Quinn, W. H., D. O. Zopf, K. S. Short and R. T. W. Kuo Yang. 1978. Historical trends and statistics of the southern oscillation, El Niño and Indonesian droughts. *U.S. Fisheries Bulletin* **76**(3): 663–78.
[**Punctuated equilibrium**]

Quinnett, Paul. 1996. *Darwin's Bass: The Evolutionary Psychology of Fishing Man*. Keokee Co Publishing, Sandpoint, Idaho, 250 pp.
[*Darwin's Bass*]

+Quoy, Jean René Constant and Joseph Paul Gaimard. 1824–25. Description des Poissons . . . Chapter IX. (Zoologie): pp. 192–401. [1–328 in 1824; 329–616 in 1825]. . . *In*: Louis Claude Desausles de Freycinet. (1824–1844). *Voyage autour du Monde, Entrepris par Ordre du Roi, Exécuté sur les Corvettes de S.M. "L'Uranie" et "La Physicienne," pendant les années 1817, 1818, 1819 et 1820*. 9 Vols. Paris, chez Phillet ainé.
[**Burrfishes (I); Eels; Duckbills; Filefishes; Jacks; Mojarras; Mullets; Parasites; Parrotfishes; Rainbow runner; Ruff; Sole**]

Quoy, Jean René Constant and Joseph Paul Gaimard. 1834. Poissons. *In: Voyage de découvertes de "l'Astrolabe," exécuté par ordre du Roi, pendant les années 1826–29, sous le commandement de M. J. Dumont d' Urville*, Vol. 3, pp. 645–720. Paris.
[**Parasites**]

+Radcliffe, Charles Bland. 1871. *Dynamics of nerves and muscle*. Macmillan & Co., London.
[9&27&&29&38 Torpedo, *Marginalia* 695; **Electric organs (II)**]

Rafinesque, Constantine Samuel. 1814. *Précis des découvertes et travaux somiologiques de Mr. C. S. Rafinesque-Schmaltz entre 1800 et 1814; ou choix raisonné de ses principales découvertes en zoologie et en botanique, pour servir d'introduction à ses ouvrages futurs*. Royale Typographie Militaire, Palerme, 55 pp.
["*Epitome of the discoveries and somiologic (= biological) works of Mr C. S. Rafinesque-Schmaltz between 1800 and 1814, or a selection of his main discoveries, to serve as introduction to his future works*"; **Pike**]

Bibliography

Randall, John E. 1973. Size of the Great White Shark (*Carcharodon*). *Science* **181**: 160–70.
 [**Great white shark**]

Randall, John E. 1987a. Introductions of marine fishes to the Hawaiian Islands. *Bulletin of Marine Science* **41**(2): 490–502.
 [**Snappers**]

Randall, John E. 1987b. Refutation of length of 11.3, 9.0, and 6.4 m attributed to the white shark, *Carcharodon carcharias*. *California Fish and Game* **73**(3): 163–8.
 [**Great white shark**]

Randall, John E. 1996. *Caribbean Reef Fishes*, 3rd edn, revised and enlarged. Tropical Fish Hobbyist Publications, Neptune City, New Jersey, 368 pp.
 [**Burrfishes (II)**]

Randall, John E. 1998. Zoogeography of shore fishes of the Indo-Pacific region. *Zoological Studies* **37**(4): 227–68.
 [**Surgeonfishes**]

Randall, John E. & Phillip C. Heemstra. 1991. Revision of Indo-Pacific groupers (Perciformes: Serranidae: Epinephelinae), with descriptions of five new species. *Indo-Pacific Fishes* no. 20: 1–332, Pls. 1–41.
 [**Eponym**]

Randall, John E., Richard L. Pyle and Robert F. Myers. 2001. Three examples of hybrid surgeonfishes (Acanthuridae). *Aqua: Journal of Ichthyology and Aquatic Biology* **4**(3): 115–20.
 [**Surgeonfishes**]

Rapp, Wilhelm Ludwig von. 1854. *Die Fische des Bodensees, untersucht und beschrieben von W. von Rapp*. Stuttgart.
 [*"Fishes of Lake Constance, studied and described by W. von Rapp"*; **Wels catfish**]

Raup, David M. 1986. *The Nemesis Affair: a story of the end of the dinosaurs and the ways of science*. Norton, New York.
 [**Extinction; Lyell**]

Reck, Günther. 1979. Una breve analisis de la pesca de bacalao en Galápagos en base a las investigaciones en puerto. *Charles Darwin Research Station Annual Report* 1979: 56–64.
 [*"A brief analysis of the fisheries for the 'bacalao' grouper* Micteroperca olfax*, based on studies at the landing places"*; **Groupers (I)**]

Reck, Günther. 1983. The coastal fisheries in the Galápagos Islands, Ecuador: description and consequences for management in the context of marine environmental protection and regional development. Doctoral Thesis, University of Kiel, 231 pp.
 [**Groupers (I)**]

Reck, Günther. 1999. Protecting the marine ecosystem around the Galápagos Islands. *In*: Daniel Pauly, Villy Christensen and Lucilla Coelho (eds.), *Proceedings of the '98 EXPO Conference on Ocean Food Webs and Economic Productivity, Lisbon, Portugal. 1–3 July 1998*, p. 52. ACP-EU Fisheries Research Report 5.
 [**Galápagos**]

Rehbock, Philip F. 1979. The early dredgers: 'naturalizing' in British Seas, 1830–1850. *Journal of the History of Biology* **12**(2): 293–368.
 [**Dredging**]

Reimchen, Thomas E. 1988. Inefficient predators and prey injuries in a population of giant stickleback. *Canadian Journal of Zoology* **66**(9): 2036–44.
[**Asymmetry; Beauty**]

Reimchen, Thomas E. 1992. Injuries on stickleback from attacks by a toothed predator (*Oncorhynchus*) and implications for the evolution of lateral plates. *Evolution* **46**(4): 1224–30.
[**Asymmetry; Beauty**]

Reimchen, Thomas E. 1997. Parasitism of asymmetrical pelvic phenotypes in stickleback. *Canadian Journal of Zoology* **75**(12): 2084–94.
[**Asymmetry; Beauty**]

Remane, Adolf. 1971. Ecology of Brackish Waters. *In*: Adolf Remane and Carl Schlieper. *Biology of Brackish Waters*, Part I, pp.1–210. E. Schweizerbart'sche Verlagsbuchhandlung, Stuttgart, and John Wiley, New York.
[2nd edition; **Caspian Sea**]

+Richardson, John. 1829–37. *Fauna Borealis-Americana; or the zoology of the Northern Part of British America: containing descriptions of the objects of natural history collected on the late northern land expeditions, under the command of Sir John Franklin, R.N.* 4 vols. Richard Bentley, London.
[Listed in Books to be read, with the comment: 2 vols. except Fishes, which read if Yarrell does not compare British with N. American; *Corrrespondence*, Vol. 4, p.469. This somewhat cryptic note may be interpreted as CD's reminder to himself that two of the volumes do not include fish ('except' can be a verb) but that he should read the remaining two if Yarrell (1836) did not provide a comparison of British and North American Fishes. We have no indication as to whether or not he ever did so. See also Richardson (1836, 1837); **Birds**]

+Richardson, John. 1836. The Fish. *In: Fauna Borealis-Americana, or the zoology of the Northern Part of British America: containing descriptions of the objects of natural history collected on the late northern land expeditions, under the command of Sir John Franklin, R.N.* Part 3: xiv + 327 pp.
[**Hawkins; Mauritius**]

+Richardson, John. 1837. Report on North American Zoology. *Reports of the British Association for the Advancement of Science, 6th Meeting, 1836*: 121–224.
[**Sounds**]

Richardson, John. 1840. Description of a collection of fishes made at Port Arthur in Van Diemen's Land. *Proceedings of the Zoological Society of London* (part 8): 25–30.
['Van Diemen's Land' is an old name for Tasmania, Australia; **Mullets**]

+Richardson, John. 1841. On some new or little known fishes from the Australian seas. *Proceedings of the Zoological Society of London* (part 9): 21–2.
[**Eels**]

+Richardson, John. 1846. Report on the Ichthyology of the Seas of China and Japan. *British Association for the Advancement of Science Report* (for 1845 meeting) **15**: 187–320.
[**Deep-sea spiny eels; Distribution; Latitude; Madeira; Range; Rattails; Submergence**]

+Richardson, John. 1856. Ichthyology. *In: Encyclopedia Britannica*, 8th edn, Vol. xii. London.
[**Eel**]

Risso, Joseph Antoine. 1810/1827. *Ichtyologie de Nice, ou Histoire Naturelle des Poissons du Département des Alpes Maritimes*. F. Schoell, Paris, xxxvi + 388 pp. + 11 plates.
["*Ichthyology of Nice, or Natural History of the Department of the Alpes maritimes*"; 1st edition 1810, 2nd 1827; **Pipefishes (II)**]

Roberts, Clive D. and Chris D. Paulin. 1997. Fish collections and collecting in New Zealand. *In*: Theodore W. Pietsch and William D. Anderson, Jr. (eds.), *Collection building in ichthyology and herpetology*, pp.207–29. American Society of Ichthyologists and Herpetologists, Special Publication Number 3.
[**Bay of Islands**]

Roberts, Tyson R. 2000. A review of the African catfish family Malapteruridae, with description of new species. *Occasional Papers in Ichthyology* **1**: 1–15.
[**Electric fishes**]

Robertson, Jack, Jacqueline McGlade and Ian Leaver. 1996. *Ecological effects of the North Sea industrial fishing industry on the availability of human consumption species: a review*. Univation, The Robert Gordon University, Aberdeen, 29 pp.
[**Capelin**]

Rojas Z, Patricia, Hector Flores G. and José I. Sepúlveda. 1985. Alimentación del bacalao de Juan Fernandez *Polyprion oxygeneios* (Bloch & Scheider 1801) (Pisces, Percichthyidae). *In*: P. Arana (ed.), *Investigaciones Marinas en el Archipiélago de Juan Fernandez*, pp.305–9. Editorial Universitaria, Santiago de Chile.
["Food of *Polyprion oxygeneios*"; **Bacalao**]

#Rolleston, George. 1870. *Forms of animal life*. Clarendon Press, Oxford.
[p. LXVII Affinities of Fishes to Dipnoi & Ganoids; LXXX CV Classification of fishes, *Marginalia* 713, with CV referring to Cuvier & Valenciennes, not '105']

Rondelet, Guillaume. 1554. *Libri de piscibus marini, in quibus veræ piscium effigies expressæ sunt*. Matthias Bonhomme, Lyons.
["*Book of marine fishes, in which are presented true images of fishes*"; facsimile edition of 1558 French translation (yes, including the figure with the baby pipe fishes...) available from Collection CTHS Sciences, no. 2, 2002, Comité des travaux historiques et scientifiques, Paris, 640 pp. **Dogfish; Pipefishes (II)**]

Rosenblatt, Richard H. and Bernard J. Zahuranec. 1967. The eastern Pacific groupers of the genus *Mycteroperca*, including a new species. *California Fish and Game* **53**(4): 228–45.
[**Groupers (I)**]

Rossiter, Margaret W. 1982. *Women Scientists in America: Struggles and Strategies to 1940*. The Johns Hopkins University Press, Baltimore and London, 439 pp.
[**Morris, Margaretta Hare (II)**]

#Rudolphi, Carl Asmund. 1812. *Beyträge zur Anthropologie und allgemeinen Naturgeschichte*. Haude & Speuer, Berlin.
["*Contributions to anthropology and general natural history*"; p.139 Bring case of F.W. Fish, difficulty in diffusion; So before Agassiz – Fish speak strongly that they have been created at many points, as same fish in distinct rivers. Glacial case makes of Fish much more difficult; Linnaeus asserts that the Pike is disseminated by Birds; 161 cases of Hybrid Fish; 184 What causes the beauty of snakes, Lizard or Newt? Salmon hook male fish different, *Marginalia* 716, 717]

Rudwick, Martin. 1974. Review of 'Darwin and his critics' by D. Hull. *Theory and Society* **1**(4): 500–3.
[**Caviller**]

Ruhlen, Merritt. 1994. *The origin of language: tracing the evolution of the mother tongue*. John Wiley and Sons, New York, 239 p.
[**Names, common**]

Rumohr, Heye. 1990. A brief history of benthos research in Kiel Bay and in the Baltic. *Deutsche Hydrografische Zeitschrift* (suppl.), ser. B, no. 22: 53–160.
[**Dredging; Möbius**]

Rupke, Nicolaas A. 1994. *Richard Owen: Victorian Naturalist*. Yale University Press, New Haven, CT, 462 pp.
[**Owen**]

Rüppel, Wilhelm Peter Eduard Simon. 1835–38. Neue Wirbelthiere zu der Fauna von Abyssinien gehörig. *In Fische des Rothen Meeres*. Frankfurt-am-Main.
[Section on *Fishes of the Red Sea* published in 1838, pp. 81–148, as: "New vertebrates belonging to the Abyssinian fauna"; **Pipefishes (I)**]

Rüppel, Wilhelm Peter Eduard Simon. 1852. *Verzeichniss der in dem Museum der Senckenbergischen naturforschenden Gesellschaft aufgestellten Sammlungen*. Vierte Abtheilung. *Fische und deren Skelette*. Frankfurt-am-Main, 40 pp.
["*Catalogue of the collections exhibited by the Senckenberg Nature Research Society*. Fourth section. *Fishes and their skeletons*"; **Mouth-brooding**]

Ruse, Michael. 1986. *Taking Darwin seriously: a naturalistic approach to philosophy*. Oxford, Basil Blackwell, xv + 303 pp.
[**Paradigm**]

Safire, William. 1999. Ockham's Razor's Close Shave. *New York Times Magazine*, January 31, p.14.
[**Occam's Razor**]

Saint-Paul, Ulrich. 1986. Potential for aquaculture of South American freshwater fishes: a review. *Aquaculture*. **54**: 205–40.
[**Creationism**]

Saint-Pierre, Jacques-Henry Bernardin de. 1773. *Voyage à l'Isle de France, à l'Isle de Bourbon, au cap de Bonne-Espérance, &c. Avec des observations nouvelles sur la nature & sur les hommes, par un officier du roi*. 2 Vols. in 1. Neuchâtel.
["*Voyage to Mauritius, Réunion, the Cape of Good Hope, etc., with new observations on nature and people, by an officer of the king*"; **Mauritius**]

#Saint-Hilaire, Auguste de. 1841. *Leçon de botanique, comprenant principalement la morphologie végétale, la terminologie, la botanique comparée,...* P.-J. Loss, Paris.
["*A course in botany, including chiefly plant morphology, terminology, comparative botany, ...*"; p.834 Important organs may vary in early stocks: hypothesis. 836 confirmed by Owen [1846] on swim bladder in vol on Fishes, *Marginalia* 735]

Sano, Mitsuhiko, Makoko Shimizu and Yukio Nose. 1984. *Food habits of teleostean reef fishes in Okinawa Island, southern Japan*. University of Tokyo Press, Tokyo, Japan, 128 pp.
[**Butterflyfishes**]

Santamaría, J. A., M. Marí-Beffa, L. Santos-Ruiz and J. Becerra. 1996. Incorporation of bromodeoxyuridine in regenerating fin tissue of the goldfish *Carassius auratus*. *Journal of Experimental Zoology* **275**: 300–7.
[**Polydactylism**]

Santos, Ricardo Serrão, Filipe Porteiro and João Pedro Barreiros. 1997. Marine Fishes of the Azores: Annotated Checklist and Bibliography. A catalogue of the Azoran Marine Ichthyodiversity. *Arquipélago – Bulletin of the University of the Azores*, supplement 1, 244 pp.
[**Azores**]

Sapp, Jan. 1994. *Evolution by association: a history of symbiosis*. Oxford University Press, Oxford, 255 pp.
[**Altruism; Non-Darwinian evolution**]

Sarhage, Dietrich and Johannes Lundbeck. 1991. *A history of fishing*. Springer-Verlag, Berlin, viii + 348 pp.
[**Herring**]

Sasal, Pierre and Christophe Pampoulie. 2000. Asymmetry, reproductive success and parasitism of *Pomatoschistus microps* in a French lagoon. *Journal of Fish Biology* **57**: 382–90.
[**Asymmetry**]

+Sauvigny, Edmé Louis Billardon, de. 1780. *Histoire naturelle des dorades de Chine, gravée par M. F. N. Martinet, accompagnée d' observations*. Paris.
["*Natural history of Chinese goldfish, with etchings by M. F. N. Martinet, along with observations*"; **Goldfish (I)**]

+Saville-Kent, William. 1873a. Permanent and temporary variation of colour in fish. *Nature* **8**: 25.
[**Wrasses**]

+Saville-Kent, Willliam. 1873b. Fish distinguished by their actions. *Nature* **8**: 263.
[**Dragonets**]

Saville-Kent, William. 1889. Preliminary observations on a natural history collection made in connection with the surveying cruise of H.M.S. "Myrmidon," at Port Darwin and Cambridge Gulf in . . . 1888. *Proceedings of the Royal Society of Queensland* **6** (5): 219–40, Pl. 13.
[**Eponym**]

Schiebinger, Londa. 1999. *Has feminism changed science?* Harvard University Press, Cambridge, MA, 252 pp.
[**Taphonomy**]

+Schiödte, Jørgen Mathias Christian. 1868. On the Development of the Position of the Eyes in Pleuronectidae. *Annals and Magazine of Natural History* (ser. 4) **1**: 378–383.
[Translated from: "Om øiestillingens udvikling hos flynderfiskene. *Naturhist. Tidsskr. Kjøbenhavn* (ser. 3) **5**: 269–275, 1868/69"; **Flatfish controversy (II)**]

Schischkoff, Gergi (ed.) 1961. *Philosophisches Wörterbuch*, 16th edn. Alfred Kröner Verlag, Stuttgart, 656 pp.
["*Philosophical Dictionary*"; **Benthos**]

Schlee, Susan. 1973. *The Edge of an Unfamiliar World: a History of Oceanography*. E. P. Dutton & Co., New York, 398 pp.
[**Plankton**]

+Schlegel, Hermann. 1843. *Essay on the physiognomy of the serpent*. Maclachlan, Stewart & Co., Edinburgh.
[p. 219 what a difficulty introduction of F.W. Eel in Otaheite & some of the Antarctic Isds, *Marginalia* 744; **Constancy**]

Schluter, Dolph, 1996. Ecological causes of adaptive radiations. *American Naturalist*. **148** (suppl.): 40–64.
[**Sticklebacks**]

Schmalhausen, Ivan Ivanovitch. 1949. *Factors of Evolution: the Theory of Stabilizing Selection*. Translated by Isadore Dordick, edited by Theodosius Dobzhansky. Blackiston, Philadelphia, xiv + 327 pp.
[**Obstinate nature**]

Schmidt-Nielsen, Knut. 1984. *Why is animal size so important?* Cambridge University Press, Cambridge, 241 pp.

[**Length**]

Schneider & Foster 1801: *see* Bloch and Schneider (1801).

Scholes, Robert. 1984. Is there a fish in this text? *College English* **46**(7): 653–64.

[**Seeing**]

Schrödinger, Edwin. 1967. *What is Life? The Physical Aspect of the Living Cell and Mind and Matter*. Cambridge University Press, Cambridge, 178 pp.

[**Oxygen**]

Schumann, D. and J. Piiper. 1966. Der Sauerstoffbedarf der Atmung bei Fischen nach Messungen and der narkotisierte Schleie (*Tinca tinca*). *Pflügers Archiv* **288**: 15–26.

["Oxygen requirement of respiration in fishes, based on measurements on narcotised Tench"; **Tench**]

Schwartz, F. J. 1972. World literature to fish hybrids, with an analysis by family, species, and hybrid. *Publications of the Gulf Coast Research Laboratory Museum* **3**: 1–328.

[**Hybrids**]

Scott, William Beverly and Edwin John Crossman. 1973. *Freshwater fishes of Canada*. Fisheries Research Board of Canada Bulletin no. 184, 966 pp.

[Reprinted 1990; **Jacks**]

Scott, William Beverly and Mildred Grace Scott. 1988. *Atlantic fishes of Canada*. Canadian Bulletin of Fisheries and Aquatic Sciences no. 219, 730 pp.

[**Cod**]

+Scrope, William. 1843. *Days and Nights of Salmon Fishing in the Tweed; with a Short Account of the Natural History and Habits of the Salmon, Instructions to Sportsmen, Anecdotes, etc*. Longman Hamilton, Adams, and Co., London.

[Listed by CD among Books to be read, and later abstracted; Burkhardt and Smith (1988), p. 468; **Homing; Hybrids; Salmon (III: Not pacific)**]

#Sedgwick, Adam. 1850. *A discourse of the studies at the University of Cambridge*. 5th edn. John Parker, London.

[The publication of the Vestiges brought out all that cd be said against the theory excellently if not too vehemently; lxv Oldest Fish highest (Book written against law of development higher & higher with which I have nothing to do; xciv . . .? Not the Fishes & Reptiles; xcvii difficulty of appearance of *Cycloids & *Ctenoids (Developed in hot ocean); But take existing fish & existing Reptiles . . .the first appearance alone ought to be chronologically in harmony with natural affinities; ci – on separation of fish & lizard; ccxvi True great classes will never run into each other – even Lepidosiren does not do that; 186 Electrical Fishes. Good account of why Fishes <Sharks> highest; 188 to 192 we here see that a Bony fish as a fish may be highest, but as part of the Vertebrata lower; 192 On coexistence of spiral valves in intestine & Bulbus arteriosus in Ganoids & so allied to Batrachians, *Marginalia* 750; **Ctenoid; Cycloid; Scales**]

Semper, Carl Gottfried. 1875. On the relationship of the vertebrata and annelida. *Annals and Magazine of Natural History* **15**: 94–5.

[**Dohrn**]

Shaw, George and F. P. Nodder. 1789–1813. *The Naturalist's Miscellany, or coloured figures of natural objects; drawn and described from nature*. London. Unnumbered pages. Pl. 253.

[**Boxfishes**]

+Shaw, John. 1836. An account of some experiments and observations on the parr, and on the ova of the salmon, proving the parr to be the young of the salmon. *Edinburgh New Philosophical Journal* **21**: 99–110.
[**Hybrids**]

+Shaw, John. 1838. Experiments on the development and growth of the fry of salmon from the exclusion of the ovum to the age of seven months. *Edinburgh New Philosophical Journal* **24**: 165–76.
[**Hybrids**]

+Shaw, John. 1840. Account of experimental observations on the development and growth of salmon-fry, from the exclusion of the ova to the age of two years. *Transactions of the Royal Society of Edinburgh* **14**: 547–66.
[**Hybrids**]

Sheets-Pyenson, Susan. 1981. Darwin's data: his reading of Natural History Journals, 1837–1842. *Journal of the History of Biology* **14**(2): 231–48.
[**Blyth;** *Marginalia*]

Sheldon, R. W. and Stephen R. Kerr. 1972. The population density of monsters in Loch Ness. *Limnology and Oceanography* **17**: 796–8.
[**Shower of fishes**]

Sheldrake, Rupert. 1988. *The presence of the past: morphic resonance and the habits of nature*. Times Books, New York, xxii + 391 pp.
[Not worth reading; **Non-Darwinian evolution**]

Shepperson, George. 1961. The intellectual background of Charles Darwin's student years at Edinburgh. *In*: M. Banton (ed.), *Darwinism and the Study of Society*, pp.17–35. Tavistock Publications, London.
[**Edinburgh**]

Sherborn, C. Davies. 1897. Notes on the dates of "The Zoology of the 'Beagle'." *Annals and Magazine of Natural History* (ser. 6) **20**(119): 483.
[*Fish;* **Jenyns (II)**]

+Silliman, Benjamin Jr. 1851. On the Mammoth Cave of Kentucky. *American Journal of Sciences and Arts*, (ser. 2) **11**: 332–9.
[**Cavefish**]

Sinclair, Michael and Per Solemdal. 1988. The development of "population thinking" in fisheries biology between 1878 and 1930. *Aquatic Living Resources* **1**: 189–213.
[**Classification**]

Singer, Peter, 2000. *A Darwinian Left: Politics, Evolution and Cooperation*. Yale University Press, New Haven, 70 pp.
[**Social Darwinism; Spencer**]

Singh, Simon. 1997. *Fermat's Enigma: The Quest to Solve the World's Greatest Mathematical Problem*. Walker & Co., New York, vxiii + 315 pp.
[*Marginalia*]

Slobotkin, Lawrence B. 1992. *Simplicity and complexity in games of the intellect*. Harvard University Press, Cambridge, MA, 266 pp.
[**Complexity**]

#Smellie, William. 1790. *The philosophy of natural history*. Elliott, Kay, Cadell & Robinson, Edinburgh and London.

[p. 513 age of some Big Birds.- Ravens & Geese lay a good many eggs, yet old livers – So with <u>Carp</u>, which lay so many eggs, *Marginalia* 758]

Smetacek, Victor. 1999. Revolution in the ocean: Victor Hensen realized that in the sea, the very small feed the very large. *Nature* **401**: 647.
[**Plankton**]

Smetacek, Victor. 2001. A watery arms race. *Nature* **411**: 745.
[Argues that the diversity of phytoplankton, and its distribution within the euphotic zone – rather than only at the water surface – are evolved responses to grazers; **Plankton**]

Smith, Bradley R. 1999. Visualizing human embryos. *Scientific American* **280**(3): 77–81.
[**Ontogeny**]

Smith, Charles Hamilton. 1852. *The natural history of the human species: its typical forms, primaeval distribution, filiations and migrations.* W. H. Lizars, Edinburgh, and Henry G. Bohn, London.
[p. 47 change in river flowing into Caspian of the Euxus; 47 cd Caspian have joined the Japan Sea; 49 Fish; 117 what an argument, *Marginalia* 766; **Hamilton Smith**]

Smith, John L. B. 1939. A living fish of Mesozoic type. *Nature* **143**: 455–6.
[**Extinction**]

Smith, R. Jan F. 1992. Alarm signal in fishes. *Reviews in Fish Biology and Fisheries*, **2**(1): 33–63.
[**Minnows**]

Smith, Sydney. 1960. The Origin of '*The Origin*' as discerned from Charles Darwin's notebooks and his annotations in the books he read between 1837 and 1842. *Advancement of Science* **16**: 391–401.
[**Blyth**]

Smith, Sydney. 1968. The Darwin Collection at Cambridge, with one example of its use: Charles Darwin and *Cirripedes*. In: *Acts of the 11th International Congress on the History of the Sciences, 24–31 August 1965*, pp.96–100. Ossolineum, Wroclaw.
[**Barnacles**]

Smith-Vaniz, William F., Bruce B. Collette and Brian E. Luckhurst. 1999. *Fishes of Bermuda: history, zoogeography, annotated checklist and identification keys*. American Fisheries Society of Ichthyologists and Herpetologists Special Publication no 4. Allen Press, Lawrence, Kansas, 424 pp.
[**Fishing; Range**]

Snodgrass, Robert Evans and Edmund Heller. 1905. Papers from the Hopkins-Stanford Galápagos Expedition, 1898–1899. XVII. Shore fishes of the Revillagigedo, Clipperton, Cocos and Galápagos Islands. *Proceedings of the Washington Academy of Sciences* **6**: 333–427.
[**Galápagos shark(s)**]

Snow, P. J., M. B. Plenderleith and L. L. Wright. 1993. Quantitative study of primary sensory neurone populations of three species of elasmobranch fish. *Journal of Comparative Neurology* **334**(1): 97–103.
[**Flying fishes**]

Sobel, Dava. 1995. *Longitude: the true story of a lone genius who solved the greatest scientific problem of his time.* Walker, New York, viii + 184 pp.
[**Beagle**]

Soto, Cristina G., John F. Leatherland and David L. G. Noakes. 1992. Gonadal histology of the self-fertilizing hermaphroditic fish *Rivulus marmoratus*. *Canadian Journal of Zoology* **70**: 2338–47.
[**Hermaphrodite**]

Southward, A. J. 1983. A new look at variation in Darwin's species of acorn barnacles. *Biological Journal of the Linnean Society* **20**: 59–72.

[**Barnacles**]

+Spencer, Herbert. 1864–67. *The Principles of Biology*. Vols. 1 and 2 bound together. Williams & Norgate, London.

[Vol. 2, p.188 Discusses one-sided fishes; 201 argues well with Amphioxus that the muscles first gave rise to Vertebrae but first of all to the Neural Spines (see quotation of Owens'); 326 How animals acquired Lungs in shallow water; 399 Struggle for existence & Law of Increase; 428 He does not understand Pangenesis; 439 Antagonism between growth & Reproduction for Pangenesis; 470 Male fish guarding nest ask further, *Marginalia* 770, 771; **Reproductive drain hypothesis; Social Darwinism; Spencer; Survival of the fittest**]

+Spix, Johann Baptist von and Louis Agassiz. 1829–31. *Selecta genera et species piscium quos in itinere per Brasiliam annis MDCCCXVII-MDCCCXX jussu et auspiciis Maximiliani Josephi I. Bavariae regis augustissimi peracto collecit et pingendos curavit Dr J. B. de Spix*, [. . .], *digessit, descripsit et observationibus anatomicis illiustravit Dr. L. Agassiz*. Monachii. Part 1: i–xvi + i–ii + 1–82, Pls. 1–48.

["*A selection of genera and species of fishes from travels in Brazil in the years 1817–1820, by the order and under the auspices of Maximilian Joseph I, the most august King of Bavaria, which were collected and painted for Dr. J. B. von Spix* [. . .], *with an account, descriptions and anatomical illustrations by Dr L. Agassiz*. Munich;" **Catfishes;** *Fish in Spirits of Wine*]

+Stark, James. 1838. On the food of the vendace, herring, and salmon. *Annals and Magazine of Natural History* **1**: 74–5.

[**Egg development**]

Stark, James. 1844. On the Existence of an Electrical Apparatus in the Flapper Skate and other Rays. *Royal Society of Edinburgh Proceedings* **2**: 1–3.

[**Electric fishes**]

Stauffer, Robert Clinton. 1960. Ecology in the long manuscript version of Darwin's *Origin of Species* and Linnaeus' *Oeconomy of Nature*. *Proceedings of the American Philosophical Society* **104**(2): 235–41.

[*Big Species Book*; **Ecology**]

Stauffer, Robert Clinton (ed.). 1975. *Charles Darwin's Natural selection: Being the second part of his big species book*. Cambridge University Press, Cambridge, 692 pp.

[*Big Species Book*]

#Steenstrup, Johann Japetus. 1845. *On the alternation of generations*. Translated by G. Busk. The Ray Society, London.

[F.W. Fish almost normally have Trematoda within eyes, *Marginalia* 786]

+Steenstrup, Johann Japetus Smith. 1863. *See* Steenstrup 1865.

+Steenstrup, Johann Japetus Smith. 1865. On the Obliquity of Flounders. *Annals and Magazine of Natural History* (ser. 3) **15**(89): 361–71.

[This contribution is a commented abstract, by Wyville Thompson, of "Steenstrup, Jean Japhet Smith 1863. Om Skjaevheden hos Flynderne, og navnlig om Vandringen af det övre Öie fre Blindsiden til Öiesiden tvers igjennend Hovedet" or translated: "On the migration of the upper eye of flounders, across, through the head, from the blind side to the eye-side," Kjöbenhaven, 1864. *Oversigt over det Kongetige Danske Videnskabernes Selskabs Forhandlinger og dets Medlemmers Arbeider i Aaret 1863*; **Flatfish controversy (II); Flounder**]

+Steenstrup, Johann Japetus Smith. 1873. Om Gjaellegitteret eller Gjaellebarderne hos Brugden (*Selachus maximus* (Gunn.)). *Oversigt over det Kongelige Danske Videnskabernes Selskabs Forhandlinger og dets Medlemmers Arbeider i Aaret 1873*. 1873–74: 47–66.

["On the gill-rakers or gill-baleen of the Basking shark *Selachus maximus* (Gunnerus, 1765)"; **Basking shark**]

Steindachner, Franz. 1867. Ichthyologische Notizen (5. Folge). *Anzeiger der philosophisch-historischen Klasse der Österreichichen Akademie der Wissenschaften Wien* **4**(14): 119–20.

["Ichthyological notes"; **Flounder, Fine**]

Steindachner, Franz. 1874. Über eine neue Gattung und Art aus der Familie der Pleuronectiden und über eine neue *Thymallus*-Art. *Anzeiger der philosophisch-historischen Klasse der Österreichichen Akademie der Wissenschaften Wien* **11**(21): 171–2.

["On a new genus and species of the Pleuronectidae Family, and on a new species of *Thymallus*"; **Eponym; Rhombus**]

Steineger, Leonhard. 1886. On the extermination of the great northern sea cow (*Rythina*). *Bulletin of the American Geographical Society* **4**: 317–28.

[**Sirenians**]

Stergiou, Konstantinos I. 1988. Feeding habits of the Lessepsian migrant *Siganus luridus* in the Eastern Mediterranean, its new environment. *Journal of Fish Biology* **33**(4): 531–43.

[**Chimaera**]

Stergiou, Konstantinos I. 1989. Capelin *Mallotus villosus* (Pisces: Osmeridae), glaciations, and speciation: a nomothetic approach to fisheries ecology and reproductive biology. *Marine Ecology Progress Series* **56**: 211–24.

[**Capelin**]

Stiassny, Melanie L. J. and Axel Meyer. 1999. Cichlids of the Rift Lakes. *Scientific American* **280**(2): 64–9.

[**Speciation**]

Stoddart, David Ross. 1962. Coral Islands by Charles Darwin, with Introduction, maps and remarks. *Atoll Research Bulletin* no. 88, 20 pp.

[Scholarly edition of notes kept at the Cambridge University Library, written by CD on Dec. 3–21, 1835 on the voyage from Tahiti to New Zealand, and forming the basis of his theory of coral reef formation; **Cocos Islands; Coral Reefs**]

Stoddart, David Ross. 1994. "This Coral Episode." *In*: Rod MacLeod and Philip F. Rehbock (eds), *Darwin's Laboratory: evolutionary theory and natural history in the Pacific*, pp. 21–48. University of Hawai'i Press, Honolulu.

[**Cocos Islands; Coral Reefs**]

Stone, Richard, 2002. Caspian ecology teeters on the brink. *Science* **295**(5554): 430–3.

[**Caspian Sea**]

Streets, Thomas Hale. 1877. Ichthyology. *In*: *Contributions to the natural history of the Hawaiian and Fannings Islands and Lower California, made in connection with the United States North Pacific surveying expedition, 1873–75. Bulletin of the United States National Museum* no. 7, pp. 43–102.

[**Surgeonfishes**]

Strickland, Hugh Edwin, John Phillips, John Richardson, Richard Owen, Leonard Jenyns, William John Broderip, John Stevens Henslow, William Edward Shuckard, George Robert Waterhouse, William Yarrell, Charles Darwin, and John Obadiah Westwood. 1843. Series of propositions for rendering the nomenclature of zoology uniform and permanent. *Annals and Magazine of Natural History* **11**: 259–75.

[Included on the Darwin CD-ROM; see **CD; Code; Names, scientific; Taxonomy; Waterhouse**]

Bibliography

Stringer, Christopher and Robin McKie. 1996. *African exodus: the origins of modern humanity*. Henry Holt and Company, New York, 282 pp.
[**Names, common**]

Suess, Eduard. 1885. *Das Antlitz der Erde*. 1. Translation by Hertha B. C. Sollas, under the direction of W. J. Sollas. Clarendon Press, Oxford.
["The Face of the Earth;" **Distribution**]

Sulak, K. J. 1990. Notacanthidae. *In*: J. C. Quero, J. C. Hureau, C. Karrer, A. Post and L. Saldanha (eds.), *Check-list of the fishes of the eastern tropical Atlantic (CLOFETA)*, Vol. 1, pp. 133–5. JNICT, Lisbon; SEI, Paris; and UNESCO, Paris.
[**Deep-sea spiny eels**]

Sulloway, Frank J. 1982. Darwin's Conversion: the *Beagle* voyage and its aftermath. *Journal of the History of Biology* **15**: 325–96.
[**Catalogue**]

Sulloway, Frank J. 1983. Further remarks on Darwin's spelling habits and the dating of *Beagle* voyage manuscripts. *Journal of the History of Biology*. **16**(3): 361–90.
[**Spelling**]

Summers, Diane and Eric Valli. 1990. The Nest Gatherers of Tiger Cave. *National Geographic Magazine*, January: 107–33.
[Also available as a book (by Eric Valli and Diane Summers, 1990) from Thames and Hudson, London; **Bird's nest soup**]

Syme, Patrick. 1821. *Werner's Nomenclature of Colours, With Additions, Arranged so as to Render it highly Useful to the Arts and Sciences, particularly Zoology, Botany, Chemistry, Mineralogy, and Morbid Anatomy. Annexed to which are examples selected from well-known objects in the animal, vegetable and mineral kingdoms*. William Blackwood, Edinburgh, and T. Cadell, London, 47 pp+ 13 colour plates.
[This book, the second edition of a work first published in 1814 by a "Flower-Painter" based in Edinburgh, and worked for "the Wernerian and Caledonian Horticultural Societies", covers 110 named "tints", and thus expands on the 79 names and descriptions originally published by Abraham Werner (e.g. "Chocolate red, is veinous blood red mixed whith a little brownish red"). Moreover, Syme included in his book 110 small (1.2 cm^2) colour panels, glued onto the appropriate plates ("Whites", "Greys", "Blacks", etc.), along with examples under columns labelled "ANIMAL", "VEGETABLE", and "MINERAL." (Most animal examples are parts of birds and insect bodies; the only fish example is "Reddish orange," as occur in "Gold Fish lustre abstracted." Some colours are not what one would expect, and "Broccoli brown" is not brown, besides having no VEGETABLE example; go figure). This small (23cm × 14cm × 1cm) book (which must have been expensive, given its handmade colour panels) is easily carried in the field, and hence CD was able to use it to standardize the colour names he used. CD's copy, much used during the voyage of the *Beagle*, is kept at Down House; **Colours; Scale**]

Tavolga, William N. 1968. *Marine animal data atlas*. Technical Report, Naval Training Device Center 1212–2, 239 pp.
[**Gurnards**]

Telles, Marcello Dantas. 1998. Modelo trofodinâmico dos recifes em franja do Parque Nacional dos Abrolhos – Bahia. Master Thesis, Universitade Federal da Bahia, Brazil, 150 pp.
["Trophodynamic model of the fringing reefs of Abrolhos National Park"; **Abrolhos**]

Templeman, Wilfred. 1948. Life History of the Caplin (*Mallotus villosus* O.F. Müller) in Newfoundland Waters. *Bulletin of the Newfoundland Government Laboratory No. 17 (Research)*, St. John's. 151 pp.
[**Capelin**]

Thewissen, J.M.G., E.M. Williams, L.J. Rose and S.T. Hussain. 2002. Skeletons of terrestrial cetaceans and the relationship of whales to artiodactyls. *Nature* **413**: 277–81.
[**Jaguar**]

+Thompson, Charles Wyville. 1841. Notes on British Char, *Salmo Umbla*, Linn., *S. Salvelinus*, Don. *Annals and Magazine of Natural History* **6**: 439–50.
[**Char**]

+Thompson, Charles Wyville. 1865. Notes on Prof. Steenstrup's Views on the Obliquity of Flounders. *Annals and Magazine of Natural History* (ser. 3) **15**(89): 361–71.
[**Flatfish controversy (II); Flounder**]

Thompson, D'Arcy Wentworth. 1917. *On growth and form*. Cambridge University Press, Cambridge.
[**Growth**]

Thompson, William. 1835. Some additions to the British fauna. *Proceedings of the Zoological Society of London* **3**: 77–82.
[**Pollan**]

+Thompson, William. 1839. On fishes: containing a notice of one species new to the British, and of others to the Irish fauna. *Annals and Magazine of Natural History* **2**: 266–73.
[CD scored in this the passage stating: "Since my account of the pollan appeared, I have been favoured by Dr. Parnell with a specimen of the *Coregonus* of Loch Lomond [...] and by Sir Wm. Jardine with one of the Ullswater species; both of which are distinct from the *Cor. Pollan*, this having not as yet been found in any of the lakes of Great Britain." The identity of these varieties, if any, could not be ascertained; **Pollan**]

Thornhill, Randy. 1986. Early history of sexual selection theory. *Evolution* **40**(2): 446–7.
[**Sexual selection**]

Thurow, Fritz. 1953. Untersuchungen über die spitz- und breitköpfigen Varianten des Flußaales. *Archiv für Fischereiwissenschaft* **9**(2): 79–97.
["Studies on the narrow- and broad-headed variants of the river eel"; **Eel**]

Todes, Daniel P. 1989. *Darwin without Malthus: the struggle for existence in Russian evolutionary thought*. Oxford University Press, Oxford, 221 pp.
[**Altruism**]

Topoff, Howard. 1997. A Charles Darwin (187th) Birthday Quiz. *American Scientist* **85**(2): 104–7.
[**Capybara**]

Tort, Patrick (ed.). 1996. *Dictionnaire du Darwinisme et de l'Evolution*. Presses Universitaires de France, Paris. Vol. 1: 1–1611; Vol. 2: 1613–3257; Vol. 3: 3258–4862.
[**Agassiz; Blyth; Carp; Catfishes; Complexity; Grant; Groupers (II); Hamilton Smith; Sex ratio; Species; Spencer; Yarrell**]

Tosteson, T. R., D. L. Ballantine and H. D. Durst. 1988. Seasonal frequency of ciguatoxic barracuda in Southwest Puerto Rico. *Toxicon* **26**(9): 795–801.
[**Barracuda**]

Traquair, Ramsay Heatley. 1865. On the asymmetry of the Pleuronectidae, as elucidated by an examination of the skeleton of the Turbot, Halibut and Plaice. *Transactions of the Linnean Society of London* **25**: 263–96.
[Session of 15 June 1865; **Flatfish controversy (II)**]

Trewavas, Ethelwynn. 1953. Sea-trout and brown trout. *Salmon and Trout Magazine* **139**: 199–215.
[**Trout**]

Tsuda, R. T. and P. G. Bryan. 1973. Food preference of juvenile *Siganus rostratus* and *S. spinus* in Guam. *Copeia* 1973(3): 604–6.
  [**Chimaera**]

Turner, William. '1866' [1867]. On a remarkable mode of gestation in an undescribed species of *Arius* (*A. Boakeii*). *Journal of Anatomy and Physiology* **1**: 78–82.
  [Published in November 1866; **Catfishes; Mouth-brooding**]

Ulanowicz, Robert E. 1986. *Growth and development: ecosystem phenomenology*. Springer-Verlag, New York, xiv + 203 pp.
  [**Complexity**]

Vaillant, Léon Louis. 1884. [No title] *In*: H. Filhol: *Explorations sous-marines – voyage du Talisman*. La Nature, Paris. February 1884 (**559**): 182–6.
  [**Rattails**]

Valenciennes, Achille. 1831–40: *See* corresponding years in Cuvier and Valenciennes.

Valenciennes, Achille. 1834–42. *Poissons* [plates only]. *In*: Orbigny (1834–47).
  [**Catfishes**]

Valenciennes, Achille. 1837–44. Ichthyologie des îles Canaries, ou histoire naturelle des poissons rapportés par Webb & Berthelot. *In*: Philip Barker Webb & Sabin Berthelot, *Histoire naturelle des îles Canaries*. Paris, 1835–1850.
  [**Ascension Island**]

Valenciennes, Achille. 1838a. Poissons des îles Canaries. *L' Institut – Journal Général des Sociétés et Travaux Scientifiques de la France et de l' Etranger* **6**(251), 18 October: 338.
  [Summary of Valenciennes (1837–44); **Ascension Island**]

Valenciennes, Achille. 1838b. Considerations générales sur l'ichthyologie de l'Atlantique. *Comptes Rendus Hebdomadaires des Séances de l' Académie des Sciences, Paris* **7**: 717–22.
  [**Ascension Island**]

+Valenciennes, Achille. 1841. Nouvelles recherches sur l'organe électrique du Silure électrique (*Malapterurus electricus* Lacépède). *Archives du Musée d'Histoire Naturelle* **2**: 43–62.
  [Also published in: *C. R. Acad. Sci. Paris*, 1840, **11**: 227–230, and *Ann. Sci. Nat.* 1840, Ser. 2: 241–244; **Electric fishes**]

Valenciennes, Achille. 1846. *See* Valenciennes (1855).

Valenciennes, Achille. 1847. Catalogue des principales espèces de poissons, rapportées de l'Amérique méridionale. Vol. 5 (part 2): 1–11. *In*: A. d'Orbigny (ed.), *Voyage dans l'Amérique méridionale*. . . .Paris.
  [See Orbigny 1834–47]

Valenciennes, Achille. 1855. Ichthyologie, pp. ii–iii and 297–351. *In*: A. du Petit-Thouars, *Atlas de Zoologie. Voyage autour du monde sur la frégate 'Vénus' pendant les années 1836–1839*. Paris.
  [Plates, including '*Serranus psittacinus*' published in 1846; **Galápagos; Groupers (I); Lizards**]

Van Valen, Leigh. 1973. A new evolutionary law. *Evolutionary Theory* **1**: 1–30.
  [**Punctuated equilibrium**]

Verne, Jules. 1997. *Histoire des grand voyages et grand voyageurs: les grand navigateurs du XVIIe siècle*. Diderot Editeurs, Paris, 636 pp.
  [Re-edition of a little known work by *the* Jules Verne; **Falkland Islands**]

Vincent, Amanda. 1994. Seahorses exhibit conventional sex roles in mating competition, despite male pregnancy. *Behaviour* **128**: 135–51.
  [**Seahorses**]

Vincent, Amanda. 1996. *The international trade in seahorses*. WWF/IUCN – TRAFFIC International, 163 pp.

[**Seahorses**]

#Virchow, Rudolf. 1860. *Cellular pathology, as based upon physiological and pathological histology*. Translated from 2nd German edition by Frank Chance. John Churchill, London.

[p. 126 Poisonous Fishes. *Marginalia* 822; it is unclear what in this text prompted CD's jotting]

Vogt, Carl. 1842. Embryologie des Salmones. Part 2. *In*: Jean Louis Rodolphe Agassiz and Carl Vogt. (1839–1842), *Histoire naturelle des poissons d'eau douce de l'Europe centrale*. Neuchâtel.

[The contents of this contribution, on the "Embryology of salmonids", were summarized by T. Huxley in a letter to CD, which he annotated (*Correspondence*, Sept. 17, 1858); however, CD does not cite Vogt, nor Karl Ernst von Baer, also mentioned in the letter, when discussing this organ in *Origin*]

Vorzimmer, Peter J. 1963. Charles Darwin and blending inheritance. *Isis* **54**(3): 371–90.

[**Lamarck;** *Marginalia;* **Words**]

Vorzimmer, Peter J. 1969. Darwin's "Lamarckism" and the "Flat-Fish Controversy" (1863–1871). *Lychnos* 1969: 121–70.

[**Flatfish controversy (II); Non-Darwinian evolution**]

#Wagner, Moritz. 1873. *The Darwinian theory of the comparative anatomy of the vertebrate animal*. Alfred Tulk (ed.). Longman, Brown, Green & Longman, London.

[Read as far as p.130 & marked thus far – & I do not think worth reading further; 217 On Electric fishes, *Marginalia* 830]

Wainwright, P. C. and R. G. Turigan. 1997. Evolution of pufferfish inflation behavior. *Evolution* **51**(2): 506–18.

[**Burrfishes (II)**]

Walbaum, Johannes Julius (ed.) 1792. *Petri Artedi Sueci Genera piscium. In quibus systema totum ichthyologiae proponitur cum classibus, ordinibus, generum characteribus, specierum differentiis, observationibus plurimis. Redactis speciebus 242 ad genera 52*. Ant. Ferdin. Röse, Greifswald, 723 pp.

["*Genera of fishes by Peter Artedi of Sweden, in which a system for all of Ichthyology is presented, with classes, orders, the characters of genera, and distinction of species, along with many observations. With 242 species collected in 52 genera.*"

Burrfishes; Greenland halibut; *Rhombus;* **Salmon (II)**]

Walcott, John. 1785. *History of British Fishes*. Manuscript.

[Cited by Fries (1839a), and Yarrell (1836), Vol. 2, pp. 327–9; appears to have been the first report of male pregnancy in *Syngnathus*; **Pipefishes (I)**]

#Walker, Alexander. 1838. *Intermarriage*. Churchill, London.

[When different variations cross, the offspring take the locomotive system from the male, because the male has the greatest desire for the female being very different – according to this, this law would be quite interfered with in a case where the ova were impregnated by the semen of the male, as in fishes & frogs, & yet we know that mule fish occur, & that it is not necessary in insects or fish that male should see female; 229 effects of desire of male nonsense. Plants & Fish &c!!; 231 plants & Fish, *Marginalia* 834, 836]

+Walker, John. 1803a. An essay on the natural, commercial, and economical history of the herring. *Prize Essays and Transactions of the Highland Society of Scotland* **2**: 270–304.

[**Hybrids**]

+Walker, John. 1803b. On the natural history of the salmon. *Prize Essays and Transactions of the Highland Society of Scotland* **2**: 346–76.

[**Hybrids**]

Bibliography

Wallace, Alfred Russel. [Read July 1, 1858]. On the tendency of varieties to depart indefinitely from the original type. *Journal of the Proceedings of the Linnean Society of London (Zoology)*. **3** (1859): 53–62.

[In which Wallace described the process now known as natural selection; **Big Species Book**; **Wallace**]

+Wallace, Alfred Russel. 1867. Mimicry, and other protective resemblances among animals. *The Westminster Review* **32**: 1–43.

[Review of contributions by Henry Walter Bates (Amazonian insects), Alfred R. Wallace (Malayan insects), Andrew Murray (Disguises in plant and animals) and Charles Darwin (4th edition of *Origin*). Wallace, whose article was not signed, concludes that females are generally better camouflaged because they are weaker and biologically more valuable, and thus require the added protection. Wallace thereby rejects sexual selection as cause for bright colours of males. CD continued to disagree, although in an earlier letter to Wallace (Feb. 18, 1867, *Calendar* no. 5404), he had conceded that the apparently high vulnerability of brightly coloured males to predation represented a difficulty to his view. The *handicap principle resolves this apparent paradox; **Colours**]

#Wallace, Alfred Russel. 1876. *The geographical distribution of animals.* 2 Vols. Macmillan & Co.

[Vol. 1, p. 463 Frogs ice – salt-water, Galaxias – without further evidence your views on which provide complications. Vol. 2, p. 465 Distribution of F.W. Fishes, *Marginalia* 838, 839]

Wallace, Alfred Russel. 1889. *Darwinism: an exposition of the theory of natural selection, with some of its applications.* The Humboldt Publishing Co., New York, Part One (of two), 332 pp.

[**Eyes; Flatfish controversy (I); Plagiarism**]

Wallace, Alfred Russel. 1890. *The Malay Archipelago: the land of the Orang-Utan and the bird of paradise. A narrative of travel with studies of man and nature,* 10th edn. MacMillan & Co., London, xvii + 515pp.

[Third page carries the following: "To Charles Darwin, Author of 'The Origin of Species' I dedicate this Book not only as a token of personal esteem and friendship but also to express my deep admiration for his Genius and his Works"; **Wallace**]

Wallace, Alfred Russel. 1905. *My life: a record of events and opinions.* New York, Dodd, Mead and Company. Vol. I, 435 pp.; Vol. II, 464 pp.

[We read on p. 22 of Vol. II: "Some of my critics declare that I am more Darwinian than Darwin himself, and in this, I admit they are not far wrong"; **Wallace**]

+Walton, Izaak. 1653. *The Compleat Angler, or the contemplative man's recreation: being the discourse of rivers, fishponds, fish and fishing not unworthy of the perusal of most anglers.* Printed by T. Maxey for R. Marriot, London.

[I used the reprint edition (1997) by Ecco Press, based on the fifth edition of 1676, the last to be revised by the author. Hopewell, New Jersey, USA, 224 pp.; **Angling;** *Darwin's Bass*; **Trout; Walton**]

+Warington, Robert. 1852. Observations on the Natural History of the Water-Snails and Fish kept in a confined and limited portion of Water. *Annals and Magazine of Natural History* (ser. 2)**10**: 273–80.

[**Sticklebacks**]

+Warington, Robert. 1855. Observations on the habits of the stickleback (being a continuation of a previous paper). *Annals and Magazine of Natural History* (ser. 2)**16**: 330–2.

[**Sticklebacks**]

+Waterhouse, George Robert. 1839. Mammalia. In: *The Zoology of the Voyage of H.M.S. Beagle, under the Command of Captain *FitzRoy, R.N. during the years 1832 to 1836.* Part II. Edited and superintendented by Charles Darwin. Smith, Elder and Co., Cornhill, London: v + 97 pp.

[Includes a geographical introduction by CD; **Waterhouse;** *Zoology*]

+Waterhouse, George Robert. 1843. Observations on the classification of the Mammalia. *Annals and Magazine of Natural History* **12**: 399–412.

[CD, in a note referring to this paper, wrote in part W. seemed baffled by *Lepidosiren*. says he wrote chiefly for Mammals a broken series (DAR 205.5:98; *Correspondence* from G. R.*Waterhouse, April 1844, n.4); the paper itself does not mention *Lepidosiren*]

Waters, Jonathan, M., Andrés López, and Graham P. Wallis. 2000. Molecular phylogenics and biogeography of Galaxiid fishes (Osteichthyes: Galaxiidae): dispersal, vicariance, and the position of *Lepidogalaxias salamandroides*. *Systematic Biology* **49**(4): 777–95.

[**Galaxiidae**]

Weber, Ernst Heinrich. 1820. De Aure et auditu hominis et animalium. Pars I: *De aure animalium aquatilium*. Leipzig.

["*On the ears and hearing of humans and animals*. Part I. *On the hearing of aquatic animals*"; not explicitly cited by CD, though he mentions Weber's organ and ossicles; **Weber**]

Wegener, Alfred. 1966. *The origin of continents and oceans*, 4th edn. Methuen, London, 248 pp.

[**Bony fishes; Distribution; Eel**]

Weinshank, Donald J., Stephan J. Ozminski, Paul Ruhlen and Wilma M. Barrett. 1990. *A Concordance to Charles Darwin's Notebooks, 1836–1844*. Cornell University Press, Ithaca and London, 739 pp.

[**Concordance**; *Notebooks*]

Weitzel, Vern (ed.) 1995. *The Journal of Syms Covington, Assistant to Charles Darwin Esq. on the Second Voyage of the* HMS Beagle, *December 1831 – September 1836*. Australian Science Archive Project, ASAPWeb: www.asap.unimelb.edu.au.

[**Covington**]

Welcomme, Robin L. 1988. International Introduction of inland aquatic species. *FAO Fisheries Technical Paper* no. 294. FAO, Rome, 328 pp.

[**Introduction; Livebearers; Wels catfish**]

Wells, Martin John. 1998. *Civilization and the Limpet*. Perseus/Helix Books, Reading, MA, 209 pp.

[**Bioluminescence**]

Went, Arthur E. J. 1972. Seventy years agrowing: a history of the International Council for the Exploration of the Sea, 1902–1972. *Rapport et Procès-verbaux des Réunions du Conseil international pour l'Exploration de la Mer* (**165**): 1–252.

[**Herring**]

Wheeler, Alwyne. 1973. Leonard Jenyns's notes on Cambridgeshire fishes. *Cambridgeshire and Isle of Ely Naturalists' Trust Annual Report*: 19–22.

[**Jenyns (I)**]

+White, Gilbert. 1789. *The Natural History of Selborne, in the County of Southampton: with engraving and an appendix*. Printed by T. Bensley for B. White and Son, London.

[A now famous book, of which young CD had the edition of 1825, known to have had much influence on him (Moore (1985), p. 460 and n. 11, p. 479), and which is occasionally cited in the work of the mature CD – though not in connection with fishes. *The Natural History of Selborne* has been kept in print to the present, through a series of new editions of which only one is listed below; **Birds**]

White, Gilbert. 1843. *The Natural History of Selborne, by the late Rev. Gilbert White, M.A.* New edition, with notes by the Rev. Leonard Jenyns, M.A., F.L.S. London.

[**Birds**]

Whitley, Gilbert Percy. 1928. Fishes from the Great Barrier Reef collected by Mr. Melbourne Ward. *Records of the Australian Museum* **16** (6): 294–304.
  [**Damselfishes; Eponym**]

Whitley, Gilbert Percy. 1958. Descriptions and records of fishes. *Proceedings of the Royal Zoological Society of New South Wales*, Volume for 1956–57: 28–51.
  [**Eponym**]

+Whitmee, R. S. 1878. On the manifestation of Anger, Fear and other Passions, in Fishes, and on the Use of their Spines. *Proceedings of the Zoological Society*, part i: 132–4.
  [*Expression*]

Whitten, Tony, Roehayat Emon Soeriaatmadja and Suraya A. Afiff. 1996. *The Ecology of Java and Bali. The Ecology of Indonesia Series*, Vol. II. Periplus Editions, Hong Kong, 969 pp. + plates.
  [**Bird's nest soup**]

Wiener, Jonathan. 1994. *The Beak of the Finch*. Random House, London, 332 pp.
  [Reports on the work by Rosemary and Peter Grant (Grant 1999); **Punctuated equilibrium**]

Wigner, Eugene. 1960. The unreasonable effectiveness of mathematics in the natural sciences. *Communications on Pure and Applied Mathematics* **13**: 1–13.
  [**Plankton**]

Wilder, Harris Hawthorne. 1923. *The History of the Human Body*. Henry Holt & Co., New York, xiv + 623 pp.
  [Chapter 13, on "The ancestry of the vertebrates", presents a detailed scenario for the transition from the 'flipped over' ancestors of annelid worms to early vertebrates, including an attempt to match their organ systems, and explicit reference to CD's work, which is contrasted to Owen's; **Dohrn; Owen**]

Williamson, Donald I. 1998. Chapter 33: Larval Transfer in Evolution. *In*: M. Syvanen and C.I. Kado (eds.), *Horizontal Gene Transfers*, pp. 436–53. Chapman & Hall, London.
  [**Non-Darwinian evolution**]

Willughby, Francis. 1686. *De historia piscium libri quatuor, jussu et sumptibus Societatis Regiae Londinensensis edititi . . .*
  ["*On the history of fish. Fourth book, published by Order and at the Expenses of the Royal Society of London, and edited by John Ray*"; reprint facsimile edition by Arno Press, New York, 1978; **Pipefishes (II)**]

Wilson, A. 1811. [Article on *Clupea*.] *In*: Rees' *Cyclopedia; or, universal dictionary of arts, sciences, and letters*. Vol. 9. No pagination. Samuel F. Bradford, Philadelphia.
  [**Shad**]

Wilson, Edward Osborne. 1980. *Sociobiology*. Belknap Press, Cambridge, MA, 366 pp.
  [**Social Darwinism**]

Wilson, Fred. 1991. *Empiricism and Darwin's science*. Kluwer Academic Publishers, Dordrecht, 353 pp.
  [**Paradigm**]

+Wilson, James. 1842. *A Voyage Round the Coast of Scotland and the Isles*. 2 Vols. Edinburgh.
  [Includes accounts on the natural history of fishes, especially Herring. Listed by CD among Books to be Read, and later summarized as poor; Burkhardt and Smith 1988, p. 469; **Variety**]

Winberg, G. G. (ed.) 1971. *Methods for the estimation of production of aquatic animals*. London, Academic Press, 175 pp.
  [**Production(s)**]

Winsor, Mary P. 1991. *Reading the Shape of Nature: Comparative Zoology at the Agassiz Museum*. University of Chicago Press, Chicago, 324 pp.
  [**Agassiz; Creationism; Jenyns (I)**]
Wischnath, Lothar. 1993. *Atlas of livebearers of the world*. T. F. H. Publications, Inc., Neptune City, NJ, 336 pp.
  [**Jenynsiinae**]
Wise, Donald U. 1998. Creationism's geologic time scale. *American Scientist* 86 (March/April): 160–73.
  [**Creationism**]
Wise, John P. 1958. The World's Southernmost Indigenous Cod. *Journal du Conseil International pour l'Exploration de la Mer* 23: 208–12.
  [**Cod**]
Wisenden, Brian D. 1999. Alloparental care in fishes. *Reviews in Fish Biology and Fisheries* 9: 45–70.
  [**Cuckoo-fish**]
Wisenden, Brian D. and R. J. F. Smith. 1997. The effect of physical condition and shoal mate familiarity on proliferation of alarm substance cells in the epidermis of fathead minnows. *Journal of Fish Biology* 50: 799–808.
  [**Minnows**]
Wisner, Robert L. 1999. Description of two new subfamilies and a new genus of hagfishes (Cyclostoma: Myxinidae). *Zoological Studies* 38(3): 307–13.
  [***Gastrobranchus***]
Wong, Kate. 2002. The mammals that conquered the seas. *Scientific American* 286(5): 71–9.
  [**Whales**]
Woodruffe, Colin, Roger McLean and Eugene Wallensky. 1990. Darwin's Coral Atoll: Geomorphology and Recent Development of the Cocos (Keeling) Islands, Indian Ocean. *National Geographic Research* 6(3): 262–75.
  [**Cocos Islands**]
#Woodward, Samuel Pickworth. 1851–56. *A rudimentary treatise of recent and fossil shells*. J. Weale, London.
  [p. 416: ... Cirripedes must now replace other animals. Hence reduced organisms now flourish, & so it is with Fish: take place of lower animals from some advantage, *Marginalia* 881]
Wootton, Robert J. 1991. *Ecology of teleost fishes*. Chapman & Hall, London, xii + 404 pp.
  [**Sunfishes**]
Wright, Robert. 1999. The accidental evolutionist: why Stephen Jay Gould is bad for evolution. *The New Yorker*, Dec. 13: 56–65.
  [R. Wright may have dubious, personal reasons for his relentless criticism of S. J. Gould, but in this essay, he expresses rather well the sense of unease that other evolutionists have expressed regarding some of Gould's positions; **Complexity; Extinction; Non-Darwinian evolution**]
Wu, Sylvia. 1984. *Cooking with Madame Wu: Yin and Yang Recipes for Health and Longevity*. McGraw-Hill, New York, 256 pp.
  [**Bird's nest soup**]
Wyman, Jeffries. 1857. On some species of fishes from the Surinam River and some conditions, heretofore unnoticed under which the eggs are developed. *Proceedings of the Boston Society of Natural History* 6: 264–9.
  [Dated September 16, 1857, issued December 18, 1859; story is on pp. 268–9; **Catfishes**]

Bibliography

Wyman, Jeffries. 1859. On some unusual modes of gestation. *American Journal of Science and Arts* (ser. 2) **27**: 5–13.

[**Catfishes**]

+Yarrell, William. 1836. *A History of British Fishes*. 2 Vols. John van Voorst, London.

[Vol. 1 cited in annotation to Bronn (1843); Vol. 2, p. 210 Flatfish; 217 Reversion; 256 Teeth of soles; 416 some [rays] have all teeth like males some sp do. are like female; G[ünther] says true male fem. Raia characters; 417 G supposes aid to claspers fin ... by double under ... catch by fins; 436 The female G says only has the thorn, *Marginalia* 883; 2nd Edition 1839; **Yarrell**]

Yarrell, William. 1839. *On the growth of the Salmon in fresh water; with six coloured illustrations of the fish of the natural size, exhibiting its character and exact appearance and various stages during the first two years*. John van Voorst, London.

[**Hybrid vigour**]

Zahavi, Amotz and Avishag Zahavi. 1997. *The handicap principle: a missing piece of Darwin's puzzle*. Oxford University Press, New York, 286 pp.

[**Handicap principle**]

Zeller, Dirk C. 1988. Short-term effects of territoriality of a tropical damselfish and experimental exclusion of large fishes on invertebrates in algal turfs. *Marine Ecology Progress Series* **44**: 85–93.

[**Damselfishes**]

Zeller, Dirk C. 1997. A long way to swim for sex (on the reef). *Research Achievements (Australian Research Council/DEETYA), Science and Technology Budget Statement 1997, Australian Federal Government*, Section 6: 6.6.

[Amazing that this title could be made part of the Australian Federal Government Budget Statement; **Groupers (II)**]

Zeller, Dirk C. 1998. Spawning aggregations: Patterns of movement of the coral trout *Plectropomus leopardus* (Serranidae) as determined by ultrasonic telemetry. *Marine Ecology Progress Series* **162**: 253–63.

[**Groupers (II); Territoriality**]

Zimmer, Carl. 1998. *At the water's edge: fish with fingers, whales with legs and how life came ashore and went back to sea*. Simon & Schuster, New York, 290 pp.

[**Climbing fish; Fish(es);** *Hox***; Jaguar; Lungfish, Australian; Whales**]

Zirkle, Conway. 1964. Some oddities in the delayed discovery of Mendelism. *Journal of Heredity* **55**(2): 65–72.

[**Mendel**]

Index to the Fishes

This index, whose first draft was kindly prepared by Dr Maria Lourdes 'Deng' Palomares, includes the common and scientific names of the fishes mentioned in this book. Most entries refer to species, generally through the pairing of a common name (current, or used by CD or his contemporaries) with a valid scientific name (*in bold italics*), or of a scientific name used by CD or L. Jenyns (*in italics*) with its valid counterpart. The updated scientific names originate largely from *FishBase; December 2003), and are likely to continue changing (see www.fishbase.org).

Some entries refer to larger groups, such as Families, or Orders. Where these have common names, or can be defined in a few words, these are paired with the scientific name as well. Single quotation marks identify uncommon or questionable usages, including misspellings, some corrected where underlined. The symbol >, when used before the second of two paired entries, implies that the second name applies to only a subset of the first group, and conversely for the < symbol. Entries with small differences in their endings (plural vs. singular, or female vs. male, '-id' rather than '-idae,' etc) are mostly grouped. Appendixes I, II and III are indexed, along with the annotations to the references, many of which include *marginalia by CD.

Aalmutter (**Zoarces viviparus**) 63
Abramis brama 21, 187
Abyssal grenadier (**Coryphaenoides armatus**) 172
Acanthaluteres spilomelanurus 76
Acanthistius brasilianus (**Plectropoma patachonica**) 101, 102
Acanthistius patachonicus (**Plectropoma patachonica**) 235, 236
'Acanthoclinus littoreus' (**Acanthoclinus fuscus**) 235
Acanthoclinus fuscus 12, 175, 233
Acanthodii (early jawed fishes) 190
Acanthopterygians 3, 4, 49, 75, 104
Acanthuridae (surgeonfishes) 199
Acanthurus humeralis (**Acanthurus olivaceus**) 199, 201, 232, 239
Acanthurus olivaceus 199, 239

Acanthurus triostegus 39, 199, 239
Acanthurus triostegus sandvicensis 199
Achiridae (American soles) 193
Achirus? 218, 239
Achirus lineatus 193, 222
African lungfishes (Protopteridae) 131, 132
Agnatha (jawless fishes) 106
Agonidae (poachers) 163
Agonopsis chiloensis 163
Agonostoma darwiniense 69
Agonostoma forsteri (**Aldrichetta forsteri**) 239
Agonostomus monticola 71
Agonus chiloensis (**Agonopsis chiloensis**) 236
Agriopus hispidus (**Congiopodus peruvianus**) 158, 238
Agriopus peruvianus (**Congiopodus peruvianus**) 158

323

Index to the Fishes

Albacore (*Thunnus alalunga*) 4
Albicore (*Thunnus alalunga*) 4, 83, 86
Aldrichetta forsteri 142
Aleuteres maculosus (*Acanthaluteres spilomelanurus*) 76, 122, 240
Aleuteres velutinus (*Nelusetta ayraudi*) 76, 123, 240
Alewife (*Alosa pseudoharengus*) 188
Alosa pectinata (*Brevoortia pectinata*) 109, 156
Alosa pseudoharengus 188
Alosa sapidissima 188
Alpine char (*Salvelinus alpinus*) 33
Amago salmon (*Oncorhynchus rhodurus*) 177
Amanses scopas 76
Amblyopsidae (<cavefishes) 31, 32, 141
'Amblyopsis' (*Amblyopsis spelaea*) 31, 32, 74, 242
Amblyopsis spelaea 32
American eel (*Anguilla rostrata*) 61
American plaice (*Hippoglossoides plattessoides*) 265
Ammocœtus (larval lamprey) 5, 43, 124
Ammodytes tobianus 179
Ammodytidae (sand-eels/sandlances) 179
'Amphioxus' (lancelets) 44, 124, 125, 274, 311
Anabas testudineus 37
Anablepidae (>Jenynsiinae) 120
anchovies (Engraulidae) 157
anchovy (>*Engraulis capensis*) 26
anchovy (>*Engraulis encrasicolus*) 26
Anglerfish (*Histrio histrio*) 161
'Anguilla' 279
Anguilla anguilla 61
Anguilla australis 63, 233, 240
Anguilla marmorata 63
Anguilla megastoma 63
Anguilla obscura 63
Anguillidae (eels) 62
Antennarius striatus (anglerfish) 161
Aphritis porosus (*Pseudaphritis porosus*) 156, 202, 223, 235
Aphritis undulatus (*Pseudaphritis undulatus*) 202, 228
Apistus niger (*Tetraroge niger*) 7
Apistus sp. 7, 123, 238

'*Aplochiton*' 279
Aplochiton taeniatus 92, 93, 202, 220
Aplochiton zebra 75, 92, 93, 221
Aplodactylidae (marblefishes) 134
Aplodactylus punctatus 94, 102, 134
Arcos poecilophthalmus 37, 38
Arcos spp. 37
Arctic char (*Salvelinus alpinus*) 33
Argentine menhaden (*Brevoortia pectinata*) 109
Arius boakeii (*Arius arius*) 31
Arius fissus (*Cathorops spixii*) 30
Armado (<catfish) 30, 222
armoured catfishes (Loricariidae) 30
Arothron stellatus 167
Arripidae (Australian 'salmon') 175
Arripis (*Centropristes*) *geogianus* (*Arripis georgianus*) 238
Arripis georgianus 102, 123, 175
Aspidontus taeniatus 154
Aspidophorus chiloensis (*Agonopsis chiloensis*) 34, 163, 229
Astyanax abramis 33, 34
Astyanax fasciatus 33, 34
Astyanax mexicanus 32
Astyanax scabripinnis 33, 34
Astyanax spp. 33
Astyanax taeniatus 33
Atherina argentinensis (*Odontesthes argentinensis*) 133, 190, 222
Atherina incisa (*Odontesthes incisa*) 156, 190, 216
Atherina microlepidota (*Basilichthys microlepidotus*) 190, 205, 228, 279
Atherinichthys argentinensis (*Odontesthes argentinensis*) 235
Atherinichthys microlepidota (*Basilichthys microlepidotus*) 239
Atherinichthys incisa (*Odontesthes incisa*) 239
Atherinidae (silversides) 190
Atlantic cod (Cod, *Gadus morhua*) 91
Atlantic salmon (*Salmo salar*) 129, 174, 177, 178, 179
Atopomycterus nychthemerus (*Diodon nictemerus*) 240
Auchenionchus microcirrhis 19

Australian lungfish (*Neoceratodus forsteri*) 131, 132
Australian ruff (*Arripis georgianus*) 175
Australian 'salmons' (Arripidae) 175

Bacalao (>*Gadus morhua*) 10
Bacalao (>*Mycteroperca olfax*) 101, 304
Bacalao (>*Polyprion oxygeneios*) 10
Bagré (<catfish) 30, 31
Balistes aculeatus (***Rhinecanthus aculeatus***) 201, 203, 232, 240
Balistes vetula 203, 214, 240
Balistidae (triggerfish) 203
Balitoridae (<loaches) 128
Banded pipefish (***Micrognathus crinitus***) 159
Bandfin scorpionfish (***Scorpaena histrio***) 181
Barbel (***Barbus barbus***) 11, 72
Barbus barbus 10
Barracuda (***Sphyraena sphyraena***) 11, 36, 195
barracudas (Sphyraenidae) 11
Barred rockfish (***Plectropoma patachonica***) 101
'Barrow cooter' (Barracuda, ***Sphyraena sphyraena***) xxiv, 11, 36, 195
Basilichthys microlepidotus 190
Basking shark (***Cetorhinus maximus***) 12, 205, 274, 312
batfishes (Ogcocephalidae)
Bathygobius lineatus 96
Batrachoididae (toadfishes) 203
Batrachus porosissimus (***Porichthys porosissimus***) 10, 203, 217
Bearded catfish (***Pseudancistrus barbatus***) 14
Belontiidae (gouramies; paradisefishes) 98
'Bering pike' 15
Betta pugnax 98
Bigeye scad (***Selar crumenophthalmus***) 116
Birdbeak burrfish (***Cyclichthys orbicularis***) 22
'black-fish' (<*Salmo salar*) 177
Black sea bream (***Spondyliosama cantharus***) 165
Blackchin tilapia (***Sarotherodon melanotheron***) 141
Blacktail snapper (***Lutjanus fulvus***) 192

Blennechis fasciatus (***Hypsoblennius sordidus***) 18, 19, 44, 229
Blennechis ornatus (***Hypsoblennius sordidus***) 18, 19, 45, 229
blennies (Blennioidei) 18, 19, 63, 202, 211
Blenniidae (combtooth blennies) 18, 19
Blennioidei 18
Blennius sanguinolentus (***Parablennius sanguinolentus***) 239
Blennius palmicornis/parvicornis (***Parablennius sanguinolentus***) 18, 26, 239
Blennius spp. 18
'*Blennius*' (misidentified eelpout) 63, 228
Blind fish (***Amblyopsis spelaea***) 31
'blind Gadus' (***Lota lota***) 62
Blotched upside-down catfish (***Synodontis nigriventris***) 58
Blue searobin (***Prionotus punctatus***) 105
'Blue shark' 47
Bluefin gurnard (***Chelidonichthys kumu***) 105
Bone-shark (***Cetorhinus maximus***) 12
bonito (>***Katsuwonus pelamis***) 20, 86
'bonitos' (<tuna) 86
bony fishes (teleosts) 20, 110, 130, 166, 180
bony-tongues (Osteoglossiformes) 139
Bovichthyidae (thornfishes) 202
boxfishes (Ostraciidae) 20
Brachydanio sp. 157
Brazilian flathead (***Percophis brasilianus***) 60
Bream (***Abramis brama***) 21, 186
Brevoortia pectinata 109, 236
Bridled burrfish (***Cyclichthis antennatus***) 22
Bridled leatherjacket (***Acanthaluteres spilomelanurus***) 76
Brill (***Scophthalmus rhombus***) 174
Bronze minnow (***Phoxinus neogaeus***) 137
Brook char (***Salvelinus fontinalis***) 33
Broom filefish (***Amanses scopas***) 76
Brown trout (***Salmo trutta***) 6, 75, 203, 204
Bull trout (***Salmo trutta***) 204
Bulldog [cod] (<***Gadus morhua***) 39
Bullseye puffer (***Sphoeroides annulatus***) 167
Bone shark (***Cetorhinus maximus***) 12
Burbot (***Lota lota***) 21

Index to the Fishes

burrfishes (Diodontidae) 10, 21, 22, 23
butterfishes (Stromateidae) 24
butterflyfishes (Chaetodontidae) 25

'*C. antennatus*' (***Cyclichthys antennatus***) 24
Calamus taurinus 165
Callichthyidae (callichthyid armoured catfishes) 30
Callichthys paleatus (***Carydoras paleatus***) 30, 234
Callionymidae (dragonets) 59
Callionymus spp. 59, 127
Callionymus dracunculus (***Callionymus lyra***) 59
Callionymus lyra 59
Callorhinchus callorhynchus 34
Callorhinchus capensis 34
Cantharus lineatus (***Spondyliosoma cantharus***) 165
Cape Verde gregory (***Stegastes imbricatus***) 51
Capelin (***Mallotus villosus***) 26, 27, 114, 179, 192
'capons' (***Scorpaena elongata, S. scrofa***) 181
Carangidae (jacks and pompanos) 116, 171
Caranx declivis (***Trachurus declivis***) 116, 123, 233, 235
Caranx georgianus (***Pseudocaranx dentex***) 116, 123, 233, 238
Caranx torvus (***Selar crumenophthalmus***) 116, 201, 232, 235
Carapidae (pearlfishes) 154
Carassius auratus 28, 97
Carassius auratus gibelio 28
Carassius carassius 27
Carassius gibelio 27, 28
Carcharinus galapagensis 91, 92
Carcharias megalodon (***Carcharodon megalodon***) 95, 135, 136
Carcharodon carcharias 100
Carcharodon megalodon 135, 278
Carcharodon rondeletti (***Carcharodon carcharias***) 100
Carp (***Cyprinus carpio***) 13, 14, 27, 28, 50, 64, 72, 123, 137, 139, 141, 180, 250, 288, 293, 310
Carp bream (***Abramis brama***) 21
carps (Cyprinidae) 13, 14, 139, 208
cartilaginous fishes (Chondrichthyes) 28, 130, 162, 172

Carydoras paleatus 30
Caspian lamprey (***Caspiomyzon wagneri***) 29
Caspiomyzon wagneri 29
catfishes (Siluriformes) 29, 30, 49, 50, 99, 141, 190, 208
catfishes (>Malapteruridae) 66
catfishes (>Mochokidae) 49
catfishes (>Pimelodidae) 30
Cathorops spixii 30
Caulolatilus princes 179
cavefish (***Astyanax mexicanus***) 32
cavefish, Northern (***Amblyopsis spelaea***) 32
cavefishes (>Amblyopsidae) xxv, 4, 31, 32, 33, 62, 74, 141
'cavern fish' (***Amblyopsis spelaea***) 242
Cephalaspidomorphi (>Petromyzontiformes >Petromyzontidae) 124
Centrarchidae (sunfishes) 198
Cephalaspis (Devonian ur-fish) 206
Ceratodontidae (<lungfishes) 131
'Ceratodus' (***Neoceratodus forsteri***) 132, 273
ceratoid anglerfishes 154
Cetonurus globiceps 172
Cetorhinus maximus 12
Chaetodon triostegus (***Acanthurus triostegus***) 199
Chaetodon auriga 25
Chaetodon setifer (***Chaetodon auriga***) 25, 38, 238
Chaetodon triostegus (***Acanthurus triostegus***) 199
Chaetodontidae (butterflyfishes) 25
Chain pickerel (***Esox reticulatus***) 116
chars (***Salvelinus*** spp.) 33, 179
Characidae (characins) 33, 34
characins (Characidae) 32, 33, 34, 90, 179
Cheilio inermis 210, 235
Cheilio ramosus (***Cheilio inermis***) 210, 226, 235
Cheilopogon xenopterus 84
Cheirodon interruptus 33, 34
Chelidonichthys kumu 105
Chelidonichthys lastoviza 105
Chilomycterus geometricus (***Cyclichthys schoepfi***) 240
'*Chilomycterus ("Diodon") antennatus*' (***Cyclichthys antennatus***) 240
Chimaera monstrosa 34, 35, 251

Chimaera/s 28, 34, 65
'Chimera'/'Chimaeroid fishes' 34, 35, 162, 172, 181
Chimaeriformes (chimaeras) 34
Chimera/chimeroid (>*Callorhinchus* sp.) 34
Chinese leatherjacket (**Nelusetta ayraudi**) 76
Chinook salmon (**Oncorynchus tshawytscha**) 177
'*Chirodon interruptus*' (**Cheirodon interruptus**) 234
Chondrichthyes (cartilaginous fishes) 28, 129
Chromidae (>cichlids and damselfishes) 35, 36, 51
Chromis crusma (Valparaiso chromis) 51
Chromis facetus (**Cichlasoma facetum**) 35, 133, 239
Chrysophrys taurina (**Calamus taurinus**) 91, 165, 231
Chub minnow (**Couesius plumbeus**) 137
Chum salmon (**Oncorynchus keta**) 177
Cichla (<Cichlidae) 35
Cichlasoma facetum 35
Cichlidae (cichlids) xxiv, 8, 35, 36, 49, 90, 95, 99, 108, 141, 148, 187, 195, 199
Ciliata mustela 126
Cirrhinus rohita (**Labeo rohita**) 28
Cirrhitus rivulatus 128
'climbing fish' (Climbing perch, ***Anabas testudineus***) 37, 83, 296
clingfishes (Gobiesocidae) 24, 37, 38
Clinus crinitus (**Auchenionchus microcirrhis**) 19, 45, 229, 235
Clupea arcuata (**Ramnogaster arcuata**) 10, 109, 217, 240
Clupea fuegensis (**Sprattus fuegensis**) 109, 202, 224, 240
Clupea harengus 109
'*Clupea (Alosa) pecttinata*' (**Brevoortia pectinata**) 217
Clupea [pseudoharengus] (**Alosa pseudoharengus**) 188
Clupea sagax (**Sardinops sagax**) 109, 157, 230
Clupea sapidissima (**Alosa sapidissima**) 188
Clupeidae (herrings and allies) 109, 188
Cnesterodon decemmaculatus 127, 234
'Cobites'/'Cobitis' (loach) 128, 129, 296
Cobitidae (<loaches) 128

Cobitis barbatula (**Barbatula barbatula**) 128, 207
'Cocoa Nuts fishes' (?) 278
'Cocos fish' (parrotfish) 155
Cod (***Gadus morhua***) xxv, 10, 39, 57, 64, 91, 141, 151, 152, 174, 179, 251, 255, 295
'cod' (various gadoids) 39, 198, 223
Cod-like fishes (Gadiformes) 27, 91, 126, 136
Coelacanth (***Latimeria chalumnae***) 74
coelacanth (***Latimeria menadonensis***) 302
Coho salmon (***Oncorhynchus kisutch***) 177
Colossoma macropopum 47
'Colitis' (*Co**bitis***) 296
combtooth blennies (Blenniidae) 18, 19
Common carp (Carp, ***Cyprinus carpio***) 72, 98, 186
'common cod' (Cod, ***Gadus morhua***) 39
Common eel (Eel, ***Anguilla anguilla***) 61, 62, 63
'common ling' (Ling, ***Molva molva***) 126
Common minnow (Minnow, ***Phoxinus phoxinus***) 137
Common silver biddy (***Gerres oyena***) 138
Common torpedo (Torpedo, ***Torpedo torpedo***) 203
conger eels (Congridae) 62, 63, 217
'Conger' 63
Conger punctus (**Maynea puncta**) 202, 225
Congiopodidae (pigfishes) 158
Congiopodus peruvianus 158
Congridae (conger eels) 62
'*Congromuraena punctus*' (**Maynea puncta**) 234
Convict surgeonfish (***Acanthurus triostegus***) 199
Cookiecutter shark (***Isistius brasiliensis***) 154
Coregonus pollan 163
Coregonus [elegans] (**Coregonus pollan**) 315
Corkwing wrasse (***Symphodus melops***) 210
Corvina adusta (**Ophioscion adustus**) 49, 133, 218, 222, 238
Coryphaenoides armatus 172
Corythoichthys flavofasciatus 159, 160
Cossyphus darwini (**Pimelometopon darwini**) 69, 91, 151, 209, 230
Cottidae (sculpins) 181
Cottus scorpius (**Myxocephalus scorpius**) 181
Crenilabrus massa (**Symphodus cinereus**) 210, 269
Crenilabrus melops (**Symphodus melops**) 210, 269

Creole perch (**Percichthys trucha**) 48, 120
croakers (Sciaenidae) 49, 102, 212
Crucian carp (**Carassius carassius**) 27, 97, 262
Cryptopsaras couesii 154
Ctenoids 49, 180, 204, 242, 300, 309
cuckoo-fish 30, 49, 154
Cuckoo wrasse (**Labrus mixtus**) 210
cusk eels (Ophidiidae) 50
Cutthroat trout (**Oncorhynchus clarki**) 177
Cyclichthys antennatus 22, 24
Cyclichthys orbicularis 22
Cyclichthys schoepfi 22
Cycloids 50, 180, 204, 242, 309
Cyclopteridae (lumpsuckers) 130
Cyclopterus lumpus 28, 130
Cyclostomes (Agnatha) 219
Cynoscion striatus 49
Cyprinidae (carps) xxi, 11, 14, 21, 26, 27, 28, 41, 50, 51, 97, 137, 156, 175, 186, 201, 279, 289
Cyprinidontidae (pupfishes) 126, 127, 187
Cypriniformes/Cyprinoids (carp-like fishes) 139, 242
'*Cyprinus*' (a carp) 250
Cyprinus auratus (**Carassius auratus**) 97, 98
Cyprinus carassius (**Carassius carassius**) 28, 250, 262
Cyprinus carpio 27, 123, 187
Cyprinus curchius (**Labeo rohita**) 28
Cyprinus cursa/-is (**Labeo rohita**) 28
Cyprinus cursis (**Labeo rohita**) 28
Cyprinus gibelio (**Carassius gibelio**) 28, 250
Cyprinus rohita (**Labeo rohita**) 28

Dace (**Leuciscus leuciscus**) 51
Dajaus diemensis (**Aldrichetta forsteri**) 123, 142, 239
damselfish (**Chromis crusma**) 51
damselfishes (Pomacentridae) 35, 51, 108, 202
Danio sp. 157
Darwin's roughy (**Gephyroberix darwinii**) 53, 69, 133
Darwin's bass (<Largemouth bass) x, 52
Deal fish (**Trachipterus arcticus**) 53

Deep-bodied pipefish (**Leptonotus blainvilleanus**), 159, 160
deep-sea spiny eels (Notacanthidae) 53
Dengat grouper (**Epinephelus goreensis**) 101
Devil's fish (**Plectropoma patachonica**) 101, 216
Diacope marginata (**Lutjanus fulvus**) 38, 102, 192, 238
Dimidiochromis compressiceps 154
'Diodon' 22, 23, 24, 98, 167, 222, 231
Diodon antennatus (**Cyclichthys antennatus**) 10, 22, 23, 214
Diodon caeruleus (**Cyclichthys orbicularis**) 22
Diodon nicthemerus 22, 123, 240
Diodon nycthemerus (**Diodon nicthemerus**) 22
Diodon rivulatus (**Cyclichthys schoepfi**) 22, 133, 222, 240
Diodontidae (burrfishes; porcupine fishes) 21
Dipneusti (lungfishes) 130
Dipnoi (lungfishes) 130, 306
Dischistodus darwiniensis 51
dogfish (>**Squalus acanthias**) 15, 57, 154, 196, 224
Dragonet (**Callionymus lyra**) 59
dragonets (Callionymidae) 31, 59
driftfishes (Nomeidae) 60
duckbills (Percophidae) 60
Dules auriga (**Serranus auriga**)102, 133, 222, 238
Dules leuciscus (**Kuhlia malo**) 102, 201, 233, 238, 279
Dules malo (**Kuhlia malo**) 238
Dusky grouper (**Epinephelus marginatus**) 101

'earth eaters' (**Geophagus** spp.) 95
Echeneidae (sharksuckers) 173
Echeneis remora (**Remora remora**) 173, 214
Eel (**Anguilla anguilla**) 56, 61, 62, 264
eel-like fishes (Anguilliformes) 53, 139
eelpouts (Zoarcidae) 63, 75
Eels (Anguillidae) 29, 50, 55, 62, 63, 108, 150, 233, 288, 308
Elagatis bipinnulata 116, 171
Electric eel (**Electrophorus electricus**) 65
Electrophorus electricus 65

'Eleginops maclovinius' (here: *Pseudaphritis porosus*) 235
Eleotridae (sleepers) 191
Eleotris gobioides (*Gobiomorphus gobioides*) 12, 191, 233, 239
elephantfishes (Mormyridae) 65, 68, 139
Engraulidae (anchovies) 157
Engraulis capensis 26
Engraulis encrasicolus 26
Engraulis ringens xiii, 114, 157, 229, 234
Entomacrodus vomerinus 19, 26
Epinephelinae (groupers) 102
'*Epinephelus ascensionis*' (prob. *Epinephelus marginatus*) 235
Epinephelus darwinensis 69
Epinephelus goreensis 101
Epinephelus labriformis 101, 102
Epinephelus marginatus 101
Eptatretus spp (<Myxinidae) 94
Esox boreus (*Esox lucius*) 242
Esox lucius 158, 186
Esox reticulatus 116, 158
Etmopterus spinax 205
Eucinostomus gula 138
Eurasian minnow (*Phoxinus phoxinus*) 137
European barracuda (Barracuda, *Sphyraena sphyraena*) 11
European eel (Eel, *Anguilla anguilla*) 61, 62
European hake (Hake, *Merluccius merluccius*) 136
European mackerel (Mackerel, *Scomber scombrus*) 53
European pike (Pike, *Esox lucius*) 158
European river lamprey (*Lampetra fluviatilis*) 124
European salmon (Salmon, *Salmo salar*) 177
Exocoetidae (flyingfishes) 84
Exocoetus exiliens (*Cheilopogon xenopterus*) 84, 229
Exocoetus volitans 92

Falkland sprat (*Sprattus fuegensis*) 75, 109
Falkland trout (*Aplochiton zebra*) 75
False cleaner fish (*Aspidontus taeniatus*) 154
fighting fishes (Belontiidae, >*Betta pugnax*) 98

filefishes (Monacanthidae) 76
Fine flounder (*Paralichthys adspersus*) 82
Finescale roughy (*Gephyroberix darwinii*) 53, 69
Fitzroya multidentata (*Jenynsia multidentata*) 234
Fivebearded rockling (*Ciliata mustela*) 126
Flat-fish (>*Platichthys flesus*) 81, 82
flatfishes (Pleuronectiformes) xx, 8, 53, 58, 79, 80, 81, 82, 100, 122, 162, 174, 193, 204, 291, 321
flatheads (Platycephalidae) 81
Flounder (*Platichthys flesus*) 64, 81, 82, 312
flounders (<Pleuronectidae) 82, 193, 287
Flying fish (*Exocoetus volitans*) 92
Flying fishes (Exocoetidae) xx, 4, 15, 20, 77, 83, 84, 86, 99, 137, 176, 206, 229, 256
Freckled driftfish (*Psenes cyanophrys*) 60
fugu (Tetraodontidae) 167
Fundulus heteroclitus 141

Gadidae (cods) 39, 226
Gadiformes (cod-like fishes) 126
'*Gadus*' (<Gadidae) 39, 62, 91, 151, 220, 225
Gadus lota (*Lota lota*) 21
Gadus molva (*Molva molva*) 126
Gadus morhua 10, 39, 91
Gadus mustela (*Ciliata mustela*) 126
Galápagos gurnard (*Prionotus miles*) 104, 105
Galápagos porgy (*Calamus taurinus*) 165
Galápagos shark (*Carcharinus galapagensis*) 91, 92
Galaxias 318
Galaxias alpinus 92, 93, 234
Galaxias attenuatus (*Galaxias maculatus*) 75, 89, 234
Galaxias maculatus 75, 89, 92, 93, 234
Galaxias smithii (*Galaxias platei*) 75
Galaxiidae 53, 55, 75, 92, 93
Gambusia spp. (Poeciliidae) 126
Ganoids/ganoid fishes xx, 74, 88, 89, 93, 94, 114, 133, 166, 180, 206, 242, 291, 293, 306, 309
Gar-pike (*Lepisosteus osseus*) 31, 94, 110, 151, 242
gar-pikes (Lepisosteidae) 94
Gastrobranchus (*Myxine*) 94

Gastrobranchus coecus (**Myxine glutinosa**) 94
Gasterosteidae (sticklebacks) 197
'Gasterosteus' (sticklebacks) 198
Gasterosteus aculeatus 197
Gasterosteus leiurus (<**Gasterosteus aculeatus**) 197, 198
Gasterosteus trachurus (<**Gasterosteus aculeatus**) 197
Gemmeous dragonet (<**Callionymus lyra**) 59
Genyroge marginata (**Lutjanus fulvus**) 238
Geophagus spp. (<Cichlidae) 35, 95, 141
Gephyroberix darwinii 53, 69, 133
Gerreidae (mojarras; silver biddies) 138
Gerres gula (**Eucinostomus gula**) 138, 175, 239
Gerres oyena 38, 138, 240
ghost pipefishes (Solenostomidae) 129, 161, 182
Giant bully (**Gobiomorphus gobioides**) 191
Giant hawkfish (**Cirrhitus rivulatus**) 128
Girardinus guppii (**Poecilia reticulata**) 104
Globefish (**Diodon nicthemerus**) 22
Globehead parrotfish (**Scarus globiceps**) 154
Globehead whiptail (**Cetonurus globiceps**) 172
'*Glyphisodon luridus*' (misidentification of **Stegastes imbricatus**) 235
goatfishes (Mullidae) 96
Gobiesocidae (clingfishes) 37, 38
gobies (Gobiidae) 96, 97, 142, 191, 211
Gobiesox marmoratus 34, 37, 38, 228, 236
Gobiesox poecilophthalmos (**Arcos poecilophthalmos**) 37, 38, 91, 231, 236
Gobiidae (gobies) 96, 142
Gobiomorphus gobioides 191
Gobius lineatus (**Bathygobius lineatus**) 91, 96, 231, 235
Gobius ophicephalus (**Ophiogobius ophicephalus**) 34, 96, 228
'*Gobiosoma ophicephalum*' (**Ophiogobius ophicephalus**) 235
Goldfish (**Carassius auratus**) xxi, 5, 27, 28, 50, 53, 62, 70, 71, 72, 97, 98, 112, 137, 138, 139, 188, 205, 250, 279, 286, 289, 308
gouramies (**Macropodus** spp.; Belontiidae) 98, 133

Grahamina capito 19
Graviceps darwini (blennid, prob. **Omobranchus lineolatus**) 69
Great white shark (**Carcharodon carcharias**) 100, 111, 136
Greenback horse mackerel (**Trachurus declivis**) 116
Greenland halibut (**Reinhardtius hippoglossoides**) 59, 100, 204, 265
grenadiers (<Macrouridae) 171
Grey wrasse (**Symphodus cinereus**) 210
groupers (Epinephelinae; <Serranidae) 49, 100, 103, 108, 187, 304
grunts (Haemulidae) 104
grunters (>Teraponidae) 202
guppies (**Poecilia** spp., Poeciliidae) 32, 70, 104, 126, 169
Guppy (**Poecilia reticulata**) 70, 104, 274
gurnards (Triglidae) 104, 105
Gymnogaster arcticus (**Trachipterus arcticus**) 251
Gymnothorax ocellatus 139, 215, 240
Gymnotus electricus (**Electrophorus electricus**) 6, 65, 67, 296

Haemulidae (grunts) 104
hagfishes (Myxinidae) 94, 106, 124, 243
hakes (Merluccidae) 136
'*Haplochiton zebra*' (**Aplochiton zebra**) 234
'*Haplochiton taeniatus*' (**Aplochiton taeniatus**) 234
Haplodactylus punctatus (**Aplodactylus punctatus**) 238
Heliases crusma (**Chromis crusma**) 51, 205, 227
'*Heliastes*'/'*Heliasus*' *crusma* (**Chromis crusma**) 239
Helotes octolineatus (**Pelates octolineatus**) 123, 203, 235
Hero facetus (**Cichlasoma facetum**) 239
Herring (**Clupea harengus**) 56, 63, 64, 109, 148, 151, 205, 320
herrings (**Clupea** spp., <Clupeidae) 55, 109
'Hippocampi' (seahorses) 160, 182
Hippocampus spp. 182, 183
Hippoglossus kingii (**Paralichthys adspersus**) 82, 122, 205

Hippoglossus pinguis (**Reinhardtius hippoglossoides**) 100, 193
Hippoglossoides platessoides 265
Histrio histrio 161
Hog-nose mullet (***Joturus pilchardi***) 71
Holacanthus darwinensis (***Chaetodontoplus duboulayi***) 69
Holocentridae (squirrelfishes; soldierfishes) 197
Holocentrus ruber (***Sargocentron rubrum***) 197
Hoplostethus atlanticus 53, 256
horse mackerels (***Trachurus***; <Carangidae) 116
Hydrocyon hepsetus (***Oligosarcus hepsetus***) 33, 133, 221, 240
'Hygrogonus' (<Cichlidae) 36
Hypsoblennius sordidus 18, 19

'Ichtus Heliodiplodokus, Fam. Heliichthinkerus' 184
Iluocoetes fimbriatus 34, 63, 228, 236
Inanga (***Galaxias maculatus***)
Indian carp (>***Labeo rohita***) 27
Indo-Pacific mackerels (***Rastrelliger*** spp.) 53
Irish pollan (***Coregonus pollan***) 163
Isistius brasiliensis 154
Istiblennius edentulus 18

Jack (***Esox reticulatus***) 116
Jack mackerel (>***Trachurus declivis***) 116
jacks (>Carangidae) 31, 171
jacks (<Salmon) 102, 116
Japanese weatherfish (***Misgurnus anguillicaudatus***) 129
Jenny mojarra (***Eucinostomus gula***) 138
Jenyns' flounder (***Pseudorhombus jenynsii***) 79
Jenynsia lineata 120, 234
Jenynsiinae (one-sided livebearers) 8, 120, 139
Jenyns's sprat (***Ramnogaster arcuata***)
Jewel moray (***Muraena lentiginosa***) 139
Joturus pilchardi 71

Katsuwonus pelamis 20
Killifish (***Fundulus heteroclitus***) 141
killifishes (Fundulidae) 120
king cod ('kongetorsk' <***Gadus morhua***) 39

Kuhlia marginata (misidentification of ***Kuhlia malo***) 238
Koi (<***Cyprinus carpio***) 123, 141

Labeo rohita 28
Labridae (wrasses) 209
Labrisomidae (<blennies) 19
'Labrus' (wrasse) 177
Labrus melops (***Symphodus melops***) 210
Labrus mixtus 210
Labrus pavo (***Thalassoma pavo***) 210
Lamnidae (white sharks) 100
Lampetra spp. 124
Lampetra fluviatilis 124
lampreys (Petromyzontidae) 124, 154, 303
Lancelets ('Amphioxus') 8, 44, 59, 74, 124, 125, 206, 296
lanternfishes (Myctophidae) 44
Largemouth bass (***Micropterus salmoides***) 52, 198
Lates darwinensis (***Lates calcarifer***) 69
Latilus jugularis (***Prolatilus jugularis***) 179, 205, 227, 238
Latilus princeps (***Caulolatilus princeps***) 91, 179, 230, 236
Latimeria chalumnae 74
Latimeria menadoensis 302
Leaf scorpionfish (***Taeniatotus triacanthus***) 181
Leafy sea dragon (***Phycodurus eques***) 160
'leather' carp (<***Cyprinus carpio***) 27
'Lebias' 279
Lebias lineata (***Jenynsia lineata***) 120, 133, 221
Lebias multidentata (***Jenynsia lineata***) 120, 139, 218
Leiognathidae (slipmouths) 16
Lemon sole (***Microstomus kitt***) 265
Lentil moray (***Muraena lentiginosa***) 139
'Lepidosiren' (***Protopterus annectens***) 5, 37, 57, 74, 88, 99, 131, 134, 206, 254, 274, 295, 296, 309, 318
Lepidosiren annectens (***Protopterus annectens***) 131
Lepidosiren paradoxa 131, 132
Lepidosirenidae (South American lungfishes) 131, 132
Lepidosteus (***Lepisosteus*** spp; gar-pikes) 151

Lepisosteus osseus 94
Lepomis gibbosus 199
Leptocephalus (eel larvae) 300
Leptonotus blainvilleanus 159, 160
Leuciscus leuciscus 51
Leuciscus phoxinus (*Phonixus phoxinus*) 187
Leviprora inops 81
'line' carp (<*Cyprinus carpio*) 27
Lined sole (*Achirus lineatus*) 193
Ling (*Molva molva*) 126
livebearers (Poeciliinae) 98, 126, 127
Liza (*Mugil liza*) 142
Liza vaigiensis 142
loaches (>Balitoridae) 128
loaches (>Cobitidae) 13, 128, 129
Longhead flathead (*Leviprora inops*) 81
Longnose gar (*Lepisosteus osseus*) 94
long-whiskered catfishes (Pimelodidae) 30
'looking-glass' carp (<*Cyprinus carpio*) 27, 250
Lophobranchii (Syngnathoidea) 129, 160, 161, 182, 277
Loricariidae (armoured catfishes) 14, 30
Lota lota 21
Lotidae (hakes and burbots) 126
Lumpfish (*Cyclopterus lumpus*) xvii, 28, 40, 61, 99, 129, 130
Lumpuckers (Cyclopteridae) 130
lungfishes (Dipnoi) 5, 130, 150, 279
Lutjanidae (snappers) 192
Lutjanus fulvus 192
Lutjanus marginatus (*Lutjanus fulvus*) 192
Lutjanus vaigiensis (*Lutjanus fulvus*) 192

Mackerel (*Scomber scombrus*) 53
Macropodus 133
Macropodus opercularis 98
Macropus (*Macropodus*) 98, 133
Macrouridae (rattails, or grenadiers) 171, 172
'*Macrourus*' (rattails) 53, 172, 198
Malacanthidae (tilefishes) 179
Malapteruridae (<catfishes) 66
Malawi eyebiter (*Dimidiochromis compressiceps*) 154

Mallotus villosus 26
Mangrove rivulus (*Rivulus marmoratus*) 108
Marblefish (>*Aplodactylus punctatus*) 94, 134
marblefishes (Aplodactylidae), 134
Marilyna darwinii 69
Masu salmon (*Oncorhynchus masou*) 177
Maynea puncta 63
Megatooth shark (*Carcharodon megalodon*) 22, 100, 135, 162
Meiacanthus lineolatus @
menhaden (<Clupeidae) 109
Menticirrhus americanus 49
Menticirrhus ophicephalus 49
merlu (*Merluccius merluccius*) 136
'merlus' (hakes) 136, 220
Merluccius gayi 136
Merluccius hubbsi 136
Merluccius merluccius 136
Mermaids 84, 99, 131, 136, 137, 157, 190, 246
'*Mesites*' 92, 279
Mesites alpinus (*Galaxias alpinus*) 92, 202, 220
Mesites attenuatus (*Galaxias maculatus*) 12, 92, 93, 233
Mesites maculatus (*Galaxias maculatus*) 92, 93, 156, 220, 226
Micrognthus crinitus 159
Micropterus dolomieui 198
Micropterus salmoides 52, 198
Microstomus kitt 265
Minnow (*Phoxinus phoxinus*) 50, 71, 137, 138, 159, 186, 187, 207, 287
minnows (<Cyprinidae) 16, 50, 137, 190
Mirror carp (<*Cyprinus carpio*) 27
Misgurnus anguillicaudatus 129
Misgurnus fossilis 129
Mochokidae (<catfishes) 49
mojarras (Gerreidae) 138
mollies (<*Poecilia* spp.) 126
Mollienesia petenensis (*Poecilia petenensis*) 127
Molva molva 126
Monacanthidae (filefishes) 76
Monacanthus auraudi (*Nelusetta ayraudi*) 240
Monacanthus scopas (*Amanses scopas*) 76

Monacanthus peronii (**Pseudomonacanthus peroni**) 76
morays (Muraenidae) 139
Mormyridae (elephantfishes) 68
Mormyrus spp. (<elephantfishes) 68, 139
Mountain mullet (*Agonostomus monticola*) 71, 272
Mozambique tilapia (*Oreochromis mossambicus*) 90
mud-fish (*Protopterus annectens*) 57, 131
'Mud Iguana' (*Protopterus annectens*) 99
mudskippers (>*Periophthalmus* spp.) 96, 142, 202
'mud-walking fish' (mudskipper) 83, 142
Mugil liza 142, 156, 217, 218, 239
Mugil sp. (mullet) 39, 142, 239
Mugilidae (mullets), 142
'mule fish' (hybrid) 112, 142, 317
mullets (Mugilidae) 29, 71, 139, 142
Mulloides flavolineatus (**Upeneus flavolineatus**) 238
Mullidae (goatfishes) 96
Muraena lentiginosa 91, 139, 231, 234
Muraena ocellata (**Gymnothorax ocellatus**) 139, 175, 240
Muraena spp. 26, 139, 201, 232, 240
Muraenidae (morays) 139
Mycteroperca olfax 101, 102, 304
'Myxina'/'Mixinidus'/'Mixinnus' (***Myxine*** spp.) 88, 106, 219
Myxine (<Myxinidae) 94
Myxine australis (**Myxine glutinosa**) 94, 106, 202, 219, 240
Myxine glutinosa 94, 106
Myxini (<Agnatha; Hagfishes) 106, 124
Myxocephalus scorpius 181

'naked' carp (*Cyprinus carpio*) 27
Narrow-headed puffer (*Sphoeroides angusticeps*) 167
Nelusetta ayraudi 76
Neoceratodus forsteri 131, 132
Network pipefish (*Corythoichthys flavofasciatus*) 159, 160
Nile tilapia (*Oreochromis niloticus*) 187

Nomeidae (driftfishes) 60
North American pike (*Esox reticulatus*) 116, 158
Northern cavefish (*Amblyopsis spelaea*) 32
Northern cod (<*Gadus morhua*) 10
Notacanthidae (deep-sea spiny eels) 53
'*Notacanthus*' (deep-sea spiny eels) 53, 172, 198
Notacanthus chemnitzii 54

Ocellated moray (*Gymnothorax ocellatus*) 139
Odontesthes argentinensis 190
Odontesthes incisa 190
Ogcocephalus darwini 69
Oligosarcus hepsetus 33
Olive rockfish (*Acanthoclinus fuscus*) 175
Oncopterus darwinii 69, 175
Oncorynchus spp. (<Pacific salmon) 177
Oncorynchus spp. (trouts) 203
Oncorhynchus gorbuscha 177
Oncorynchus keta 177
Oncorhynchus kisutch 177
Oncorhynchus masou 177
Oncorhynchus nerka 177
Oncorhynchus rhodurus 177
Oncorynchus tshawytscha 177
One-sided livebearer (*Jenynsia lineata*) 120
Ophidiidae (cusk eels) 50
Ophidium (Ophidiidae) 50, 194
Ophioblennius atlanticus 18, 26
Ophiocara darwiniensis (**Ophiocara porocephala**) 69
Ophiogobius ophicephalus 96
Ophioscion adustus 49
Orange roughy (*Hoplostethus atlanticus*) 53, 256
Orangespot surgeonfish (*Acanthurus olivaceus*) 199
Oreochromis mossambicus 90
Oreochromis niloticus 187
Ornate wrasse (*Thalassoma pavo*) 210
Orthopristis cantharinus 104
Osmeridae (smelts and capelin) 27, 192
Osmerus eperlanus 192
Osteoglossidae (bony-tongues) 139
Osteoglossiformes (bony-tongues) 139
Ostraciidae (boxfishes) 20
Ostracion meleagris (**Ostracion punctatus**) 20, 232

Ostracion punctatus 20, 201, 240
Otolithus analis 49, 157, 235
Otolithus guatucupa (*Cynoscion striatus*) 49, 133, 222, 238
'*Otolithus augusticeps*' (*Otolithus analis*) 230

Pacific red sheephead (*Pimelometopon darwini*) 209
Pacific salmon (<*Oncorhynchus* spp.) 116, 174, 177
Parablennius sanguinolentus 18
Paralabrax albomaculatus 100, 102
Paralichthyidae (<flatfishes) 82
Paralichthys adspersus 82
Parona signata 116
Paropsis signata (*Parona signata*) 116, 156, 218, 235
parrotfishes (Scaridae) xxiv, 22, 35, 36, 46, 47, 48, 108, 154, 155, 165
Parupeneus trifasciatus 96
Patagonian redfish (*Sebastes oculatus*) 181
'Peacock Labrus' (*Thalassoma pavo*) 210
pearlfishes (Carapidae) 154
Pelates octolineatus 202
Perca flavescens 156
Perca fluviatilis 156
Perca laevis (*Percichthys trucha*) 48, 156, 226
Perch (*Perca fluviatilis*) 156, 180, 201
Perch-likes (Perciformes) 3, 100
'Perch, True' (*Percichthys trucha*) 279
perches (Percidae) 80, 156, 242
Percichthyidae (temperate basses) 48
Percichthys laevis (*Percichthys trucha*) 235
Percichthys trucha 47
Percidae (perches) 156
Percophidae (duckbills)
Percophis brasilianus 60, 151, 155, 215, 222, 238
Percopsidae (trout-perches) 60, 204
Percopsis 204, 242
Perilampus perseus (*Danio* or *Brachydanio* sp.) 156, 157, 289
Periophthalmus spp. (<mudskippers) 142
Peruvian anchoveta (*Engraulis ringens*) xiii, 114, 157, 169

Peruvian weakfish (*Otolithus analis*) 49
'Pescado del Rey' (prob. Pejerrey, *Odontesthes bonariensis*) 223
Petromyzon marinus 124
Petromyzontidae (lampreys) 124
Petromyzontiformes 124
Petroscirtes fasciatus (*Hypsoblennius sordidus*) 235
Phoxinus phoxinus 137
Phucocoetes latitans 63, 75, 221, 236
Phycodurus eques 160
Phyllopterix taeniolatus 160
Pigfish (*Congiopodus peruvianus*) 158
pigfishes (Congiopodidae) 158
Pike (*Esox lucius*) xxi, 21, 22, 53, 68, 137, 138, 158, 159, 162, 186, 207, 253, 287, 303, 306
Piked dogfish (*Squalus acanthias*) 15, 57, 197
pilchards (<Clupeidae) 109
Pilot fish (*Remora remora*) 268
Pimelodidae (long-whiskered catfishes) 30
Pimelodella gracilis 30
Pimelodus exsudans (*Rhamdella exsudans*) 30, 175, 215, 239
Pimelodus gracilis (*Pimelodella gracilis*) 30, 175, 215, 239
Pimelometopon darwini x, 53, 69, 151, 209
Pinguipedidae (sandperches) 179
Pinguipes brasilianus 179
Pinguipes chilensis 179, 205, 227, 238
Pinguipes fasciatus (*Pinguipes chilensis*) 155, 179, 216, 235
Pink salmon (*Oncorhynchus gorbuscha*) 177
pipefishes (<Syngnathidae) 129, 141, 159, 160, 161, 174, 182, 211, 306
piranha (<Characidae) 33, 34
placoderms/placoids 28, 180, 291
plagiostomous fishes (cartilaginous fishes) 35, 162, 172
Plagusia ? 193, 218, 239
Plagusia brasiliensis (*Symphurus tessellatus*) 193
Plagusia fasciatus 193
Plagusia lineatus 193
Plaice (*Pleuronectes platessa*) 162
Platessa ? 217
Platessa flesus (*Platichthys flesus*) 82

Platessa orbignyana (**Paralichthys orbignyanus**) 10, 79, 123, 217
Platichthys flesus 81
Platycephalidae (flatheads) 81
Platycephalus inops (**Leviprora inops**) 81, 123, 236
Platycephalus laevigatus 81
Plecostomus barbatus (**Pseudancistrus barbatus**) 14
Plectropoma patachonica 101, 102, 155, 215, 216
Plesiopidae (roundheads) 175
Pleuronectes pinguis (**Reinhardtius hippoglossoides**) 100
Pleuronectes platessa 162
Pleuronectes (*Platessa*) (not identified) 239
Pleuronectidae (<flatfishes) 53, 79, 80, 162, 193, 204, 313
Pleuronectiformes 100, 162
Pleuronectoids (flatfishes) 53, 81
poacher (>*Agonopsis chiloensis*) 163
poachers (Agonidae) 163
Poecilia decemmaculata (**Cnesterodon decemmaculatus**) 127, 133, 221
Poecilia reticulata 104
Poecilia spp. 126, 279
Poecilia unimaculata (**Poecilia vivipara**) 127, 175, 215, 240
Poecilia vivipara 127
Poeciliidae (livebearers, mosquitofishes) 126, 127
Poeciliinae (livebearers) 126
Pollan (**Coregonus pollan**) 179, 315
Polyprion oxygeneios 10, 306
Pomacentridae (damselfishes) 51
Pomacentrus darwiniensis (**Dischistodus darwiniensis**) 51, 69
Pomotis (<Centrarchidae) 36, 199
porcupine fish (Diodontidae) 21
porgies (Sparidae) xxiv, 49, 108, 154, 165
Porichthys porosissimus 203, 236
Pot-bellied leatherjacket (**Pseudomonacanthus peroni**) 76
Prionodes fasciatus (**Serranus psittacinus**) 49, 91, 102, 231
Prionotus spp. 105
Prionotus miles 91, 105, 236
Prionotus punctatus 105, 175, 215, 238

Prionotus 'ruber' (**Prionotus miles**) 104, 230
Pristipoma cantharinum (**Orthopristis cantharinus**) 91, 104, 232, 235
'*Pristipoma miles cantharinum*' (**Orthopristis cantharinus**) 230
Prolatilus jugularis 179
Protopteridae (<Lungfishes) 131
Protopterus spp. (African lungfishes) 131
Protopterus annectens 131
Prussian carp (**Carassius gibelio**) 28, 71
Psenes cyanophrys 60
Psenes leucurus (**Psenes cyanophrys**) 60, 238
Pseudancistrus barbatus 14
Pseudaphritis porosus 202
Pseudaphritis undulatus 202
Pseudocaranx dentex 116
Pseudomonacanthus peroni 76
Pseudorhombus jenynsii 79
Pseudoscarus globiceps (**Scarus globiceps**) 239
Pseudoscarus lepidus (**Scarus globiceps**) 235
Pseuudupeneus prayensis 96
Pterichthys (not a fossil flying fish) 206
Pterois volitans 181
puffers (Tetraodontidae) 166
Pumpkinseed sunfish (**Lepomis gibbosus**) 199
pupfishes (Cyprinidontidae) 126

Queen triggerfish (**Balistes vetula**) 203

Rabbitfish (>**Chimaera monstrosa**) 34
rabbitfish (>Siganidae) 34
'Raia' (ray) 321
Raia batis (**Dipturus batis**) 172
Raia clavata (**Raja clavata**) 172
Raia maculata (**Raja montagui**) 172
Rainbow runner (**Elagatis bipinnulata**) 116, 171
Rainbow trout (**Oncorhynchus mykiss**) 177
Rajidae (Rays and Skates) 191
Raja batis 191
Ramnogaster arcuata 109
'rascasse' (<Scorpaenidae) 181
Rastrelliger spp. 53
rattails (<Macrouridae) 171, 172

Index to the Fishes

rays (<Rajidae) xxi, 28, 35, 65, 66, 67, 84, 130, 162, 172, 173, 191, 203, 217, 223, 286, 321
Red clingfish (**Arcos poecilophthalmos**) 37
Redcoat squirrelfish (**Sargocentron rubrum**) 197
Redlip blenny (**Ophioblennius atlanticus**) 18
Reinhardtius hippoglossoides 100, 265
Remo flounder (**Oncopterus darwinii**) 175
Remora (**Remora remora**) 154, 173
Remora remora 173
'resident trout' (**Salmo trutta fario**) 203
Rhamdella exsudans 30
Rhinecanthus aculeatus 203
Rhombus(<Scophthalmidae) 174, 217, 239
Rhombus darwinii (**Oncopterus darwinii**) 69
Ribbonfish (**Trachipterus arcticus**) 53, 251
Rio de la Plata one-sided livebearer (**Jenynsia multidentata**) 139
Rippled rockskipper (**Istiblennius edentulus**) 18
Rivulus marmoratus 108, 173
Roach (**Rutilus rutilus**) 50, 175, 201
'roncador' (croaker) 212
roughies (<Trachichthyidae) 53
'rough-tailed stickleback (G. trachurus)' (<**Gasterosteus aculeatus**) 197
roundheads (Plesiopidae) 175
Ruee (**Labeo rohita**) 28
Rusty blenny (**Parablennius sanguinolentus**) 18
Rutilus rutilus 175

Salarias atlanticus (**Ophioblennius atlanticus**) xix, 18, 26, 214, 239
Salarias quadricornis (**Istiblennius edentulus**) 18, 39, 236
Salarias spp. 18
Salmo (>**Salmo**) 129, 179, 203, 204
Salarias vomerinus (**Entomacrodus vomerinus**) 19, 26, 214, 239
Salmo eriox (**Salmo trutta**) 204
Salmo lycaodon (**Oncorhynchus nerka**) 177
Salmo salar 112, 113, 143, 177, 178, 204
Salmo trutta 75, 112, 203, 204
Salmo trutta fario 203
Salmo trutta trutta 203

salmon, Atlantic (**Salmo salar**) 6, 55, 64, 80, 111, 112, 113, 114, 116, 129, 150, 151, 174, 177, 178, 179, 184, 186, 204, 205, 207, 250, 278, 293, 303, 306
salmon, Pacific (**Oncorhynchus** spp.) 6, 55, 114, 116, 129, 177
Salmonidae (salmonids) xxi, 26, 27, 33, 113, 163, 177, 179, 203, 204, 242, 289, 317
'salmons (-ids)' (actually: Characidae) 114, 179, 223, 279
Salvelinus alpinus 33
Salvelinus fontinalis 33, 143
Salvelinus sp. 33
Salvelinus umbla 33
'sammon' (**Salmo salar**) 64
sand-eels (<Ammodytidae) 43, 179
sandlance (**Ammodytes tobianus**) 179
sandperches (Pinguipedidae) 179
sardines (<Clupeidae) 109
Sardina sagax (**Sardinops sagax**) 236
Sardinops sagax 109
Sargocentron rubrum 197
Sarotherodon melanotheron 141
'Sauroid fish' 132
sawbellies (<Trachichthyidae) 53
scads (<Carangidae) 116
'scaly mermaid' (**Lepidosiren**) 131
Scaridae (parrotfishes) 154
Scarus chlorodon (**Scarus prasiognathos**) 39, 154
Scarus globiceps 154, 201, 232
Scarus lepidus (**Scarus globiceps**) 154, 201, 232
Scarus prasiognathos 154
Scarus spp./'Scari' xxiv, 39, 48, 154, 155, 165, 189, 239
Sciaena adusta (**Ophioscion adustus**) 238, 240
Sciaena aquila (**Argyrosomus regius**) 194
Sciaenidae (croakers) 49
Scomber scombrus 53
Scophthalmidae (<flatfishes) 174, 204
Scophthalmus maximus 204
Scorpaena azorica 9, 270
Scorpaena elongata 181
Scorpaena histrio 91, 181, 231, 236
Scorpaena scrofa 181

Index to the Fishes

Scorpaenidae (scorpionfishes) 181
scorpionfishes (>Scorpaenidae) 181
scorpionfishes (>Tetrarogidae) 7
sculpins (Cottidae) 181
Sea lamprey (*Petromyzon marinus*) 124
'sea pike' (*Merluccius merluccius*) 136
seahorses (<Syngnathidae) xx, 129, 141, 161, 174, 182, 183, 285
searobins (Triglidae) 104
'sea-scorpion' (*Myxocephalus scorpius*) 181
'sea-trout' (*Salmo trutta*) 32, 203
Sebastes darwini (*Sebastes capensis*) 69
Sebastes oculata (*Sebastes oculatus*) 94, 205, 227
Sebastes oculatus 94, 181, 238
Sebastodes darwini (*Sebastes capensis*) 280
Selacean/Selachian (shark) 58, 93, 166, 268, 274
Selachus maximus (*Cetorhinus maximus*) 312
Selar crumenophthalmus 116
Seriola bipinnulata (*Elagatis bipinnulata*) 38, 39, 116, 171, 238
Seriolichthys ("*Seriola*") *bipinnulatus* (*Elagatis bipinnulata*) 238
Serranidae (groupers) 102
Serranus (>*Serranus*) 102, 303
Serranus spp. (groupers; Epinephelinae) 102, 103, 108, 255
Serranus albo-maculatus (*Paralabrax albomaculatus*) 91, 100, 102, 232, 236
Serranus aspersus (*Epinephelus marginatus*) 26, 101
Serranus fasciatus (*Serranus psittacinus*) 102
Serranus '*galapagensis*' (*Mycteroperca olfax*) 101, 231
Serranus goreensis (*Epinephelus goreensis*) 26, 101, 214, 238
Serranus labriformis (*Epinephelus labriformis*) 91, 101, 102, 230, 236
Serranus olfax (*Mycteroperca olfax*) 91, 101, 102, 231, 236
Serranus psittacinus 102, 316
Shad (*Alosa sapidissima*) 148, 188
shads (Alosinae) 6
sharksuckers (Echeneidae) 173

shark(s) xxi, 3, 10, 11, 15, 24, 28, 31, 34, 35, 43, 44, 47, 57, 65, 78, 84, 94, 95, 100, 109, 110, 111, 125, 130, 135, 136, 138, 141, 145, 152, 154, 162, 164, 166, 172, 173, 179, 180, 188, 189, 214, 216, 217, 268, 275, 278, 291, 293, 296, 303, 309
Sheephead grunt (*Orthopristis cantharinus*) 104
Shortfin eel (*Anguilla australis*) 63
Shorthorn sculpin (*Myxocephalus scorpius*) 181
Siganidae (<rabbitfish) 34
Siluridae (<freshwater catfishes) 30, 208, 230
Siluriformes (catfishes) 29
Siluri/Silurus/siluroid/ (catfishes) 13, 14, 30, 190, 208, 221
Silurus glanis 190, 208
silver biddies (Gerreidae) 138
'silver fish' (*Carassius* sp.) 97, 112
silversides (Atherinidae) 90, 190
'Siren' (Mud iguana; lungfish) 99
Singapore parrotfish (*Scarus prasiognathos*) 154
Skate (*Dipturus batis*)
skates (<Rajidae) 172, 191, 220, 233, 275
Skipjack (*Katsuwonus pelamis*) 20
sleepers (Eleotridae) 191
slimeheads (<Trachichthyidae) 53
slipmouths (Leiognathidae) 16
Smallmouth bass (*Micropterus dolomieui*) 198
Small-toothed flounder (*Pseudorhombus jenynsii*) 79
Smelt (*Osmerus eperlanus*) 64, 192
smelts (Osmeridae) 27
'smooth-tailed stickleback (*G. leiurus*)' (<*Gasterosteus aculeatus*) 198
snailfishes (Liparidae) 129
Snakehead king croaker (*Menticirrhus ophicephalus*) 49
snappers (Lutjanidae) 192
soapfishes (Grammistinae) 24
Sockeye salmon (*Oncorhynchus nerka*) 177
soldierfishes (Holocentridae) 197
Sole (>*Solea solea*) 81, 204
Solea jenynsii (*Pseudorhombus jenynsii*) 236
Solea solea 193
Soleidae (<soles) 193

Index to the Fishes

Soles (<Soleidae) 193, 321
Solenostoma spp. 161
Solenostomidae (ghost pipefishes) 161
soles (Achiridae; Soleidae) 100, 193
Sordid dragonet (<***Callionymous lyra***) 59
South American lungfish (***Lepidosiren paradoxa***) 131, 132
South American pilchard (***Sardinops sagax***) 109
South American 'salmon' (Characidae) 179
'southern cod-fish' (cod-like fish) 198
Southern king croaker (***Menticirrhus americanus***) 49
Sparidae (porgies) 165
Sparrow flying fish (***Cheilopogon xenopterus***) 84
'Spari' (actually *Scarus* spp.) 155
Sparus spp. xxiv, 48, 49, 154, 155, 165
Sphoeroides angusticeps 167
Sphoeroides annulatus 167
Sphyraena sphyraena 11
Sphyraenidae (barracudas) 11
'spiegel-carpe' (Spiegelkarpfen) (<***Cyprinus carpio***) 27
Spiny eel (***Notacanthus chemnitzii***) 54
Spondyliosoma cantharus 165
Spotted robust triplefin (***Grahamina capito***) 19
sprats (<Clupeidae) 109
Sprattus fuegensis 75
Squalidae (dogfish sharks) 196
'Squalus' (shark, dogfish) 11, 196, 216, 217, 218
Squalus acanthias 15, 57, 197
Squalus elephas (***Cetorhinus maximus***) 12
Squalus maximus (***Cetorhinus maximus***) 11, 274
Squalus rhinoceros (***Cetorhinus maximus***) 12
Squalus spinax (***Etmopterus spinax***) 11, 205
Squaretail mullet (***Liza vaigiensis***) 142
squirrelfishes (Holocentridae) 197
stargazers (Uranoscopidae) 65
Starry grouper (***Epinephelus labriformis***) 101
Starry toadfish (***Tetrodon aerostaticus***) 167
Steelhead trout (***Oncorhynchus mykiss***) 255
Stegastes imbricatus/a 26, 51, 235
Stickleback (***Gasterosteus aculeatus***) xxv, 174

sticklebacks (Gasterosteidae) 184, 197, 198, 190, 198
Streaked gurnard (***Chelidonichthys lastoviza***) 105
Striped burrfish (***Cyclichthys schoepfi***) 22
Striped perch (***Pelates octolineatus***) 203
Stromateidae (butterfishes) 24
Stromateus maculatus (***Stromateus stellatus***) 24, 34, 223, 228, 238
Stromateus stellatus 24
sturgeon 130, 180, 296
'sucking fish' (***Hypsoblennius sordidus***) 229
'sucking fish' (***Gobiesox marmoratus***) 37, 228
'sucking fish' (***Remora remora***) 173, 214
surgeonfishes (Acanthuridae) 199
sunfishes (Centrarchidae) 114, 184, 198, 199
Syluri, awful (***Silurus glanis***) 269
Symphodus cinereus 210
Symphodus melops 210
Symphurus tessellates 193
Syngnathidae (pipefishes and seahorses), 159, 182
Syngnathoidea 129, 277
'Syngnath(o)us' (pipefish) 108, 160, 161, 182, 229, 317
Syngnathus acicularis (***Leptonotus blainvilleanus***) 159, 160, 205, 228, 235
Syngnathus conspicillatus (***Corythoichthys flavofasciatus***) 159, 160, 201, 232, 235
Syngnathus crinitus (***Micrognathus crinitus***) 10, 159, 214, 235
Syngnathus ophidion (***Nerophis ophidion***) 161
Synodontis multipunctatus 49
Synodontis nigriventris 58
Synodontis petricola 49

'Taeniatole austral' (scorpionfish) 181
'taeniato<u>te</u>s' (scorpionfishes) 181
Taeniatotus triacanthus 181
Tambaqui (***Colossoma macropopum***) 47
'tanche' (***Tinca tinca***; wrasses) 87
Teleost/telostean (bony fishes) xx, xxi, 20, 34, 93, 110, 129, 166, 201, 203, 300
temperate basses (Percichthyidae) 48

Tench (***Tinca tinca***) 50, 72, 87, 97, 151, 175, 186, 201, 202, 309
Ten-spotted livebearer (***Cnesterodon decemmaculatus***) 127
Teraponidae (tigerperches) 202
'*Tetragonopterus*' 33, 279
Tetragonopterus abramis (***Astyanax abramis***) 33, 34, 223, 239
Tetragonopterus interruptus (***Cheirodon interruptus***) 33, 34, 133, 221
Tetragonopterus rutilus (***Astyanax fasciatus***) 33, 34, 223, 234
Tetragonopterus scabripinnis (***Astyanax scabripinnis***) 33, 34, 175, 215, 234
Tetragonopterus taeniatus (***Astyanax taeniatus***) 33, 175, 215, 239
Tetraodontidae (puffers) 166
Tetraroge niger 7
Tetrarogidae (scorpionfishes) 7
Tetrodon aerostaticus (***Arothron stellatus***) 167, 240
Tetrodon aerostaticus (misidentified as ***Sphoeroides angusticeps***) 167, 230
Tetrodon angusticeps (***Sphoeroides angusticeps***) 91, 167, 240
Tetrodon annulatus (***Sphoeroides annulatus***) 91, 167, 231, 236
Tetrodon darwinii (***Marilyna darwinii***) 69, 254
Tetrodon hispidus (***Tetrodon implutus***) 236
Tetrodon implutus 39, 167, 236
Tetrodon stellatus (***Arothron stellatus***) 240
Teuthis sandvicensis (***Acanthurus triostegus sandvicensis***) 199
Thalassoma pavo 210
Thymallus sp. (grayling, Salmonidae) 313
Thornback (***Raja clavata***) 172
thornfish (***Pseudaphritis undulatus***) 202
thornfishes (Bovichthyidae) 202
Threadfin butterflyfish (***Chaetodon auriga***) 25
threefin blennies (Tripterygiidae)
Three-spined stickleback (***Gasterosteus aculeatus***) 197
Thunnus alalunga 4
tigerperches (Teraponidae) 202

tilefishes (Malacanthidae)
Tinca vulgaris (***Tinca tinca***) 187
Tinca tinca 201
toadfishes (Batrachoididae) 203
Torpedinidae (electric rays) 203
Torpedo (<Torpedinidae) 6, 65, 67, 191, 203, 288, 293, 296, 303
Torpedo torpedo 203
Trachichthyidae (roughies, sawbellies and slimeheads) 53
Trachichthys darwinii (***Gephyroberix darwinii***) 53, 69
Trachypterus arcticus (***Trachipterus arcticus***) 53, 80, 81
Trachipterus arcticus 53
Trachurus declivis 116
triggerfishes (Balistidae) 76, 203
Trigla (>***Chelidonichthys lastoviza***) 105, 193
Trigla kumu (***Chelidonichthys kumu***) 12, 105, 233, 240
Trigla lineata (***Chelidonichthys lastoviza***) 105
Triglidae (gurnards and searobins) 104, 105
Triplewart seadevil (***Cryptopsaras couesii***) 154
Tripterygiidae (threefin blennies) 19
Tripterygion capito (***Grahamina capito***) 12, 19, 233
'*Tripterygium nigripinne*' (***Grahamina capito***) 235
Tropical two-wing flying fish (***Exocoetus volitans***) 92
Trout (>***Salmo trutta***) xvii, 32, 33, 48, 112, 129, 150, 179, 184, 186, 203, 204, 205, 207, 251, 255, 288
trout-perches (Percopsidae) 204
tuna (-like) 20, 31
Turbot (***Scophthalmus maximus***) 81, 151, 174, 204
'Turbot' (***Reinhardtius hippoglossoides***) 100, 204

Umbrina arenata (***Menticirrhus americanus***) 49, 156, 217, 222, 238
Umbrina ophicephala (***Menticirrhus ophicephalus***) 45, 49, 229

Index to the Fishes

'Umbrinas (*Sciaena aquila*)' 194
Upeneus (or ***Mulloidichthys***) ***flavolineatus*** 38, 96, 238
Upeneus prayensis (***Pseudupeneus prayensis***) 26, 96, 214, 238
Upeneus trifasciatus (***Parupeneus trifasciatus***) 201, 232, 238
upside-down catfishes (Mochokidae) 49, 58
Uranoscopidae (stargazers) 65

Velvet belly (***Etmopterus spinax***) 205

Weatherfish (***Misgurnus fossilis***) 129
Wels catfish (***Silurus glanis***) 114, 208, 269
Weedy sea dragon (***Phyllopterix taeniolatus***) 160
West African goatfish (***Pseudupeneus prayensis***) 96
West African lungfish (***Protopterus annectens***) 131

'white fish' 51
white sharks (Lamnidae) 100
White trevally (***Pseudocaranx dentex***) 116
Whitespotted sandbass (***Paralabrax albomaculatus***) 100
Whitetip flying fish (***Cheilopogon xenopterus***) 84
wrasses (Labridae) 35, 69, 87, 103, 108, 151, 187, 209, 210

Xiphophorus helleri (Green swordtail) 127
Xiphorhamphus jenynsii (***Oligosarcus hepsetus***) 240

Yellow perch (***Perca flavescens***) 156
Yellow-eye mullet (***Aldrichetta forsteri***) 142
Yellowstripe goatfish (***Upeneus*** [or ***Mulloidichthys***] ***flavolineatus***) 96

Zoarces viviparus 63
Zoarcidae (eelpouts) 63